Ultrasonics of High-T_c and Other Unconventional Superconductors

PHYSICAL ACOUSTICS

Volume XX

Physical Acoustics

Volume XX

R.N. THURSTON AND ALLAN D. PIERCE, Series Editors

CONTRIBUTORS TO VOLUME XX

S. ADENWALLA
D. P. ALMOND
S. BHATTACHARYA
BRAGE GOLDING
J. B. KETTERSON
VLADIMIR Z. KRESIN
MOISES LEVY
J. D. MAYNARD
M. J. MCKENNA
A. MIGLIORI
BIMAL K. SARMA
KEUN JENN SUN
WILLIAM M. VISSCHER
MIN-FENG XU
Z. ZHAO

Ultrasonics of High-T_c and Other Unconventional Superconductors

Edited by MOISES LEVY

UNIVERSITY OF WISCONSIN
MILWAUKEE, WISCONSIN

PHYSICAL ACOUSTICS
Volume XX

ACADEMIC PRESS, INC.
Harcourt Brace Jovanovich, Publishers

Boston San Diego New York
London Sydney Tokyo Toronto

This book is printed on acid-free paper. ∞

COPYRIGHT © 1992 BY ACADEMIC PRESS, INC.

All rights reserved.
No part of this publication may be reproduced or
transmitted in any form or by any means, electronic
or mechanical, including photocopy, recording, or
any information storage and retrieval system, without
permission in writing from the publisher.

ACADEMIC PRESS, INC.
1250 Sixth Avenue, San Diego, CA 92101-4311

United Kingdom Edition published by
ACADEMIC PRESS LIMITED
24–28 Oval Road, London NW1 7DX

ISSN 0893-388X

ISBN 0-12-477920-4

PRINTED IN THE UNITED STATES OF AMERICA

92 93 94 95   BB   9 8 7 6 5 4 3 2 1

This book is dedicated to the seventeen students who obtained their Ph.D.'s under my supervision and without whom this work would not have been possible, and to my wife who provided the encouragement and incentive required to complete it.

Contents

CONTRIBUTORS xiii
PREFACE xv

1

Ultrasonic Attenuation in Conventional Superconductors

MOISES LEVY

1. INTRODUCTION	1
2. TEMPERATURE DEPENDENCE	2
3. MAGNETIC FIELD DEPENDENCE	13
ACKNOWLEDGMENT	21
REFERENCES	21

2

Sound Propagation and Collective Modes in Superfluid ^{3}He

BIMAL K. SARMA, J. B. KETTERSON, S. ADENWALLA, AND Z. ZHAO

1. INTRODUCTION	23
2. NORMAL STATE OF ^{3}HE	27
3. SUPERFLUID ^{3}HE	33
4. COLLECTIVE MODES	43
5. EXPERIMENTAL RESULTS	53
6. CONCLUSIONS	101
ACKNOWLEDGMENTS	102
REFERENCES	103

3

Sound Propagation in the Heavy Fermion Superconductors

BIMAL K. SARMA, MOISES LEVY, S. ADENWALLA AND J. B. KETTERSON

1. INTRODUCTION	108
2. SOUND PROPAGATION IN CONVENTIONAL SUPERCONDUCTORS	114
3. UNCONVENTIONAL SUPERCONDUCTIVITY	118
4. HEAVY FERMION SYSTEMS	140
5. UPt_3	143
6. UBe_{13}	176
7. URu_2Si_2	179
8. $Ce\,Cu_6$	181
9. CONCLUSIONS	185
ACKNOWLEDGMENTS	185
REFERENCES	185

4

Ultrasonic Attenuation in the Magnetic Superconducting System $Er_{1-x}Ho_xRh_4B_4$

KEUN JENN SUN AND MOISES LEVY

1. INTRODUCTION	191
2. EXPERIMENTAL DETAILS AND PHYSICAL PROPERTIES	194
3. COMMON ATTENUATION BEHAVIOR	196
4. SPECIAL ATTENUATION BEHAVIOR	198
5. SUMMARY	230
ACKNOWLEDGMENTS	232
REFERENCES	233

5

Ultrasonic Propagation in Sintered High-T_c Superconductors

MOISES LEVY, MIN-FENG XU, BIMAL K. SARMA AND KEUN JENN SUN

1. INTRODUCTION	237
2. ATTENUATION AND VELOCITY IN $La_{2-x}Sr_xCuO_4$	238
3. ORDINARY SINTERED $YBa_2Cu_3O_{7-\delta}$	243
4. ORIENTED $YBa_2Cu_3O_{7-\delta}$	254

Contents ix

5. SOUND PROPAGATION IN $GdBa_2Cu_3O_{7-\delta}$ AND $ErBa_2Cu_3O_{7-\delta}$	271
6. BiSrCaCuO AND TℓBaCaCuO SUPERCONDUCTING COMPOUNDS	274
7. SOUND PROPAGATION IN $Ba_{1-x}K_xBiO_3$	280
8. SUMMARY	289
ACKNOWLEDGMENT	292
APPENDIX A CRYSTAL STRUCTURE OF HIGH-T_c SUPERCONDUCTORS	292
APPENDIX B SELECTED VELOCITIES AND ELASTIC CONSTANTS FOR HIGH-T_c SUPERCONDUCTORS	295
APPENDIX C TEMPERATURE POSITION OF ATTENUATION PEAKS FOR SINTER-FORGED $YBa_2Cu_3O_7$ SAMPLES	297
APPENDIX D ACTIVATION ENERGIES FOR RELAXATION TIMES ASSOCIATED WITH RELAXATION ATTENUATION PEAKS	297
REFERENCES	298

6

Sound Velocity Studies of Ceramic High-Temperature Superconductors

S. BHATTACHARYA

1. INTRODUCTION	303
2. THEORY	306
3. SOUND VELOCITY IN CERAMIC SAMPLES	311
4. RESULTS NEAR T_c	328
5. MAGNETIC FIELD DEPENDENCE	337
6. CONCLUSIONS	342
REFERENCES	345
BIBLIOGRAPHY	346

7

Acoustic Studies of Single-Crystal High-T_c Superconductors

BRAGE GOLDING

1. INTRODUCTION	349
2. THE HIGH-T_c SUPERCONDUCTORS	351
3. ACOUSTIC METHODS FOR SMALL SINGLE CRYSTALS	352
4. SINGLE-CRYSTAL EXPERIMENTS AND RESULTS	357
5. SUMMARY AND OUTLOOK	376

| ACKNOWLEDGMENTS | 378 |
| REFERENCES | 378 |

8

Ultrasonic Measurements of Elastic Constants in Single Crystals of La_2CuO_4

J. D. MAYNARD, M. J. MCKENNA, A. MIGLIORI AND WILLIAM M. VISSCHER

1. INTRODUCTION	381
2. DEVELOPMENT OF THE SMALL-SAMPLE RESONANT ULTRASOUND TECHNIQUE	382
3. MEASUREMENTS ON SINGLE-CRYSTAL SAMPLES OF LA_2CUO_4	397
4. MEASUREMENTS IN SUPERCONDUCTING $LA_{1.86}SR_{0.14}CUO_4$	401
ACKNOWLEDGMENTS	407
REFERENCES	407

9

A Rationalisation of the Diversity in the Elastic Response of Polycrystalline Superconducting Oxides

D. P. ALMOND

1. INTRODUCTION	409
2. REVIEW OF INITIAL WORK	410
3. A COMPARISON OF THE ULTRASONIC CHARACTERISTICS OF SUPERCONDUCTING AND NON-SUPERCONDUCTING MATERIAL	418
4. HIGH-TEMPERATURE ANOMALIES	428
5. CONCLUSIONS	429
ACKNOWLEDGMENTS	431
REFERENCES	431

10

High-T_c Superconductivity and Ultrasonics—Theoretical Aspects

VLADIMIR Z. KRESIN

| 1. INTRODUCTION | 435 |
| 2. SOUND ATTENUATION AND CONVENTIONAL SUPERCONDUCTIVITY | 436 |

Contents xi

3. Exotic Superconductors (Organic Materials; Heavy Fermions)   440
4. High-T_c Oxides   441
5. Ultrasonic Attenuation in High-T_c Oxides   446
   Acknowledgment   452
   References   452

Index   455
Contents of Volumes in this Series   461

Contributors

Numbers in parentheses indicate the pages on which the authors' contributions begin.

S. ADENWALLA (23, 107)
Northwestern University
Physics Department
2145 Sheridan Road
Evanston, IL 60208

D. P. ALMOND (409)
School of Materials Science
University of Bath
Claverton Down
Bath, BA2 7AY, United Kingdom

S. BHATTACHARYA (303)
NEC Research Institute
4 Independence Way
Princeton, NJ 08540

BRAGE GOLDING (349)
Department of Physics and
   Astronomy
Michigan State University
East Lansing, Michigan 48824-1116

J. B. KETTERSON (23, 107)
Northwestern University
Physics Department
2145 Sheridan Road
Evanston, IL 60208

VLADIMIR Z. KRESIN (435)
Materials and Chemical Science
   Division
Lawrence Berkeley Laboratory
University of California
Berkeley, CA 94720

MOISES LEVY (1, 107, 191, 237)
Department of Physics
P.O. Box 413
University of Wisconsin—Milwaukee
Milwaukee, WI 53201

J. D. MAYNARD (381)
Department of Physics
The Pennsylvania State University
University Park, PA 16802

M. J. MCKENNA (381)
Department of Physics
The Pennsylvania State University
University Park, PA 16802

A. MIGLIORI (381)
Los Alamos National Laboratory
PO Box 1663, MS k-764
Los Alamos, New Mexico 87545

BIMAL K. SARMA (23, 107, 237)
Department of Physics
University of Wisconsin—Milwaukee
Milwaukee, WI 53201

KEUN JENN SUN (191, 237)
Department of Physics
The College of William and Mary
Williamsburg, VA 23185

WILLIAM M. VISSCHER (381)
Los Alamos National Laboratory
Los Alamos, New Mexico 87545

MIN-FENG XU (237)
Department of Physics
University of Wisconsin—Milwaukee
Milwaukee, WI 53201

Z. ZHAO (23)
Northwestern University
Physics Department
2145 Sheridan Road,
Evanston, IL 60208

Preface

In 1911 Kammerlingh Onnes (1991) discovered superconductivity in mercury at a temperature near 4 K. Seventy-five years later, an unprecedented set of discoveries led to the birth of high-T_c superconductivity. Mainly because of the efforts of Bednorz and Muller (1986) and Chu (Wu et al., 1987), superconductivity above liquid nitrogen temperatures has become a reality.

The first microscopic theory for superconductivity was given by Bardeen, Cooper, and Schrieffer (BCS) (Bardeen et al., 1957), and it invoked electron–phonon interaction as being responsible for the pairing mechanism. The measurement of ultrasonic attenuation in the superconducting state was one of the experimental techniques that provided definitive confirmation for the validity of the BCS theory. Both attenuation and velocity measurements have proven to be essential in the discovery of new effects and new phases in both conventional and unconventional superconductors. Therefore, it appears appropriate at this time to publish a volume on the acoustic study of the high-T_c superconductors. Since these new superconductors are very much different from the conventional superconductors that can be reasonably well explained by the BCS theory, we have included three chapters on some unconventional superconducting systems: superfluid ^{3}He, heavy Fermion superconductors, and magnetic reentrant superconductors. We hope that their inclusion will set the groundwork for our study of the high-T_c superconductors, which is covered in the succeeding five chapters. The last chapter will attempt to provide a theoretical understanding of the different mechanisms that may be responsible for superconductivity in these novel superconducting systems, and for the contributions that sound measurements have made and could make to our understanding of these systems.

The authors of most of the chapters in this book assume that the readers possess some familiarity with sound attenuation in conventional BCS superconductors. Therefore, Chapter I summarizes the principal results that have been observed in these systems as functions both of temperature and of magnetic field.

Chapter II is on the ultrasonic study of superfluid ^{3}H3. Ever since the publication of the BCS theory in 1957, explaining the physics of s-wave superconductors, we have been looking for a p-wave superconductor. In 1972, this was discovered in ^{3}He (Osheroff et al., 1972), where, because of the large hard core repulsion, the pairing is in the $\ell=1$ (p-wave), $s=1$ (triplet) state. This system is by far the best studied, both theoretically and experimentally, of the unconventional superconductors. Surprisingly, the theory can be explained on the basis of the BCS theory modified for triplet p-wave pairing, which includes strong coupling effects at higher pressures. The interaction is spin–spin. The order parameter is a complex tensor, and many superfluid phases exist in the P–T–H planes (pressure, temperature, magnetic field). Because of the low superfluid transition temperature (2.5 mK), the pair breaking energy is only 180 MHz in frequency units. Because of the complex nature of the order parameter, many collective mode states exist in the energy gap, and as these couple to density, they can be observed by ultrasonic methods. There is a very large absorption and dispersion of the sound waves as one goes through these modes. Ultrasonic techniques have been developed to study these modes, and absorption in excess of 1,000 dB/cm and group velocities less than 25 m/s have been measured. The chapter includes experimental techniques and data, as well as the necessary theory.

Chapter III is on the ultrasonic study of the heavy-Fermion superconductors. These materials are characterized by a very large electronic heat capacity (100–1,000 times that of copper), and surprisingly, some of these have a superconducting ground state. A reasonable theory for these systems is lacking, though it seems likely that there may be many things in common between these and superfluid ^{3}He, viz., a higher angular momentum pairing. Most of the ultrasonic measurements are on UPt$_3$. The attenuation is very much different from that of a conventional BCS superconductor such as niobium or vanadium, showing evidence of an anisotropic energy gap. UPt$_3$ also shows evidence of multiple phases in the H–T plane, the first evidence of which came from ultrasonic measurements. Spin fluctuations seem to play a dominant role in the pairing mechanism, and the study of these systems may reveal their pairing mechanisms that may prove useful in building a theoretical understanding of the high-T_c superconductors.

Chapter IV is on the magnetic reentrant superconductors. These form an interesting system. The ground state seems to be a magnetic state, over which lies a superconducting state. Again, ultrasonic attenuation has proved to be a useful tool to study the subtle interplay among magnetism, electromagnetic screening, and superconductivity. Measurements have been performed on the $Er_{1-x}Ho_xRh_4B_4$ system, where, as the concentration of Ho is changed from zero to one, the system goes from a reentrant superconductor with a superconducting

Preface xvii

transition at 8.9 K and a ferromagnetic transition at around 1 K, to a purely magnetic system with a ferromagnetic transition at 6.9 K. In fact, in the concentration range $0 \leq x \leq 0.3$, there is a coexistence temperature range between superconductivity and a sinusoidal modulated antiferromagnetic state. Attenuation measurements have been performed on several members of this system. The most striking feature is observed for $0.6 \leq x \leq 0.9$ ($x = 0.9$ is the upper value for the observation of superconductivity in this system). In this range the attenuation increased in the superconducting state, as opposed to the expected decrease for a BCS superconductor. In $ErRh_4B_4$, attenuation measurements in a magnetic field exemplify the importance of electromagnetic screening for this reentrant superconducting system. In addition, for $0 \leq x \leq 0.3$, the boundaries of the coexistence region can be easily identified by features in the attenuation measurements. Some of the high-T_c superconductors exhibit reentrant behavior, and comparisons between future measurements on these high-T_c superconductors and the present results may provide insight into the mechanisms that are producing superconductivity and are responsible for the interaction with sound waves.

The next five chapters are on measurements with sound waves on the high-T_c superconductors. The first of these, Chapter V, reviews the experimental data on the sintered oxide superconductors. Most of these measurements are done by pulsed ultrasonics, and the samples are made by a shake-and-bake method. Some of these samples are relatively well oriented, being either grown by a sinter-forged technique, or grown in the presence of a strong external magnetic field in the case of ions with strong paramagnetic moments; a large anisotropy with propagation direction is seen in the elastic constants, as inferred from velocity measurements. Typical results are a stiffening of the lattice at the superconducting transition, and the observation of several attenuation peaks. Most of these attenuation peaks may be due to some kind of relaxation mechanism that could be associated with the excitations that are supposed to be responsible for the high transition temperature of these superconductors. However, in the thallium superconducting compounds, peaks appear in the vicinity of T_c that do not shift with frequency and therefore may be truly associated with the superconducting phase transition. A possible mechanism for this effect could be phonon interaction with superconducting fluctuations associated with the phase transition.

Chapter VI covers the temperature and magnetic field dependence of the velocity and elastic constants in sintered high-T_c superconductors. It concentrates mainly on velocity measurements on the 40 K La–Sr system. From a detailed thermodynamic analysis of both the longitudinal and transverse velocity, the author finds that the main elastic modulus involved at the transition is the shear modulus. Anomalies in the velocity in applied magnetic fields have been interpreted as evidence of a multiple phase diagram in the H–T plane, showing their strong unconventional nature akin to that in superfluid ^{3}He, and in the heavy fermion superconductor UPt$_3$.

Chapters VII and VIII cover both sound absorption and sound dispersion measurements on single crystals of high-T_c superconductors. Since the single crystals are small, particularly for sound measurements, novel experimental approaches have been developed for determining the attenuation and the elastic constants in these small crystals. Chapter VII discusses vibrating reed techniques at audio frequencies, bulk resonance techniques at rf frequencies, and plane wave propagation techniques at microwave frequencies. This chapter covers some of the most beautiful velocity data that have been obtained at microwave frequencies at the phase transition of the high-T_c superconductors. Since the measurements are performed on single crystals, it becomes possible to try to separate the different mechanisms and effects that contribute to the real and imaginary parts of the elastic constants, such as electrons, defects, tunneling, gaplessness, and flux lattice interactions.

The authors of Chapter VIII have perfected an intriguing new small-sample resonant ultrasound technique that uses thin piezoelectric films or weakly coupled lithium niobate transducers. By continuously varying the frequency at a particular temperature, the resonant frequencies of a large number of modes may be obtained, from which all of the independent elastic constants of submillimeter-sized single crystals may be determined from a single experimental setup. Using this technique the authors have observed minima of the shear moduli near the structural phase transition of $La_{1-x}Sr_xCu O_4$ with $x = 0.14$ and 0.1.

Chapter IX discusses both the effect of oxygen on superconducting properties and the response of sound to these additions. An important technique in the study of conventional superconductors was the ability to destroy superconductivity by applying an external magnetic field in excess of H_{c2}. This allowed one to estimate the contribution due to superconductivity. For the high-T_c materials, fields in excess of H_{c2} or approaching this are difficult to achieve at present, but it is possible to destroy superconductivity by varying the oxygen content in the high-T_c cuprates, although this is accompanied by conduction electron density changes, and perhaps even structural changes. The author has measured the velocity and attenuation in such an oxygen-modified system. Although it is difficult to separate those effects that are due to varying oxygen content from those that are intrinsic to the high-T_c superconductors, the author has attempted to distinguish effects that are due to the superconductivity itself from those that are due to the complex lattice structure of these highly interesting crystallographic systems.

Chapter X provides a theoretical foundation for sound measurements in the superconducting state. The author emphasizes the effects of multigap structures and gap anisotropy on sound attenuation in the superconducting state of the

Preface

cuprate superconductors. These effects will be particularly important in the high-T_c superconductors, since the coherence length is intrinsically smaller than the electron mean free path in these superconductors, a fact that then precludes anisotropies of the energy gap from being averaged out by collisions.

There appears to be incontrovertible evidence that lattice stiffening is occurring below T_c according to the work reported in Chapter VI, and some lattice softening right at T_c according to Chapter VII. Although measurements of attenuation below T_c have not yet detected the expected decrease or at least change in attenuation that should accompany a superconducting transition in a non-reentrant or reentrant superconductor, several measurements, Chapter V and IX, have observed attenuation maxima near T_c that do not shift with frequency, indicating that they are actually associated with the transition and could be produced by superconducting fluctuations. Other attenuation maxima have also been observed in these high-T_c systems whose origin appears to be relaxation processes. These relaxation processes may be associated with the excitations that some theoretical models are postulating as the quanta that are responsible for the attractive interaction that produces Cooper pairs and, therefore, superconductivity in these high-T_c systems.

The editor hopes that the timely publication of the monographs included in this book will lead to its considerable use in reference libraries, and will help to formulate theoretical models that will explain the presence of superconductivity in the high-T_c and unconventional superconductors.

The guest editor wishes to express his appreciation to the other authors of the chapters of this volume who have devoted themselves to produce comprehensive and up-to-date reviews in a timely manner, and to Kathleen Jackson for patiently preparing these and numerous other manuscripts and to Joseph Herro for preparing the diagrams in Chapter I. He also wishes to express his gratitude to the numerous colleagues who have encouraged and enlightened him, starting with his thesis advisor Professor Isadore Rudnick; his experimental colleagues, Reynold Kagiwada, Peter Wyder, Klaus Andres, Elias Burstein, and Bimal Sarma; his theoretical colleagues, Masashi Tachiki, Kazumi Maki, Michael Revzen, Charles Kuper, the late Theodore Holstein, Vladimir Z. Kresin, and Richard Sorbello; and his grant funding officers, the late Max Swerdlow, AFOSR, Harold Weinstock, AFOSR, and Logan Hargrove, ONR.

Bardeen, J., Cooper L. N., and Schrieffer, J. R. (1957). *Phys. Rev.* **108,** 1175–1204.
Bednorz, J. G. and Muller, K. A. (1986). *Z. Phys.* **B64,** 189–193.
Onnes, H. Kammerlingh (1911) Akad. van Wetenshappen (Amsterdam) **113,** 818. U.S. Center for Research Libraries Chicago, Monographs, D-803.
Osheroff, D. D., Richardson, R. C., and Lee, D. M. (1972). *Phys. Rev. Lett.* **28,** 855–858.
Wu, M. K., Ashburn, J. R., Torng, C. J., Hor, Ph. H., Meng, R. L., Gao, L., Huang, Z. J., Wang, Y. Q., and Chu, C. W. (1987). *Phys. Rev. Lett.* **58,** 908–910.

—1—
Ultrasonic Attenuation in Conventional Superconductors

MOISES LEVY
Department of Physics, University of Wisconsin—Milwaukee, Milwaukee, Wisconsin

1. Introduction 1
2. Temperature Dependence 2
   2.1. Transverse Waves 4
   2.2. Longitudinal Waves 7
3. Magnetic Field Dependence 13
   3.1. Dirty Type II Superconductors 16
   3.2. Clean Type II Superconductors 19
   Acknowledgment 21
   References 21

1. Introduction

The transition into the superconducting state is a second-order phase transition. Therefore, the first-order derivatives of the Gibbs free energy are continuous across the transition, but the second-order derivatives are discontinuous. This implies that there should be a discontinuity in the specific heat, the susceptibility, the thermal expansion coefficient, and the elastic constants. The jump in the specific heat was first found experimentally by Keesom and Kok (1932). The discontinuity in the susceptibility that leads to perfect diamagnetism and that is now identified as the Meissner–Ochsenfeld effect was discovered experimentally in 1933 (Meissner and Ochsenfeld, 1933). In fact, it was the discovery of these two effects that led to the conclusion that the superconducting transition in the absence of a magnetic field was a second-order phase transition, and that spurred investigators to attempt to find the discontinuities in the mechanical constants. Although their bimetallic strips of tin and a non-superconducting metal, brass, were not sufficiently sensitive to detect the change in thermal expansion coefficient, Lasarew and Sudovstov (1949) were able to observe changes in volume between the two states that corresponded to the first-order phase transition across the critical magnetic field boundary in the temperature vs. magnetic field phase diagram of a type I superconductor. It was not until 1961 that Andres and Rhorer (1961) were able to detect a change in the thermal expansion coefficient of lead. They observed a decrease on the order of a few parts in 10^8. In 1954, J. K. Landauer succeeded in measuring the difference in elastic constants between the normal and superconducting states of tin. He observed a decrease of a few ppm. Therefore, it wasn't until 1961 that all the second-order thermodynamics discontinuities were experimentally established for the superconducting transition in the absence of a magnetic field.

In 1953, Bommel and Olsen had attempted to detect the decrease in velocity in the superconducting state of lead. The sensitivity of their apparatus, which operated in the MHz frequency range, was not sufficient to measure the decrease in velocity. However, they noticed (Bommel, 1960) that the amplitude of the sound waves appeared to decrease and increase as the temperature was lowered. These observations gave Bommel (1960) the incentive to perform his now-famous ultrasonic attenuation measurements in a single crystal of lead. The consequences of these measurements, which provided the impetus for subsequent ultrasonic attenuation investigators both in the normal state and the superconducting state, will be discussed in Section 2.

The magnetic field dependence of the attenuation in dirty and clean type II superconductors will be presented in Section 3.

2. Temperature Dependence

The first measurements of sound attenuation in the normal state and in the superconducting state were performed by Bommel (1954) in a single crystal of Pb. Shortly thereafter, Pippard (1955) produced an elegant model to describe the interaction between the sound waves and the electrons in the normal state, which we will call electron–phonon interaction. He found that the attenuation was directly dependent upon the square of the radial frequency ω and on the electron mean free path ℓ when $q\ell < 1$, where q is the phonon wavevector $2\pi/\lambda$, and λ is the phonon wavelength. These two facts had already been established experimentally by Bommel. In addition, both Bommel and Pippard determined that when $q\ell > 1$, the attenuation was dependent on the first power of ω and independent of ℓ.

Although Pippard was the first to develop a complete model for electron–phonon interaction, Mason (1955), using a viscous interaction between the electrons and the lattice, had previously found the result for $q\ell < 1$, and Morse (1955), the result for $q\ell > 1$. In 1957, Bardeen, Cooper, and Schrieffer (BCS) (Bardeen et al., 1957) derived the now-famous expression for the ratio of the attenuation in the superconducting state α_s to that in the normal state α_n:

$$\frac{\alpha_s}{\alpha_n} = \frac{2}{e^{\Delta/kT} + 1}, \quad (1)$$

where 2Δ is the temperature-dependent superconducting energy gap, k is Boltzmann's constant, and T is the temperature. The BCS result was derived using the Golden Rule method, similar to that used by Morse, for longitudinal waves in the $q\ell > 1$ limit.

The same relation has been obtained for both longitudinal and transverse waves in the $q\ell < 1$ limit, and a slightly modified relationship has been obtained for transverse waves in the $q\ell > 1$ limit. We now will briefly describe the Boltzmann transport method that was used for obtaining these results (Levy, 1963).

The Boltzmann transport equation in one dimension for the electron–phonon interaction process may be written as follows:

$$\frac{\partial f}{\partial t} + v_z \frac{\partial f}{\partial z} + a \, \text{grad}_v f = \left(\frac{\partial f}{\partial t}\right)_{\text{coll}} = -\frac{f - f_{eq}}{\tau}, \quad (2)$$

where the first-order linear expansion term has been used for the collision term, τ is the relaxation time, v_z is the electron velocity in the z direction, and a is the acceleration. In the normal state, f would be the Fermi distribution function of the normal electrons. In the superconducting state, since the energy of the phonons is not sufficient to break up a Cooper pair, the distribution function should be that for the thermally activated superconducting quasiparticle excitations (Bardeen et al., 1957):

$$f = \frac{1}{e^{E/kT} + 1}, \quad (3)$$

where the energy of the quasiparticles is $E = (\varepsilon^2 + \Delta^2(T))^{1/2}$, ε is the kinetic energy of the quasiparticles referred to the Fermi level E_F, and $2\Delta(T)$ is the temperature dependent energy gap. For a single instantaneous distortion of the lattice followed by a relaxation to equilibrium, one could assume that f_{eq} would be the value of the distribution function for the undisturbed metal. However, this assumption would be improper for a sound wave, since during the actual process, the equilibrium electron distribution would be the one that would follow, completely in phase, the distortions produced by the sound wave. Thus, for longitudinal waves that produce a change in density n and impart a velocity to the lattice $\mathbf{u}$, $f_{eq} = f_{\mathbf{u},n}$, where $f_{\mathbf{u},n}$ is the distribution function for electrons having an average velocity $\mathbf{u}$ in the z direction and undergoing a small density change n. For transverse waves, $f_{eq} = f_{\mathbf{u}}$, since no density changes are produced by transverse waves.

2.1. TRANSVERSE WAVES

This Boltzmann transport approach appears to yield results for transverse waves in the superconducting state that are valid for all values of $q\ell$. Therefore, these will be presented first.

The procedure for obtaining the attenuation coefficient α would be to calculate the amount of energy loss produced by the collision term. In effect, one is finding the irreversible transfer of energy from the electron part of the system to the thermal bath. In both the BCS (1957) and Mason (1955) approaches, the transfer of energy from the lattice to the electrons is found by calculating the rate at which electrons absorb phonons. Obviously, the electrons would then have to return their excess energy to the thermal bath. In the Pippard (1955) and the Boltzmann transport approach (Levy, 1963), we concentrate on this latter part of the process.

In the normal state, the acceleration term in Eq. (2) is produced by the electromagnetic forces that are induced by the phase lag between the lattice current and the electron current. Through Maxwell's equations both magnetic and electric fields will be set up. However, in the superconducting state, the magnetic fields will be shielded by the Meissner–Ochsenfeld effect, and we can assume that the acceleration force, for transverse waves, is zero. Thus, Eq. (2) becomes

$$\frac{\partial f}{\partial t} + v_z \frac{\partial f}{\partial z} = -\frac{f - f_{\mathbf{u}}}{\tau}, \qquad (4)$$

where the equilibrium value of E in $f_{\mathbf{u}}$ is given by

$$E = [(\tfrac{1}{2} m(\mathbf{v} - \mathbf{u})^2 - E_{\mathrm{F}})^2 + \Delta^2(T)]^{1/2}. \qquad (5)$$

The electrons have to relax to a Fermi surface that is centered about the lattice velocity $\mathbf{u}$. This is the origin of the phonon drag term $\tfrac{1}{2}m(\mathbf{v} - \mathbf{u})^2$ in Eq. (5), where m is the electron mass. Now, one can linearize Eq. (4) by taking the first Taylor expansion term of $f_{\mathbf{u}}$ about its value in the undisturbed metal f_0, and also by setting $f = f_0 + \gamma$, where γ is the instantaneous change of the distribution function, which is assumed to be small. Thus,

$$f_{\mathbf{u}} = f_0 - \frac{mvu \sin\theta \cos\phi}{\left[1 + \dfrac{\Delta^2(T)}{\varepsilon^2}\right]^{1/2}} \frac{\partial f_0}{\partial E},$$

where θ is the angle between the electron velocity and the propagation direction, and ϕ is the azimuthal angle, measured from the polarization direction. And, since all variables associated with the wave are multiplied by $e^{i(\omega t - qz)}$, then

$$\gamma = \frac{mvu\sin\theta\cos\phi}{[1 + i\omega\tau - iqv\tau\cos\theta]\left[1 + \frac{\Delta^2(T)}{\varepsilon^2}\right]^{1/2}} \frac{\partial f_0}{\partial E},$$

and after neglecting terms in $\omega\tau$ since $q\ell = \left(\dfrac{v_F}{v_t}\right)\omega\tau$, where the electron mean free path $\ell = v_F\tau$, v_F is the Fermi velocity, v_t is the transverse sound velocity, and v_F is usually between 300 and 1,000 times larger than v_t, one obtains

$$\left(\frac{\partial f}{\partial \tau}\right)_{coll} = \frac{iqmv^2u\sin\theta\cos\theta\cos\phi}{\tau[1 - iqv\tau\cos\theta]\left[1 + \frac{\Delta^2(T)}{\varepsilon^2}\right]^{1/2}} \frac{\partial f_0}{\partial E}.$$

The rate of energy loss due to collisions of the electrons with the lattice is given by (Blount, 1959)

$$Q = \int \left\langle H\left(\frac{\partial f}{\partial t}\right)_{coll}\right\rangle d\mathbf{v}. \tag{6}$$

After performing the averages per cycle and doing the integral over velocity space, one finds

$$Q = \frac{mu^2 N}{\tau}[1 - g]f_0(\Delta),$$

where N is the density of electrons in the normal state, and

$$g = \frac{3}{2}\frac{1}{(q\ell)^2}\left[\left(\frac{(q\ell)^2 + 1}{q\ell}\right)\tan^{-1}q\ell - 1\right].$$

A plot of g as a function of $q\ell$ is given in Fig. 1.

The attenuation coefficient is given by

$$\alpha = \frac{Q}{\tfrac{1}{2}\rho u^2 v_{sound}},$$

which is the ratio of the rate of energy loss divided by the energy in the sound wave and by the sound velocity in order to convert the attenuation per unit time to that per unit distance. Thus, the attenuation coefficient for transverse waves in the superconducting state is given by

$$\alpha_{st} = \frac{2mN}{v_t\rho\tau}[1 - g]f_0(\Delta), \tag{7}$$

and, for $q\ell < 1$,

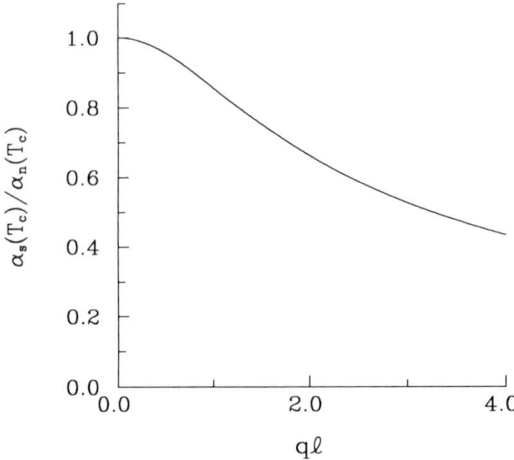

FIG. 1. The ratio of the attenuation of transverse waves in the superconducting state to that in the normal state α_{st}/α_{nt} at the superconducting transition temperature T_c plotted as a function of $q\ell$. This is just a plot of g as a function of $q\ell$.

$$\alpha_{st} = \frac{2 m N (q\ell)^2}{5 v_t \rho \tau} f_0(\Delta)$$

For normal metals $\Delta = 0$ and $f_0(0) = \frac{1}{2}$. Therefore,

$$\alpha_{nt} = \frac{m N (q\ell)^2}{5 v_s \rho \tau},$$

which is the result obtained by Pippard (1955) in the $q\ell < 1$ limit. Thus,

$$\frac{\alpha_{st}}{\alpha_{nt}} = 2 f_0(\Delta) = \frac{2}{e^{\Delta/kT} + 1}, \qquad (8)$$

which shows the validity of the BCS (1957) result for transverse waves in the $q\ell < 1$ in the superconducting state. And, by using the expression derived by Pippard (1955) for transverse waves that is valid for all values of $q\ell$ in the normal state,

$$\alpha_{nt} = \frac{1 - g}{g} \frac{N m}{\rho v_t \tau},$$

we find

$$\frac{\alpha_{st}}{\alpha_{nt}} = 2 g f_0(\Delta). \qquad (9)$$

1. Ultrasonic Attenuation in Conventional Superconductors

Upon looking at the values of g as a function of $q\ell$ in Fig. 1, it can be seen that Eq. (8) is a direct consequence of Eq. (9), since $g \simeq 1$ for $q\ell < 1$. For values of $q\ell > 1$, Eq. (9) predicts a drop in attenuation at T_c, which has been observed (Clayborne and Morse, 1964). It also predicts that the residual attenuation will follow the BCS result and that for $q\ell \gg 1$, the residual attenuation will be $Nm/(v_t\rho\tau)$, which is the result that Pippard obtains in the normal state for the case when the skin depth becomes larger than the wavelength. It is interesting to note that in the superconducting state, this result is achieved because the superconducting currents screen out the ionic currents, while in the normal state, this occurs because the electron currents do not respond to the ionic currents. Thus, in the former case the electron currents are large, but since the resistance is zero the dissipation due to this process is also zero. In the latter case, the screening currents tend to zero because the electrons have no time to respond, and therefore the dissipation is also zero.

2.2. Longitudinal Waves

A similar procedure can be followed to obtain the ratio of the attenuation coefficients for longitudinal waves in the $q\ell < 1$ limit. In this case the Boltzmann transport equation becomes

$$\frac{\partial f}{\partial t} + v_z \frac{\partial f}{\partial t} + \mathbf{a}\ \text{grad}_v f = -\frac{f - f_{\mathbf{u},n}}{\tau}, \tag{10}$$

and

$$f_{\mathbf{u},n} = f_0 - \frac{mvu\cos\theta - mnv_F^2/3N)}{\left[1 + \frac{\Delta^2(T)}{\varepsilon^2}\right]^{1/2}} \frac{\partial f_0}{\partial E}.$$

Now, the problem is to determine the forces that produce acceleration in the superconducting state. In the normal state, electric fields are set up by both the density gradients and the electrostatic charges that accumulate because of the phase lag between the lattice ionic current and the electron current. However, in the superconducting state it can be assumed that the charge accumulation will be shorted out by the superconducting currents. We are still left with the acceleration produced by the density gradients. The change in Fermi energy for a spherical Fermi surface produced by a density change n is given by

$$\Delta E_f = \frac{mnv_F^2}{3N},$$

and, therefore, the acceleration is given by

$$a = -\frac{1}{m}\frac{\partial E_f}{\partial z} = iq\frac{v_F^2}{3}\frac{n}{N}\cos\theta.$$

Again, let $f = f_0 + \gamma$ and neglecting $\mathrm{grad}_v \psi$ one obtains from Eq. (10)

$$\gamma = -\frac{mvu\cos\theta + (mnv_F^2/3N)(1 - iqv\tau\cos\theta)\frac{\partial f_0}{\partial E}}{(1 + i\omega\tau - iqv\tau\cos\theta)\left[1 + \frac{\Delta^2(T)}{\varepsilon^2}\right]^{1/2}}.$$

We can solve for n by using the continuity equation, $\nabla \cdot J_e + \frac{\partial}{\partial t}(N + n) = 0$, to find that the electron current density $J_e = env_e$. Since there are no space charges set up, this must cancel the lattice current density eNu. Therefore, $n = Nu/v_\ell$, where the subscript ℓ refers to longitudinal waves.

After neglecting terms in $\omega\tau$, we have

$$\left(\frac{\partial f}{\partial t}\right)_{\mathrm{coll}} = \frac{ium[v\cos^2\theta - v_F^2/3v]}{\tau}\left[\frac{1}{qv\tau - i\cos\theta}\right]\left[1 + \frac{\Delta^2(T)}{\varepsilon^2}\right]^{-1/2}\left(\frac{\partial f_0}{\partial E}\right).$$

The rate of heat dissipation may be computed according to Eq. (6). Following the same procedure as outlined for transverse waves and neglecting higher-order terms, one finds for $q\ell < 1$

$$\alpha_{s\ell} = \frac{8mN}{15\rho v_\ell \tau}(q\ell)^2 f_0(\Delta)$$

and

$$\frac{\alpha_{s\ell}}{\alpha_{n\ell}} = 2f_0(\Delta) = \frac{2}{e^{\Delta/kT} + 1}.$$

Thus, we have found that the BCS result is valid for longitudinal waves for all values of $q\ell$, and for transverse waves when $q\ell < 1$. But, for transverse waves when $q\ell > 1$, the BCS result has to be multiplied by g as given in Fig. 1, which then gives a drop in attenuation at T_c followed by a residual attenuation portion that also obeys the BCS temperature dependence. For completeness, the temperature dependence of α_s/α_n for Ta is given in Fig. 2, and the resulting temperature dependent energy gap is shown in Fig. 3 (Levy et al., 1963). Superimposed on this figure is the normalized BCS temperature-dependent energy gap. Thus, these measurements are very powerful in that they can provide the complete energy gap function for a superconductor by measurements made throughout the bulk of a sample.

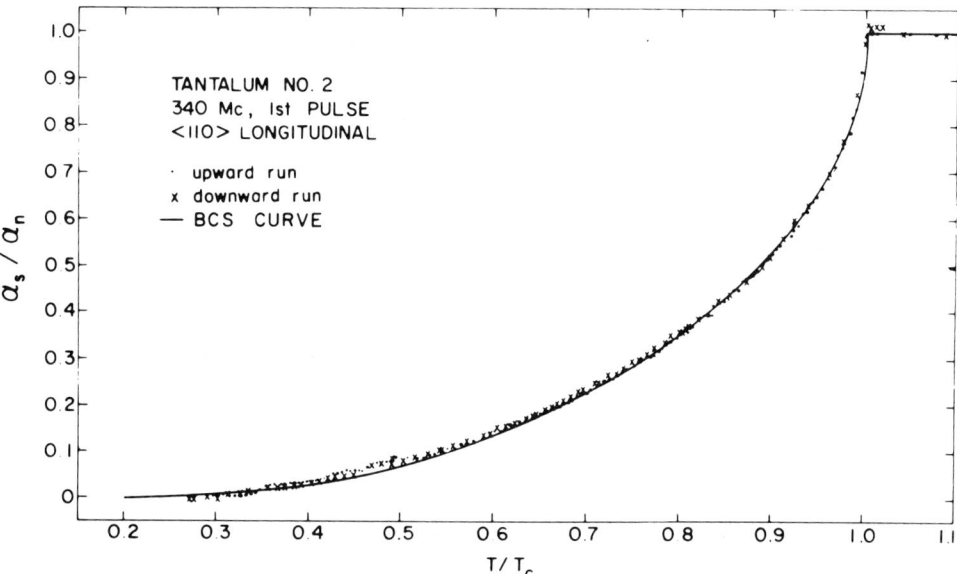

FIG. 2. Normalized attenuation versus reduced temperatures in the superconducting state of Ta obtained with 340 MHz longitudinal waves propagating along the $\langle 110 \rangle$ direction. The solid line is plotted according to the BCS relation (Eq. (1)) with a zero temperature energy gap of $2\Delta = 3.5\, kT_c$ and a $T_c = 4.42$ K. (After Levy, et al., 1963.)

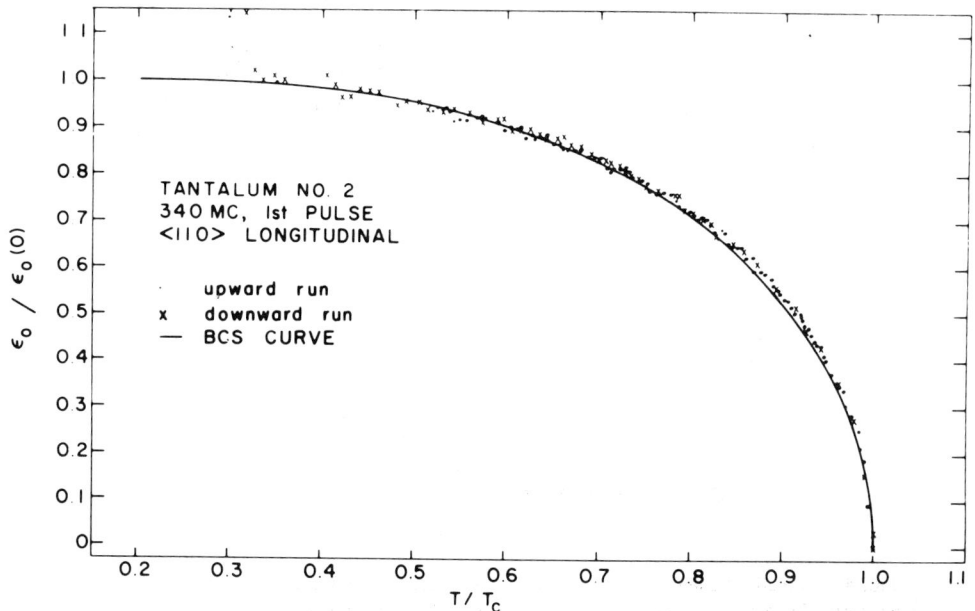

FIG. 3. Normalized energy gap versus reduced temperature obtained from data in Fig. 2 by solving the BCS relation, Eq. (1), for Δ. The solid line is a plot of the BCS temperature-dependent normalized energy gap $\Delta(T)/\Delta(0)$. (After Levy, et al., 1963.)

BCS predicted a zero temperature energy gap $2\Delta(0)$ equal to $3.52\,kT_c$; this is the value that was used for plotting the solid theoretical curve in Fig. 2. Several measurements have found values for the zero temperature energy gap that were different from the BCS value. If it is assumed that the temperature dependence of the normalized energy gap follows the solid curve shown in Fig. 3, then it is possible to plot complete attenuation curves even for such values. This is done in Fig. 4 for three different values of $2\Delta(0)/kT_c$: 4.00, 3.50, and 3.0. It is evident that as $2\Delta(0)$ increases, the attenuation drops off more rapidly below T_c. Thus, it is possible to obtain values for $2\Delta(0)$ just from the data close to T_c by superimposing the normalized experimental data over the theoretical curves. Obviously, $2\Delta(0)$ may also be determined from low-temperature data by plotting $\log\,([2\alpha_n/\alpha_s] - 1)$ as a function of $1/T$ and taking the asymptote of the low-temperature slope, particularly since Δ is almost constant at low temperatures.

The BCS result has been obtained for an isotropic energy gap as schematically shown in Fig. 5a. However, in the heavy Fermion superconductors and in superfluid ^{3}He, it is possible to have anisotropies in the superconducting energy gap wherein the gap may vanish at poles, the axial state (Fig. 5b) or along an equatorial line, the polar state (Fig. 5c). This vanishing of the energy gap will necessarily yield a different temperature dependence for the attenuation of sound waves in the superconducting state. A three-dimensional schematic of the energy gap near a pole is shown in Fig. 6a. It is assumed that the gap vanishes linearly. Thermally excited electrons will occupy this cone up to an energy equal to kT. Therefore, the number of electrons contained in this volume of depth kT will be proportional to $(kT)^3$. And since the attenuation is proportional to the number

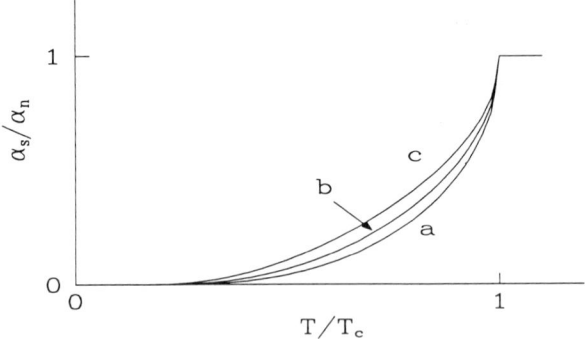

FIG. 4. Normalized attenuation versus reduced temperature using the BCS relation, Eq. (1). The three curves have been plotted for different values of the zero temperature superconducting energy gap $2\Delta/kT_c$; 4.0 (a), 3.5 (b), 3.0 (c). It is assumed that the normalized energy gap follows the BCS temperature dependence shown in Fig. 3.

1. Ultrasonic Attenuation in Conventional Superconductors

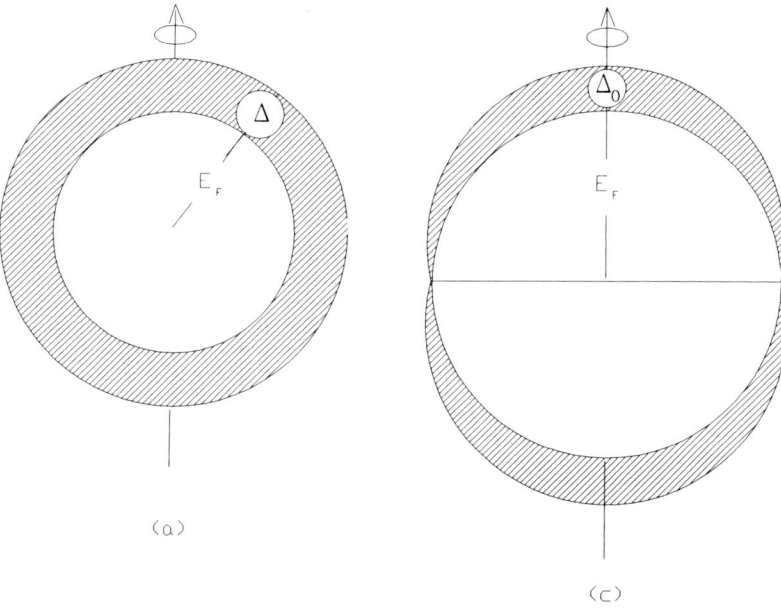

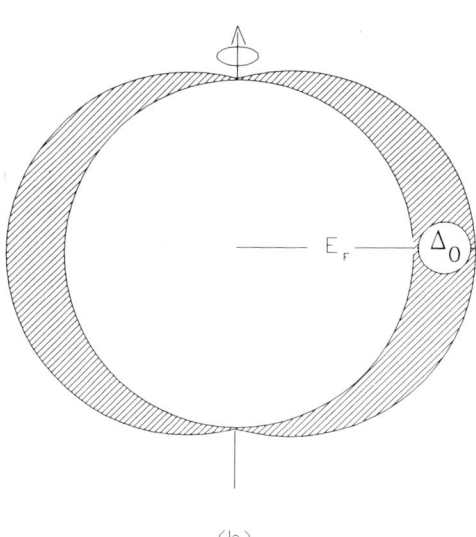

FIG. 5. Schematic drawing of the energy gap of (a) an isotropic state (b) an axial state with nodes at the two poles, and (c) a polar state wherein the gap vanishes on an equatorial line. In these drawings Δ is not drawn to scale.

of excited electrons, it should also be proportional to T^3. On the other hand, the temperature dependence for a polar state should be T^2. This may be seen by unfolding the region about the equatorial node, producing a trough of length $2\pi E_F$, Fig. 6b. Again it will be filled to a depth kT by thermally excited electrons, and the number of electrons contained in this volume will be proportional to $(kT)^2$ and therefore the attenuation will be proportional to T^2. As will be seen in Chapter III, these are the temperature dependences that have been observed in the heavy Fermion superconductor UPt$_3$ with longitudinal waves.

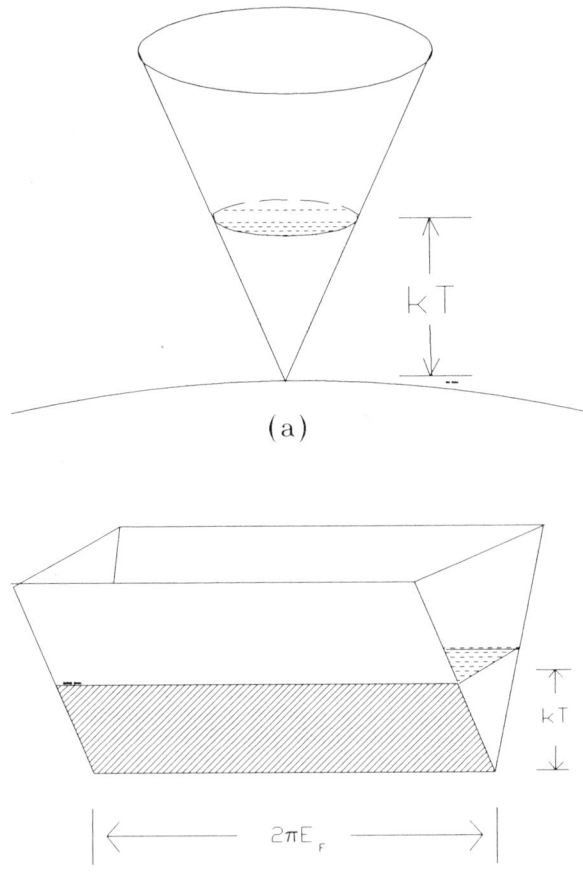

FIG. 6. (a) Expanded schematic of a cone around a pole. (b) Unfolded schematic of the gap on an equatorial line. The shaded volumes of depth kT are occupied by thermally excited electrons.

3. Magnetic Field Dependence

Now, let us turn our attention to the effect of a magnetic field on the attenuation in the superconducting state. Before doing this, however, we must briefly discuss the effect of a magnetic field on the superconducting state. Superconductors can be broadly divided into type I and type II. Type I superconductors exhibit a full diamagnetic or Meissner–Ochsenfeld state throughout their superconducting phase. A magnetization curve is shown for a type I superconductor in Fig. 7a. A phase diagram is shown in Fig. 8a. As seen in Fig. 7a, the sample is completely diamagnetic up to a thermodynamic critical field H_c, whereupon it abruptly loses its diamagnetism, allows the magnetic field B to penetrate throughout its volume, and becomes normal. Thus, since the magnetization is discontinuous at H_c, a superconductor undergoes a first-order phase transition along this boundary, with a latent heat equal to $H_c^2/8\pi$ and a change in volume as discussed in the introduction. Within a few percent, the critical magnetic field follows a parabolic temperature dependence, $H_c = H_0(1 - t^2)$, where $t = T/T_c$.

A magnetization curve for a type II superconductor is shown in Fig. 7b. A sample is completely diamagnetic up to a lower critical field H_{c1}, whereupon magnetic field lines in the form of flux lines, whose value is $\phi_0 = hc/2e = 2 \times 10^{-7}$ gauss-cm^2, start to penetrate, making the sample only partially diamagnetic up to the upper critical field H_{c2}, where the bulk of the sample becomes normal. Surface superconductivity will persist up to $H_{c3} = 1.7\, H_{c2}$. The transitions along H_{c1}, H_{c2}, and H_{c3} are second-order, since the magnetization as a function of magnetic field is continuous along these boundaries, but its slope is discontinuous.

The superconducting phase above H_{c1} is known as the mixed phase. The ratio of the London penetration length $\lambda_L = (mc^2/4\pi Ne^2)^{1/2}$ to the coherence length $\xi_0 = 2\hbar v_F/\pi \Delta(0)$ determines whether a superconductor is type I or type II. In these expressions, c is the velocity of light and $\hbar$ is Planck's constant. The London penetration length determines how far a magnetic field will penetrate into a superconductor before it is screened out by superconducting currents, while the coherence length determines who far a disturbance of the superconducting wave function or order parameter Δ will persist before superconductivity is recovered. The Ginzburg–Landau parameter is defined as

$$\kappa = \frac{1}{\sqrt{2}} \frac{\lambda}{\xi}.$$

The transition between a type I and a type II superconductor occurs at $\kappa = 1/\sqrt{2}$.

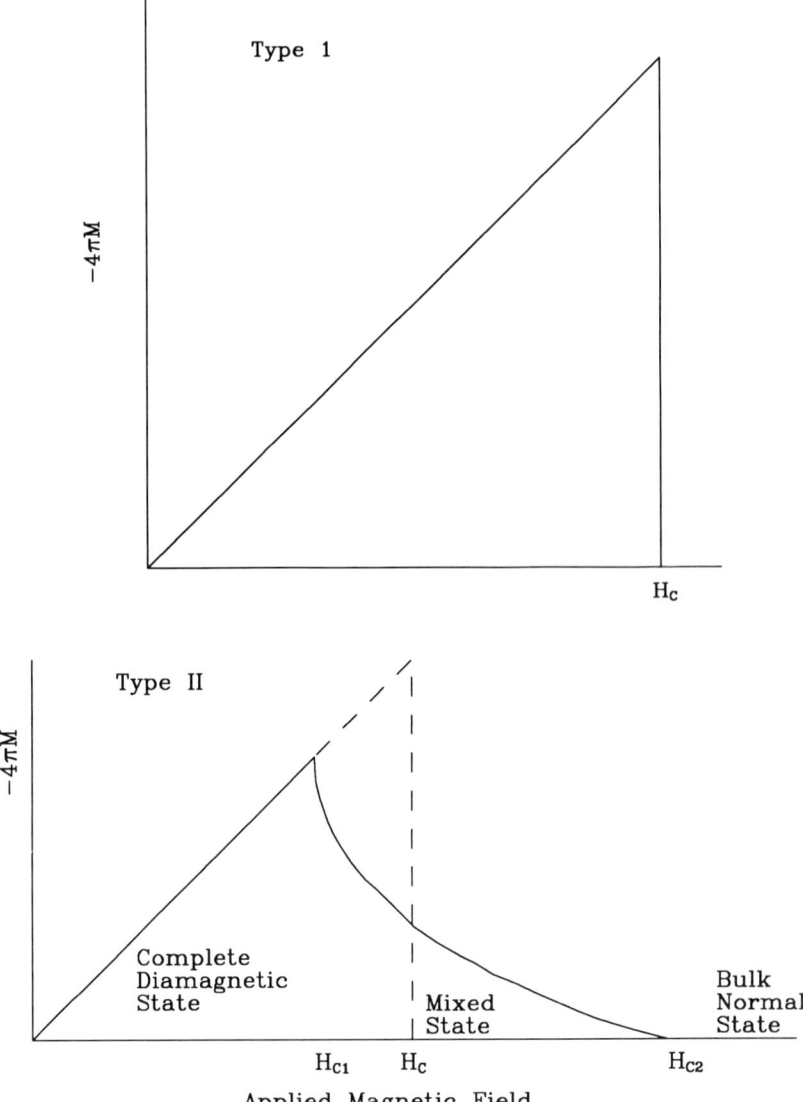

FIG. 7. (a) Magnetization curve for a Type I superconductor. Below the critical field H_c, the bulk superconductor is fully diamagnetic. Above H_c, it is in the normal state. (b) Magnetization curve for a Type II superconductor. Below the lower critical field, H_{c1}, the sample is fully diamagnetic. Between H_{c1} and the upper critical field H_{c2}, the sample is in the mixed state, where flux lines penetrate the sample; however, it still exhibits superconducting properties. Above H_{c2}, the bulk of the sample is in the normal state. The thermodynamic critical field H_c is obtained by setting the area under the solid curve equal to that under the dashed curve.

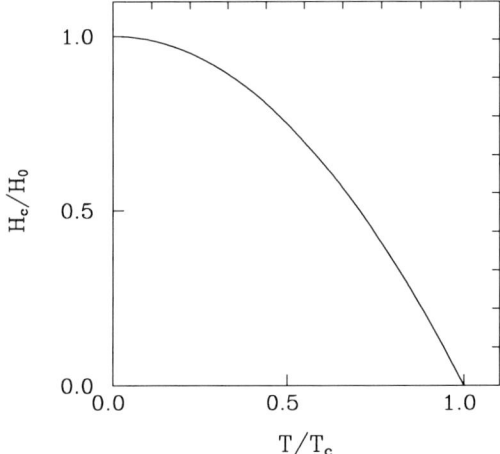

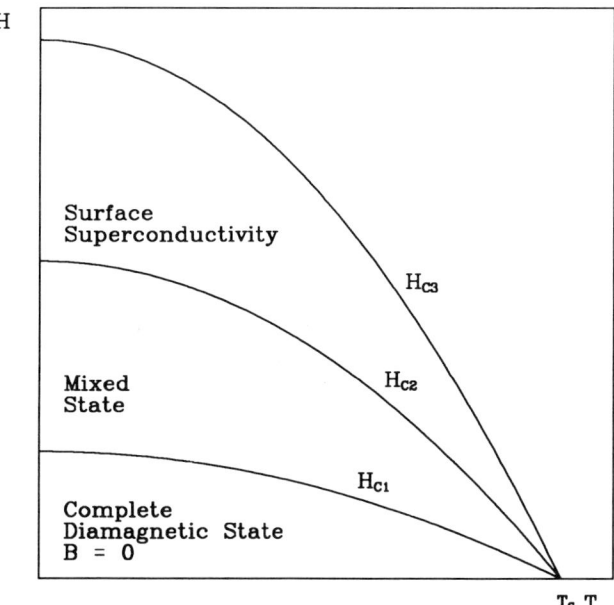

FIG. 8. (a) Phase diagram for a Type I superconductor. The thermodynamic critical field is plotted according to $H_c = H_0(1 - t^2)$, where $t = T/T_c$. In the superconducting state, $B = 0$. (b) Phase diagram for the Type II superconductor. The specimen may exhibit surface superconductivity up to H_{c3}.

A type I superconductor has $\kappa < 1/\sqrt{2}$, and a type II superconductor, $\kappa > 1/\sqrt{2}$. A type I superconductor undergoes a phase transition at the thermodynamic critical field H_c when the energy required to exclude the magnetic field from the sample volume $H_c^2/8\pi$ becomes equal to the decrease in energy due to superconductivity $\frac{1}{2}N_0\Delta^2$, where N_0 is the density of electrons per unit energy at the Fermi level. A type II superconductor undergoes a phase transition at H_{c1} when the energy required to quench superconductivity within a flux line of radial dimensions ξ is comparable to the magnetic energy contained in a cylinder of radius λ_L. Thus, to a first approximation

$$H_{c1} = \frac{H_c}{\sqrt{2}\,\kappa},$$

where H_c is the thermodynamic critical field obtained by integrating the area under Fig. 4b and setting it equal to $H_c^2/8\pi$. The bulk of a type II superconductor becomes normal at H_{c2} when flux lines of dimension ξ overlap at

$$H_{c2} = \sqrt{2}\,\kappa\, H_c.$$

The transition at H_{c2} is second-order, and the magnetization has a linear dependence near H_{c2}, $-4\pi M \simeq H_{c2} - H$. In addition, a distinction should be made between clean and dirty type II superconductors. In a clean type II superconductor, the electron mean free path is longer than the coherence length $\ell > \xi_0$, and the effective coherence length is just ξ_0, while in a dirty type II superconductor the electron mean free path is shorter than the coherence length $\ell < \xi_0$ and the effective coherence length is given by $\xi = \sqrt{\ell\xi_0}$. Both λ and ξ are temperature-dependent, going to infinity at T_c as $(T_c - T)^{-1/2}$. However, κ is only slightly temperature-dependent. It may increase by less than 26% as the temperature is lowered from T_c to 0 K (Eilenberger, 1967).

After this brief introduction, we can discuss the attenuation in type I and type II superconductors. In the absence of a magnetic field, both should follow the BCS relation. In a magnetic field, the attenuation in a type I superconductor should just remain constant up to the critical field where it should rapidly increase up to its normal state value.

3.1. Dirty Type II Superconductors

We shall only discuss the behavior of the attenuation close to H_{c2} for type II superconductors. In Fig. 9a, we have sketched the square of the order parameter, Δ^2, as a function of position in the mixed state of a dirty type II superconductor. The square of the order parameter, Δ^2, is proportional to the square of the superconducting wave function, $|\psi|^2$, which is also proportional to the magnet-

1. Ultrasonic Attenuation in Conventional Superconductors

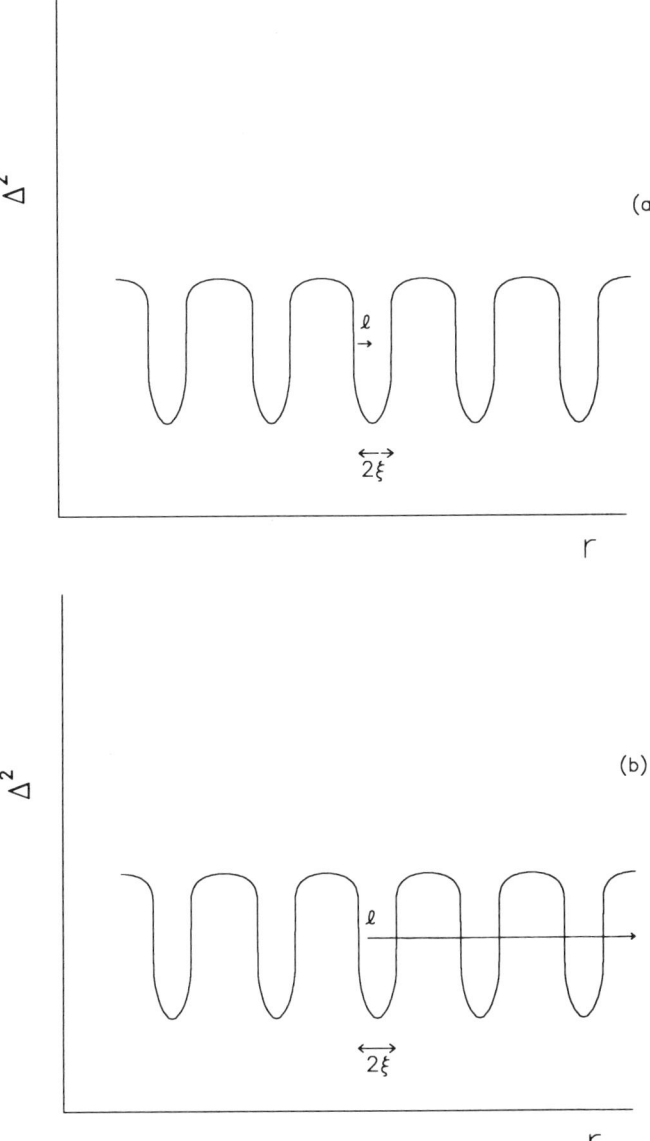

FIG. 9. Schematic of the order parameter Δ as a function of position r in the mixed state of a Type II superconductor. The order parameter vanishes at the center of the flux lines. The square of the order parameter Δ^2 is proportional to the superconducting density n_s. The diameter of the flux lines is given by twice the coherence distance, 2ξ, approximately. (a) For a dirty Type II superconductor, the electron mean free path ℓ is smaller than the coherence distance ξ, and an electron may undergo several collisions before escaping the flux line. (b) For a clean Type II superconductor, ℓ is larger than ξ, and an electron takes a spatial average of the order parameter.

ization. At the center of each flux line Δ^2 vanishes, since the magnetic flux has quenched superconductivity within a radius ξ. Since $\ell < \xi$, we have also indicated a typical mean free path inside one of the flux lines. Thus, electrons inside a flux line will suffer several collisions before escaping into the superconducting region. And, electrons in the superconducting region will also have several collisions before entering a flux line. Therefore, electrons inside a flux line will contribute the energy loss associated with the normal state, while those in the superconducting region contribute that of the superconducting state.

Thus, the attenuation in the mixed state for a dirty type II superconductor should be proportional to the fraction of normal regions, which should be proportional to one plus the magnetization since $-4\pi M \simeq |\psi|^2 \simeq |\Delta|^2$:

$$\frac{\alpha_s(H)}{\alpha_n} \simeq 1 + 4\pi M,$$

which close to H_{c2} yields

$$\frac{\alpha_n - \alpha_s(H)}{\alpha_n} \simeq -4\pi M \simeq H_{c2} - H. \qquad (11)$$

Therefore, the attenuation should be linearly dependent on the applied magnetic field close to H_{c2}. In Fig. 10 we have plotted the attenuation of surface

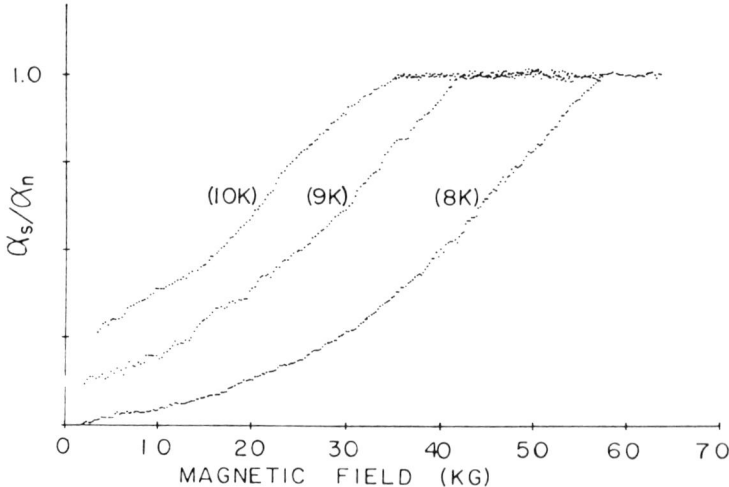

FIG. 10. Normalized attenuation of 700 MHz surface acoustic waves versus applied magnetic field for a Nb_3Sn film on a $LiNbO_3$ substrate. Data were taken at constant temperatures of 8 K, 9 K, and 10 K. This is an example of the behavior of a dirty Type II superconductor, wherein the attenuation decreases linearly below $H_{c2}(T)$ as shown in Eq. (11). (After Fredricksen et al., 1979. © 1979 IEEE.)

1. Ultrasonic Attenuation in Conventional Superconductors

acoustic waves traveling through a film of Nb_3Sn as a function of magnetic field (Fredricksen *et al.*, 1979). It can clearly be seen that the attenuation is linearly dependent on the magnetic field close to H_{c2}.

3.2. CLEAN TYPE II SUPERCONDUCTORS

Now, let us turn our attention to clean type II superconductors. In Fig. 9b we have indicated a typical mean free path traversing several flux lines, since $\ell > \xi_0$ in this case. As opposed to the previous case shown in Fig. 9a, the electrons sample several regions during each mean free path. And we can postulate that

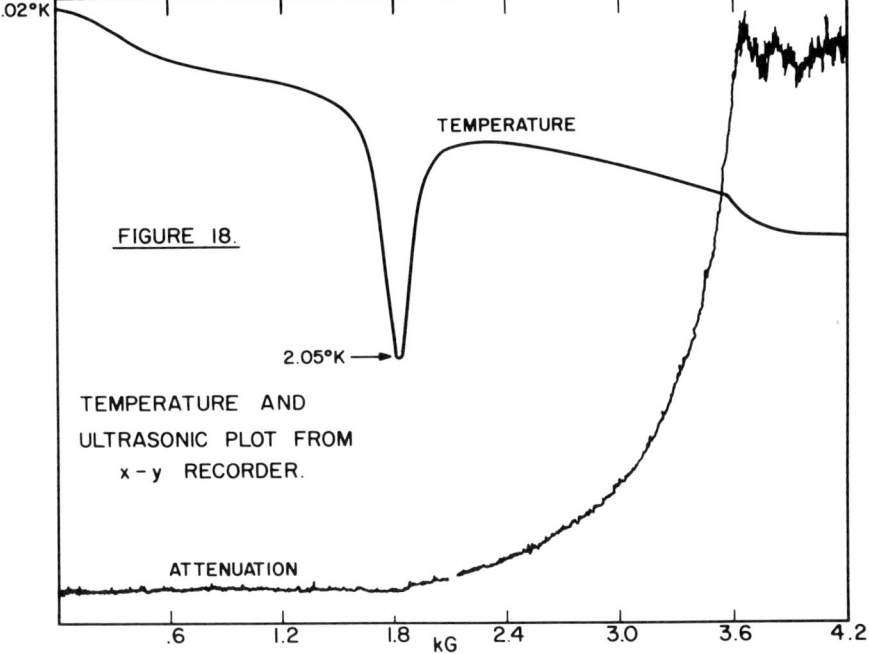

FIG. 11. Raw ultrasonic attenuation data versus magnetic field for a Nb single crystal obtained with 104 MHz longitudinal waves at 2 K. This is an example of the behavior of a clean Type II superconductor. The attenuation drops rapidly below $H_{c2} = 0.36\ T$, as would be expected for a square root dependence. In this experimental run, the temperature of the sample was monitored simultaneously with the attenuation. The 0.03 K temperature spike at about 1.8 kG is associated with the influx of magnetic flux lines at H_{c1}. The motion of the flux lines should cause an increase in temperature, while the adiabatic quenching of superconductivity within the flux lines should decrease the temperature. For this sample the exothermic process is larger than the endothermic process, resulting in a temperature rise. The subsequent decrease is produced by the sample coming to equilibrium with the thermal bath. (After Kagiwada, 1966.)

the electrons sample a spatially averaged energy gap or order parameter before each collision. So we can return to the BCS expression and replace Δ by a spatially averaged $\langle\Delta\rangle$ that is proportional to $M^{1/2}$. For small values of $\langle\Delta\rangle$ we can expand Eq. (1):

$$\frac{\alpha_s}{\alpha_n} = \frac{2}{e^{\Delta/kT} + 1} \simeq \frac{2}{1 + \langle\Delta\rangle/kT + 1} \simeq 1 - \frac{\langle\Delta\rangle}{2kT},$$

$$\frac{\alpha_n - \alpha_s}{\alpha_n} \simeq \langle\Delta\rangle \simeq (-4\pi M)^{1/2} \simeq \sqrt{H_{c2} - H}. \quad (12)$$

Therefore, the attenuation close to H_{c2} for a clean type II superconductor should be proportional to the square root of the difference in field. In Fig. 11 we have plotted the attenuation of bulk waves through single crystal Nb as a function of H (Kagiwada, 1966) and in Fig. 12 as a function of $\sqrt{H_{c2} - H}$ (Kagiwada *et al.*, 1967). The straight-line fit in Fig. 12 shows that this square root relationship is followed.

This concludes our necessarily brief summary of the attenuation effects that can be observed in classical BCS superconductors. However, even in these systems, there are numerous effects that have not been mentioned, and indeed

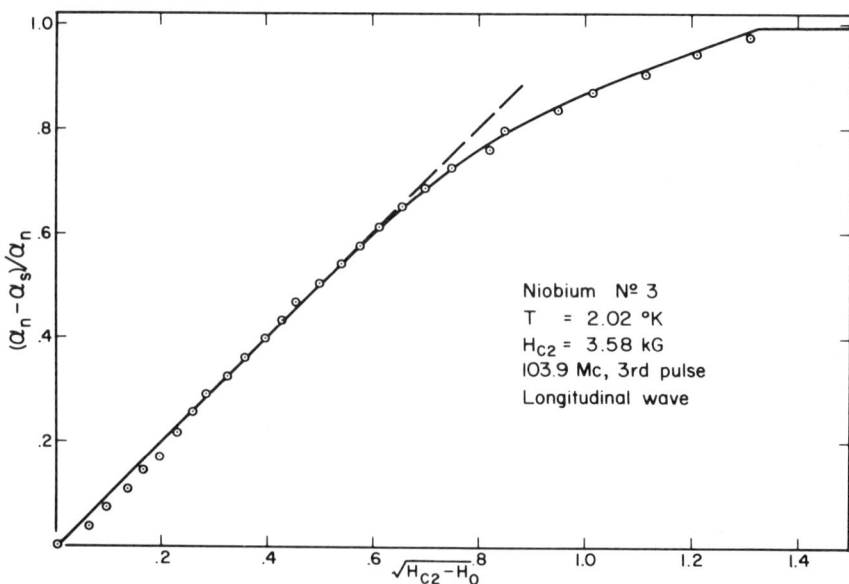

FIG. 12. Analysis of data for a clean Type II superconductor shown in Fig. 8. This is a plot of the normalized difference in attenuation $(\alpha_n - \alpha_s)/\alpha_n$ versus the square root of the difference in magnetic field $\sqrt{H_{c2} - H}$. The straight-line fit near the origin demonstrates the validity of Eq. (12). (After Kagiwada *et al.*, 1967.)

even remain unexplained. There are two band effects in the normal state, which may give rise to attenuation drops in the $q\ell > 1$ limit even for longitudinal waves in the superconducting state (Robinson and Levy, 1970). There are unexplained large apparent energy gaps found in clean niobium (Carsey *et al.*, 1971), mercury (Thomas *et al.*, 1966) and lead (Deaton, 1966). There is excess attenuation in the superconducting state caused by dislocation unpinning due to a reduction of electron viscosity in the superconducting state (Tittman and Bommel, 1965), and there is a linear attenuation region close to H_{c2} even for clean type II superconductors due to a gapless density of state (Carsey and Levy, 1971). Several articles covering some of these and other topics concerning superconductivity have been published in previous volumes of this series on physical acoustics (Gottlieb *et al.*, 1970; Rayne and Jones, 1970; Suenaga and Galligan, 1972; Testardi, 1973; Eisenmager, 1976; Testardi, 1977).

ACKNOWLEDGMENT

This work was supported by the Office of Naval Research.

References

Andres, K., and Rohrer, H. (1961). *Helv. Phys. Acta* **34**, 398–400.
Bardeen, J., Cooper, L. N., and Schrieffer, J. R. (1957). *Phys. Rev.* **106**, 1175–1204.
Blount, E. I. (1959). *Phys. Rev.* **114**, 418–436.
Bommel, H. E. (1954). *Phys. Rev.* **96**, 220–221.
Bommel, H. E. (1960). Private communication.
Bommel, H. E., and Olsen, J. L. (1953). *Phys. Rev.* **91**, 1017–1018.
Carsey, F., Kagiwada, R., Levy, M., and Maki, K. (1971). *Phys. Rev.* **B4**, 854–862.
Carsey, F., and Levy, M. (1971). *Phys. Rev. Let.* **27**, 853–856.
Clayborne, L. T., and Morse, R. W. (1964). *Phys. Rev.* **136**, A893–A905.
Beaton, B. C. (1966). *Phys. Rev. Lett.* **16**, 577–581.
Eilenberger, G. (1967). *Phys. Rev.* **153**, 584–598.
Eisenmenger, W. (1976). In "Physical Acoustics," Vol. 12. Academic Press, New York, pp. 79–153.
Fredricksen, H. P., Salvo, H. L., Jr., Levy, M., Hammond, R. H., and Geballe, T. H. (1979). *Proceedings 1979 IEEE Ultrasonics Symposium* (79 CH1482-9; J. de Klerk and B. R. McAvoy, eds.). IEEE, New York, pp. 435–438.
Gottlieb, M., Garbury, M. and Jones, C. K. (1970). In "Physical Acoustics," Vol. 7. Academic Press, New York, pp. 1–49.
Kagiwada, R. S. (1966). Ph.D. Thesis, University of California, Los Angeles (unpublished).
Kagiwada, R., Levy, M., Rudnick, I., Kagiwada, H., and Maki, K. (1967). *Phys. Rev. Lett.* **18**, 74–76.
Keesom, W. H. and Kok, J. A. (1932). *Commun. Phys. Lab. Univ. Leiden*, No. 221e.
Landauer, J. K. (1954). *Phy. Rev.* 94.
Lasarew, B. G., and Sudovstov (1949). *Dokl. Akad. Nauk S.S.S.R.* **69**, 345.
Levy, M. (1963). *Phys. Rev.* **131**, 1497–1500.
Levy, M., Kagiwada, R., and Rudnick, I. (1963). *Phys. Rev.* **132**, 2039–2046.

Mason, W. P. (1955). *Phys. Rev.* **97**, 557–558.
Meissner, W. and Oshsenfeld, R. (1933). *Naturwissenschaften* **21**, 787.
Morse, R. W. (1955). *Phys. Rev.* **97**, 1716–1717.
Pippard, A. B. (1955). *Phil. Mag.,* Series 7, **46**, 1104–1114.
Rayne, J. A., and Jones, C. K. (1970). In "Physical Acoustics," Vol. 7. Academic Press, New York, pp. 149–218.
Robinson, D. A., and Levy, M. (1970). *Phys. Rev. Lett.* **24**, 1238–1242.
Suenaga, M., and Galligan, J. M. (1972). In "Physical Acoustics," Vol. 9. Academic Press, New York, pp. 1–34.
Testardi, L. R. (1973). In "Physical Acoustics," Vol. 10. Academic Press, New York, pp. 193–296.
Testardi, L. R. (1977). In "Physical Acoustics," Vol. 13. Academic Press, New York, pp. 29–47.
Thomas, R. L., Wu, H. C., and Tepley, N. (1966). *Phys. Rev. Lett.* **17**, 22–24.
Tittman, B. R., and Bommel, H. E. (1965). *Phys. Rev. Lett.* **14**, 296–298.

—2—
Sound Propagation and Collective Modes in Superfluid ^{3}He

BIMAL K. SARMA
Department of Physics, University of Wisconsin–Milwaukee, Milwaukee, Wisconsin

J.B. KETTERSON, S. ADENWALLA, Z. ZHAO*
Northwestern University, Evanston, Illinois

1. Introduction 23
   1.1. Collective Modes in a Many-Body System 25
2. Normal State of ^{3}He 27
   2.1. Fermi Liquid 27
   2.2. Zero Sound 31
3. Superfluid ^{3}He 33
   3.1. Experimental Properties of Superfluid ^{3}He 33
   3.2. Superfluid Phase Diagram 36
   3.3. Order Parameter of Superfluid ^{3}He 39
4. Collective Modes 43
   4.1. Hydrodynamic Modes 44
   4.2. Order Parameter Collective Modes 46
5. Experimental Results 53
   5.1. Experimental Techniques 53
   5.2. Superfluid ^{3}He-B 68
   5.3. Superfluid ^{3}He-A 92
6. Conclusions 101
   Acknowledgments 102
   References 103

1. Introduction

Superfluid ^{3}He is the only established example of unconventional superfluidity. The BCS theory explained metallic superconductivity on the basis of singlet Cooper pairing arising from a weak attractive interaction mediated by the electron-phonon interaction. There has been much speculation on other

---

*Present address: Department of Physics, Lyman Lab of Physics, Harvard University, Cambridge, MA 02138

pairing mechanisms. In addition, pairing in higher angular momentum and triplet spin states was investigated.

The ^{3}He nucleus has a spin $\frac{1}{2}$, and hence obeys Fermi statistics. In many respects it is similar to the electron system, at low temperatures, except for the absence of electronic charge. In analogy with the BCS picture, two ^{3}He atoms can form a bound pair in the presence of an attractive interaction. However, because of the strong hard core repulsion of the helium atoms, pairing in the zero angular momentum state (an s state with a large amplitude at small particle separations) is not possible. The lowest angular momentum state is thus $L = 1$ (p-wave), and because of the requirements of an antisymmetric pair wave function (^{3}He atoms being Fermi particles), the spin state will be $S = 1$ (spin triplet). Thus the total angular momentum, J, can be $J = 0$, 1, or 2, leading to a more complicated order parameter structure. The number of degrees of freedom for the order parameter has $(2L + 1) \times (2S + 1) = 9$ complex coefficients, and taking into account both the real and imaginary parts for these coefficients, this amounts to 18 degrees of freedom.

Experimentally it is found that there are multiple superfluid phases. Only two exist at zero magnetic field, in the P–T plane: the A-phase and the B-phase. The A-phase is stable only at relatively high pressures, in excess of 21 bars. The B-phase is the stable low-temperature, low-pressure state. However, on applying a magnetic field, the A-phase can be stabilized at low pressures: A narrow wedge of A-phase appears between the normal phase and the B-phase. This wedge of A-phase widens as the field is increased, and by 6 kilogauss the B-phase is completely suppressed. A new phase, the A_1 phase, which is stable only in a magnetic field, exists near the normal superfluid phase boundary.

^{3}He is a quantum liquid: The large zero-point motion overshadows the weak van der Waals attraction, and as a result ^{3}He (as well as ^{4}He) remains a liquid all the way down to absolute zero. At zero temperature, a pressure in excess of 34 bars is required to solidify ^{3}He (25 bars for ^{4}He). In contrast to ^{4}He, ^{3}He, because of its odd number of nucleons (two protons and one neutron), has a spin $\frac{1}{2}$ and is governed by Fermi statistics. As a result, the heat capacity of ^{3}He decreases linearly with temperature. The low-temperature thermal properties of the solid are dominated by the nuclei, which have an entropy $R\ell n 2$ well above the ordering temperature. At very low temperatures where the exchange energy of the nuclear spins becomes of the order of $k_B T$, the spins order (in an antiferromagnetic state), and the entropy drops dramatically. The exchange coupling in ^{3}He is large because of the large zero-point motion. The resulting spin–spin interaction is of the order of 1 mK, instead of the typical 1 μK (arising from dipole–dipole interactions) encountered in nonquantum solids. From thermo-

2. Sound Propagation and Collective Modes

dynamics (the Clausius–Clapeyron equation), the melting curve has a minimum at the point where the lattice entropy (γT) is equal to the spin entropy of the nuclei ($R\ell n2$). This occurs at a temperature of 316 mK and a pressure of 29.3 bar. Below this temperature the slope of the melting curve dP/dT is negative, and thus, applying a pressure to a liquid–solid mixture of ^{3}He will cause cooling. This phenomenon was predicted by Pomeranchuk (1950) and successfully applied to cool a mixture of solid and liquid ^{3}He into the superfluid phase by the Cornell group (Osheroff *et al.*, 1972a). The results were initially interpreted as an ordering transition in the solid phase. The experiment consisted of measuring the pressure (traced on a chart recorder) as a function of time, as the mixture was being cooled. They observed distinct changes in the slope at 2.6 mK, and again at 1.9 mK, which were identified as phase transitions. However, in a subsequent NMR experiment (Osheroff *et al.*, 1972b), these slope changes were found to be associated with the liquid phase, and it was clear that they were observing the superfluid transition.

There are many excellent reviews discussing superfluid ^{3}He: We mention only a few. The first reviews were by Leggett (1975) and Wheatley (1975), followed by the book edited by Bennemann and Ketterson (1978). See especially the article by Lee and Richardson (1978). Recently there have been the theoretical review by Vollhardt and Wölfle (1990) and the book edited by Halperin and Pitaevskii (1990). There are two recent review articles dealing with the ultrasonic aspects by Dobbs and Saunders (1991) and Zhao *et al.* (1991). Because of the restricted nature of this article we shall concentrate mostly on the experiments performed at Northwestern, with special emphasis on the collective mode spectrum of the superfluid, which provides some of the strongest support for unconventional superfluidity.

1.1. COLLECTIVE MODES IN A MANY-BODY SYSTEM

When disturbed, a many-body system relaxes back to equilibrium. The relaxation time, τ, depends on the interactions and is typically very fast, of the order of 10^{-14} s. There are a restricted class of excitations where the system behaves in a collective manner: When disturbed, the excitations persist for a time much longer than τ. Such excitations are called collective modes. A study of the collective mode response yields information about the many-body interactions of the system.

The collective modes may be classified in terms of the physical principles that generate them: conservation laws, spontaneously broken symmetries, and mean fields. The familiar hydrodynamic collective modes of fluids and solids arise because these systems obey the local conservation laws of mass, momentum, and energy (solids have additional hydrodynamic modes arising from the broken translational symmetry). The conservation equations define the dynamic variables of the system and govern the behavior of these variables as they undergo periodic variations about some equilibrium value. These deviations are the collective modes. For example, the hydrodynamic sound modes are collective modes based on the conservation laws. A fluid can support longitudinal sound modes, which are compressional waves, the velocity being given by $c_1 = \sqrt{(B/\rho)}$, where B is the bulk modulus (or the inverse of the compressibility). The restoring forces are the many interatomic collisions. As the temperature is lowered, the collisions decrease (increasing the viscosity), and the damping increases. In ^{3}He liquid this mode is called the first sound mode. Since liquids do not support a shear stress, there is no shear velocity mode, as would be present in the case of a solid. (The shear waves are a direct manifestation of the broken translational symmetry.)

Many systems when cooled make a transition into a more ordered state. The symmetry in the ordered state is lower. For example, in a ferromagnet the crystal symmetry is lost at the Curie temperature where the system chooses a specific direction along which to magnetize. This symmetry is spontaneously broken at the transition that leads to a second-order transition. The ground state is then degenerate with respect to these symmetry changes, and small periodic oscillations in this symmetry breaking variable about some equilibrium value may exist. In our example of the ferromagnet, the direction of the magnetization axis may vary slowly in space and time, which is referred to as a spin wave. These oscillations are called Goldstone collective modes. They may be hydrodynamic or nonhydrodynamic in nature depending on the local equilibrium conditions.

In the mean field approach, the complex particle–particle interactions in the system are approximated by the interaction of each particle with the mean field generated by the distribution of the other particles. The mean field produces a "rigidity" in the system that resists large deviations from the equilibrium distribution; small oscillations about the equilibrium configuration are interpreted as collective modes. In the Fermi liquid regime in normal ^{3}He, at very low temperatures, such a mode, called the zero sound mode, becomes possible. This is a collisionless mode, with a velocity different from c_1, and the restoring force comes from the mean field due to the strong interactions between the atoms.

All three types of collective modes exist in superfluid ^{3}He. The zero sound mode will be discussed in Section 2.2. The other collective modes will be discussed in Sections 4.1 and 4.2. Many of these superfluid collective modes have an energy of the order of the energy gap, and since some of these couple to density variations, it has been possible to study them with ultrasonic methods.

2. Sound Propagation and Collective Modes

2. Normal State of ^{3}He

2.1. FERMI LIQUID

Liquid ^{3}He is an example of a strongly interacting Fermi liquid. Because of the nuclear spin ($s = \frac{1}{2}$), the system obeys Fermi statistics; however, because of the large mass (as compared to the electron system), the Fermi temperature is very low (of the order of 1 K or lower, depending on the pressure). Landau put forth a phenomenological model: The excitations of the strongly interacting ^{3}He atoms are replaced by an identical number of more weakly interacting excitations called quasiparticles; the excitations behave as a Fermi gas in many respects, having the same temperature dependence for the thermodynamic properties. The strong interactions are absorbed in a (renormalized) effective mass, and through the introduction of a small number of phenomenological parameters called Landau parameters. For liquid ^{3}He the ratio of the effective mass to the bare mass, m^*/m, varies from three at the saturated vapor pressure (s.v.p.) to six at the melting pressure (m.p.). The Fermi momentum, which is given by the same expression as for a noninteracting gas, is

$$p_F = \hbar(3\pi^2 n)^{1/3}, \tag{1}$$

where n is the number density of ^{3}He atoms. However, both the Fermi energy and Fermi velocity are lowered because of the enhanced mass. An additional feature (of the theory) is the presence of a new normal state collective mode called zero sound. This was predicted by Landau (1957) and first observed by Abel *et al.* (1965).

For a noninteracting Fermi gas, the particles obey the Fermi distribution,

$$n_p = \frac{1}{e^{(\varepsilon(p)-\mu)/k_B T} + 1}, \tag{2}$$

where μ is the chemical potential or Fermi energy with

$$\varepsilon_F = \frac{p_F^2}{2m} \tag{3}$$

and

$$v_F = \left(\frac{\delta \varepsilon}{\delta p}\right)_{p_F} = \frac{p_F}{m}. \tag{4}$$

The density of states, N_F, per unit energy and unit volume at the Fermi surface (for both spins) is given by

$$N_F = \frac{2}{(2\pi\hbar)^3} \frac{4\pi p_F^2}{\left(\dfrac{d\varepsilon}{dp}\right)_{p_F}} \quad (5)$$

$$= \frac{mk_F}{\pi^2\hbar^2}. \quad (6)$$

The heat capacity is linear in T, and the Pauli paramagnetic susceptibility is temperature-independent.

$$C_V = \frac{\pi^2}{3} k_B^2 N_F T, \quad (7)$$

$$\chi_N^0 = \frac{1}{4} \gamma^2 \hbar^2 N_F; \quad (8)$$

γ is the gyromagnetic ratio $2\mu/\hbar^2$, where μ is the magnetic moment (for electrons this is the Bohr magneton, μ_B, and for ^{3}He it is the ^{3}He nuclear moment μ_3).

In the Landau–Fermi liquid model, there is a one-to-one correspondence between the noninteracting particles and the "quasi-particles" as the interaction is switched on. As a result the "quasi-particles" are like ^{3}He atoms with a back flow arising from the surrounding atoms resulting in an effective mass, m^*. The "quasi-particles" behave similarly to the free particles: They obey the same Fermi statistics. At $T = 0$, all energy states up to the Fermi level are filled (note p_F remains the same), and at a finite T some of the states above ε_F become occupied, which are called quasi-particles, leaving unoccupied states below ε_F, called quasi-holes. Alternatively, one refers to the hole and particle states collectively as quasi-particles. One may describe the system with the distribution function $n(p,\sigma)$ or $\delta n(p,\sigma)$, the deviation of the distribution from the equilibrium value.

In addition to a change in the effective mass, there is also an effective interaction between the "quasi-particles." A local change in the "quasi-particle" distribution $\delta n_{k\alpha\beta}(\mathbf{r}, t)$ causes a change in the local "quasi-particle" energy

$$\delta\varepsilon_{k\alpha\beta}(\mathbf{r}, t) = \sum_{k'\alpha'\beta'} f_{k\alpha\beta;k'\alpha'\beta'}(\mathbf{r}, t)\, \delta n_{k'\alpha'\beta'}(\mathbf{r}, t), \quad (9)$$

where $f_{k\alpha\beta k'\alpha'\beta'}$ is the Fermi liquid interaction function introduced by Landau (1956, 1957). The excitations of interest usually involve states near the Fermi surface. The deviation can then be approximated as a δ function evaluated at p_F. The angular dependence is expanded in Legendre functions in the form

2. Sound Propagation and Collective Modes

$$f_{k\alpha\beta;k'\alpha'\beta'} = N_F^{-1} \sum_{\ell=0}^{\infty} P_\ell(k,k') \left[F_\ell^s \delta_{\alpha\beta}\delta_{\alpha'\beta'} + F_\ell^a \sum_{\mu=1}^{3} (\sigma_\mu)_{\alpha\beta} (\sigma_\mu)_{\alpha'\beta'} \right]. \quad (10)$$

F_ℓ^s and F_ℓ^a are the dimensionless spin symmetric and spin antisymmetric Landau parameters. It is sufficient to keep only the first few terms, typically $\ell = 0, 1$, and 2. This Landau interaction describes the shift in energy of a specific quasiparticle arising from the mean field (molecular field) produced by the other quasiparticles. Following Vollhardt and Wölfe (1990), one obtains

$$\delta\varepsilon_{k\alpha\beta} = N_F^{-1} \left[F_0^s \delta n \delta_{\alpha\beta} + F_0^a \frac{2}{\hbar} \sum_\mu (\sigma_\mu)_{\alpha\beta} S_\mu \right] \quad (11)$$

$$+ \frac{1}{3\rho} F_1^s \frac{m}{m^*} \hbar \mathbf{k} \cdot \mathbf{g} \delta_{\alpha\beta}, \quad (12)$$

where

$$\delta n = \sum_{k\alpha} \delta n_{k\alpha\alpha} \quad \text{(particle density)}, \quad (13)$$

$$S_\mu = \frac{\hbar}{2} \sum_{k\alpha\beta} (\sigma_\mu)_{\alpha\beta} \delta n_{k\alpha\beta} \quad \text{(spin density)}, \quad (14)$$

$$\mathbf{g} = \sum_{k\alpha} \hbar \mathbf{k} \delta n_{k\alpha\alpha} \quad \text{(momentum density)}. \quad (15)$$

Since ^{3}He is a fluid with translational invariance, it obeys a Galilean invariance principle; in metals, this symmetry is lost because of the discrete nature of crystal symmetry.

Thermodynamic Properties

The quasiparticle distribution function is given by

$$n_{k\alpha} = \left[\exp\left(\frac{\varepsilon_{k\alpha} - \mu}{k_B T}\right) + 1 \right]^{-1}. \quad (16)$$

The derivation of the various thermodynamic functions is treated in many texts. The heat capacity is linear in temperature and given by

$$C_N = -\frac{\pi^2}{3} N_F k_B^2 T = \frac{m^* k_F}{3\hbar^2} k_B^2 T, \quad (17)$$

where

$$\frac{m^*}{m} = 1 + \frac{1}{3}F_1^s. \qquad (18)$$

Since experimentally m^*/m ranges from 3 to 6, F_ℓ^s ranges from 6 (at s.v.p.) to 15 (at m.p.), an effect of the strong correlations.

The spin susceptibility is

$$\chi_N = \frac{\chi_N^0}{1 + F_0^a}. \qquad (19)$$

It is enhanced by a factor of ~ 4 over that of a free Fermi gas of mass m^* leading to $F_0^a \sim -0.75$, and is very weakly pressure-dependent. The compressibility is given by

$$K_N = \frac{1}{n^2}\frac{\delta n}{\delta \mu} = \frac{1}{n^2}\frac{N_F}{1 + F_0^s}, \qquad (20)$$

and the sound velocity c_1 by

$$c_1^2 = (\rho K_N)^{-1} = \frac{1}{3}(1 + F_0^s)\left(1 + \frac{1}{3}F_1^s\right)v_F^2. \qquad (21)$$

Table I lists data on the pressure dependence of the Fermi liquid parameters.

The Landau parameters F_0^s, F_1^s, F_0^a may thus be obtained from the heat capacity, sound velocity, and magnetic susceptibility. The higher-order Landau parameters must be deduced from various transport properties. F_1^a may be obtained from

TABLE I

FERMI–LIQUID PARAMETERS FOR ^{3}He. FROM HALPERIN AND VAROQUAUX (1990).

P (bar)	v (cm³/mol)	v_f (m/s)	c_1 (m/s)	$(c_0-c_1)/c_1$	C/nRT (K^{-1})	m^*/m	T* (K)	F_0^s	F_1^s	F_0^a
0	36.818	59.03	183.1	0.03734	2.78	2.80	0.359	9.3	5.4	−0.70
3	33.934	53.82	227.0	0.02329	2.98	3.16	0.306	15.9	6.5	−0.72
6	32.036	49.77	260.1	0.01708	3.16	3.48	0.276	22.5	7.4	−0.73
9	30.716	46.58	286.2	0.01366	3.32	3.77	0.256	29.1	8.3	−0.74
12	29.716	44.00	308.0	0.01150	3.48	4.03	0.239	35.5	9.1	−0.75
15	28.888	41.83	327.2	0.00997	3.62	4.28	0.224	41.8	9.9	−0.75
18	28.168	39.92	344.6	0.00881	3.77	4.53	0.212	48.3	10.6	−0.76
21	27.541	38.17	360.7	0.00787	3.92	4.77	0.204	55.1	11.3	−0.76
24	27.012	36.53	375.6	0.00709	4.06	5.02	0.198	62.2	12.1	−0.76
27	26.572	35.00	389.5	0.00643	4.21	5.26	0.192	69.6	12.8	−0.75
30	26.170	33.63	402.8	0.00587	4.36	5.50	0.184	77.2	13.5	−0.75
33	25.677	32.52	416.2	0.00542	4.50	5.74	0.179	84.7	14.2	−0.75
34	25.456	32.23	421.0	0.00529	4.55	5.81	0.180	87.1	14.4	−0.75

2. Sound Propagation and Collective Modes

the spin diffusion measurements in a magnetic field, and F_2^s from the first sound–zero sound velocity difference.

2.2. Zero Sound

The transport properties can be derived from the kinetic equation (the Boltzmann transport equation), which is an integro-differential equation given by

$$\partial_t n_{k\alpha\beta} + \sum_{\alpha'} \nabla_{\mathbf{k}} \varepsilon_{k\alpha\alpha'} \cdot \nabla_{\mathbf{r}} n_{n\alpha'\beta} - \sum_{\alpha'} \nabla_{\mathbf{r}} \varepsilon_{k\alpha\alpha'} \cdot \nabla_{\mathbf{k}} n_{k\alpha'\beta} = I\{n_{k\alpha\beta}\}; \quad (22)$$

here $I\{n_{k\alpha\beta}\}$ accounts for the effects of collisions. This gives the time evolution of the local quasiparticle distribution function. Linearizing, we obtain

$$(\omega - \mathbf{v_k} \cdot \mathbf{q})\delta n_{k\alpha} + \mathbf{v_k} \cdot \mathbf{q}\frac{\partial n_k}{\partial \varepsilon_k}\partial \varepsilon_{k\alpha} = iI\{\delta n_{k\alpha}\}, \quad (23)$$

where

$$\delta\varepsilon_{k\alpha} = \frac{2}{(2\pi\hbar)^3} \int f(\mathbf{k},\mathbf{k}')\delta n(\mathbf{k}')d^3k'. \quad (24)$$

From hydrodynamics one obtains a collective mode in the form of ordinary longitudinal sound, where the velocity and the attenuation are given by (Landau and Lifshitz, 1959)

$$c_1^2(\rho) = \left(\frac{\delta P}{\delta \rho}\right)_s = (\rho K)^{-1} \quad (25)$$

and

$$\alpha_1 = \frac{2\omega^2}{3c_1^3\rho}\eta, \quad (26)$$

where K and η are the compressibility and viscosity, respectively.

In the hydrodynamic limit, $\omega\tau \ll 1$, there are many collisions within one cycle. As one cools the liquid, the Pauli principle limits the number of collisions possible. Both the number of excited ^{3}He atoms, and the number of possible empty excited states to scatter into, scale as k_BT; thus, the rate is proportional to T^2 and τ to T^{-2}. From kinetic theory the viscosity then goes as T^{-2}. As the temperature is lowered we eventually cross the "boundary" $\omega\tau \cong 1$ and enter the collisionless regime, $\omega\tau \gg 1$.

In the extreme collisionless limit, one drops the collision integral term. The solution of the kinetic equation (with $I\{n_{k\alpha\beta}\} = 0$) gives the zero sound modes. Longitudinal zero sound is the best-defined mode. Including the effect of collisions, one finds that the attenuation is frequency-independent and goes as T^2. Other modes, corresponding to the separate angular momentum, may also exist that are not so easily observed. These include transverse zero sound and (in a magnetic field) spin waves.

There is a fundamental difference between the structure of the solutions to the hydrodynamic equation and the collisionless kinetic equation: Hydrodynamics involves partial differential equations in space and time, which on Fourier transforming give linear algebraic equations in $\mathbf{v}$, P, etc. The kinetic equation is an integro-differential equation, which on linearizing and Fourier transforming becomes a linear integral equation in the particle distribution function $\delta n(p)$. For a discussion of these solutions see Pines and Nozieres (1966).

In the collisionless limit, one finds

$$c_0 = c_1 \left\{ 1 + \frac{2}{15} \frac{m^*}{m} \left(1 + \frac{1}{5} F_2^S \right) \left(\frac{v_F}{c_1} \right)^2 + O\left[\left(\frac{v_F}{c_1} \right)^4 \right] \right\} \tag{27}$$

and

$$\alpha_0 = \frac{2}{15} \frac{m^*}{m} \frac{1 + \frac{1}{5} F_2^S}{v_F \tau_z} \left(\frac{v_F}{c_1} \right)^3 \left\{ 1 + O\left[\left(\frac{v_F}{c_1} \right)^2 \right] \right\}, \tag{28}$$

where the relaxation time τ_z is related to the viscous relaxation time τ_η as

$$\tau_z = \frac{\tau_\eta}{\left(1 + \frac{1}{5} F_2^S \right)}. \tag{29}$$

The crossover from first sound to zero sound takes place at about 10 mK for frequencies of order 10 MHz. In the zero sound mode, there is no frequency dependence of the attenuation. The zero sound velocity is about 3.7% larger than c_1 at s.v.p. and 0.5% larger at the melting pressure. In the superfluid phase, the zero sound velocity evolves back to the first sound velocity.

Figure 1 shows data for the first sound–zero sound crossover and the change in sound velocity (Abel et al., 1966). As the temperature is lowered below 100 mK, the attenuation increases as the collision rate decreases. However, around 10 mK a crossover takes place into the collisionless or zero sound region. The sound velocity increases to c_0, and the attenuation starts dropping as T^2. At the superfluid transition there is a sharp peak in the attenuation due to pair breaking and absorption into collective modes. This will be discussed in Section 5.

2. Sound Propagation and Collective Modes

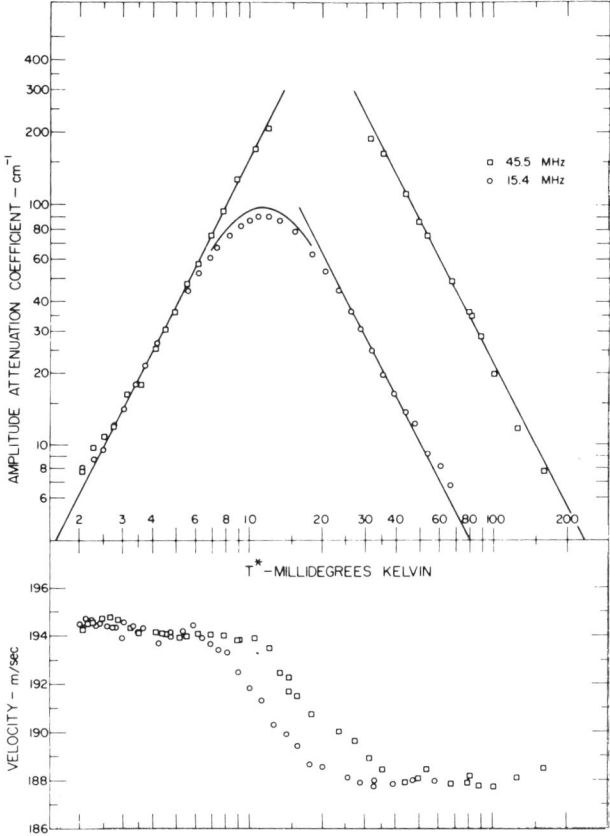

FIG. 1. Zero sound attenuation and velocity data of Abel *et al.* (1966).

3. Superfluid ^{3}He

3.1. EXPERIMENTAL PROPERTIES OF SUPERFLUID ^{3}HE

As mentioned earlier, the study of superfluid ^{3}He was possible only after improvements involving several cooling techniques: the dilution refrigerator, nuclear cooling, and Pomeranchuk (compressional) cooling. One of the major problems at these low temperatures is to establish an absolute temperature scale. In the initial period after the discovery of superfluidity in ^{3}He, there was a large discrepancy ($\sim$10%) in temperature between the various laboratories. To overcome these difficulties, all data reported are on a reduced temperature scale,

T/T_c, wherever possible. A critical evaluation of the temperature scale has been done by Greywall (1986). Only the phase diagram will be shown on an absolute temperature scale (the Greywall scale).

In the normal state the heat capacity displays a linear temperature dependence, as expected for a Fermi system. However, theory predicts an additional $T^3 \log T$ term in the heat capacity, arising from strong spin fluctuations. (A similar $T^3 \log T$ term was later found in the heavy fermion superconductor UPt_3.) As the temperature is lowered, there is a heat capacity anomaly at T_c, similar to that at the superconducting transition. Detailed measurements of the heat capacity have been performed by Alvesalo et al. (1980, 1981), Roach et al. (1982), and Greywall (1986), for all pressures. There is a sharp discontinuity at T_c, and $\Delta C/C_N$ can be as large as 2 at the melting pressure. The discontinuity becomes smaller with decreasing pressure and approaches the weak coupling BCS value of 1.43 for pressures below 10 bar. Figure 2 shows the heat capacity data in zero magnetic field at the melting pressure (Halperin et al., 1976). In the superfluid phase the specific heat shows a power law behavior

$$C(T) \simeq C(T_c) \left(\frac{T}{T_c}\right)^3, \qquad (30)$$

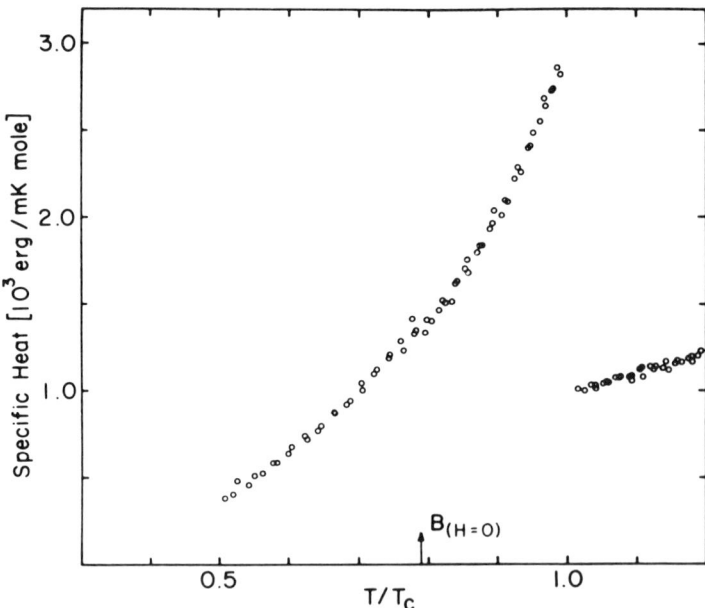

FIG. 2. Specific heat jump at the superfluid transition in zero field. The B-transition is indicated by the arrow. After Halperin et al. (1976).

2. Sound Propagation and Collective Modes

which is expected for the ABM state (A-phase). In the B-phase (at very low temperatures, or at low pressures), the temperature dependence should be exponential $C(T) \propto e^{-\Delta/kT}$. The A–B transition is a first-order transition, with a small latent heat $\ell_{AB} = 15.4$ erg/mol; $T_{AB} = 1.932$ mK and $(C_B - C_A)/C_A = 0.091$ on the melting curve. A small amount of supercooling is seen, and both the A and B transition can be identified on a chart recorder trace if the pressure is swept out in time as a mixture of solid and liquid ^{3}He is cooled at a constant rate. The Clausius–Clapeyron equation gives a volume change of $1:10^8$ at the AB transition (the A-phase has the higher density).

On the application of a magnetic field there is a new phase, the A_1 phase. Both the N–A_1 and A_1–A_2 transitions are second-order. The heat capacity shows a step at both these transitions (see Fig. 3).

Magnetization. The odd number of particles in the ^{3}He nucleus results in a nuclear magnetic moment. The resulting magnetization can be measured either by nmr techniques, or by standard magnetization techniques (preferably employing a sensitive squid detector). In the normal state, there is a temperature-independent Pauli susceptibility. In the A phase the magnetic susceptibility is the same as in the normal state (actually there is a very small increase). In the B phase, the susceptibility is reduced and drops to a value 0.35 that of the normal state value. In the B-phase there are three spin states, only two of which contribute

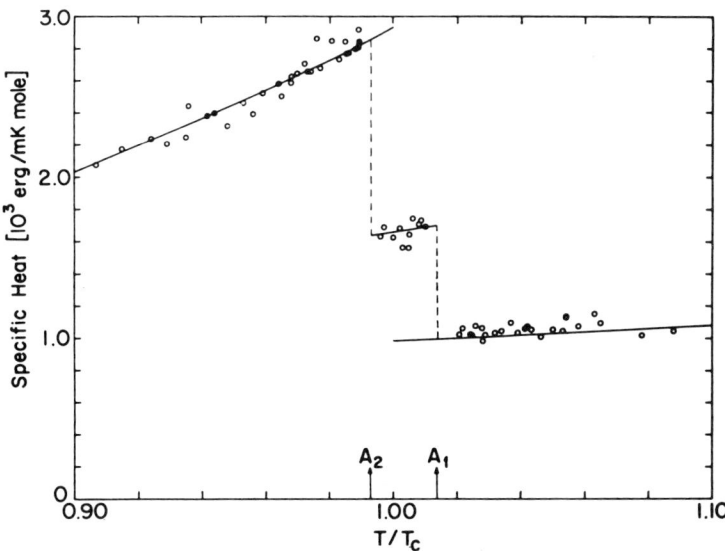

FIG. 3. Specific heat data in a field of 8.8 kG at the A_1 and A_2 transitions ($T_{A2} < T_{A1}$). After Halperin et al. (1976).

to the magnetic susceptibility. However, because of the Fermi liquid interactions, the susceptibility is further reduced to the above value.

Sound Attenuation and Velocity. Near the superfluid transition and the AB transition, peaks are observed in the sound attenuation. The peaks arise from the existence of collective modes in the superfluid and will be discussed at length in later sections. The velocity also drops from the normal state zero sound value, c_0, to the first sound value, c_1.

Viscosity and Normal Fluid Density. Both vibrating wire and torsional oscillator measurements in the superfluid have been performed. The shift in resonant frequency and the damping of the vibrations can be converted into a viscosity and a density fraction of the normal-fluid component. Briefly the results are: (i) in the normal phase $\eta_{shear} \sim T^{-2}$, (ii) a rapid drop occurs in viscosity below $T_c \sim (T_c - T)^{1/2}$, and (iii) there is a smooth decrease of the normal fluid density to very small values below 0.5 T_c (Alvesalo *et al.*, 1973, Johnson *et al.*, 1975).

3.2. SUPERFLUID PHASE DIAGRAM

There are two superfluid phases in zero magnetic field; see Fig. 4. The normal-to-superfluid transition (NA and NB) is second-order, and the A–B transition is

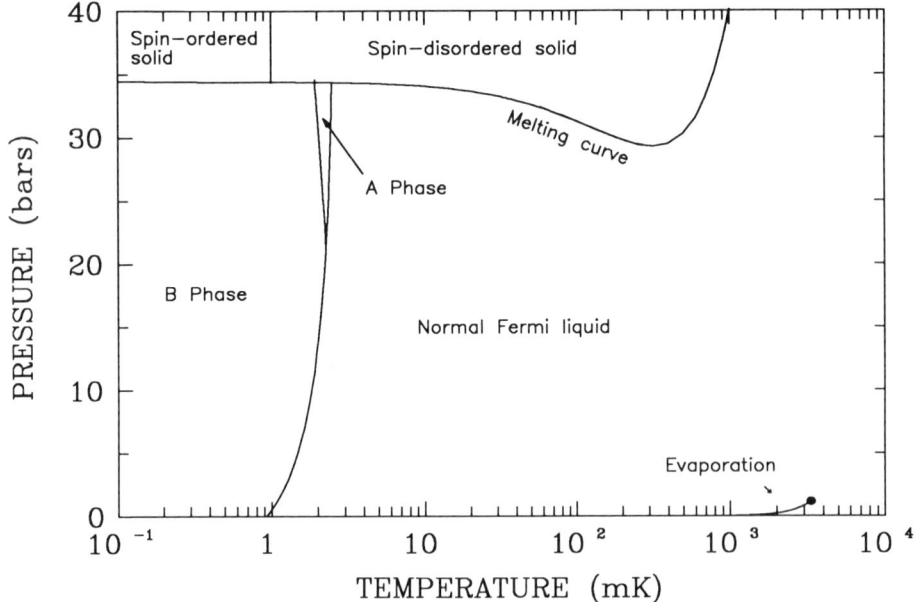

FIG. 4. ^{3}He phase diagram in zero magnetic field.

2. Sound Propagation and Collective Modes

first-order. The NA, NB, and AB lines meet at a "polycritical point" with $P_{PCP} = 21.22$ bar and $T_c(P_{PCP}) = 2.273$ mK in the revised temperature scale (Greywall, 1986). The pressure–temperature coordinates of the superfluid and the A–B transition are given in Table II.

A magnetic field has a significant effect on the phase diagram (see Fig. 5):

(i) A new phase, the A_1 phase, appears between the normal and the A or B phase for all pressures. The polycritical point disappears.

(ii) The A phase (now called the A_2 phase) is stabilized at high fields down to zero pressure, and the B-phase is completely suppressed for magnetic fields in excess of 0.6 Tesla.

Rotating ^3He. In some respects a rotating superfluid (both ^{3}He and ^{4}He) is equivalent to a type II superconductor in a magnetic field. When the speed is above the critical velocity (which is very small), vortex lines nucleate. The rotational (circulation) flux is quantized, and the lines order in a hexagonal closed-pack lattice. The circulation, is given by

$$\kappa = \oint_c \mathbf{v} \cdot d\mathbf{l} = \oint (\nabla \times \mathbf{v}) \cdot d\mathbf{s} = \frac{\hbar}{2m} \oint_c \nabla \phi \cdot d\mathbf{l} = n\kappa_0, \quad (31)$$

TABLE II

P–T Coordinates for the Superfluid Phase Diagram. From Greywall (1986).

P (bars)	T_c (mK)	T_{A-B} (mK)	P (bars)	T_c (mK)	T_{A-B} (mK)
0	0.929	—	19	2.209	—
1	1.061	—	20	2.239	—
2	1.181	—	21	2.267	—
3	1.290	—	21.22	2.273	2.273
4	1.388	—	22	2.293	2.262
5	1.478	—	23	2.317	2.242
6	1.560	—	24	2.339	2.217
7	1.636	—	25	2.360	2.191
8	1.705	—	26	2.378	2.164
9	1.769	—	27	2.395	2.137
10	1.828	—	28	2.411	2.111
11	1.883	—	29	2.425	2.083
12	1.934	—	30	2.438	2.056
13	1.981	—	31	2.451	2.027
14	2.016	—	32	2.463	1.998
15	2.067	—	33	2.474	1.969
16	2.106	—	34	2.486	1.941
17	2.143	—	34.338	2.491	—
18	2.177	—	34.358	—	1.932

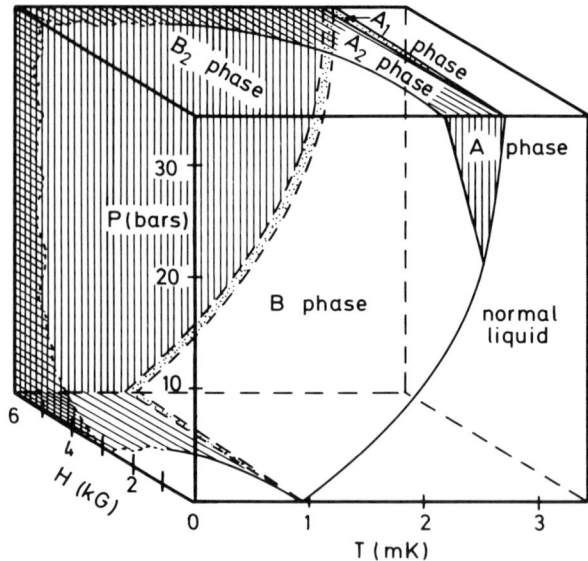

FIG. 5. ^{3}He phase diagram in the H–P–T plane. After Vollhardt and Wölfle (1990).

where $\kappa_0 = h/m_4 = 0.997$ cm^{2}s^{-1} for superfluid ^{4}He, and $\kappa_0 = h/2m_3 = 0.662$ cm^{2}s^{-1} for superfluid ^{3}He. Because superfluid ^{3}He undergoes Cooper pairing, there is an additional factor 2 in the flux quantum. In ^{4}He the vortices are of a simple nature. They are axisymmetric with a singularity on the axis, and a very small core size (~ 1 Å). However in ^{3}He, because of the higher-order (unconventional) pairing there is more than one kind of vortex. Extensive theoretical work on the vortices has been done by Thuneberg (1986) and Salomaa and Volovik (1987) (see also Vollhardt and Wölfle, 1990).

There are three topologically different types of vortices in ^{3}He-A: (i) singular vortices with an odd quantum of circulation, (ii) continuous coreless vortices with even integer circulation, and (iii) singular vortices with a half-integer circulation. The singular, singly quantized vortex has two different cores; a "hard" core with a radius of the order of the coherence length $\xi \sim 100$ Å, inside which the order parameter deviates from its bulk ^{3}He-A value, and a "soft" core with a radius of the order of the dipolar healing length $\xi_D = 6$ μm. All vorticity in the liquid is localized inside the soft core. The continuous vortices have no singular core, but may acquire a soft core with localized vorticity.

Because of the isotropic energy gap in ^{3}He-B, one might expect the vortex line to be similar to that in superfluid ^{4}He. Of the theoretically many possible vortex types, in the B-phase, only two have stable structures: an antisymmetric

2. Sound Propagation and Collective Modes

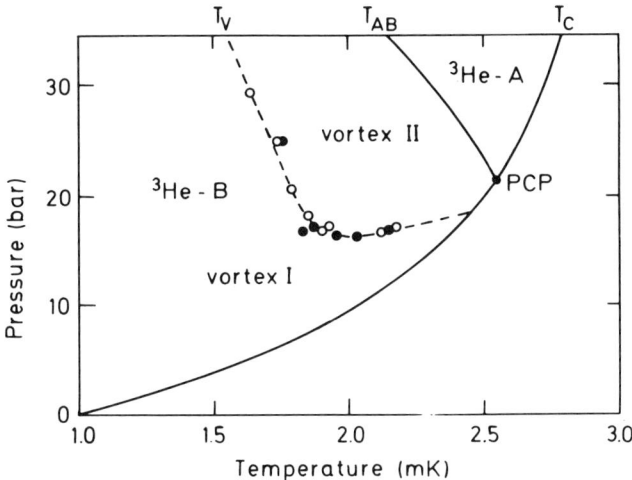

FIG. 6. Phase diagram of rotating superfluid ^{3}He. After Vollhardt and Wölfle (1990).

v-vortex core at high pressures (≥ 20 bar) and the double core vortex at low pressures. The lattice has hexagonal symmetry in both cases. A phase transition is seen (in the P–T plane) between these two states in NMR experiments. The phase diagram of rotating ^{3}He is shown in Fig. 6.

Of the five possible axisymmetric vortices at high pressures, the v-vortex is the energetically favored: The core is a superfluid and has a strong magnetic moment which can be studied by NMR. At low pressures, a nonaxisymmetric vortex has a lower energy. This transition between the two types of vortices, at a temperature $T_v \sim 0.6 T_c$ (Ikkala *et al.*, 1982; Hakonen *et al.*, 1983; Pekola *et al.*, 1984), shows up as a discontinuity in the spin-wave resonant frequencies.

Vortices in superfluid ^{3}He are difficult to see. In superfluid ^{4}He the vortices can be imaged via trapped ions that are released at the surface. In superconductors one can use decoration techniques (Bitter patterns), neutron diffraction, or STM. In superfluid ^{3}He, information on vortices has to be obtained indirectly. The probes that have been used are NMR, ultrasonics (see Section 5.2.6), and negative ions.

3.3. Order Parameter of Superfluid ^{3}He

The first successful microscopic theory of superconductivity was the BCS (Bardeen *et al.*, 1957) theory which describes a superconductor in terms of pairs of electrons having equal and opposite momenta (Cooper pairs) (Cooper, 1956).

Cooper calculated the energy of a pair of electrons interacting via an attractive potential in the presence of a filled Fermi sea. He showed that it was always energetically favorable for the electrons to form a bound pair (rather than condense into the Fermi sea), irrespective of how weak the interaction was. The Cooper problem depended only on the statistics of the particles involved and could be applied to any system of fermions with an attractive potential. (For an introduction to the BCS theory, see deGennes, 1989, or Tinkham, 1975.)

In most superconductors, the electrons pair in an $\ell = 0$ state with opposite spins (i.e., spin singlet). The source of the attractive interaction is the electron–phonon interaction. In liquid ^{3}He, because of short-range repulsion, an $\ell = 0$ state was ruled out. The experimentally observed magnetic response requires triplet pairing. Since the overall wave function must be antisymmetric, the angular momentum state must be antisymmetric, i.e., $\ell = 1, 3, 5$, etc. It is generally accepted that ^{3}He atoms pair in an $\ell = 1, s = 1$ state.

A number of texts on ^{3}He and spin triplet pairing exist, and we refer the reader to these for details (Leggett, 1975; Anderson and Brinkman, 1978; Vollhardt and Wölfle, 1990).

For spin triplet pairing a general wave function can be written as

$$\Psi = \Psi_{\uparrow\uparrow}|\uparrow\uparrow\rangle + \Psi_{\downarrow\downarrow}|\downarrow\downarrow\rangle + \frac{\Psi_{\uparrow\downarrow+\downarrow\uparrow}}{\sqrt{2}}|\uparrow\downarrow + \downarrow\uparrow\rangle. \quad (32)$$

The spin wave functions $|\uparrow\uparrow\rangle, |\downarrow\downarrow\rangle$, and $|\uparrow\downarrow + \downarrow\uparrow\rangle$ can be written in the form

$$\langle S_1 S_2 | \uparrow\uparrow \rangle = \begin{pmatrix} 1 & 0 \\ 0 & 0 \end{pmatrix}, \quad (33a)$$

$$\langle S_1 S_2 | \downarrow\downarrow \rangle = \begin{pmatrix} 0 & 0 \\ 0 & 1 \end{pmatrix}, \quad (33b)$$

$$\langle S_1 S_2 | \uparrow\downarrow + \downarrow\uparrow \rangle = \frac{1}{\sqrt{2}} \begin{pmatrix} 0 & 1 \\ 1 & 0 \end{pmatrix}. \quad (33c)$$

Then

$$\Psi = \begin{pmatrix} \Psi_{\uparrow\uparrow} & \Psi_{\uparrow\downarrow+\downarrow\uparrow} \\ \Psi_{\uparrow\downarrow+\downarrow\uparrow} & \Psi_{\downarrow\downarrow} \end{pmatrix}. \quad (34)$$

Also,

$$\langle S_1 S_2 | \uparrow\uparrow \rangle = \frac{1}{2}(I + \sigma_z), \quad (35a)$$

$$\langle S_2 S_2 | \downarrow\downarrow \rangle = \frac{1}{2}(I - \sigma_z), \quad (35b)$$

2. Sound Propagation and Collective Modes

$$\langle S_1 S_2 | \uparrow\downarrow + \downarrow\uparrow \rangle = \frac{1}{\sqrt{2}} \sigma_x, \tag{35c}$$

and

$$\Psi = \frac{\Psi_{\uparrow\uparrow}}{2}(I + \sigma_z) + \frac{\Psi_{\downarrow\downarrow}}{2}(I - \sigma_z) + \frac{\Psi_{\uparrow\downarrow + \downarrow\uparrow}}{2}\sigma_x, \tag{36}$$

$$I = \sigma_y \sigma_y, \qquad \sigma_z = -i\sigma_x \sigma_y, \qquad \sigma_x = i\sigma_z \sigma_y, \tag{37}$$

$$\Psi = \frac{(\Psi_{\uparrow\uparrow} + \Psi_{\downarrow\downarrow})}{2i} i\sigma_y \sigma_y + \frac{(\Psi_{\downarrow\downarrow} - \Psi_{\uparrow\uparrow})}{2} i\sigma_x \sigma_y \tag{38}$$

$$+ \left(\frac{\Psi_{\uparrow\downarrow + \downarrow\uparrow}}{2}\right) i\sigma_z \sigma_y.$$

Define a vector $\mathbf{d}(\hat{k})$ with components

$$d_x = \frac{\Psi_{\downarrow\downarrow} - \Psi_{\uparrow\uparrow}}{2}, \tag{39a}$$

$$d_y = \frac{\Psi_{\uparrow\uparrow} + \Psi_{\downarrow\downarrow}}{2i}, \tag{39b}$$

$$d_z = \frac{\Psi_{\uparrow\downarrow + \downarrow\uparrow}}{2}; \tag{39c}$$

then

$$\Psi = \sum_{j=1}^{3} d_j (i\sigma_j \sigma_y), \tag{40}$$

or

$$\Psi = \begin{pmatrix} -d_x + id_y & d_z \\ d_z & d_x + id_y \end{pmatrix}. \tag{41}$$

The vector $\mathbf{d}(\hat{\mathbf{k}})$ is a function of the momentum $\hat{\mathbf{k}}$ and contains information about the orbital part of the order parameter. An orbital wave function with $\ell = 1$ can be expanded in the spherical harmonics Y_ℓ^m or equivalently in the cartesian coordinates k_x, k_y, and k_z:

$$k_x = \sin\theta \cos\phi, \qquad \propto Y_{1,1} + Y_{1,-1}, \tag{42a}$$

$$k_y = \sin\theta \sin\phi, \qquad \propto Y_{1,1} - Y_{1,-1}, \tag{42b}$$

$$k_z = \cos\theta, \qquad \propto Y_{1,0}. \tag{42c}$$

We expand $\mathbf{d}(\hat{\mathbf{k}})$ as

$$d_j = \sum_{i=1}^{3} A_{ji}k_i, \qquad (43)$$

$$\Psi = \sum_{i,j} A_{ji}k_i\,(i\sigma_j\sigma_y). \qquad (44)$$

The matrix $\overleftrightarrow{A}$ is a (3×3) matrix with nine complex elements.

There are many possible choices for the order parameter. On minimizing the Landau–Ginzburg free energy and using the weak coupling values for the Landau–Ginzburg parameters, the most stable state for ^{3}He is

$$\overleftrightarrow{A} = \Delta \begin{pmatrix} 1 & 0 & 0 \\ 0 & 1 & 0 \\ 0 & 0 & 1 \end{pmatrix}, \qquad (45)$$

and $\qquad (45a)$

$$\Psi = k_x(i\sigma_x\sigma_y) + k_y(i\sigma_y\sigma_y) + k_z(i\sigma_z\sigma_y).$$

This is the B-phase, also known as the Balian–Werthamer (BW) state (Balian and Werthamer, 1963), which is stable at zero field at pressures up to $\sim$21 bar. All three spin species $\uparrow\uparrow$, $\downarrow\downarrow$, and $\uparrow\downarrow$ are present. The square of the gap is given by

$$(\Psi\Psi^+) = |\mathbf{d}(\mathbf{k})|^2 + i\sum_\mu [\mathbf{d}(\mathbf{k}) \times \mathbf{d}^*(\mathbf{k})]_\mu \sigma_\mu. \qquad (46)$$

For unitary states $\mathbf{d}(\mathbf{k}) \times \mathbf{d}^*(\mathbf{k}) = 0$. The gap for the B-phase is

$$(\Psi\Psi^+)^{1/2} = \Delta(k_x^2 + k_y^2 + k_z^2)^{1/2}, \qquad (47)$$

which is independent of $\hat{k}$, i.e., the gap is isotropic in k-space. In many respects the BW state is similar to an s-wave state in a conventional superconductor, because of the isotropic energy gap. The BW gap is shown schematically in Fig. 7a.

At pressures over 21 bar, strong coupling corrections lower the energy of the A-phase relative to the B-phase. The A-phase (also known as the Anderson–Brinkman–Morel [Anderson and Brinkman, 1973] or ABM state) is given by

$$\overleftrightarrow{A} = \Delta \begin{pmatrix} 1 & i & 0 \\ 0 & 0 & 0 \\ 0 & 0 & 0 \end{pmatrix}, \qquad (48)$$

$$\Psi = \Delta(k_x + ik_y)(i\sigma_x\sigma_y) \propto Y_{1,1}. \qquad (49)$$

2. Sound Propagation and Collective Modes

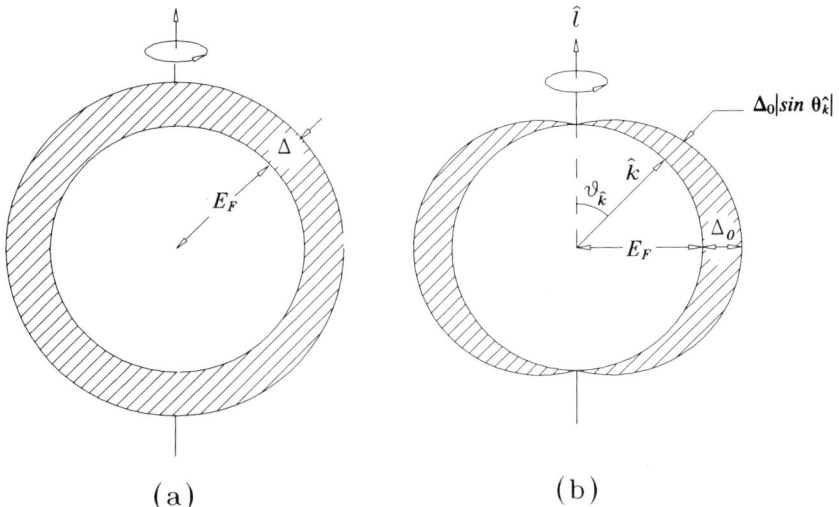

FIG. 7. Schematic drawing of the energy gap for (a) the B-phase (isotropic energy gap) and (b) the A-phase (the gap vanishes at two points).

Only the ↑↑ and ↓↓ spin species are present. The gap is given by

$$(\Psi\Psi^+)^{1/2} = \Delta(k_x^2 + k_y^2)^{1/2} \sim \Delta \sin \theta. \tag{50}$$

The gap is maximum at the equator and goes to zero at the poles, as shown in Fig. 7b. The $\hat{x}$, $\hat{y}$, and $\hat{z}$ axes form an orthogonal triad in orbital space, with the gap going to zero along $\hat{z}$. (These three directions are also referred to as $\hat{m}$, $\hat{n}$ and $\hat{\ell}$.)

4. Collective Modes

The collective modes can be classified into two groups:

(i) The gapless modes: Here the frequency goes to zero as the wavevector goes to zero (long wavelength limit). These are the Goldstone modes and arise because of a broken symmetry.

(ii) Modes with a gap in the energy spectrum: These are coherent oscillations of the Cooper pairs.

4.1. Hydrodynamic Modes

4.1.1. Sound Modes

In the superfluid phase some of the hydrodynamic modes are similar to modes in the normal phase; in addition, there are some notable exceptions. The hydrodynamics below T_c can be understood on the basis of the two-fluid model: a normal fluid component and a superfluid component with no viscosity; the ratio of the two can be determined from torsional oscillator measurements. There are two soundlike modes as in superfluid ^{4}He: first sound and second sound. In the first sound mode, both the normal and superfluid component move in phase with $v_s = v_n$. In second sound, the two fluid components move out of phase, with the total mass current density

$$\mathbf{g} = \rho_n \mathbf{v}_n + \rho_s \mathbf{v}_s = 0. \tag{51}$$

The second sound mode, which is an entropy wave, is very difficult to see in ^{3}He, as the remaining entropy of ^{3}He at these temperatures is very much smaller than in ^{4}He. The second sound velocity is of the order of 5 cm/s. The dispersion of second sound is given by

$$\omega^2 = c_2^2 q^2 (1 - i\omega\tau_2), \tag{52}$$

where the velocity c_2 and the relaxation time, τ_2, are given by

$$c_2 = \frac{s^2 T}{\rho C_v} \left(\frac{\rho_s}{\rho_n}\right), \tag{53}$$

$$\tau_2 = \frac{C_v}{s^2 T} \left[\frac{4}{3}\eta + \rho^2 \zeta_3 + \frac{\rho\kappa}{C_v}\left(\frac{\rho_n}{\rho_s}\right)\right]; \tag{54}$$

here s, ρ are the entropy and mass density, C_v the heat capacity, ρ_s and ρ_n the superfluid and normal fluid densities. The transport coefficients η, ζ_3, and κ are, respectively, the shear viscosity, second viscosity, and thermal conductivity (see Landau and Lifshitz, 1959).

There are other ingenious methods of generating "sound-like" oscillations to determine the superfluid parameters, such as the superfluid density. In restricted geometries (for example, in narrow pores or channels, as in a Vycor matrix), the normal fluid component is "clamped" to the surface because of viscosity, and the oscillating motion of the superfluid is called fourth sound. The dispersion relation and velocity of fourth sound are given by

$$\omega^2 = c_4^2 q^2 (1 - i\omega\tau_4), \tag{55}$$

2. Sound Propagation and Collective Modes

$$c_4^2 = c_1^2 \frac{\rho_s}{\rho}, \tag{56}$$

and

$$\tau_4 = \rho \frac{\zeta_3}{c_1^2}. \tag{57}$$

Fourth sound has been observed in superfluid ^{3}He (Yanof and Reppy, 1974; Kojima et al., 1974). The ^{3}He was cooled by adiabatic demagnetization of powdered CMN. The experiments involved detection of longitudinal resonance in a cylinder packed with CMN. Kojima et al.'s results for ρ_s/ρ_n for various pressures are shown in Fig. 8. Note that the temperature scale is in the CMN magnetic temperature, T^*. Temperatures are obtained by assuming the susceptibility of CMN follows a Curie-like law. The low temperature data shows a steep rise caused by the distortion of the temperature scale due to magnetic ordering of the CMN, which takes place around 1 mK.

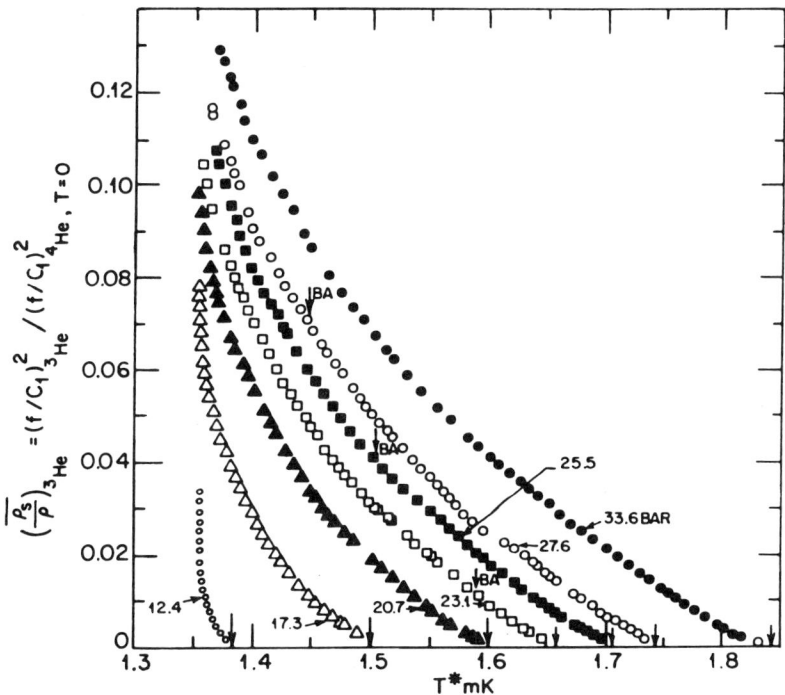

FIG. 8. Results of fourth sound measurements for $\rho_s\rho$ by Kojima et al. (1974).

4.1.2. Spin-Wave Modes

Since the ^{3}He Cooper pairs carry spin, there are well-defined hydrodynamic spin-wave modes in both ^{3}He-A and B that are not present in the normal phase. These arise from the spontaneously broken rotational symmetry in spin space. Theoretical discussions of spin waves in ^{3}He are given by Combescot (1974). Spin waves are more difficult to study experimentally because there is no analogue to a sound transducer as used in sound absorption measurements. However, hydrodynamic (and collisionless) spin-waves have been generated and observed by NMR techniques (Osheroff et al., 1977; Masuhara et al., 1984). Bogacz and Ketterson (1986) have proposed a novel method of generating and detecting spin-waves by using a meander-line transducer.

4.2. ORDER PARAMETER COLLECTIVE MODES

In superfluid ^{3}He, the Cooper pairs have an internal structure the excitation of which leads to collective modes of the order parameter. These can be visualized as certain "oscillations" of the order parameter that lead to eigenfrequencies with energies lying between 0 and 2 Δ. Both types of modes (gapped and gapless) are present. The theoretical derivation of these collective modes is very complicated and beyond the scope of this article. For a detailed discussion of sound propagation and collective modes, see Wölfle (1977, 1978) and Vollhardt and Wölfle (1990). For a phenomenological treatment of sound and collective modes in ^{3}He-B, see Zhao et al. (1991). A group theoretical treatment has been given by Volovik and Khazan (1983). A description of the theoretical results and a detailed discussion of the experiments is given by Halperin and Varoquaux (1990) and Dobbs and Saunders (1990).

For p-wave superfluidity, the order parameter has 18 degrees of freedom $(2 \times [2l + 1] \times [2s + 1] = 18)$ (arising from the real and imaginary components of a 3 × 3 matrix), from which it follows that there are 18 collective modes, some of which are degenerate.

One approach to obtaining the collective modes in the superfluid is to start from the kinetic equation (Section 2) and look at the solution for small oscillations of the order parameter about its equilibrium value.

The density operator in the superfluid phase is represented by a 2 × 2 matrix

$$n_{\mathbf{k}} = \begin{pmatrix} f_{\mathbf{k}} & g_{\mathbf{k}} \\ g_{\mathbf{k}} & 1 - f_{-\mathbf{k}} \end{pmatrix}, \tag{58}$$

where

$$f_{\mathbf{k},\alpha\beta} = \langle a_{\mathbf{k}\alpha}^+ a_{-\mathbf{k}\beta} \rangle, \tag{59}$$

2. Sound Propagation and Collective Modes

and

$$g_{\mathbf{k},\alpha\beta} = \langle a_{\mathbf{k}\alpha} a_{-\mathbf{k}\beta} \rangle; \qquad (60)$$

since $f_{\mathbf{k}}$ and $g_{\mathbf{k}}$ are themselves 2×2 matrices in spin space, the entire problem involves the solution of a 4×4 matrix kinetic equation. The quasiparticle energy is replaced by

$$\varepsilon_k = \begin{vmatrix} \varepsilon_k & \Delta_{\mathbf{k}} \\ \Delta_{\mathbf{k}}^+ & -\varepsilon_{-\mathbf{k}} \end{vmatrix}. \qquad (61)$$

The solution of the linearized kinetic equation gives the energy eigenvalues of the order parameter, which are the collective modes. The q dependence gives the dispersion of the collective modes.

4.2.1. ^{3}He-B Order Parameter

The equilibrium gap parameter given previously can be written as (omitting the spin dependence)

$$d_\mu^0(\hat{\mathbf{k}}) = \Delta \, \hat{k}_\mu. \qquad (62)$$

The fluctuations (real and imaginary parts) are written as

$$\frac{1}{2}[\delta d_\mu(\hat{\mathbf{k}};q,\omega) \pm \delta d_\mu^*(\hat{\mathbf{k}};-q,-\omega)] = \begin{cases} \sum_j D'_{\mu j} \hat{k}_j & \text{for } (+) \\ \sum_j D''_{\mu j} \hat{k}_j & \text{for } (-) \end{cases}, \qquad (63)$$

where $(+)$ represents the real components, and $(-)$ represents the imaginary components. The solution of the kinetic equation yields 18 collective modes of the order parameter, which are listed in Table III and graphically represented in Fig. 9. The ^{3}He-B collective modes can be classified as $J = 0, 1, 2$, both real and imaginary. The calculations are done with the weak coupling theory and in the collisionless limit (see Vollhardt and Wölfle, 1990). Both the $J = 2^+$ (rsq) and the $J = 2^-$ (sq) modes show a fivefold Zeeman splitting in the presence of a magnetic field.

We now summarize the mode structure.

The Goldstone modes or gapless modes. There are four modes with $\omega \propto q$:

$J = 0^-$ corresponds to oscillations of the phase of the order parameter and couples strongly to density fluctuations and is the zero sound mode.

$J = 1$, $J_z = \pm 1, 0$ involve relative rotations of spin and orbital variables and correspond to the spin waves. The spin wave velocity is less than the zero sound velocity.

TABLE III

COLLECTIVE MODES OF THE B-PHASE. FROM VOLHARDT AND WÖLFLE (1990).

Mode	J, K	m_j	ω^2	Coupling to observables	Magnetic-field splitting
	0^+	0	$4\Delta^2$	Density, $O(\omega/E_f)$	—
Spin waves	1^+	0 ± 1	0	Spin density, $O(q^0)$	$\omega_L m_j\, g_{1+}$
Real squashing mode	2^+	0 ± 1 ± 2	$8/5\,\Delta^2$	Spin density, $O(q)^2$ Density, $O(q^2\,\omega/E_f)$	$\omega_L\, m_j\, g_{2+}$
Sound	0^- 1^-	0 0 ± 1	0 $4\Delta^2$	Density, $O(q^0)$ Spin density, $O(\omega/E_f)$	$\omega_L\, m_j\, g_{1-}$
Squashing mode	2^-	0 ± 0 ± 0	$12/5\,\Delta^2$	Density, $O(q^2)$; Spin density, $O(q^2\,\omega/F_F)$	$\omega_L\, m_j\, g_{2-}$

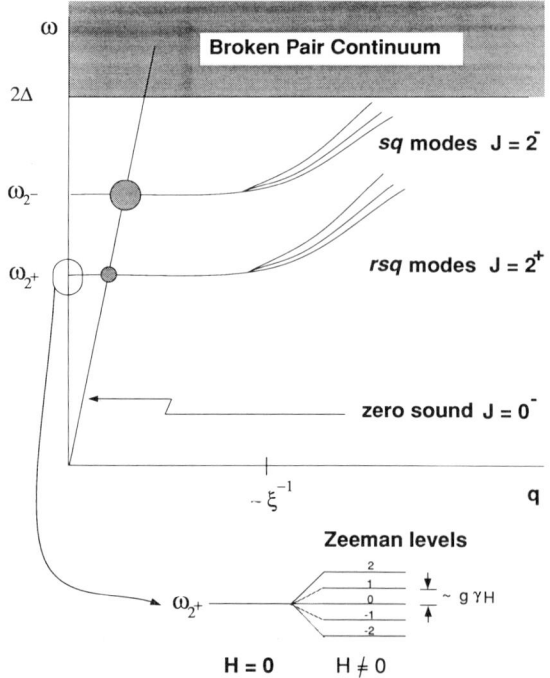

FIG. 9. The pair–excitation spectrum for ^{3}He-B. From Sauls and McKenzie (1991).

2. Sound Propagation and Collective Modes

Pair-vibration modes. There are two sets of $J = 2$ modes. The $J = 2$ modes were first calculated by Vdovin (1963). The $J = 2^-$ and 2^+ modes are called the (imaginary) squashing mode and the real squashing mode, and represent oscillatory quadrupolar modes in which the imaginary and real spherical gap parameter are "squashed." The real component, $J = 2^+$, has the eigenfrequencies $\hbar\omega_2^+ = \sqrt{(8/5)}\Delta$, and the imaginary component $J = 2^-$ has the eigenfrequency $\hbar\omega_2^- = \sqrt{(12/5)}\Delta$.

The other four modes $J = 0^+$ and 1^- are at the pair breaking edge, $\hbar\omega = 2\Delta$. They couple, respectively, to spin-density or zero sound (density) variations.

All these modes are weakly dispersive; the leading term goes as $\sim q^2$ for low q, becoming linear at high q. In a magnetic field these modes undergo a Zeeman splitting that has been observed (see Section 5.2). For small magnetic fields the linear splitting of the $J = 2$ modes is given by

$$\omega_{2\pm}^{(m_J)}(T,H) = \omega_{2\pm}(T) + m_J g_{2\pm}(T)\omega_L(T,H), \tag{64}$$

where $m_J = 0, \pm 1, \pm 2$. The g-factor and the (renormalized) Larmor frequency are given by

$$g_{2\pm}(T) = \frac{1}{12}\left[1 \pm \frac{1 - Y_0}{\lambda(\omega_{2\pm})}\right], \tag{65}$$

$$\omega_L(T,H) = \frac{\gamma H}{1 + F_0^a(\tfrac{2}{3} + \tfrac{1}{3}Y_0)}; \tag{66}$$

F_0^a is a Fermi liquid parameter, Y_0 is the Yoshida function, and λ is a "coupling constant" introduced by Wölfle (1977). The Yoshida function (see Vollhardt and Wölfle, 1990) is a measure of the density of the thermal excitations on the Fermi surface.

4.2.2. Collective Modes in ^{3}He-A

In the axial state the equilibrium gap parameter is given by (omitting the spin dependence)

$$d_\mu^0(\hat{\mathbf{k}}) = \left(\frac{8}{3}\right)^{1/2}\Delta_0 \hat{d}_\mu^0 Y_{11}(\hat{\mathbf{k}}), \tag{67}$$

and expanding the order parameter fluctuations in spherical harmonics (d_μ^0 is taken along the z-axis),

$$\delta d_\mu(\hat{\mathbf{k}}) = \sum_{m=0,\pm 1} \delta d_{\mu m} Y_{1m}(\hat{\mathbf{k}}). \tag{68}$$

The procedure is the same as in the B-phase case. However, there is a major difference. Since the gap vanishes at the poles, pair breaking takes place at all frequencies. As a consequence, the determinant which arises in the solution does not have any zeroes (except when ω = 0), and strictly speaking, there are no well-defined modes, but resonances. We shall still refer to these solutions as modes.

The collective modes for the A-phase are listed in Table IV. Three of the collective modes, referred to as the normal flapping, the clapping, and the superflapping modes, (where the names clapping and flapping were coined by Saslow) are predicted to couple to zero sound (density oscillations) (Serene, 1974; Wölfle, 1977). The effect of a magnetic field was studied by Tewordt and Schopohl (1979a). A graphical representation of the A-phase collective modes is shown in Fig. 10. The normal flapping mode may be visualized as a flapping motion of $\hat{n}$ and $\hat{m}$ about the fixed directions of $\hat{\ell}$ (see Fig. 11). For rapid oscillations of $\hat{\ell}$, the quasiparticles do not have time to relax. A fixed distribution of thermal quasiparticles will support oscillations of the $\hat{\ell}$ vector. As $T \to 0$, the number of quasiparticles diminishes, and the frequency of this mode tends

TABLE IV

COLLECTIVE MODES OF THE A-PHASE. FROM VOLHARDT AND WÖLFLE (1990).

Mode	$|L_z-I|$	$|S_z|$	Weak-coupling frequency ω $T \to T_c$	$T \to 0$	Coupling
Sound	0	0			Density, $O(q^0)$
Spin waves		1	$\propto v_F q$	$3^{-1/2} v_F q$	Spin density, $O(q^0)$
None		0		Density, $O(\omega/E_F)$	
		1		Spin density, $O(\omega/E_F)$	
Orbital waves	1	0	$\omega_{orb} \propto i^{1/2} q^2$	$\propto iq^3$	Density, $O(q^2\omega/E_F)$
Flapping mode					Density, $O(q^2)$
			$\omega_{nfl} = (4/5)^{1/2} \Delta_0(T)$	$\propto T$	
					Spin density, $O(q^2)$
Flapping modes		1			
			$\omega_{sfl} = 2 \Delta_0(T)$	$1.56 \Delta_0(0)$	Spin density, $O(q^2\omega/E_F)$
					Density, $O(q^2\omega/E_F)$
	2	0			
					Density, $O(q^2)$
Clapping modes			$1.23 \Delta_0(T)$	$1.22 \Delta_0(T)$	
					Spin density, $O(q^2)$
		1			
					Spin density, $O(q^2\omega/E_F)$

2. Sound Propagation and Collective Modes

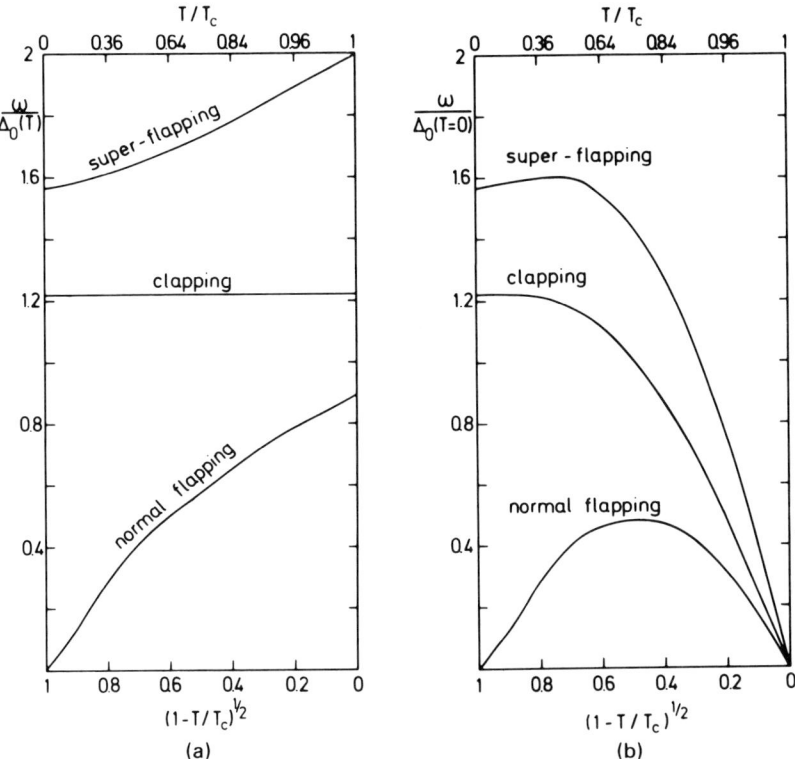

FIG. 10. Collective modes in the A-phase. The left figure is normalized to the temperature-dependent gap, and the right to the zero temperature gap. After Vollhardt and Wölfle (1990).

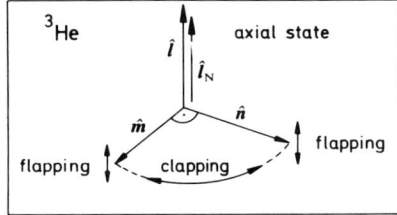

FIG. 11. A pictorial representation of the order parameter collective modes in the A-phase. From Vollhardt and Wölfle (1990).

to zero. The other two modes are associated with internal vibrations of the order parameter. The superflapping mode is interpreted as a rapid flapping motion of $\hat{n}$ and $\hat{m}$ against the equilibrium orientation $\hat{\ell}$. The third mode is an oscillation of $\hat{n}$ and $\hat{m}$ against each other—a clapping mode. For the clapping mode, $\hbar\omega_{cl} \simeq 1.23\Delta_0(T)$, and the frequency scales as the gap.

In the A-phase the gap can be approximated by the equation

$$\Delta_0(T) = \Delta_0 k_B T_c \tanh\left\{\left(\frac{\pi}{\Delta_0}\right)\left(\frac{\Delta C_N}{C}\right)^{1/2}\left(\frac{T_c}{T} - 1\right)^{1/2}\right\}, \qquad (69)$$

with $\Delta_0 = 2.03$, and $\Delta C_N/C = 2.0$ (Δ_0 is the maximum of Δ). The effect of a magnetic field on these modes is shown in Fig. 12. The pair breaking effects in the A-phase play a dominant role. Since the gap vanishes at the poles,

$$|\Delta(\hat{k})| = \Delta_0 \sin\theta, \qquad (70)$$

pair breaking, via absorption, will take place at all frequencies, and hence there will not be a sharp pair-breaking edge.

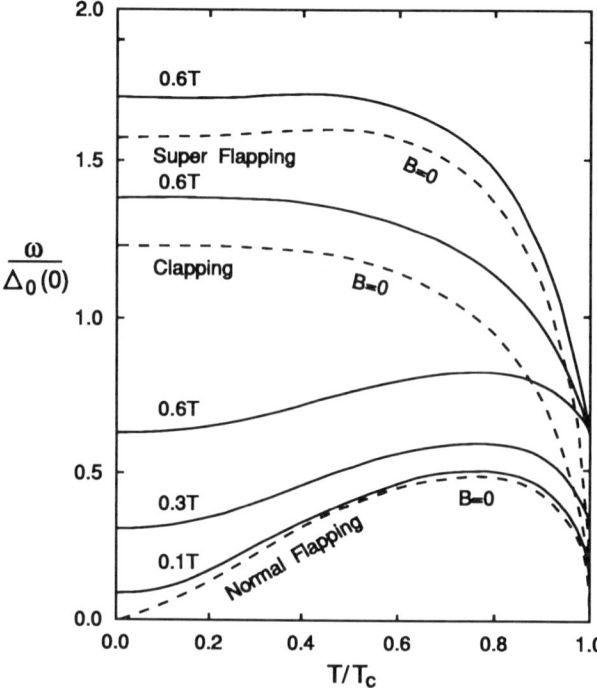

FIG. 12. A-phase collective modes in a magnetic field. After Tewordt and Schopohl (1979a).

2. Sound Propagation and Collective Modes

The dispersion of the normal-flapping mode is given by

$$\omega_{nfl}^2(q) = \omega_{nfl}^2(q=0) + v_F^2(\mathbf{q}\cdot\hat{\ell})^2. \tag{71}$$

For the flapping mode,

$$\omega \sim (v_F q)^2 \quad \text{as } T \to T_c. \tag{72}$$

The dispersion relations are also given in Table IV.

Because of the anisotropic nature of the A-phase order parameter and energy gap, the sound attenuation and velocity are also anisotropic.

$$\alpha(\beta) = \alpha_\parallel \cos^4\beta + 2\alpha_c \cos^2\beta \sin^2\beta + \alpha_\perp \sin^4\beta, \tag{73}$$

$$c_0 - c(\beta) = \Delta c_\parallel \cos^4\beta + 2\Delta c_c \cos^2\beta \sin^2\beta + \Delta c_\perp \sin^4\beta, \tag{74}$$

where β is the angle between $\hat{\mathbf{q}}$ and $\hat{\ell}$ and c_0 is the zero sound velocity. The maximum of the attenuation is when $\hat{\mathbf{q}} \| \hat{\ell}$, and the minimum is at an angle $\beta = \cos^{-1}(1/\sqrt{3}) = 55°$. The first attenuation measurements were done by Paulson *et al.* (1973) and Roach *et al.* (1975a,b).

5. Experimental Results

In this section we review the experimental results of the sound experiments in superfluid ^{3}He.

5.1. EXPERIMENTAL TECHNIQUES

There are many special experimental techniques that needed to be developed to study the superfluid properties of ^{3}He. We shall just mention a few of these. As discussed earlier, there are collective modes of the order parameter, whose energies lie between 0 and 2Δ above the Fermi surface. Since T_c is between 1 and 2.5 mK, according to the BCS model, the maximum value of $2\Delta(0) = 3.5\, k_B T_c$ is about 180 MHz, which is easily accessible by sound. Figure 13 (upper) shows a sketch of the temperature dependence of 2Δ (pair breaking) and the two other collective modes ($J = 2$) for superfluid ^{3}He-B. The temperature dependence of the gap function $\Delta(T)$ may be obtained from the BCS theory (Mühlschlegel, 1959) and is given in Table V.

In a typical experiment the transducer would be resonated (pulsed or cw) at an odd harmonic of its fundamental frequency. Consider the attenuation of sound at a fixed frequency, ν, as the temperature is reduced to T_c and below. Since Δ is temperature-dependent, this is analogous to keeping the temperature fixed and varying the frequency. Above T_c the attenuation is relatively small (and

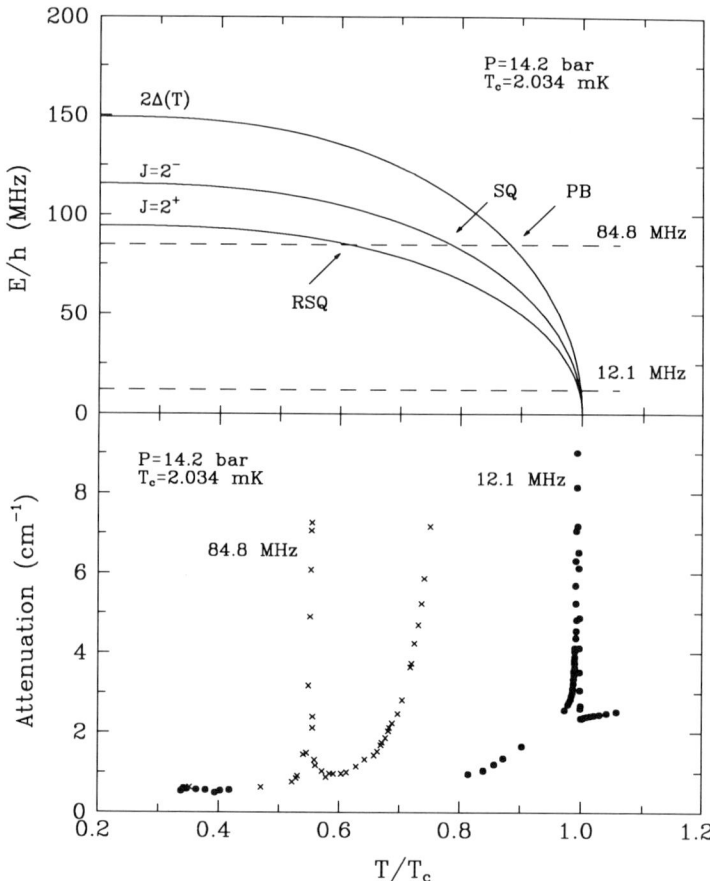

FIG. 13. Schematic of the principle of sound measurements of the collective modes of ^{3}He-B (upper), and actual attenuation data (lower), showing the attenuation peaks.

frequency-independent), being in the collisionless limit. At T_c the liquid goes superfluid, and Cooper pairs are formed. However, since the sound frequency (energy) $h\nu$ is larger than 2Δ, the sound quanta can break these pairs, giving rise to an absorption or attenuation increase. As the temperature is lowered, the number of pairs increases, and the attenuation is increased. This mechanism turns off when the sound quantum becomes smaller than the pair breaking energy, 2Δ; there is a relatively abrupt drop (edge) in the attenuation at this point, designated PB. At a somewhat lower temperature the sound can excite the squashing or the real squashing mode, resulting in large attenuation peaks, points SQ and RSQ. Associated with these peaks are large dispersions of the phase velocity and a large dip in the group velocity. The amplitude of these attenuation

2. Sound Propagation and Collective Modes

TABLE V

MÜHLSCHLEGEL TABLES (ADAPTED FROM B. MÜHLSCHLEGEL, 1959).

T/T_C	Δ/Δ_0
0	1
0.1	1.000000
0.15	0.999998
0.20	0.999875
0.25	0.999186
0.30	0.99711
0.35	0.9928
0.40	0.9850
0.45	0.9733
0.50	0.9569
0.55	0.9351
0.60	0.9070
0.65	0.8719
0.70	0.8288
0.75	0.7760
0.80	0.7110
0.85	0.6303
0.90	0.5141
0.92	0.4749
0.94	0.4148
0.96	0.3416
0.98	0.2436
0.99	0.1730
0.995	0.1226
1.0	0

peaks and their widths depend on the coupling of the sound to the collective modes and the (temperature dependent) relaxation time. Any fine structure in these collective modes—for example, a splitting in an applied magnetic field—results in a splitting in the attenuation peak as the temperature is varied (provided this splitting is of the order of or larger than the width of the attenuation peak). A typical attenuation curve obtained in the pulsed transmission mode is shown in Fig. 13 (lower). The BCS temperature dependence of 2Δ (pair-breaking) and the two $J = 2$ collective modes (squashing and real squashing) are shown for ^{3}He-B at a pressure of 14.2 bar (having a T_c value of 2.034 mK). The resonances (absorption peaks) for the two frequencies 12.1 MHz and 84.8 MHz (seventh harmonic) are shown as the intersection of the horizontal lines at the respective frequency with the three curves. For 12.1 MHz, because of the closeness of the three curves, all three features merge together, and show up as one attenuation peak just below T_c. For the 84.2 MHz data, the rsq mode is clearly resolved and shows up at $\sim 0.55T_c$. The attenuation from the pair breaking and the real squashing mode at 84.2 MHz are too high to be completely resolved in the transmission experiments (the Northwestern experiments could measure up to 10 cm^{-1}). However, in the cw impedance measurements, these high attenuation features could be completely resolved, and such a trace is shown later in Fig. 16. From Fig. 13, one notes that the rsq mode occurs at a much lower temperature than predicted by theory, which corresponds to a frequency of 15% lower than the ideal $\sqrt{(8/5)}\Delta$ predicted for the rsq mode. The discrepancy can be accounted for by introducing strong coupling effects into the BCS theory.

Another way of sweeping through these resonances is to depressurize continuously. This technique was developed simultaneously by the Northwestern and Cornell groups. Since T_c depends on the pressure, changing the pressure at a fixed temperature changes Δ and the value of all the other collective modes. Using this technique the group velocity of zero sound near the squashing mode was measured by Movshovich *et al.* (1988) at very low temperatures (where the energy gap varies very slowly with temperature and it is awkward to sweep through the spectrum using temperature as the variable) while the liquid was depressurized at a rate slower than 0.5 bar/hour. The group velocity was observed to decrease by more than a factor of 15 in these experiments (they measured a velocity as low as ~ 25 m/s). The same technique was later used to identify the substates of the squashing mode in a magnetic field (Movshovich *et al.*, 1988). The pressure sweeping technique was also used by Adenwalla *et al.* 1989) to study both of the $J = 2$ collective modes and the pair breaking edge, over a wide range of pressures (see Section 5.2).

5.1.1. Cryogenics

Several cryogenic advances preceded the discovery of superfluid ^{3}He. (For reviews on cryogenic techniques, see Lounasmaa, 1974; Richardson and Smith, 1988.) Of these the most important was the development of the dilution refrigerator. This is the only continuous refrigerator that produces temperatures less than 5 mK. A mixture of liquid ^{3}He and ^{4}He is employed. The phase diagram of this mixture is quite interesting. Below 0.8 K there is a phase separation: Two phases appear. The concentrated phase, which is mostly ^{3}He, is lighter, and floats on top of the dilute phase, which is low in ^{3}He. Because of quantum mechanical effects, even at absolute zero, there is a finite solubility of ^{3}He in ^{4}He; below 100 mK it is constant at about 6%. At low temperatures, the ^{3}He in the dilute phase behaves like a Fermi gas, with an entropy proportional to T^2. The entropy of the concentrated ^{3}He is also proportional to T^2; however, being more concentrated, and hence more degenerate (on a per mole basis), the coef-

2. Sound Propagation and Collective Modes

ficient is smaller. By a suitable design the ^{3}He in the dilute phase is "pumped" out; to maintain equilibrium, some of the ^{3}He in the concentrated solution dissolves in the dilute phase. This process is analogous to evaporation of liquid ^{3}He into the dilute phase, and since the molar pumping rate is held constant, the cooling power (involving the difference in the enthalpies) is proportional to T^2. This process can be compared to the cooling power of a ^{3}He refrigerator, where the cooling power drops off exponentially (with the vapor pressure). The "pumped" ^{3}He is recovered (at room temperature) and returned to the refrigerator, so that the cooling is a continuous process. The outgoing liquid is used to precool the returning liquid through heat exchangers. Commercial refrigerators can now produce a minimum temperature of ~5 mK with circulation rates up to 2,000 μmoles/s. The state-of-the-art dilution refrigerator at Leiden (Frossatti) can produce a (continuous) base temperature of below 2 mK, making possible the study of ^{3}He superfluid without using a second cooling stage.

The dilution refrigerator is generally used only as a precooling stage. Secondary refrigeration is obtained from Pomeranchuk cooling or adiabatic demagnetization of a suitable spin-containing material. Pomeranchuk or compressional cooling was first used by Wheatley's group; the Cornell group employed it in the experiments that led to the discovery of superfluidity in ^{3}He. Below 316 mK, the melting curve of ^{3}He has a negative slope. On adiabatic compression of a mixture of liquid and solid ^{3}He, liquid is converted to solid, and the temperature drops. The compression is applied by using a liquid ^{4}He driven bellows system. The design has to comply with the fact that at these temperatures ^{4}He solidifies at 25 bars, whereas the ^{3}He is liquid at a pressure over 30 bars (solidifying close to 34 bar). One also has to minimize any heating from the compression process. The pressure of the ^{3}He mixture fixes the temperature; however, the experiments have to be confined to the melting curve, unless one uses it as an indirect cooling method (i.e., the ^{3}He experimental cell is cooled by the ^{3}He in the Pomeranchuk cell through heat exchangers, which can limit the low temperature achieved). Because of the ordering transition in solid ^{3}He, the melting curve flattens out at 1 mK, and thus the lowest temperature that can be achieved is 1 mK.

Adiabatic demagnetization (generally of copper) has proved to be the most powerful technique to obtain temperatures below 1 mK. The weak nuclear paramagnetism of copper is employed. A dilution refrigerator is used to cool a bundle of copper (wires or plates, to reduce eddy current heating when the magnetic field is changed) in a large magnetic field (usually 8 Tesla) to temperatures of the order of 10–15 mK. The superconducting solenoids are suitably compensated, with a fast roll-off to limit the spillover magnetic field in the vicinity of the experiment and the mixing chamber to a manageable value. The heat of magnetization is removed by the dilution refrigerator (typically this takes 24 to 48 hours); the two stages are then thermally isolated (using a heat switch), and a slow (adiabatic) demagnetization is performed in which the copper bundle cools, along with the experiment. The external heat leak into the experimental ^{3}He can be as low as 1 nW, and in most of these systems temperatures below 1 mK can be maintained for periods of the order of days. Many temperature sweeps (up and down) can be achieved by either magnetizing or demagnetizing slowly (this is nearly reversible), instead of employing a heater. Other paramagnetic systems have also been employed for magnetic cooling. One of these is the salt CMN (cerium magnesium nitrate). This is an electronic paramagnet; hence, the magnetic moment is much larger than that of copper. The advantages are that only a small magnetic field is necessary (1 Tesla is more than enough), and one can start with a much higher precooling temperature. However, the large magnetic moment causes CMN to order antiferromagnetically at a temperature of 2 mK, thus limiting the minimum temperature possible. Despite this, in the early experiments two groups did manage to cool into the superfluid phase and perform sound experiments (Paulson et al., 1973; Roach et al., 1975a and b). The salt is in the form of a "pill" (which has been pressed from very fine powder) and is placed in contact with the liquid. Because of this geometry, the experiments were limited to near zero field, but pressures up to the melting curve were possible. Another magnetic refrigerant that is popular is PrNi$_5$. This has a hyperfine enhanced nuclear magnetic moment that, like CMN, gives a much higher cooling power with fewer constraints on the starting parameters (precooling temperature and initial magnetic field), but it has an ordering transition at 0.3 mK. The metallurgical properties of this material are rather cumbersome; however, it has been widely used as a precooling stage for a two-stage demagnetization refrigerator.

To isolate the dilution refrigerator thermally from the next lower cooling stage, a heat switch is employed. Mechanical heat switches have been employed effectively by Roach et al. (1975a). It is more common to employ a superconducting heat switch made out of zinc, tin, or aluminum. When a metal becomes a superconductor, the thermal conductivity can be significantly lowered, and the ratio $\kappa_s/\kappa_n \sim T^2$. For tin below 100 mK, this is a very effective method of thermal isolation. The tin can be switched between its superconducting and normal state by energizing a small solenoid (H_c = 300 gauss). Aluminum heat switches need a smaller magnetic field and are more robust than the tin heat switches. A second important design consideration is heat transfer between liquid ^{3}He and metals. Although the Fermi excitations (electrons or ^{3}He quasiparticles) effectively transfer heat within the metal or liquid, only excitations common to both systems may transfer energy across a boundary. Two such excitations are

2. Sound Propagation and Collective Modes

mechanical (phonon) and magnetic fluctuations. Because of the very low acoustic impedance of liquid helium, there is a large mismatch in the acoustic impedance at the boundary of liquid helium and a metal container. This results in a large thermal boundary resistance called the Kapitza resistance which goes as $\sim 1/T^3$. This becomes prohibitively large in the mK range, and very large surface-area heat exchangers are used. In the case of ^{3}He, however, the magnetic interactions (possibly involving a coupling between impurities in copper and ^{3}He nuclei) improve the thermal heat transfer, and at temperatures below 1 mK the boundary resistance goes only as $1/T$.

5.1.2. Thermometry

The measurement of temperature at these low temperatures plays an important role. Temperatures are usually obtained by measuring the susceptibility of a suitable paramagnetic moment and using a modified Curie law. Either the magnetization of powdered LCMN (La-diluted cerium magnesium nitrate) is measured with a SQUID detector, or the (nuclear) susceptibility of Pt (powder or fine wires) is measured by pulsed NMR methods. At 1 mK the Pt has a long spin lattice relaxation time, $T_1 \sim 30$ s, and one must separate each temperature measurement by several minutes. It is common to use a platinum thermometer to calibrate the LCMN thermometer at these temperatures, and then use the high resolution and fast response of the LCMN thermometer.

The ^{3}He melting curve and the superfluid transition provide a readily available temperature scale (Greywall, 1986) to calibrate a thermometer. The scale has been established by careful measurements, and earlier controversy among the various temperature scales (Helsinki, La Jolla, and Cornell) has now been settled. T_c on the melting curve is now 2.49 mK (as compared to 2.75 mK on the Helsinki and the Cornell scales and 2.6 mK on the La Jolla scale). Secondary thermometers are calibrated against the superfluid transition; T_c may be varied with pressure, and most liquid probes display a distinct signature at T_c. The Greywall T_c values are listed in Table II.

Knowing the absolute temperature is crucial. At low pressures, the gap at $T = 0$ is given (in the weak coupling theory) by the BCS value of $\Delta(0) = 1.76 \, k_B T_c$, which may be measured by ultrasonic techniques. At high pressures, strong coupling corrections modify the above expression. If the absolute temperature is not known, it is not possible to determine the strong coupling corrections. Prior to the Greywall scale, there had been many erroneous identifications of the collective mode substates, especially close to 2Δ, as the accepted T_c values were 10% higher.

5.1.3. Ultrasonics

Various pulsed and cw acoustic techniques have been employed to study superfluid ^{3}He. In one respect the design of sound experiments in ^{3}He is rather different from that for a typical metallic or dielectric sample. In liquid ^{3}He the density (0.083 g/cm^3) and sound velocity ($\sim$300 m/s) are very much smaller than those in quartz or in a metal, making the acoustic impedance very much different. The liquid helium does not load the transducers very well, resulting in a high Q and a long ring-down (up) time. Thus, to get the full amplitude, the pulse widths have to be larger than 5 μs (typically 10 μs). However, path lengths of 4–6 mm are adequate to resolve the transmitted pulse from the electric feedthrough, since the sound velocity is also small. Also, there is no need to make bonds between the sample and the transducer. The transmitting and receiving transducers are maintained parallel by placing a cylindrical hollow (quartz) spacer between them. The ends of the quartz spacer are polished optically flat and parallel, and are also castellated in a way that minimizes any direct sound transmission through the spacer. This also provides for adequate flow of liquid for thermal equilibrium. A typical insertion loss for pulsed transmission measurements is 80 dB. The two coax lines are specially designed to prevent heat leak down the center conductor, at the same time maintaining a 50-Ω impedance. The center conductor is often a superconducting wire such as Nb–Ti. The coaxial feedthroughs into the helium cell have to be leak-tight at high pressures.

Since the measurements are done at very low temperatures, only low levels of power can be used and, in the superfluid phase, one pulses the system periodically, giving sufficient time for the heat generated to be removed. In the Northwestern pulsed experiments (Mast *et al.*, 1980), the measurements were done with a heterodyne spectrometer (Gibson *et al.*, 1981); the schematics of this system are shown in Fig. 14. Pulsed sound experiments were performed at the fundamental and odd harmonics of 12 MHz quartz transducers separated by a 6-mm path length. Bursts of 32 pulses each 5 μs wide were transmitted within an interval of less than 2 s. The envelope of the received rf pulse was calculated by taking the square root of the sum of the squares of the in-phase and 90°-phase shifted, signal-averaged, outputs of a phase detector. This was fitted by nonlinear least squares with a Gaussian. The peak height determined the attenuation, the peak position determined the group velocity, and the rf phase at the peak determined the phase velocity. Absolute attenuations and group velocities were inferred from the pulse echo trains easily observed at all frequencies at the lowest temperatures. After cooling to a minimum temperature, data were taken at all frequencies during the warm-up. With the total heat leak less than 2 nW, temperatures below 2 mK could be maintained for more than four days.

2. Sound Propagation and Collective Modes

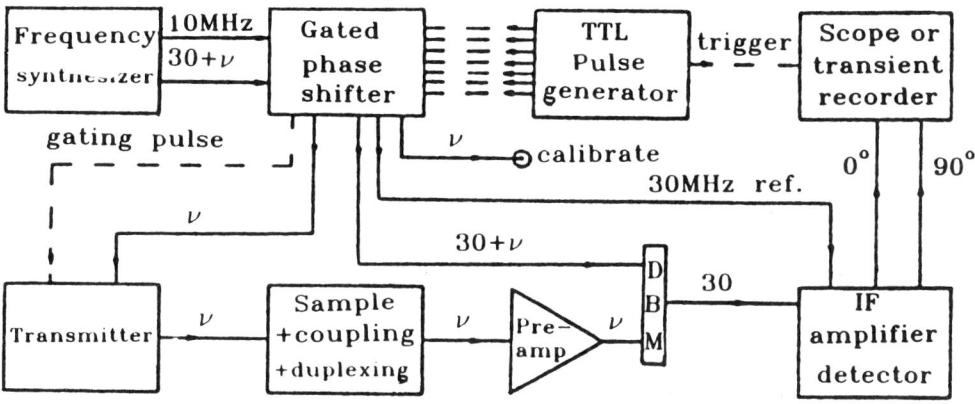

FIG. 14. Block diagram of heterodyne spectrometer. After Gibson et al. (1981).

In the vicinity of a collective mode, the attenuation is very high. High attenuation regimes can be studied by making the path length small, and introducing a quartz delay line to produce sufficient acoustic time delay to time-separate the transmitted and received signals. This method was first used (in liquid helium) by Abel et al. (1966) to detect longitudinal zero sound, and recently by Saunders et al. (1983, 1990) and Ling et al. (1987) to study the high attenuation regime near the pair-breaking and squashing mode of ^{3}He.

An alternative technique is the cw acoustic impedance technique, which has been widely used by the Northwestern group. We shall discuss this method in some detail here. It is problematic to determine the absolute velocity and attenuation with the cw acoustic impedance techniques, which is a disadvantage relative to pulsed methods. In the pulsed mode it is possible to time-separate the effects whose origin are purely electrical (transmission lines and detector electronics) from those that are primarily acoustical (the liquid or liquid transducer system). The single-ended cw technique senses the impedance of the transducer and, indirectly, the liquid. The block diagram of the electronics is shown in Fig. 15.

A frequency modulated signal at the resonant frequency of the transducer is passed through a fixed attenuator and applied to one arm of a quadrature hybrid. A reference signal derived from the same source is connected to the second arm, after being sent through a variable attenuator and delay line (phase shifter). The third arm is connected to the transducer (in contact with the liquid), and the fourth arm to a low-noise rf amplifier, the output of which goes to an amplitude modulation detector. The signal generator is locked to a synchronizer to stabilize

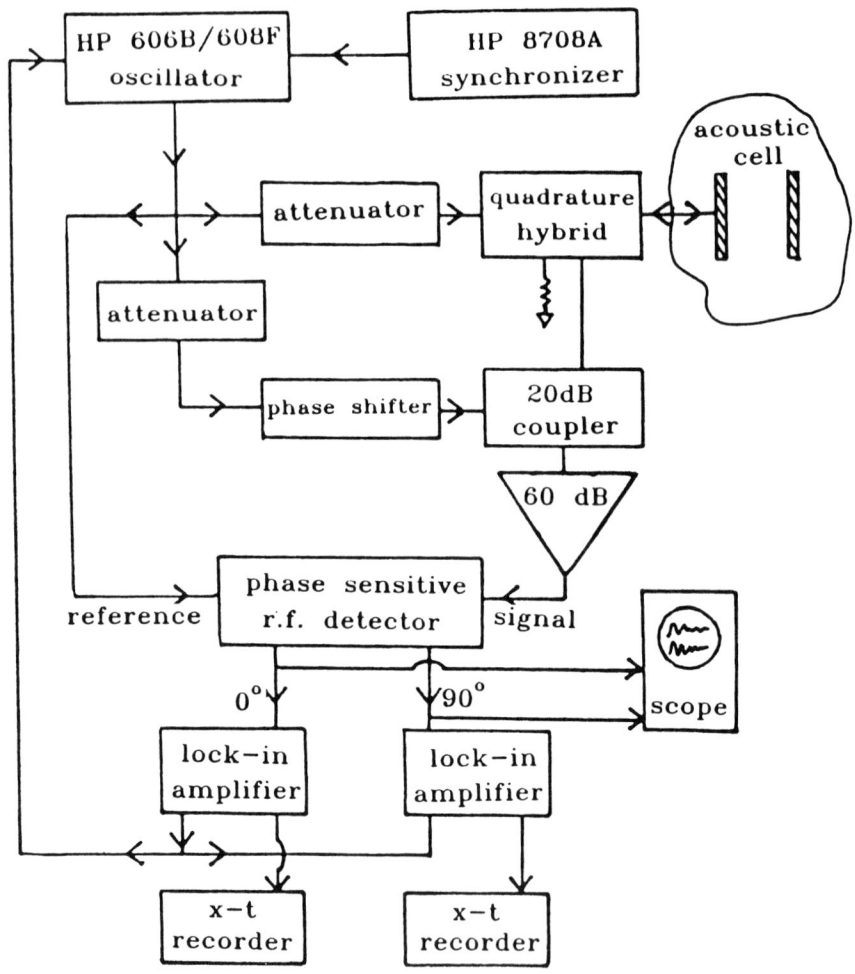

Fig. 15. Block diagram of the Northwestern cw impedance spectrometer. After Shivaram et al. (1986).

the frequency of the output signal. The signal is frequency modulated by the synchronizer using a reference signal from a lock-in amplifier. Because of the relatively high Q of a ^{3}He-loaded transducer, a frequency-modulated signal applied at the resonant frequency will be converted to an amplitude-modulated signal. The output from the fourth arm of the quadrature hybrid will then consist of an FM part (the applied signal) and an AM part (the signal reflected back from the transducer), both being modulated by the same frequency. The detector will detect only the AM signal, and the output of the detector

2. Sound Propagation and Collective Modes

will be a signal at the frequency of the modulating signal. This signal is then fed to the lock-in.

By this method the actual impedance of the liquid is not measured. There are many elements (the coaxes, the transducer, and their coupling), each with its own frequency-dependent transfer characteristics. However, once the bridge is balanced (nulled) at any suitable point, then following the output of the fourth arm, one can pick up any change in the liquid impedance as any of the following parameters are varied: temperature, pressure, or magnetic field. Typical traces are shown in Fig. 16. There are two advantages of the cw impedance method: (i) the bandwidth is very much narrower than the pulsed technique, and thus has a high resolution in picking up the splitting of the collective modes; (ii) the amplitude of the sound waves is much lower, resulting in less heating. This

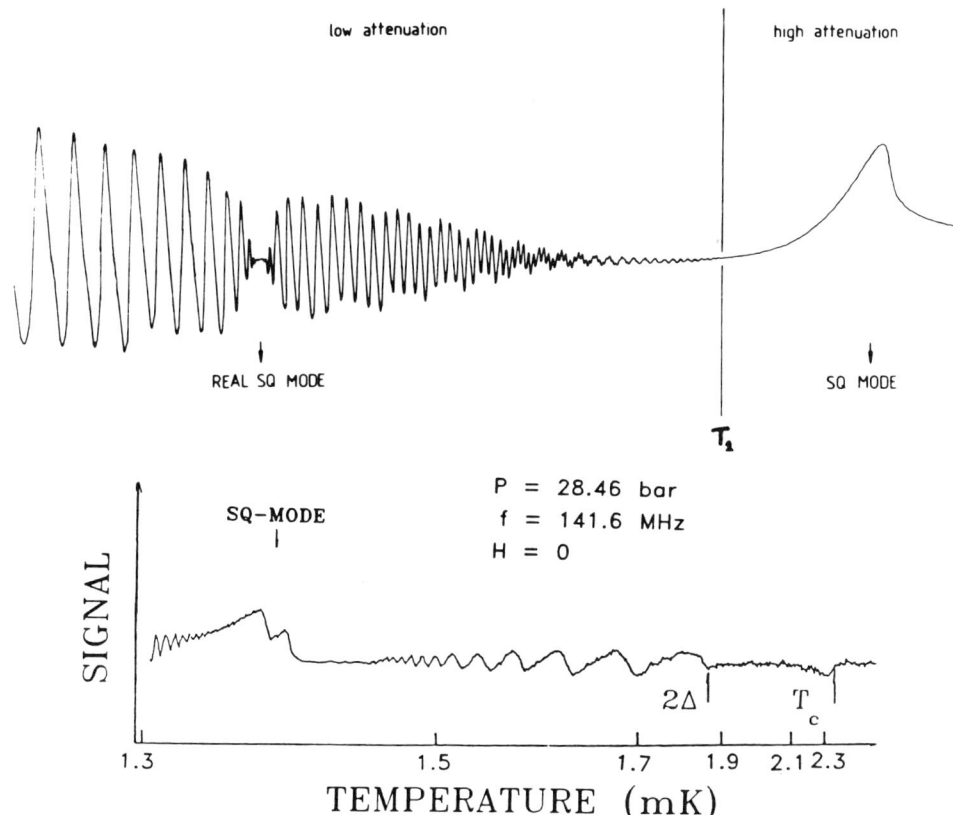

FIG. 16. Typical traces from FM modulated impedance measurements: (a) sq and rsq modes in the 4 mm path length cell (after Shivaram et al., 1986) and (b) sq mode, pair-breaking edge and T_c in the 180 μm path length cell (after Adenwalla et al., 1989).

lower amplitude also helps to minimize amplitude dependent (nonlinear) effects in the liquid.

5.1.4. Cell Design

In this section we briefly review some of the ^{3}He experimental cells and unusual cryogenic features of a few laboratories.

Figure 17 shows the Argonne cell (Ketterson et al., 1975) made out of epoxy (Epibond 100A). The vacuum seal was accomplished with a soap–glycerine seal. The heat exchanger in the cell involved a stack of thin gold foils precooled through a mechanical heat switch connected to the mixing chamber. The refrigerant was a pressed pellet of CMN.

The Northwestern cryostat used a copper nuclear stage (19 moles of copper made from wires of diameter 0.25 mm) with a sintered copper powder heat exchanger (34 m^2 surface area) which cooled 12–16 cm^3 of liquid ^{3}He; the stage was precooled through a tin heat switch. In the first cell (Mast et al., 1980), ^{3}He was cooled to 0.41 mK. The LCMN thermometer was extensively calibrated against a Pt NMR thermometer and the superfluid transition (Helsinki scale), since below 0.7 mK the LCMN showed departures from Curie law. In later experiments the measurements were performed in a magnetic field. The sound cell was located more remotely from the heat exchanger with which it communicated via a 3 mm liquid column. This removed the cell from the fringing field of the demagnetization magnet. At these temperatures the thermal conductivity of liquid ^{3}He is as good as that of copper. The low-temperature experimental setup is shown in Fig. 18, and the sound cell is shown in Fig. 19. A split housing holds the transducers and spacer assembly with $\hat{\mathbf{q}}$ perpendicular to the cryostat axis. Two superconducting magnets were available to apply a magnetic field perpendicular to $\hat{\mathbf{q}}$ (a 3 T solenoid) and parallel to $\hat{\mathbf{q}}$ (a 0.6 T split pair). Two LCMN towers, one on top of the cell (and thus outside the magnetic field) and the second just above the heat exchanger, allowed one to estimate the temperature of the ^{3}He in the sound path in the presence of possible thermal gradients. Various spacers were used. The earlier experiment used a 6 mm quartz spacer, and the later experiments had either a 4 mm quartz or a 4 mm Macor spacer. All spacers were castellated. The short path cell ($d = 190$ μm) was made from a gold-plated tungsten wire, coiled into a single loop and placed between the transducers.

The London group (Saunders et al., 1983, 1990) used a short path cell (250 μm) and a quartz delay line. Their cell is shown in Fig. 20. They were able to resolve structure in the high attenuation regime, close to 2Δ and the sq mode.

Fraenkel et al. (1989) have developed an acoustic cell using a pair of

2. Sound Propagation and Collective Modes

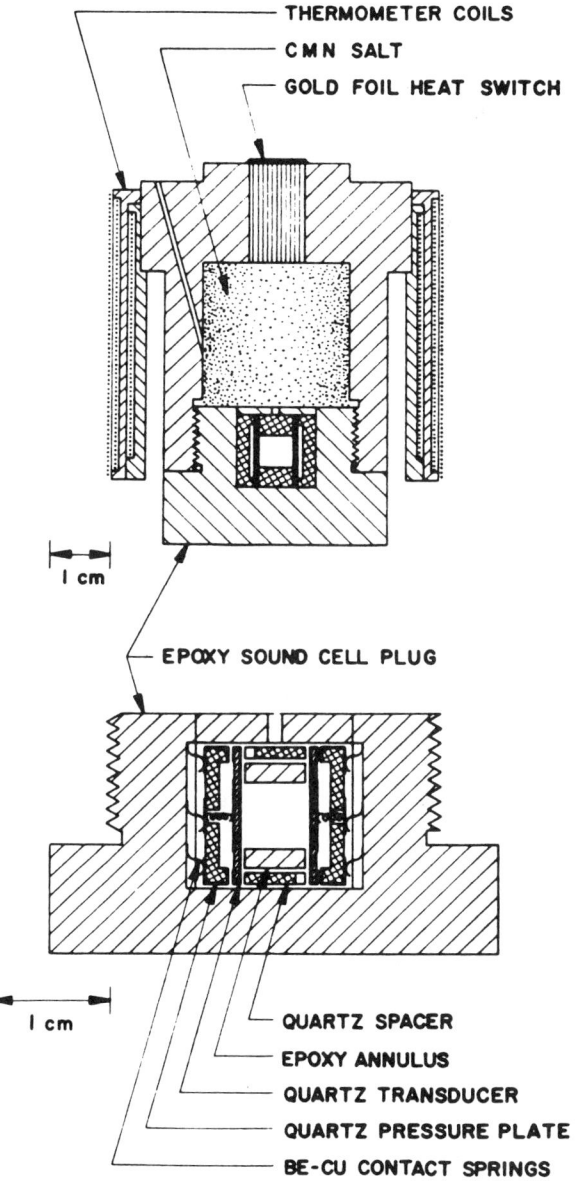

FIG. 17. Epoxy sound cell of Roach et al. (1975a). After Ketterson et al. (1975).

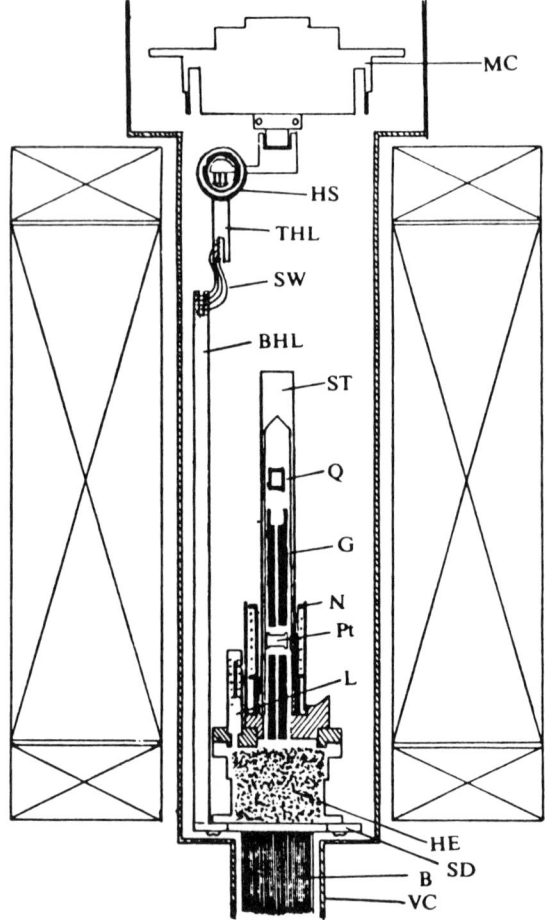

FIG. 18. Low-temperature experimental setup. MC, mixing chamber; HS, heat switch; THL, top heat link; BHL, bottom heat link; SW, silver wires; B, copper bundle; ST, silver tube; G, graphite; Q, quartz transducers; SD, silver disk; Pt, platinum, H_1 coil; N, niobium shield; L, LCMN tower; HE, heat exchanger; VC, vacuum can. After Shivaram et al. (1986).

nonresonant piezoelectric polyvinylidene fluoride (PVDF) plastic film transducers. A schematic of their cell is shown in Fig. 21. The composite transducers are separated by a 1.6 mm quartz spacer ring; a torsional oscillator functioned as a thermometer by measuring the normal fluid fraction, ρ_n/ρ, of the ^{3}He liquid inside the cell. The plastic film transducers have the advantage of a very wide bandwidth (which permits a very rapid rise time and consequently allows a short path length). The lower acoustic impedance of a plastic film and the fact that it

2. Sound Propagation and Collective Modes

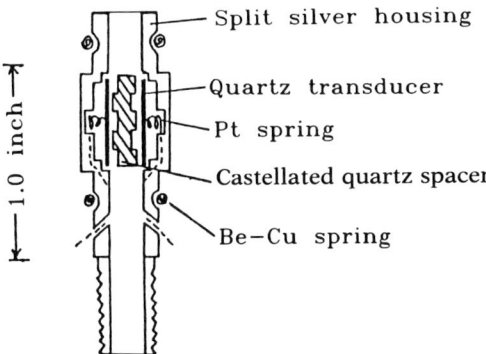

FIG. 19. Sound cell of the Northwestern group. After Shivaram et al. (1986).

is thin allow a rapid transfer of energy to the liquid (ring down) and hence a large bandwidth.

The Finnish group has done sound measurements in rotating ^{3}He-B and observed coupling to the rsq collective mode from rotational effects. Both the Northwestern (Adenwalla, 1989) and the Cornell groups have used a depressurizing technique to sweep through the collective modes.

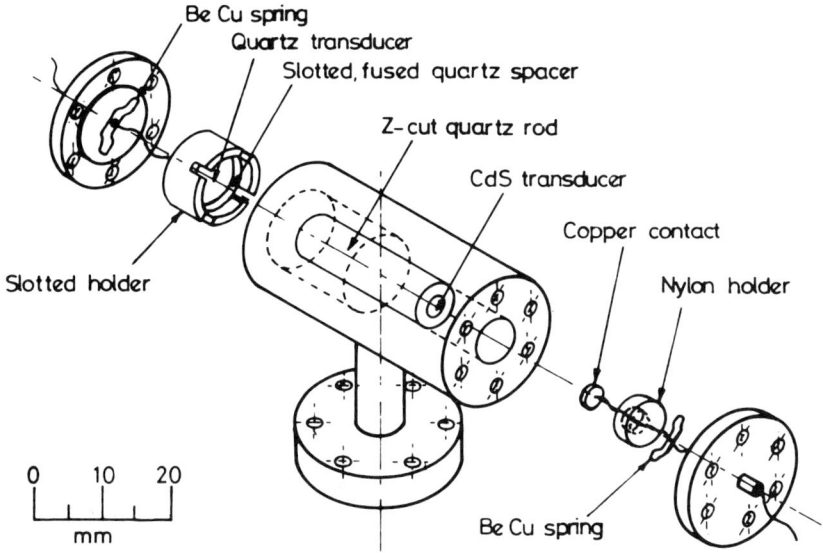

FIG. 20. Sound cell of the London group. After Saunders et al. (1983).

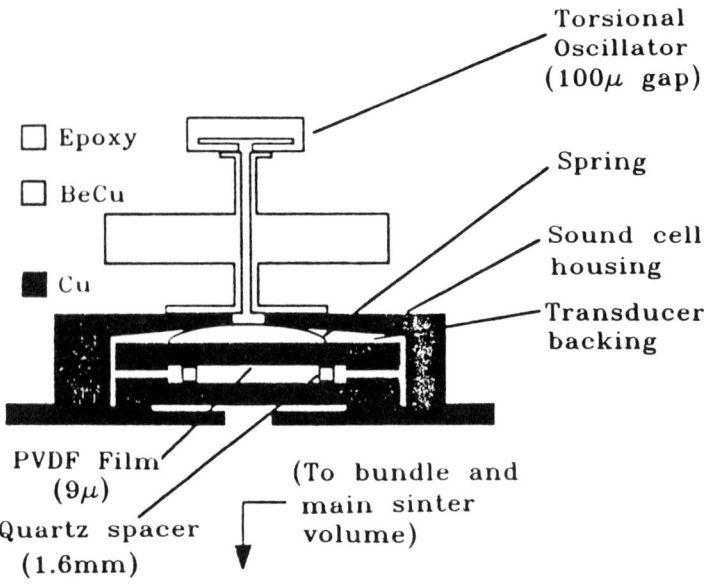

FIG. 21. Sound cell of the Cornell group using a plastic piezoelectric transducer, PVDF. After Fraenkel et al. (1989).

5.2. Superfluid ^{3}He-B

5.2.1. Early Experiments on Superfluid ^{3}He-B

The first ultrasonic measurements in superfluid ^{3}He-B were performed by Paulson et al. (1973) and Roach et al. (1975a) using a CMN demagnetization cell. Figure 22 shows the data of Roach et al. (1975a) taken at a pressure of 21.00 bar, just below the polycritical point. Measurements were done at a high pressure to ensure the highest T_c, since the cooling power of CMN is very limited at these temperatures. At the superfluid transition there is a sharp drop in the zero sound velocity (approaching the first sound value c_1 as $T \rightarrow 0$). The attenuation increases rapidly due to pair breaking and absorption into the sq collective mode. The change in velocity associated with the first sound–zero sound transition may be thought of as arising from the stiffening of the liquid caused by a change in shape (as well as size) of the Fermi surface, which cannot relax when $\omega\tau > 1$; however, the appearance of a gap suppresses the shape change, which is the qualitative reason why the sound velocity approaches the first sound value as $T \rightarrow 0$.

2. Sound Propagation and Collective Modes

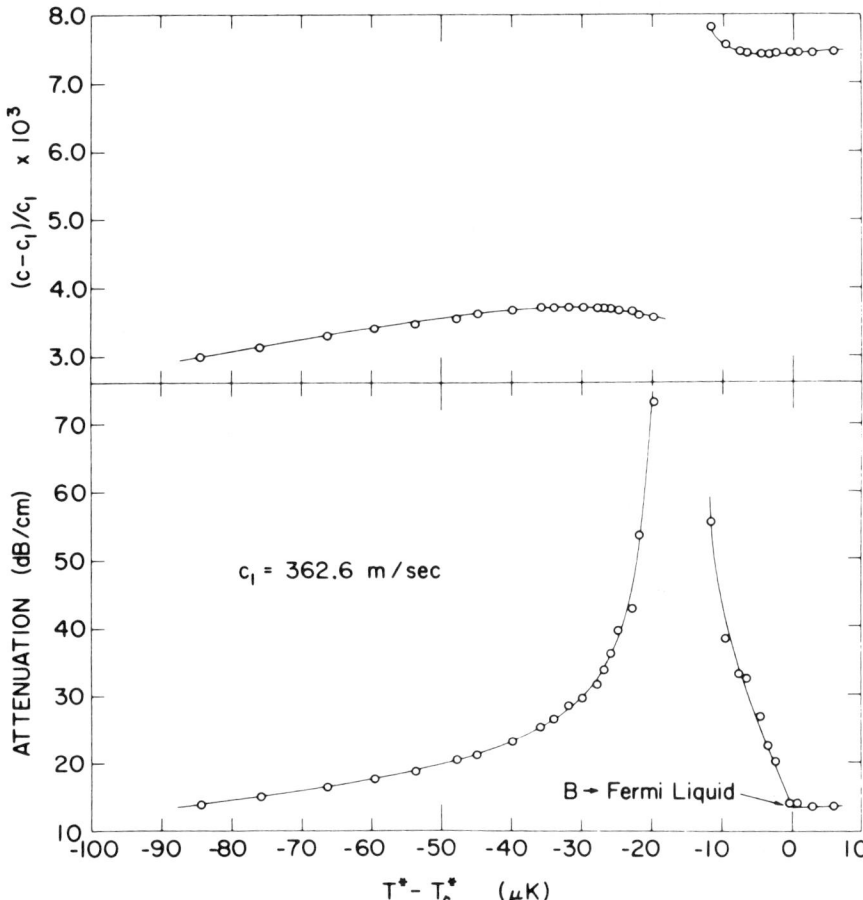

FIG. 22. Zero sound results of the superfluid transition in the B-phase. After Roach et al. (1975a).

5.2.2. The Real Squashing Mode (2^+)

The real squashing mode is a collective mode corresponding to the $J = 2$ real part of the order parameter, and occurs at a frequency $\hbar\omega = \sqrt{(8/5)}\Delta(T)$. Initially this was expected not to couple to zero sound, but only to spin waves.

Experiments by Mast et al. (1980) and Giannetta et al. (1980) showed a very sharp and narrow absorption peak at $\omega = 1.09\ \Delta$ (15% lower than that predicted for the rsq mode). The Northwestern experiments were done on the previously described copper nuclear demagnetization apparatus. Pulsed transmission measurements were made of the attenuation, group velocity, and changes in the phase velocity. The observed sharp peak in the sound attenuation

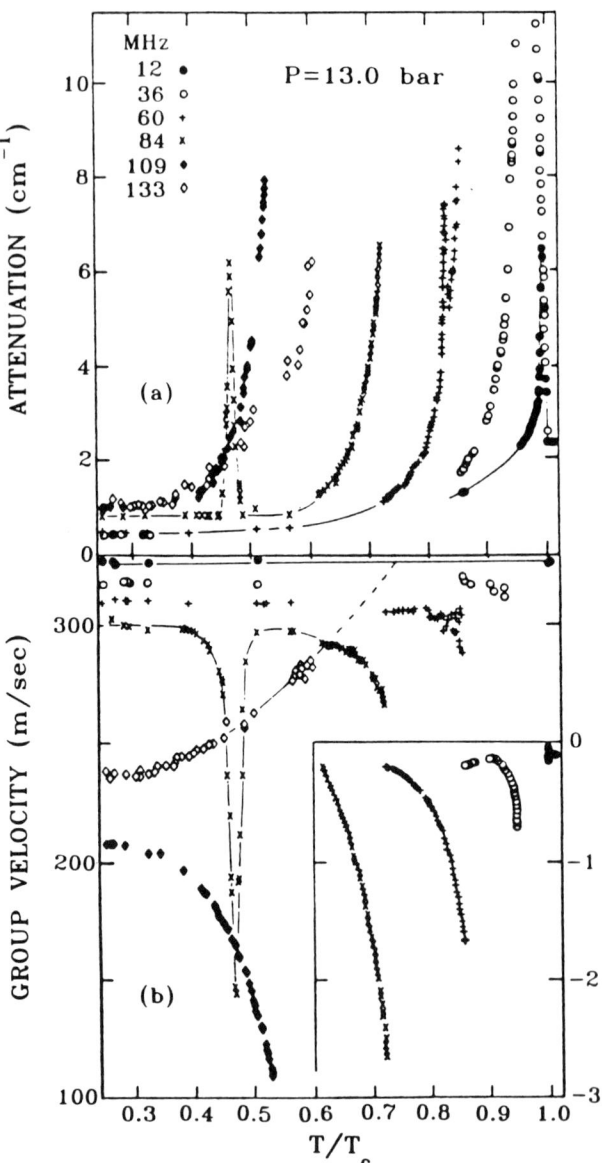

FIG. 23. (a) Attenuation and (b) group velocity at the rsq mode. After Mast et al. (1980).

2. Sound Propagation and Collective Modes

was accompanied by a sharp drop in the group velocity; the details are show in Fig. 23. The measurements were done at a pressure of 13 bar ($T_c = 1.981$ mK) at six frequencies. A new resonance was seen, which is clearly resolved for the 60 and 84 MHz data. Also, at temperatures where the attenuation increases sharply, it is observed that the group velocity of the acoustic pulse decreases by as much as a factor of 3 below that of the velocity of zero sound.

It was quickly realized that the presence of an asymmetry between the energy of particle and hole states located asymmetrically about the Fermi surface would provide the needed coupling mechanism to this newly observed collective mode. Magnetic fields applied parallel to $\hat{\mathbf{q}}$ should have no effect. However, on applying a field perpendicular to $\hat{\mathbf{q}}$, Avenel et al. (1980) observed a fivefold splitting of the mode, which unambiguously defined its $J = 2$ character. Their data are shown in Fig. 24, where the single peak in zero field undergoes a fivefold Zeeman splitting. They obtained a splitting of 360 μKT^{-1} (or 10 MHzT^{-1}), close to the theoretical value of 10 MHzT^{-1} predicted by Tewordt and Schopohl (1979b). Theory predicts that the coupling of sound to these substates should be proportional to the spherical harmonics

$$\Lambda_{2\pm}(M_J,\phi) \propto \begin{cases} \frac{1}{4}(1 - 3\cos^2\phi)^2 & M_J = 0 \\ \frac{3}{2}\sin^2\phi \cos^2\phi & M_J = \pm 1 \\ \frac{3}{8}\sin^4\phi & M_J = \pm 2 \end{cases} \qquad (75)$$

Avenel et al. (1980) measured the angular dependence and, from the intensities of the spectral lines, deduced the coupling strengths. The anisotropy in the

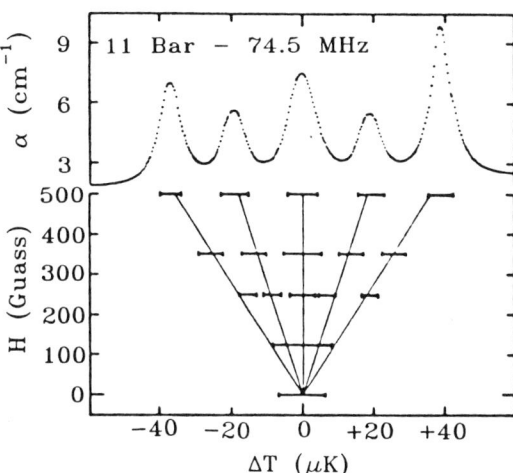

FIG. 24. Fivefold splitting of the rsq attenuation peak (Avenel et al., 1980).

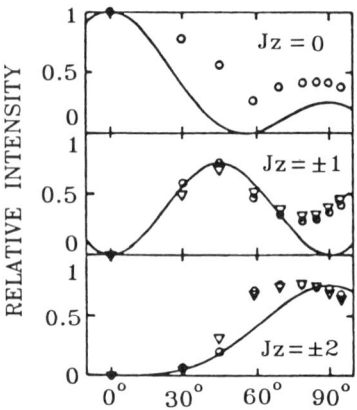

FIG. 25. Anisotropy of the fivefold splitting of the rsq attenuation peak. The angle (x-axis) is measured between the applied magnetic field and the sound propagation direction. (Avenel et al., 1980).

coupling strengths is plotted in Fig. 25 and agrees very well with the theoretical predictions for the $J = 2^+$ mode. These observations of a fivefold splitting and the anisotropy of the coupling strengths were conclusive evidence that this new peak was the $J = 2^+$ rsq mode.

The observed frequency of the rsq collective mode is lower than predicted. A compilation of data for various frequencies and pressures by Adenwalla et al. (1988) is shown in Fig. 26.

The cw method gave a much higher resolution, and many details of the magnetic field splitting of the rsq mode and other highly attenuating modes could be studied. Figure 27 is a typical trace showing the splitting in a magnetic field. A nonlinear evolution of the rsq mode was observed in higher fields (up to 180 mT) by Shivaram et al. (1983); see Fig. 28. The distribution of the various components became asymmetric about the central state above 30 mT, and when the field was increased to about 140 mT a mode crossing occurred between the $J_z = -1$ and the $J_z = -2$ states (the separate $J_z = +1$ and $J_z = +2$ components were not resolvable above 120 mT). In addition, the central $J_z = 0$ state shifted to lower frequency as the field was increased. (See Fig. 29.)

Schopohl et al. (1983) explained this nonlinear phenomenon as a result of a distortion of the B-phase energy gap in a magnetic field. It has been referred to as the Paschen–Back effect. Figure 30 is a schematic of the gap distortion in a magnetic field, and Fig. 31 shows the field dependence of the two components of the gap.

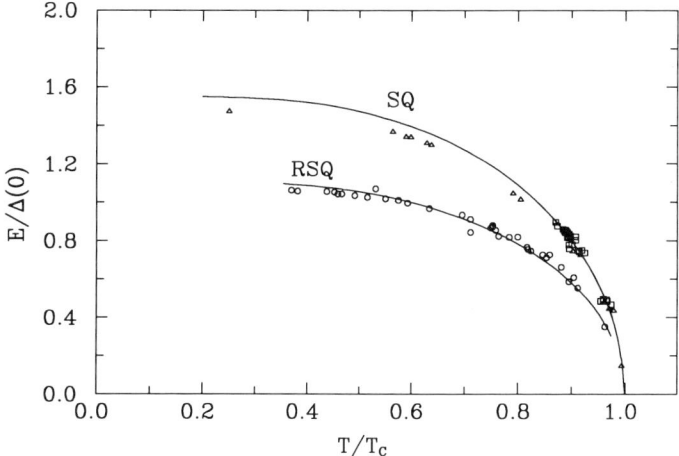

FIG. 26. A compilation of data for the rsq and sq modes. After Adenwalla et al. (1988).

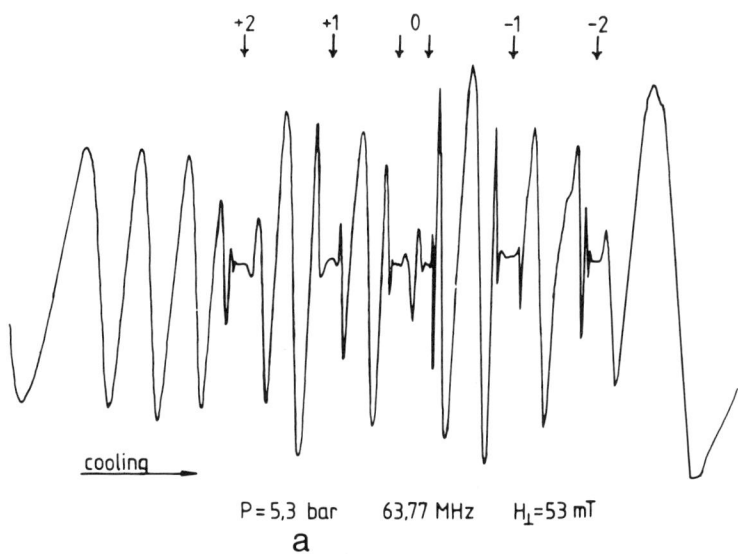

FIG. 27. Trace from impedance spectrum showing the fivefold splitting of the rsq mode and the doublet $J_Z = 0$. After Shivaram et al. (1986).

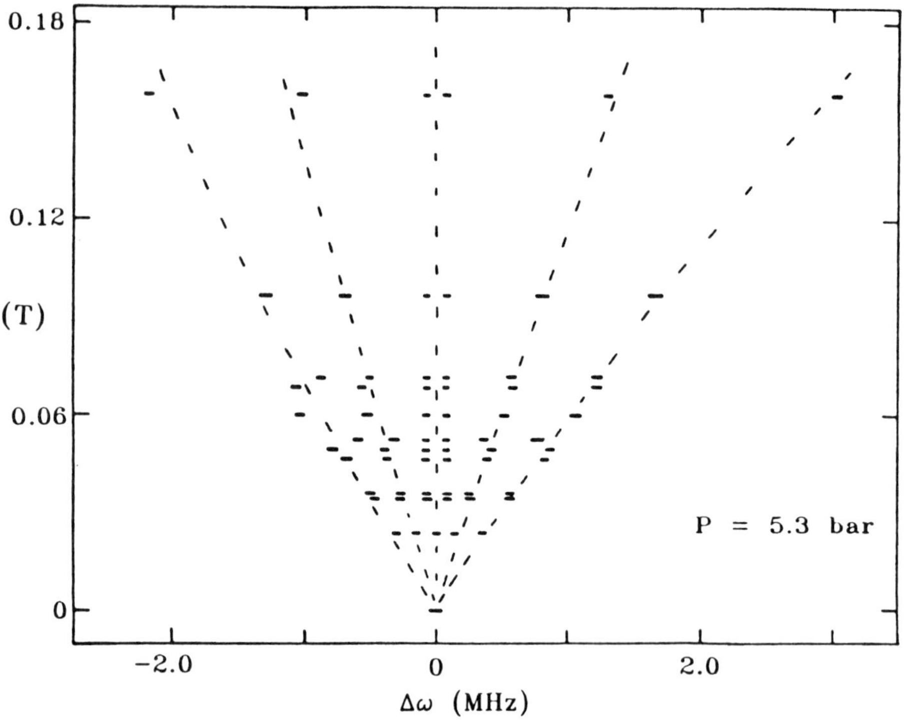

FIG. 28. Evolution of the Zeeman fivefold splitting of the rsq mode. The doublet splitting of the $J_z = 0$ mode is also shown. After Shivaram *et al.* (1986).

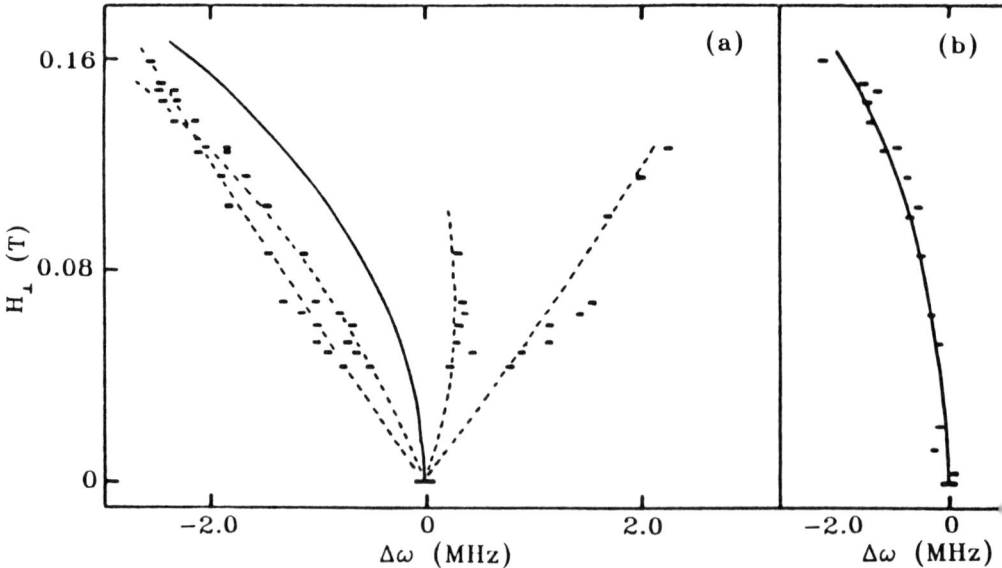

FIG. 29. Nonlinear evolution of the fivefold splitting (Paschen–Bach effect). (Shivaram *et al.*, 1986).

2. Sound Propagation and Collective Modes

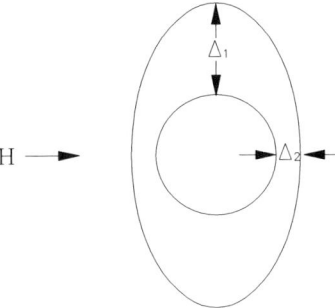

FIG. 30. Gap distortion of the B-phase in a magnetic field.

5.2.3. Squashing Mode (2⁻)

The imaginary component $J = 2^-$ squashing mode, which should occur at $\sqrt{(12/5)}\Delta$, couples strongly to sound. The coupling strength is over 100 times larger than for the rsq mode. At higher frequencies, the attenuation becomes enormously large; the peak becomes very broad and merges with the attenuation

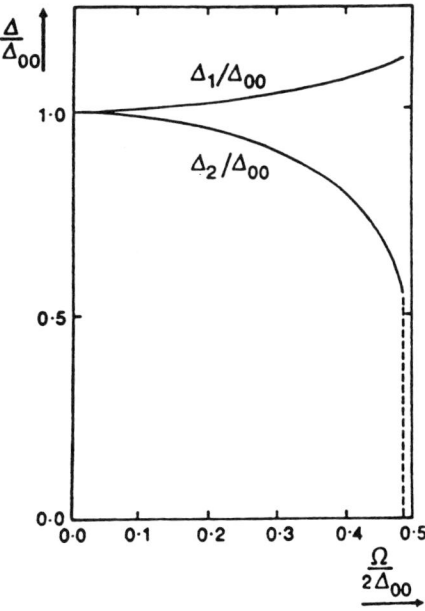

FIG. 31. The B-phase gap distortion as a function of magnetic field. (Tewordt and Schopohl, 1979b).

peak coming from the pair breaking. As a result, it is difficult to locate the peak position and hence the sq mode frequency.

Calder *et al.* (1980) developed a mathematical technique that allowed them to locate the position of the peak by analyzing the dispersion of the group velocity in the vicinity of the sq and rsq modes. They modeled the group velocity behavior by a phenomenological description of the coupling of sound to a collective mode, first developed by Wolfle (1976), and they were able to infer the collective mode frequency. They called this method group velocity spectroscopy (GVS). We now sketch this theory.

Associated with each collective mode of the order parameter is a frequency

$$\omega_j = a_j \Delta(T); \tag{76}$$

the a_j are constants (of order unity) and have a weak dependence on pressure and temperature, and the index j runs over all the collective modes. We introduce a phenomonological equation of motion for a distortion of the order parameter, δ, having the form of a damped harmonic oscillator with quadratic dispersion:

$$\ddot{\delta}_j + \frac{1}{\tau_j}\dot{\delta} + \omega_j^2 \delta_j - c_j^2 \nabla^2 \delta = \gamma \nabla^2 \rho; \tag{77}$$

γ is a constant, which gives the coupling to the (second derivative of the) density. Combining this with the hydrodynamic relation

$$\ddot{\rho} = \nabla^2 P \tag{78}$$

and an "equation of state"

$$P = P(\rho, \delta), \tag{79}$$

one obtains a dispersion relation of the sound in the vicinity of the collective mode from which the expressions for the phase velocity, attenuation, and group velocity are obtained. These expressions are

$$2\frac{c^2}{c_1^2} = 1 + \left[1 + \frac{8\lambda_j \eta}{(\eta^2 + b^2)}\right]^{1/2}, \tag{80}$$

$$\alpha = \frac{\lambda_j \left(\frac{c_1}{c}\right)^2}{\eta^2 + b^2 + 4\lambda_j \left(\frac{c_1}{c}\right)^2 \eta}, \tag{81}$$

and

2. Sound Propagation and Collective Modes

$$\frac{v_g}{c_1} = \frac{1 + 2\lambda_j \left(\frac{c_1}{c}\right)^4 \frac{\eta}{\eta^2 + b^2}}{1 + 2\lambda_j \left(\frac{c_1}{c}\right)^4 \frac{[\eta^2 + b^2(\eta - 1)]}{(\eta^2 + b^2)^2}}, \quad (82)$$

where $\eta = 1 - \omega_j^2/\omega^2$ and $b^{-1} = \omega\tau$.

Absolute group velocities (accurate to better than 1%) could be obtained at the lowest temperatures. At a given pressure, v_g and v_p can be fitted by least squares to yield ω_{sq} and λ_{sq}; see Fig. 32 and Fig. 33. Calder *et al.* (1980) obtained $\omega_{sq} = 1.50 \, \Delta_{BCS}(0)$ and $\lambda_{sq} = 0.010$ at 13.1 bars.

Avenel *et al.* (1981) refined the preceding theory by adding two boundary conditions at a solid (transducer) interface, to describe the acoustic impedance of superfluid ^{3}He. A pulsed acoustic impedance technique was employed by Avenel *et al.* (1981) in their ring-down experiment using a sound frequency $f = 104.3$ MHz at $P = 13.5$ bar. By fitting both the real and imaginary parts

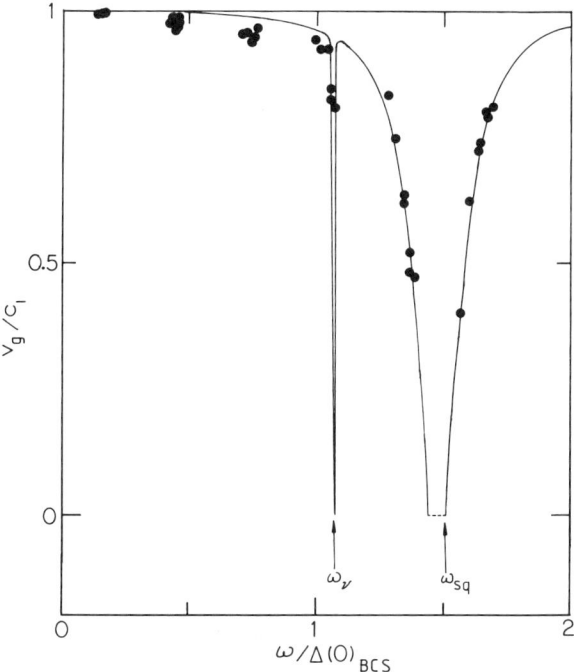

FIG. 32. Group velocity at the rsq and sq modes. The solid line is a fit from the GVS expressions (Calder *et al.*, 1980).

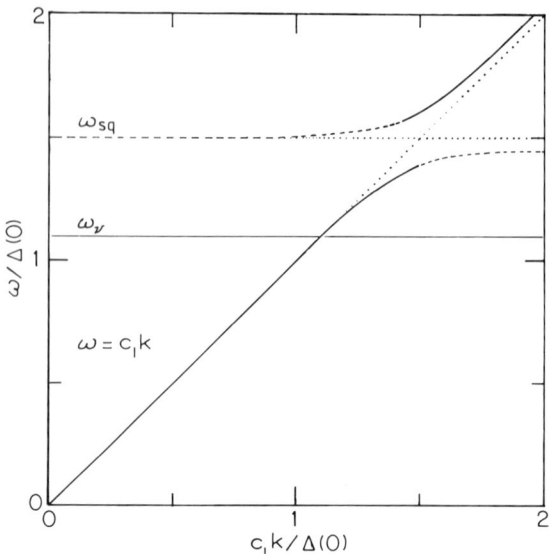

FIG. 33. Mode crossing of the phase velocity at the rsq and sq modes. (Calder et al., 1980).

of the detected acoustic impedance, the squashing mode frequency, the coupling constant, the relaxation time of the order parameter, the accommodation length, and the phase velocity of the acoustic sound were obtained (see Fig. 34). A detailed theoretical study of the propagation characteristics near a collective mode was carried out by Avenel et al. (1983) and Varoquaux et al. (1986). They performed numerical studies that clarified the operational applicability of the concepts of phase, signal, and group velocity.

Because of the very large attenuation (100 cm^{-1}) and the width of the sq mode peak, and the small Lande g-factor (typically 10 times smaller than that for the rsq mode), it is difficult to observe the Zeeman splitting of the sq mode. The problems were overcome by Movshovich et al. (1988). The liquid was cooled to 0.4 mK in a high magnetic field (460 mT) in order to maximize the splitting and minimize the line width (quasiparticle scattering induces line broadening and is smaller at lower temperatures). A conventional pulse transmission technique was employed to measure the group velocity spectrum (while the liquid was depressurized slowly), which, when fit to a theoretical model (Varoquaux et al., 1986), yielded the field split positions of the different J_z components of the sq mode; see Fig. 35. The nonlinear shift of the strongly coupled $J_z = 0$ component was observed earlier by Meisel et al. (1983b) using the acoustic impedance technique.

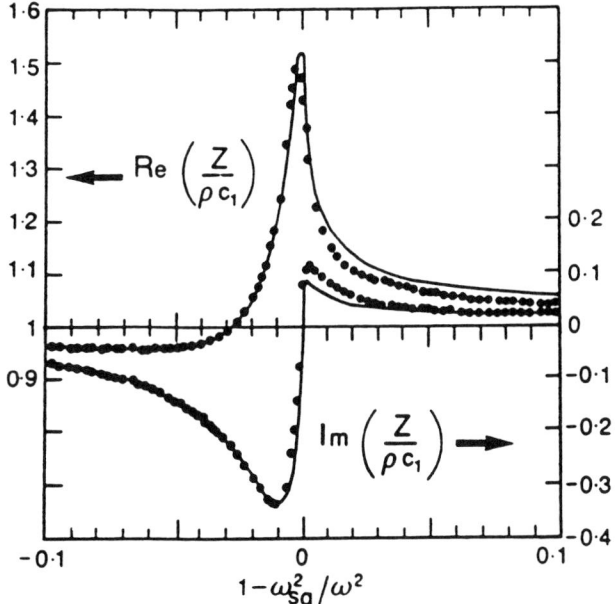

FIG. 34. Transmission impedance data of Avenel *et al.* (1983).

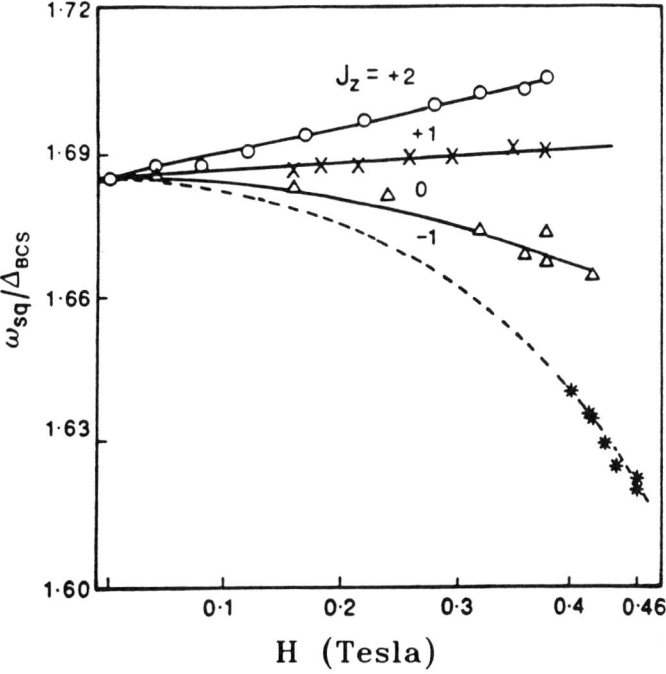

FIG. 35. Field splitting of the sq mode. After Movshovich *et al.* (1988).

5.2.4. Pair Breaking

The B-phase gap is isotropic, and at low pressures the value is given by the weak coupling BCS value,

$$\Delta(0) = 1.76 k_B T_c. \tag{83}$$

When the frequency of the sound waves is greater than 2Δ, Cooper pairs can be broken, and this gives rise to a very large attenuation. When $\hbar\omega = 2\Delta(T)$, the pair-breaking ceases and the attenuation drops abruptly, defining the pair-breaking edge. However it is difficult to observe the pair-breaking edge, since immediately below this the attenuation rises again because of absorption into the sq mode.

The first indication of the pair-breaking edge was seen in the pulsed transmission measurements of Giannetta et al. (1980), at 5.3 bar and 60 MHz. Later, Daniels et al. (1983), using their quartz delay line and a short path length of 250 μm, were able to resolve completely the pair-breaking edge from the sq mode. Meisel et al. (1983c) used the pulsed zero sound method at low pressures (P ~ 1 bar) to resolve the PB edge from the sq mode. They also detected an abrupt change in the group velocity at the PB edge.

Extensive measurements of the pair-breaking edge have been performed by Adenwalla et al. (1989) using the Northwestern cw impedance technique and a short path cell. A typical trace from the impedance spectrometer is shown in Fig. 16. As the temperature is lowered one sees the signature at T_c (there is a change in impedance because of the abrupt increase in attenuation coming from the breaking of pairs). About 0.2 to 0.4 mK below T_c, there is a step in the trace that is the PB signature. Below this, oscillations appear in the trace arising from the approach of the squashing mode (facilitated by the low attenuation). As the temperature is lowered, the oscillations die out as the attenuation rises near the sq mode.

Their measurements were done in zero field at pressures ranging from 3 to 28 bar, and at frequencies of 64.3, 90.1, 141.6, and 167.4 MHz; the path length in the cell was 390 μm. Their data are plotted in Fig. 36 and are well represented by the weak coupling-plus model and the Greywall temperature scale. (Greywall, 1986).

Application of a magnetic field has two effects on the pair-breaking edge: (i) it results in a Zeeman shift of the single particle excitation energies, and (ii) it results in a distortion of the B-phase energy gap. The Zeeman shift is the same as that for the sq or the rsq modes obtained earlier. The effect of the magnetic field is to lower the longitudinal gap, $\Delta_\parallel$, and increase the transverse energy gap, $\Delta_\perp$ (parallel and perpendicular to the field). The smaller of the two

2. Sound Propagation and Collective Modes

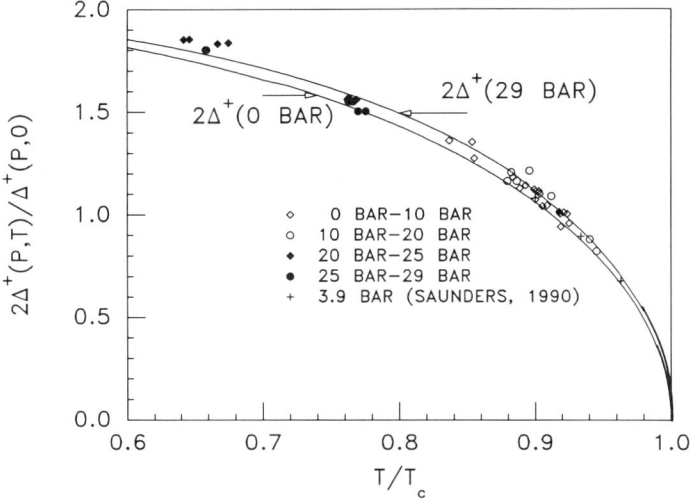

FIG. 36. Frequency of the pair-breaking edge (Adenwalla et al., 1989).

energy gaps will result in pair breaking and cause a shift of the pair-breaking edge to a lower frequency by an amount $c\Omega^2/\Delta_0(P,T)$, where the coefficient c goes to $\frac{1}{2}$ as $T \to 0$. The net result in the presence of a magnetic field is

$$\hbar\omega_{PB} = 2\Delta_0(P,T) - \Omega - \frac{c\Omega^2}{\Delta_0(P,T)}. \tag{84}$$

This prediction has been qualitatively confirmed by Movshovich et al. (1990).

Daniels et al. (1983) see an additional structure at $\omega = 1.83\Delta$ for the data at 44.2 MHz and 2.4 bar. They see some excess attenuation between the pair breaking and sq mode, which they initially identified with two modes $J = 0^+$ and $J = 4^-$. However, making use of the Greywall temperature scale, T_c is reduced by $\sim 10\%$. It was then reasonable to identify this new feature with the $J = 1^-$ mode, which is a 2Δ mode and initially was not expected to couple to sound. Schopohl and Tewordt (1984) interpreted the structure as arising from the $J = 1^-$ gap mode, which can couple to sound in the presence of a magnetic field through the induced distortion of the B-phase gap. The $J = 1^-$ mode is degenerate with the pair-breaking edge, $2\Delta(T)$, in zero magnetic field. At finite fields the $J_z = \pm 1$ components do couple to sound; the frequencies are given by

$$\omega_{J,J_z} = \omega^-_{1,\pm 1} = 2\Delta(H) \pm g\Omega_L, \tag{85}$$

where g is the g-factor and Ω_L the Larmor frequency. The $J_z = -1$ mode moves down below the gap and is now observable. In a magnetic field the coupling is proportional to the gap distortion $(\Delta_1 - \Delta_2)^2$, and hence rises strongly with increasing field.

Schopohl and Tewordt (1984) also predicted that the $J_z = -1$ gives an ordinary resonance in the attenuation, whereas the $J_z = +1$ component should show an anti-resonance or a dip in the attenuation relative to the background attenuation. This line shape is similar to the Fano resonance (Fano, 1961) seen in atomic physics. Here this effect arises from the quantum interference between the $J_z = +1, J = 1$ mode and the continuum. The net effect is a peak–antipeak structure (see Fig. 37), increasing rapidly with applied field. Ling et al. (1987) have observed the splitting of the $J_z = \pm 1$ components and compared it with theory. They find the splitting

$$\delta\left(\frac{\Delta}{\Delta_0}\right) \equiv \frac{\Delta(t^-) - \Delta(t^+)}{\Delta_0}, \tag{86}$$

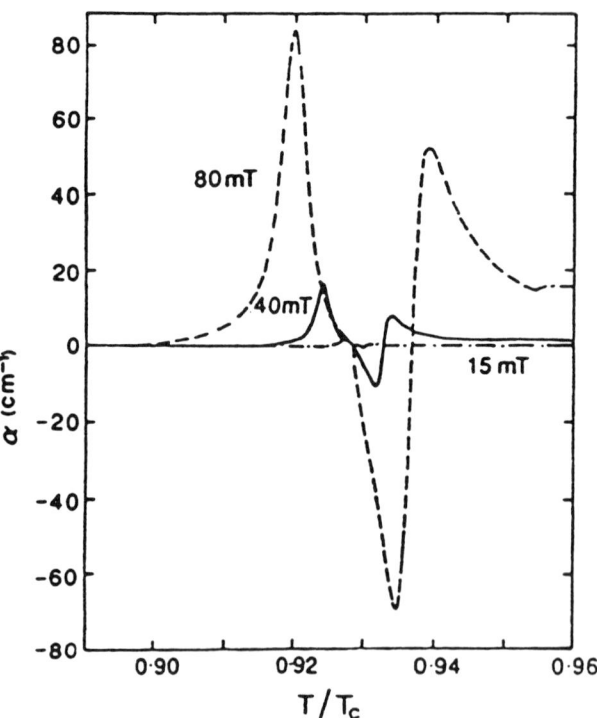

FIG. 37. Theoretical calculation for the peak antipeak structure (Fano resonance) (Schopohl and Tewordt, 1984).

2. Sound Propagation and Collective Modes

to be linear in applied field where t^- and t^+ are the reduced temperatures of the peak and antipeak, respectively. From this data they obtain an experimental g-factor of 0.40 and 0.41, in excellent agreement with the theoretical value of 0.39, which is obtained using the Greywall (1986) F_0^a values. The measurements of the splitting of the $J_z = \pm 1$ components with field show an unexpected asymmetry that may be due to gap distortion and mode–mode coupling.

5.2.5. Additional Structure

In addition to the linear and nonlinear Zeeman splitting of the $J = 2$ modes, there is some additional structure that shows up in the high-resolution cw impedance measurements. Vdovin (1963) had predicted, and it was later recalculated by Brusov and Popov (1980), that the degeneracy of the collective mode is lifted at finite vector $\mathbf{q}$, since various J_z components have different dispersion relations and, in analogy with the Stark effect, it splits into three levels: $J_z = 0$, $J_z = \pm 1$ and $J_z = \pm 2$. This was observed for the rsq mode by Shivaram et al. (1982). The observed and predicted splitting is proportional to J_z^2, and the coupling to the three modes may arise from superflow effects (Sauls and Serene, 1984).

Brusov and Popov (1980) have done extensive calculations on the dispersion corrections to all the 18 modes in the B-phase at $T = 0$. A schematic of the dispersion-induced mode splitting (for both rsq and sq modes) is shown in Fig. 38. The zero sound vector cuts the three modes $J_z = 0$, $J_z = \pm 1$ and $J_z = \pm 2$ at ω_0, ω_1, and ω_2, which satisfy the relation

$$\frac{(\omega_0 - \omega_1)}{(\omega_0 - \omega_2)} = \frac{1}{4}. \tag{87}$$

In this article we only quote the results for the rsq mode. Shivaram et al. (1986) see a splitting in their cw impedance measurement, involving three absorption features, labeled A, B, and C, at a pressure of 4.88 bar, $T/T_c = 0.453$, and a magnetic field of 7.8 mT. There was a strong absorption for the $J_z = 0$ component (feature A), and two very weak absorption signals corresponding to $J_z = \pm 1$ (feature B) and $J_z = \pm 2$ (feature C). These were observed in a temperature sweep, at a transducer frequency of 63 MHz (the fifth harmonic), and the separations AB = 5.9 µK and AC = 24.8 µK are in good agreement with the theoretical prediction of AB/AC = $\frac{1}{4}$. The weak coupling of the two nonzero J_z modes is due to the presence of a small magnetic field, superflow, or wall effects. It was difficult to observe this splitting at lower frequencies, as the lines are now much closer.

Using a very short path cell ($d \sim 380$ µm), Zhao et al. (1990) were able to

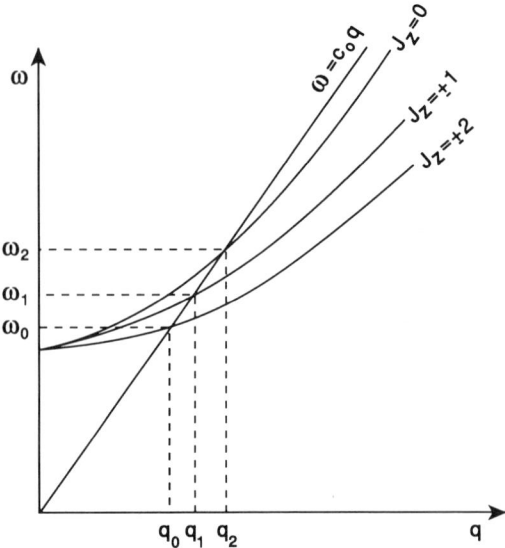

FIG. 38. Dispersion-induced mode splitting—theory. The ($\omega_i - q_i$) labels the energies and the wave vectors at which the zero sound mode $\omega = c_0 q$ crosses the various J_Z modes.

resolve a splitting of the sq mode. Figure 39 shows the experimental trace, along with a conventional nonlinear least-square fit to a Lorentzian line shape used to simulate this phenomenon (measured at 19.2 bar in zero field). Measurements were performed at frequencies of 115.8 MHz and 141.6 MHz, and studied in the pressure range from 19.2 to 27.7 bars in zero magnetic field and at a single pressure of 27.3 bars in a magnetic field (perpendicular to $\hat{q}$) up to 1.3 kG. The doublet splitting of the squashing mode was observed in both zero field and finite magnetic field; the splitting was unaffected by the field (up to 1.3 kG). The observed doublet splitting of the squashing mode is strongly pressure- and thermal-gradient–dependent, but is independent of magnetic field. The splitting increased with an increase in pressure or temperature.

Shivaram et al. (1982, 1986) observed a doublet splitting of the $J_z = 0$ state of the real squashing mode above a threshold magnetic field ($\mathbf{q} \perp \mathbf{H}$). In the trace of Fig. 27, the central $J_z = 0$ state shows a doublet splitting. The separation δT (or $\delta \nu = (d\nu/dT)\delta T$) between the two center states was found to be independent of magnetic field, and varied between 5 and 15 μK, depending on the temperature and the pressure. The presence of the central doublet structure was systematically studied (Shivaram et al., 1986) at a number of pressures and temperatures. The feature was found to be very reproducible. Considerable hysteresis was observed in the evolution of the two center states. The threshold field

2. Sound Propagation and Collective Modes

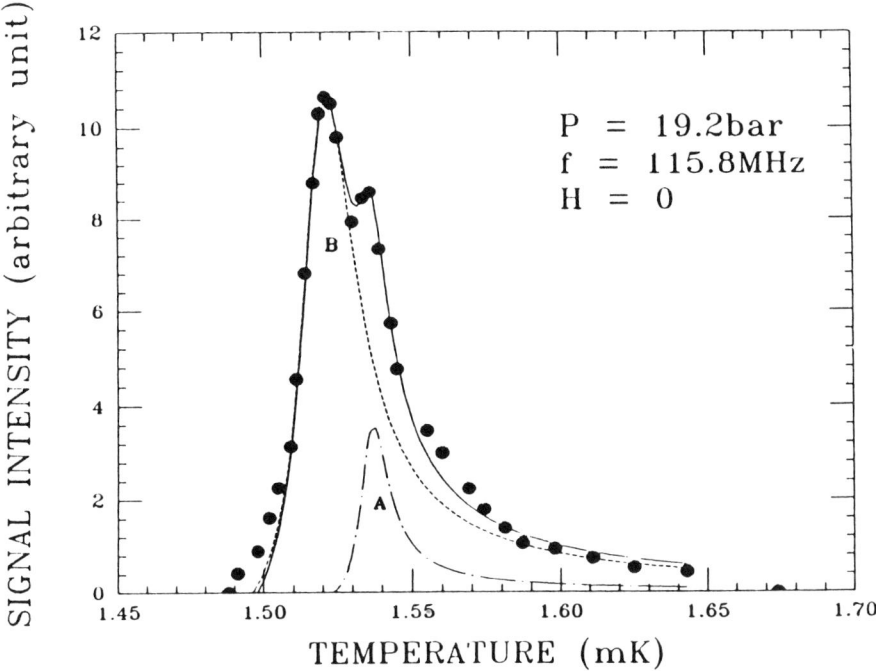

FIG. 39. Dispersion splitting of the sq mode. Solid line is Lorentzian fit to the data (Zhao et al., 1990).

at which the $J_z = 0$ doublet appears is dependent on whether the liquid was cooled into the superfluid state in a magnetic field or not.

This phenomenon has been attributed to a texture effect on the dispersion of the rsq mode (Volovik, 1984; Brusov, 1985; Fishman and Sauls, 1988). The texture introduces an additional term into the dispersion equation,

$$\omega^2 = (\omega_0 + \gamma J_z H)^2 + (c_1^2 + c_2^2 \cos^2 \theta) q^2, \qquad (88)$$

where $\gamma J_z H$ is the Zeeman term and $c_1^2 q^2$ the dispersion term. The $\cos \theta$ measures the texture effect. The angle θ defines the direction of quantization axis in the B-phase and is given by $\theta = \cos^{-1}(\hat{\mathbf{q}} \cdot \mathbf{R} \cdot \hat{\mathbf{H}})$, where $\mathbf{R}$ is the order parameter rotation matrix. This angle depends on the texture in the superfluid and is expected to be 90° in the bulk region for $\hat{\mathbf{q}} \cdot \mathbf{R} \cdot \hat{\mathbf{H}}$ geometry, the texture being aligned with the field. Near the walls or transducers, $\theta = 0°$. The effect of the texture changing from $\theta = 0°$ near the transducer to $\theta = 90°$ in the bulk (the sound wavelength must be smaller than the magnetic bending length ξ of the texture, $q\xi > 1$) gives two separate regions of the fluid where the angular dependent collective mode frequency is extremal, the difference being given by

$$\Delta\omega = \frac{q^2 c_2^2}{2\omega_0}. \tag{89}$$

The textural splitting has been seen only for the $J_z = 0$ state. The splitting of the other states is small and not observable. No splitting was seen for $\hat{q} \parallel \hat{H}$, and this is consistent with the above model since $\theta = 0°$ everywhere.

5.2.6. Sound Propagation in Rotating ^{3}He-B

Rotating ^{3}He superfluid offers a very rich subject for study. Rotating a superfluid is analogous to applying a magnetic field to a type II superconductor. The rotation axis Ω acts as an additional orienting axis that can now compete with the orienting effects of an applied magnetic field. Many types of vortex states are possible for both the A and B superfluids (see Section 3.2); for a review, see Salomaa and Volovik (1987).

The first sound experiments in rotating ^{3}He-B were performed by Salmelin *et al.* (1989). They used conventional pulsed transmission techniques to determine the attenuation and a phase sensitive technique to determine velocities. The sound cell consisted of a pair of X-cut quartz transducers separated by 4 mm, with a fundamental frequency of 8.895 MHz; pulse widths of 5 µs or 12 µs were used. The cryostat was rotated about its axis at speeds up to 4 rad/s. Sound was propagated along Ω, and a magnetic field of either 250 G or 320 G was applied along Ω and q.

The rotating cryostat (ROTA 2) works as follows. The cryostat is mounted on an air bearing, with the electronics rotating on a synchronous platform. After precooling the nuclear stage in a magnetic field to a suitable low temperature, all pumping lines are disconnected. The dilution refrigerator and the 1.5 K ^{4}He pot then operate on charcoal adsorption pumps that are in the helium bath, in a single-cycle mode. Rotation periods of 24 to 48 hours are possible before it is necessary to reconnect to the external pumping system. For each fixed angular velocity, the temperature is allowed to vary slowly, allowing the various substate frequencies to pass into resonance with the zero sound.

For $H \parallel \hat{q}$, the splitting of the rsq mode should not be observed, since the nonzero J_z states do not couple to sound in zero field. Rotating the ^{3}He-B

2. Sound Propagation and Collective Modes

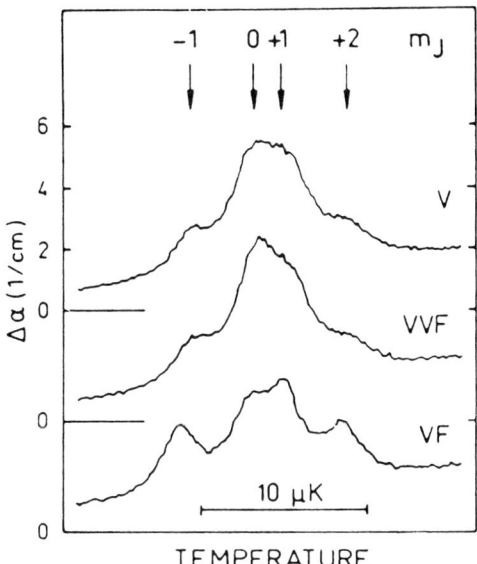

FIG. 40. The effect of rotation on the rsq mode attenuation (Salmelin et al., 1989).

sample changes the coupling strengths of the substates: the $m_J = \pm 2, \pm 1$ side peaks now become visible. These data are shown in Fig. 40 and Fig. 41. Two types of vortex states could be generated in ^{3}He-B. When the rotation was started before the sample had been cooled to the B-phase, an equilibrium vortex lattice—the V state—was created. In this case the $m_J = 0$ and $m_J = +1$ resonances merge into one central prominent peak. When rotation was started well below T_c, an equilibrium vortex state (the VF state) was obtained: The side peaks then couple more strongly to the zero sound. The VF spectra were always obtained with very slow accelerations (0.004 rad/s^2). At higher values (0.008 or 0.015 rad/s^2), intermediate spectra (VVF) were sometimes observed.

Figure 41 shows the effect of rotation on the rsq mode attenuation when $H \parallel q \parallel \Omega$. When $\Omega = 0$, only the $m_J = 0$ substate is observed. When the rotation is started, the $m_J \neq 0$ peaks appear; the splitting is 150 µK/T for $\Delta m_J = 1$. At higher pressures the spectral lines are better resolved as the ratio of the Zeeman splitting to the natural width is larger. The authors also observed the doublet splitting at $m_J = 0$ at high rotation speed of 4.00 rad/s.

The effect of rotation is to create a difference between the normal and superfluid velocities, $v_n - v_s$. This acts as a new orienting axis perpendicular to $\hat{q}$.

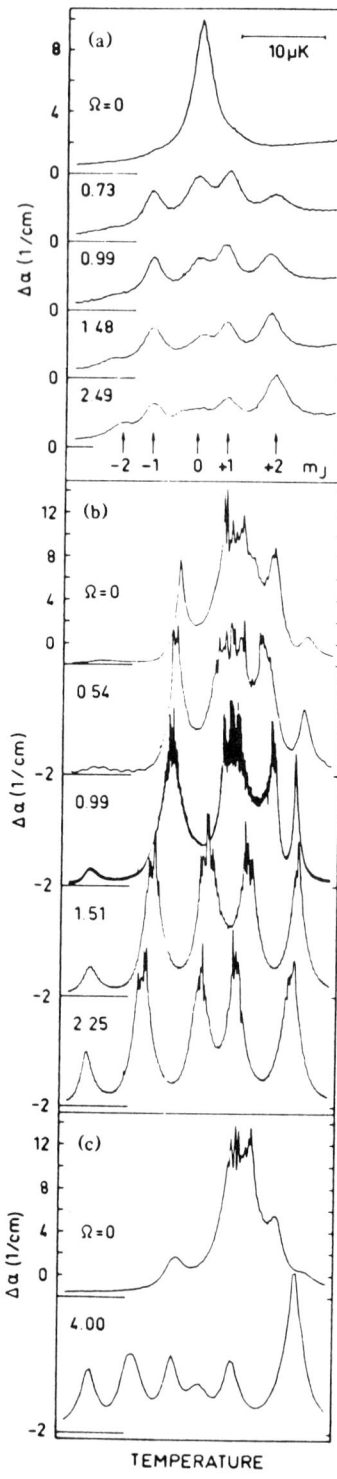

2. Sound Propagation and Collective Modes

5.2.7. Nonlinear Effects

Many nonlinear acoustic phenomena are expected in superfluid ^{3}He, because of the intrinsic complexity of the order parameter. For a discussion of this phenomenon, see Halperin and Varoquaux (1990), and for a theoretical description see McKenzie and Sauls (1990). Many of these nonlinear acoustic phenomena are analogous to nonlinear optical excitations of atoms and molecules.

The first measurement of acoustic nonlinear effects in the sound properties of the B-phase came from the Cornell group (Polturak et al., 1981), where the measurements were done at high pressures, between 24 and 29 bars. In the study of the rsq mode, these authors observed a saturation of the zero sound attenuation that was accompanied by a pulse breakup at high input power levels in the vincinity of the collective modes. They interpreted these results on the basis of soliton propagation, but there were many discrepancies. This effect was not seen at the lower pressure measurements (15 bar) of both the Orsay and the Northwestern groups, even though the power levels of the pulses were comparable. More experiments are needed to obtain a clear picture of nonlinear sound propagation near the rsq mode.

The possiblity of harmonic generation in acoustic waves in superfluid ^{3}He has been considered by Serene (1984), and similar frequency mixing effects have been considered by McKenzie and Sauls (1989). According to McKenzie and Sauls, the cleanest nonlinear effects are likely to be stimulated Raman scattering of zero sound and two-phonon absorption by the rsq ($J = 2^+$) mode.

The $J = 2^+$ resonance mode may be treated like an elementary excitation, the "real squashon." The three-body interaction involving the real squashon could be either a two-phonon excitation of the real squashon, or the decay of a phonon into a squashon and a phonon (see Fig. 42). For this there are two requirements: (i) a satisfaction of the conservation theorems,

$$\nu_3 = \nu_1 + \nu_2 \quad \text{and} \quad \mathbf{q}_3 = \mathbf{q}_1 + \mathbf{q}_2, \tag{90}$$

and (ii) nonzero matrix elements or a favorable selection rule. The particle-hole asymmetry (provided by the parametric field of the pumped wave phonons) makes this transition possible for the real squashing mode.

The first experiment showing this parametric effect has been performed in Helsinki by Torizuka et al. (1991) (see Fig. 43). They measured the attenuation and phase velocity of a low-intensity wave (the signal wave with frequency and wave vector $\omega_s, \mathbf{q}_s$, and intensity E_s) in the presence of a co-propagating high-intensity wave (the pump wave with ω_p, $\mathbf{q}_p$ and E_p). If the signal wave is at a

---

FIG. 41. Finnish data from the rotation cryostat (Salmelin et al., 1989).

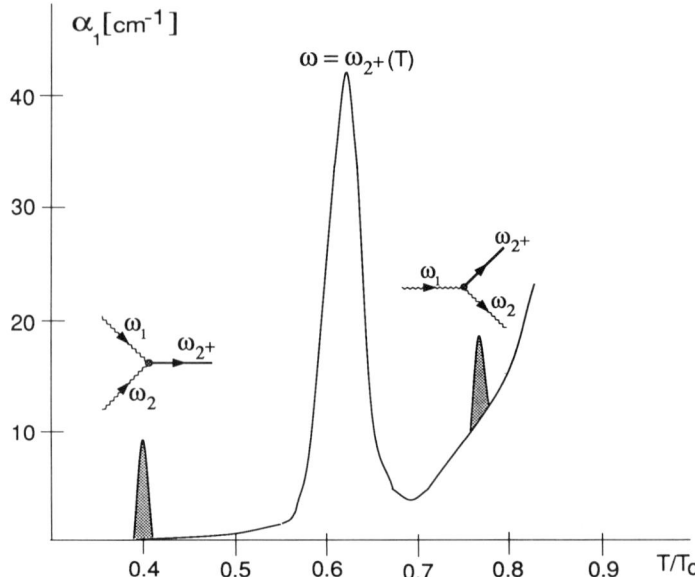

FIG. 42. Signal wave absorption spectrum for the three-body interactions involving real squashon. See text for details (Sauls and McKenzie, 1991).

higher frequency than the pump wave ($\omega_s > \omega_p$, $E_s \ll E_p$) a two-phonon absorption is expected to be seen at the temperature T^+ and T^- such that

$$\hbar(\omega_s \pm \omega_p) = \sqrt{\frac{8}{5}} \Delta(T^\pm). \tag{91}$$

The interaction of these peaks should increase linearly with the power of the pump wave.

According to Sauls and McKenzie (1991), the change in attenuation and the velocity are given by

$$\frac{\Delta\alpha}{\alpha} \simeq \frac{\Delta c}{c} \simeq \frac{|\Lambda|^2}{(1+F_0^s)^2} \frac{\Delta}{\Gamma} \frac{U_p}{U_c}, \tag{92}$$

where U_c is the superfluid condensation energy, U_p the energy density of the pump wave, $1/\Gamma$ the lifetime of the mode, and λ the coupling strength, which has been calculated from theory. Both low temperatures (sharper line width) and low pressures (smaller $2F_0^s$ and U_c) are favorable to see these peaks.

Measurements were done on the ROTA 2 cryostat in Helsinki, with a path length of 4 mm, using the pulsed transmission technique. The pump and signal waves were produced by two separate generators. The fundamental frequency

2. Sound Propagation and Collective Modes

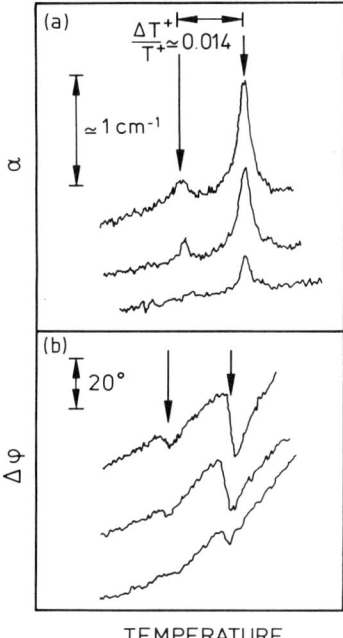

FIG. 43. Data on attenuation and velocity from Torizuka et al. (1991).

was at $\nu_s = 8.9$ MHz. The pump frequency was the third or the fifth harmonic at $\nu_p = 26.8$ or 44.7 MHz. Pulses at these two frequencies were simultaneously sent to the transducer through a directional bridge. The pulse width was 12 μs. Only the signal wave was monitored by the receiver system. The maximum averaged heating power of $E_p \sim 1.5$ nW was estimated to produce a temperature gradient of $\delta T = 13$ μK at a temperature $T = 1.0$ mK.

Figure 43 shows the attenuation peaks and velocity shifts at three different power levels. The T^+ peaks were seen at temperatures in good agreement with those predicted. The heights of the attenuation peaks scale linearly with the energy of the pump wave, as expected. However, there is an additional satellite peak, even in zero magnetic field.

The splitting into the main peak and small satellite peak originates from the existence of both parallel and antiparallel sound waves, as originally suggested by Volovik (1984). Suppose the pump and signal waves have not reached the receiver crystal; in this case the temperature, T_m^+, of the rsq mode is given by

$$h(f_s + f_p) = \sqrt{\frac{8}{5}} \Delta(T_m^+) + \frac{7}{15} [(q_s + q_p)v_F]^2. \qquad (93)$$

When part of the pump and signal waves is reflected, the antiparallel sound waves can also interact, and the temperature, T_s^+, of the mode is given by

$$h(f_s + f_p) = \sqrt{\frac{8}{5}} \Delta(T_s^+) + \frac{7}{15} [(q_s + q_p)v_F]^2. \tag{94}$$

Putting the numbers in, one obtains that the splitting, $\Delta T = T_m^+ - T_s^+$, is $\Delta T/T_c \sim 0.017$, in agreement with the experiment. This is also independent of magnetic field.

The relative magnitudes of the two peaks are also consistent with this model: Because of the shorter period of the interaction for the $\mathbf{q}_p \uparrow \downarrow \mathbf{q}_s$ with respect to $\mathbf{q}_p \uparrow \uparrow \mathbf{q}_s$ case, and because of nonperfect reflection at the receiver crystal, we expect that the low-temperature ($\mathbf{q}_p \uparrow \downarrow \mathbf{q}_s$) peak is weaker than the higher-temperature ($\mathbf{q}_p \uparrow \uparrow \mathbf{q}_s$) peak.

5.3. Superfluid ^{3}He-A

Sound measurements in the ^{3}He-A phase are complicated for a variety of reasons. The situation is more complicated than in the B-phase, because of orbital anisotropy. However, the spin and orbital parts of the order parameter are decoupled (except for the effects of the weak dipole interaction), and the order parameter collective modes have less structure. Textural effects are very important in the A-phase, leading to large anisotropies and geometry dependencies. Also, the A-phase is stable only in a small region in the P–T plane, close to T_c and the melting curve. Thus, quasiparticle scattering is higher, leading to the broadening of the resonances; also, strong coupling corrections have to be made. One may apply a large field to stabilize the A-phase at low pressures and temperatures: this, however, splits the A-phase into the A_1 and A_2 phases, and also splits the collective mode resonances, and these effects have to be taken into account. Pair breaking takes place at all temperatures and frequencies—i.e., there is finite absorption (down to absolute zero) and line broadening. Also, since the gap is anisotropic, this leads to an additional anisotropy.

5.3.1. Zero Sound Interactions and Textures

The first study was done by Lawson *et al.* (1973, 1974) and Paulson *et al.* (1973) in a Pomeranchuk cell along the melting curve, and later, extensive measurements of the anisotropy were obtained by Roach *et al.* (1975b). The sound propagation properties are anisotropic, depending on the angle β between the sound propagation vector $\hat{\mathbf{q}}$ and the preferred direction $\hat{\ell}$. The angular de-

2. Sound Propagation and Collective Modes

pendence of the attenuation $\alpha(\beta)$ and the velocity $c(\beta)$ have been given earlier (see Eqns. 73 and 74),

$$\alpha(\beta) = \alpha_\parallel \cos^4\beta + 2\alpha_c \sin^2\beta \cos^2\beta + \alpha_\perp \sin^4\beta, \quad (95a)$$

$$c_0 - c(\beta) = \Delta c_\parallel \cos^4\beta + 2\Delta c_c \cos^2\beta \sin^2\beta + \Delta c_\perp \sin^4\beta, \quad (95b)$$

where expressions for α_i and c_i have ben obtained by Wölfle and Koch (1978). For sound propagating along $\hat{\ell}$, none of the flapping and clapping modes is excited. For $\hat{q}$ perpendicular to $\hat{\ell}$, only the clapping mode is excited: A compressional wave perpendicular to $\hat{\ell}$ will lead to a local widening and narrowing of the angle between the preferred directions $\hat{m}$ and $\hat{n}$, which is the clapping mode, while leaving the angle with respect to $\hat{\ell}_N$ unchanged.

Roach et al. (1975b) have measured the temperature dependence of the attenuation and velocity shift in the A-phase at 26.0 bar for the angles $\psi = 0$, 30, 60, and 90° and for a field of 27 G. The anisotropy was quite apparent at 9 G and essentially saturated at 18 G. Figure 44 shows the attenuation and velocity shift in the A-phase at 26.0 bar in a field of 18 G and at a temperature of 48 μK below T_c, where a detailed angular study was made. The data exhibit an angular dependence that is very close to the predicted form of Eq. (95).

In Fig. 45 are plotted the data of Paulson et al. (1976, 1977), along with the numerical calculations of Wölfle and Koch (1978). The theory includes both pair breaking and collision effects. The peaks associated with the clapping mode ($\alpha_\perp$) and the normal flapping mode (α_c) are easily seen. The super flapping mode is masked by the pair-breaking effects.

Extensive textural anisotropies exist in the A-phase. There are forces acting on both the $\hat{\ell}$ and the $\hat{d}$ vectors, tending to align them in a suitable direction. Forces acting on $\hat{\ell}$ are due to the following causes:

- wall effects—$\hat{\ell}$ orients perpendicular to the walls;
- superfluid currents orient $\hat{\ell}$ parallel to v_s (superflow currents may arise due to heat flow from temperature gradients);
- elastic bending energy of the $\hat{\ell}$ vector field;
- dipolar coupling to $\hat{d}$.

Similarly, there are forces tending to align the $\hat{d}$ vector. In addition to the dipolar coupling to $\hat{\ell}$, a magnetic field tends to align $\hat{d}$ perpendicular to $\mathbf{H}$, because of the anisotropy in the susceptibility.

A variety of textures may appear depending on the experimental situation in the test cell. The dipole field is of the order of 25 G. For fields in excess of 25 G, the $\hat{d}$-vector will lock in a plane perpendicular to $\mathbf{H}$, and the $\hat{\ell}$-vector will

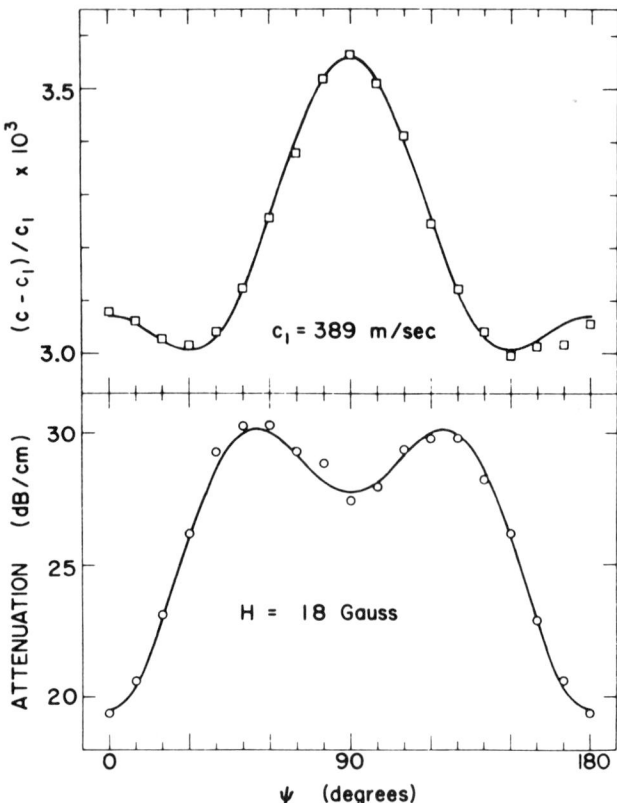

FIG. 44. The angular dependence of the attenuation and velocity shift in superfluid ^{3}He-A. After Roach et al. (1975b).

tend to align with $\hat{\mathbf{d}}$—the order parameter is said to be dipole-locked. If $\hat{\ell}$ is forced into a different orientation, then $\hat{\ell}$ will heal back from the wall-locked case to the dipole-locked case over a length scale called the dipole length, $\zeta_d \sim 10\ \mu m$.

Because of the uncertainty or instability of the A-phase textures, arising chiefly from an uncontrolled superflow (heat flow), there are often irreproducibilities in the propagation of sound, which can be very frustrating. The USC group, Bates et al. (1984, 1986) and Bozler and Gould (1987), have performed extensive measurements on the study of textures, using ultrasonic methods. Using a bellows assembly and a superleak, they created a well-controlled superflow velocity and monitored the textural changes by sound propagation. Note that the sound wavelength at 30 MHz is of the order of the dipole healing length, ζ_d. When they increased the magnetic field, at constant v_s, they observed a significant change

2. Sound Propagation and Collective Modes

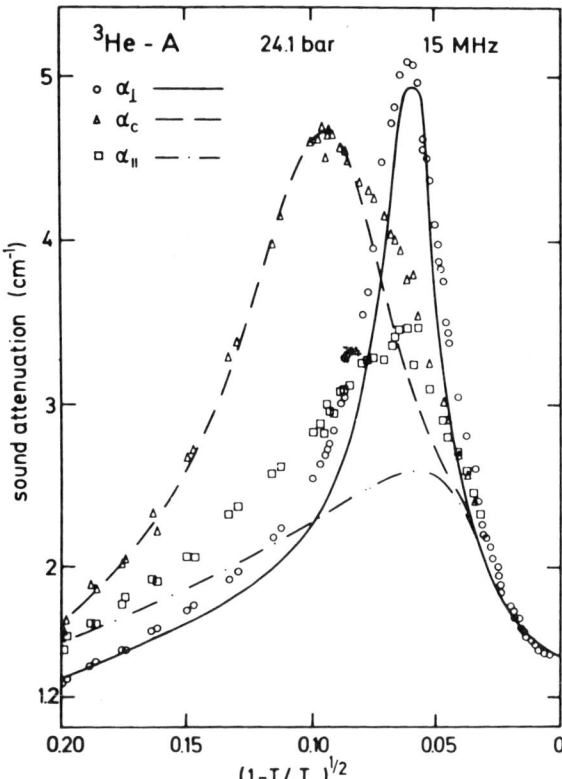

FIG. 45. Temperature dependence of the three coefficients $\alpha_\parallel$, α_c, and $\alpha_\perp$ of attenuation of zero sound in ^{3}He-A at 33.5 bar and 25 MHz. Experimental results from Paulson *et al.* (1977), and theoretical curves from Wölfle (1978). After Vollhardt and Wölfle (1990).

in the attenuation, α, at certain threshold values of H. They deduced a textural phase diagram in the v_s–H plane, showing the existence of different textures. Their observed phase diagram differs appreciably from that calculated, and thus it is difficult to identify the textural nature of these phases. (For a more detailed account refer to Bozler, 1990.)

5.3.2. Pair Breaking

Because of the nodes in the gap in ^{3}He-A, pair breaking is possible at all temperatures (below T_c) and at all frequencies. As one cools into the superfluid phase the attenuation initially increases with the superfluid fraction; eventually, it decreases as the region in the Fermi surface over which it is possible to break pairs diminishes. Theoretical calculations (Serene, 1974; Ebisawa and Maki, 1974; Ashida and Nagai, 1983) show that there is a cusplike behavior (for $q \parallel \ell$), similar to van Hove–like singularities in the phonon spectrum in solids. In a magnetic field, Schopohl *et al.* (1985) predict a splitting—a double cusp structure arising from the difference in energy of the spin up and spin down states. The calculations of Ashida and Nagai (1983) are shown in Fig. 46.

Features corresponding to this have been observed by Meisel *et al.* (1983a) and independently by Avenel and Varoquaux (1984). Meisel *et al.* (1983a) observed a new feature in their impedance measurement at $\hbar\omega \sim 1.8\ \Delta_A(T)$, over a wide range of pressures and temperatures. Measurements were done at 38 MHz from 0.9 to 16.8 bar and at 64 MHz from 5 to 12 bar. A typical impedance trace is shown in Fig. 47, where the pair-breaking feature and a new feature can be seen. Their initial interpretation was to identify this new feature with the superflapping mode, which has not yet been observed. All the North-

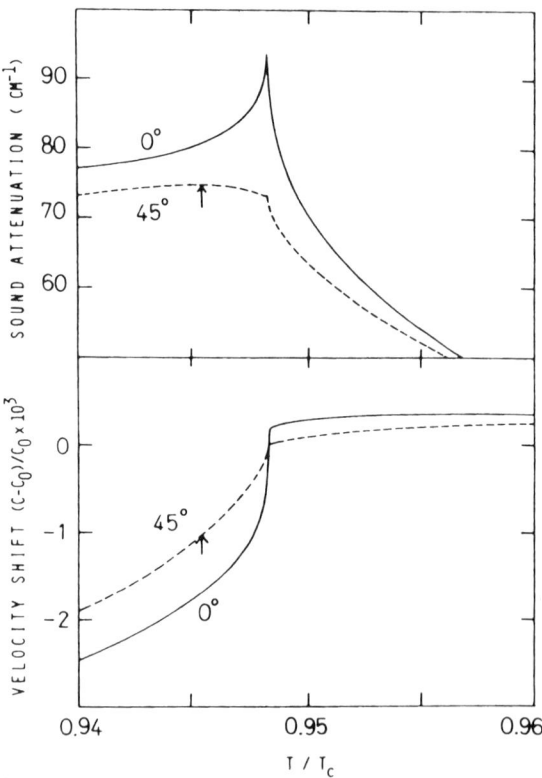

FIG. 46. Theoretical calculations for the $J = 1^-$ mode (Ashida and Nagai, 1983).

2. Sound Propagation and Collective Modes

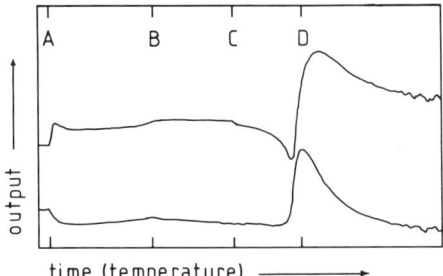

FIG. 47. Impedance trace showing the $J = 1^-$ mode. Initially this was identified as a pair-breaking mode. After Meisel et al. (1983a).

western data prior to 1984 used the Helsinki temperature scale ($T_c = 2.75$ mK on the melting curve), which is 10% higher than the revised temperature scale due to Greywall (1986). The correction to the temperature scale would place this new feature exactly at the pair-breaking frequency, and this new feature is now believed to arise from pair breaking.

The data of Avenel and Varoquaux (1984) are shown in Fig. 48 and are in qualitative agreement with those predicted by theory. Their measurements were done in the pulsed impedance mode, with a strong magnetic field (2 kG) applied

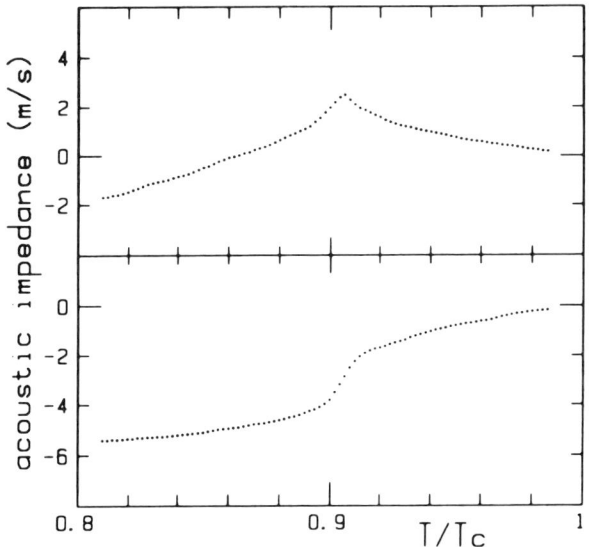

FIG. 48. The temperature dependence of the real (upper) and imaginary (lower) parts of the acoustic impedance in ^{3}He-A. After Avenel and Varoquaux (1984).

along $\hat{\mathbf{q}}$ (1 bar, 44.7 MHz). Because of its peculiar shape, they called it the Chinese hat. However, it is to be noted that, since the real part of the longitudinal impedance in a viscoelastic fluid is related to the velocity, they see a cusp in the velocity, whereas the theory predicts a cusp only for the attenuation. The authors attribute this discrepancy to the possibility that the acoustic impedance is not probing the bulk ^{3}He, and a better model for the coupling between the transducer and superfluid is required.

Meisel et al. (1983a) studied the effect of a magnetic field on this cusp feature for fields up to 1.5 kG for $\mathbf{H} \parallel \mathbf{q}$ and up to 7.2 kG for $\mathbf{H} \perp \mathbf{q}$. They saw a very weak dependence (if any) on field, in contrast to the predictions of Schopohl et al. (1985).

Ling et al. (1990) have measured the temperature dependence of the pair-breaking attenuation (P = 29.3 bar, frequency ~44–84 MHz) with pulsed transmission using their quartz delay line technique (^{3}He path length ~250 μm). To achieve a uniform texture with $\hat{\mathbf{q}} \parallel \hat{\ell}$, they put $\mathbf{H} \perp \hat{\mathbf{q}}$. Their results are in overall agreement with the theory of Wölfle and Koch (1978). However, they do not see the cusplike behavior; this may be due to one of two reasons: smearing due to quasiparticle scattering, or instrumental broadening from the pulsed measurements.

5.3.3. Clapping Mode

The strongest coupling of sound to the clapping mode occurs with $\hat{\mathbf{q}} \perp \hat{\ell}$. This may be achieved by applying $\mathbf{H} \parallel \mathbf{q}$. In the early sound data, in the superfluid phase of Paulson et al. (1977), the clapping mode was identified with the peak of $\alpha_\perp$ (see Fig. 45). Later Israelsson et al. (1986) made a number of measurements of the clapping mode frequency, ω_{c1}, over the pressure range 11 to 33 bar at 21.3 MHz. They observed ω_{c1} to be always significantly less than the theoretical value of 1.23 Δ (see Section 4.2.2) and concluded that there must be significant f-wave (ℓ = 3) mixing with the p-wave in the pairing. However, quasiparticle collisions can significantly lower the clapping mode frequency (Einzel et al., 1987), and this was certified in the measurements of Ling et al. (1990). Attenuations as high as 100 cm^{-1} could be measured with their quartz delay line cell, and they studied the clapping mode for frequencies ranging from 24 to 94 MHz at a pressure of 29.3 bar ($\hat{\mathbf{q}} \parallel \mathbf{H}$). At these higher frequencies (where the clapping mode lies is at a lower temperature), the quasiparticle scattering has a smaller effect on the clapping mode resonance. Figure 49 shows the data of Ling et al. (1990). Their data agree very well with the weak coupling theory of Wölfle and Koch (1978), which includes collisions and corrections due to F_2^s, but does not use any f-wave interactions. From their data they extract a gap (in the A-phase)

2. Sound Propagation and Collective Modes

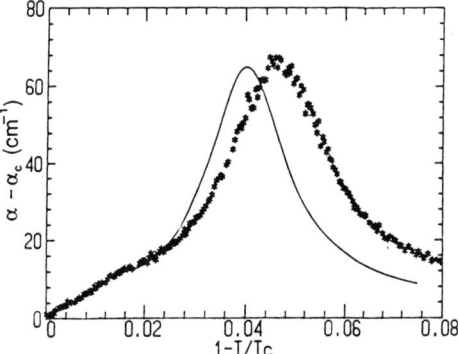

FIG. 49. Data of Ling et al. (1990) showing the clapping mode resonance in the A-phase (at 54.0 MHz and 29.3 bar), compared with the theory of Wölfle and Koch (1978).

of $\Delta_0(0) = 2.64\, k_B T_c$ and $\omega_{c1}/\Delta_0 = 1.15$. This is still less than the weak coupling value of 1.23; the difference can be accounted for by the strong coupling corrections.

In a magnetic field the superfluid A-phase splits into the A_1 phase and the A_2 phase, and a double peak in the attenuation is seen. Sagan et al. (1984) and Israelsson et al. (1984) have studied the A_1 and the A_2 phases as a function of magnetic field. The clapping mode peaks in the A_1 and A_2 phases were observed by Lawson et al. (1975), and their data are shown in Fig. 50. The phase diagram showing the A_1 and A_2 phases obtained from attenuation measurements is shown in Fig. 51 (Sagan et al., 1984).

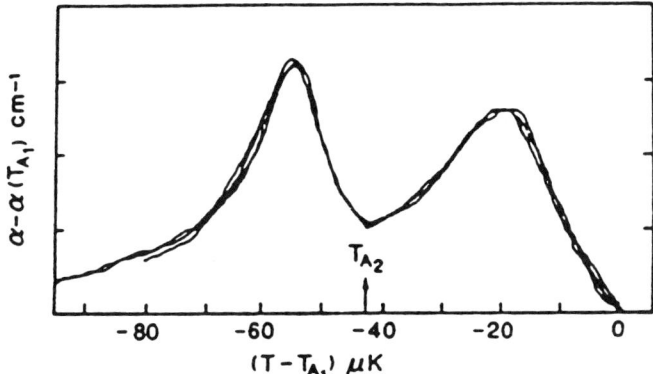

FIG. 50. Splitting of the clapping mode in a magnetic field (Lawson et al., 1975).

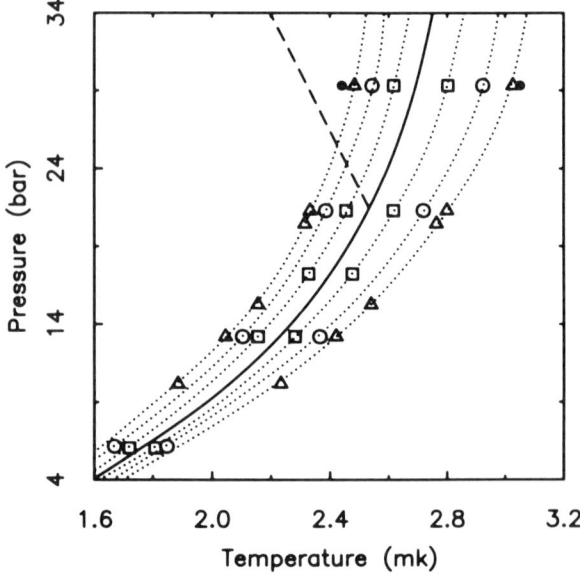

FIG. 51. A_1, A_2 phase diagram obtained from attenuation measurements. After Sagan *et al.* (1984).

5.3.4. Normal Flapping Mode

From the angular dependence of the attenuation, α, Paulson *et al.* (1977) obtained the attenuation component α_c near T_c. Since the flapping mode frequency dips down at lower temperatures, this mode can couple to zero sound at two temperatures. The reappearance of this mode has been observed by Ling *et al.* (1989, 1990). A large magnetic field (0.35 to 0.45 Tesla) was applied to stabilize the A phase at sufficiently lower temperatures. Maximum coupling to the flapping mode is obtained by orienting the texture at 45°. The large line width of the flapping mode can be easily distinguished from the sharp clapping mode data near T_c. To quantitatively explain their data, substantial f-wave corrections have to be applied. Figure 52 is the data from Ling *et al.* (1989) and shows the clapping mode (the sharper peak), the broad flapping mode, both near T_c, and the reentrant flapping mode at $\sim 0.5 T_c$. To explain these data, a substantial f-wave correction has to be applied. At sufficiently high frequencies, the resonant coupling to the normal flapping mode must disappear, and this is verified in Ling's measurements, which were done at frequencies up to 73.7 MHz. The splitting of the normal flapping mode in a high magnetic field (5 Tesla) has been observed by deVegvar *et al.* (1986).

2. Sound Propagation and Collective Modes

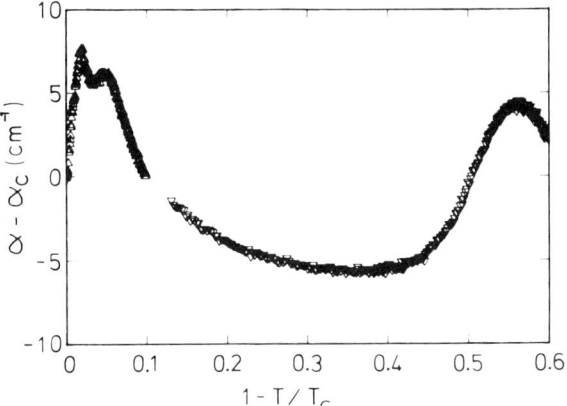

FIG. 52. Normal flapping resonances and clapping resonance at 34.2 MHz. Data of Ling *et al.* (1989). From Halperin and Varoquaux (1990).

5.3.5. Rotating ^{3}He-A

Vortices in ^{3}He-A in low magnetic fields has also been studied by ultrasonics (Pekola *et al.*, 1990). Experiments were done with various combinations of rotating speeds and magnetic fields. Extensive hysteresis was seen in the sound amplitude, as a function of both magnetic field and rotating speed, from which information on the stability of the various vortex types was inferred. They observe a first-order phase change between the vortex types, with dissimilar sound absorption taking place at a field of the order of the dipolar field H_d.

Two types of vortices were prominent. They saw a phase transition between dipole-unlocked high-field vortices and dipole-locked low-field vortices. From these measurements, they obtain an experimental phase diagram of vortex textures in ^{3}He-A in low magnetic fields. Below H_c the stable phase is the dipole-locked state, the low-field vortex. Above H_{c1} the stable phase is the dipole-unlocked state. Between H_c and H_{c1}, both vortex states may exist, but the low-field vortex is metastable. Pekola *et al.* (1990) obtain a value of 1.5 mT for H_c and 3.8 mT for H_{c1}.

6. Conclusions

In the nearly two decades since the discovery of superfluid ^{3}He much has been learned about unconventional superfluidity. The rapid development of the subject is due to three principal reasons:

(i) Much of the theoretical framework for p-wave superfluidity/superconductivity had been worked out in the 1960s and could be readily applied to explain the new results.

(ii) The experiments were necessarily of a very high quality, since they had to be carefully planned and performed: A typical experiment takes one to two years in designing and troubleshooting.

(iii) Pure ^{3}He is readily available (i.e., there are no serious sample problems).

Ultrasonics has proved to be a very powerful probe for the study of p-wave superfluidity and the spectroscopy of the collective modes. Identifying and labeling the various modes and substates has involved remarkable ingenuity in which both theorists and experimentalists have played their respective roles. Resoluton (revision) of the temperature scale has helped in identifying the states, especially those close to 2Δ.

There are many cross-disciplinary comparisons for superfluid ^{3}He. The anisotropy and textural effects are akin to those in liquid crystals. Volovik (1991) has done a critical comparison of superfluid ^{3}He properties with analogous effects in cosmology and high-energy physics.

The closest comparison of importance to us is with unconventional superconductivity in UPt_3. Both have strong Fermi-liquid characteristics, spin–spin interactions, and a $T^3 \log T$ heat capacity. The symmetry classification of all possible superfluid states in ^{3}He has proved useful for a similar classification to the heavy fermion and the high-T_c superconductors.

Very rich and complex vortex structures and textures are possible in superfluid ^{3}He, including half-quantum and doubly quantized vortices. Somewhat similar vortex structures have been suggested for the case of UPt_3, where a phase transition may have been observed between two possible vortex core states.

ACKNOWLEDGMENTS

The work at University of Wisconsin—Milwaukee was supported by the Office of Naval Research, and that at Northwestern University by the National Science Foundation under grant DMR-89-07396. We would like to thank Dr. Pekola for discussing the Helsinki ultrasonic experiments on superfluid ^{3}He. We thank William Halperin, David Mast, Ian Calder, John Owers-Bradley, Bellave Shivaram, and Mark Meisel for earlier collaborations, Jim Sauls for theoretical discussions, and S. W. Lin for his help in preparing the manuscript.

References

Abel, W. R., Anderson, A. C., and Wheatley, J. C. (1966). *Phys. Rev. Lett.* **17**, 74.
Adenwalla, S., Zhao, Z., Ketterson, J. B., and Sarma, B. K. (1988). *J. Low Temp. Phys.* **76**, 1.
Adenwalla, S., Zhao, Z., Ketterson, J. B., and Sarma, B. K. (1989). *Phys. Rev. Lett.* **63**, 1811.
Alvesalo, T. A., Anufriyev, Yu. D., Collan, H. K., Lounasmaa, O. V., and Wennerström, P. (1973). *Phys. Rev. Lett.* **30**, 1962.
Alvesalo, T. A., Haavasoja, T., Manninen, M. T., and Soinne, A. T. (1980). *Phys. Rev. Lett.* **44**, 1076.
Alvesalo, T. A., Haavasoja, T., and Manninen, M. T. (1981). *J. Low Temp. Phys.* **45**, 373.
Anderson, P. W., and Brinkman, W. F. (1973). *Phys. Rev. Lett.* **30**, 1108, 1973.
Anderson, P. W., and Brinkman, W. F. (1978). "Theory of Anisotropic Superfluidity in ^{3}He," in "The Physics of Liquid and Solid Helium," Part II (K. H. Bennemann and J. B. Ketterson, eds.). Wiley, New York, p. 177–286.
Ashida, M., and Nagai, K. (1983). *Prog. Theor. Phys.* **70**, 1672.
Avenel, O., Varoquaux, E., and Ebisawa, E. (1980). *Phys. Rev. Lett.* **45**, 1952.
Avenel, O., Piche, P., Saslow, W. M., Varoquaux, E., and Combescott, R. (1981). *Phys. Rev. Lett.* **47**, 803.
Avenel, O., Rouff, M., Varoquaux, E., and Williams, G. A. (1983). *Phys. Rev. Lett.* **50**, 1591.
Avenel, O., and Varoquaux, E. (1984). In "Proc. LT-17" (U. Eckern, A. Schmid, W. Weber, and H. Wuhl, eds.). North-Holland, Amsterdam, p. 771–2.
Balian, R., and Werthamer, N. R. (1963). *Phys. Rev.* **131**, 1553.
Bardeen, J., Cooper, L. N., and Schrieffer, J. R. (1957). *Phys. Rev.* **108**, 1175.
Bates, D. M., Ytterboe, S. N., Gould, C. M., and Bozler, H. M. (1984). *Phys. Rev. Lett.* **53**, 1574.
Bates, D. M., Ytterboe, S. N., Gould, C. M., and Bozler, H. M. (1986). *J. Low Temp. Phys.* **62**, 177.
Bennemann, K. H., and Ketterson, J. B. (1978). "The Physics of Liquid and Solid Helium." Part II, John Wiley and Sons, New York.
Bogacz, S. A., and Ketterson, J. B. (1986). *Phys. Rev. Lett.* **57**, 591.
Bozler, H. M., and Gould, C. M. (1987). *Can. J. Phys.* **65**, 1476.
Bozler, H. M. (1990). *Textures in Flowing Superfluid ^{3}He-A*, In "Helium Three" (W. P. Halperin and L. P. Pitaevskii, eds.). Elsevier Science Publishers, North Holland, Amsterdam, p. 695–726.
Brusov, P. N. (1985). *J. Low Temp. Phys.* **58**, 265.
Brusov, P. N., and Popov, V. N. (1980). *Sov. Phys. JETP* **51**, 1217.
Calder, I. D., Mast, D. B., Sarma, B. K., Owers-Bradley, J. R., Ketterson, J. B., and Halperin, W. P. (1980). *Phys. Rev. Lett.* **45**, 1866.
Combescot, R. (1974). *Phys. Rev. Lett.* **33**, 946.
Cooper, L. N. (1956). *Phys. Rev.* **104**, 1189.
Daniels, M. E., Dobbs, E. R., Saunders, J., and Ward, P. L. (1983). *Phys. Rev.* **B27**, 6988.
deGennes, P. G. (1989). "Superconductivity of Metals and Alloys." Addison-Wesley, New York.
deVegvar, P. G. N., Morshovich, R., Ziercher, E. L., and Lee, D. M. (1986). *Phys. Rev. Lett.* **57**, 1028.
Dobbs, E. R., and Saunders, J. (1991). "Ultrasonic Spectroscopy of the Order-Parameter Collective Modes of Superfluid ^{3}He" in "Progress in Low Temperature Physics," Vol. XIII, p. 91 (D. F. Brewer, ed.) North Holland, Amsterdam.
Ebisawa, H., and Maki, K. (1974). *J. Low Temp. Phys.* **32**, 19.
Einzel, D., Hirschfeld, P., and Wölfle, P. (1987). *Phys. Rev. Lett.* **58**, 1383.
Fano, V. (1961). *Phys. Rev.* **124**, 1866.

Fishman, R. A., and Sauls, J. A. (1988). *Phys. Rev. Lett.* **61**, 2871.
Fraenkel, P. N., Keolian, R., and Reppy, J. D. (1989). *Phys. Rev. Lett.* **62**, 1126.
Giannetta, R. W., Ahonen, A., Polturak, E., Saunders, J., Zeise, E. K., Richardson, R. C., and Lee, D. M. (1980). *Phys. Rev. Lett.* **45**, 262.
Gibson, A. A. V., Owers-Bradley, J. R., Calder, I. D., Ketterson, J. B., and Halperin, W. P. (1981). *Rev. Sci. Inst.* **52**, 1509.
Greywall, D. S. (1986). *Phys. Rev.* **B33**, 7520.
Hakonen, P. J., Krusius, M., Salomaa, M. M., Simola, J. T., Bunkov, Yu. M., Mineev, V. P., and Volovik, G. E. (1983). *Phys. Rev. Lett.* **51**, 1362.
Halperin, W. P., Archie, C. N., Ramussen, F. B., Alvesalo, T. A., and Richardson, R. C. (1976). *Phys. Rev.* **B13**, 2124.
Halperin, W. P., and Varoquaux, E. (1990). *Order Parameter Collective Modes in Superfluid ^{3}He, in* "Helium Three." (W. P. Halperin and L. P. Pitaevskii, eds.). Elsevier Science Publishers, North Holland, Amsterdam
Ikkala, O. T., Volovik, G. E., Hakonen, P. J., Bunkov, Yu. M., Islander, S. T., and Kharadze, G. A. (1982). *Sov. Phys. JETP Lett.* **35**, 338.
Israelsson, U. E., Crooker, B. C., Bozler, H. M., and Gould, C. M. (1984). *Phys. Rev. Lett.* **53**, 1943.
Israelsson, U. E., Crooker, B. C., Bozler, H. M., and Gould, C. M. (1986). *Phys. Rev. Lett.* **56**, 2383.
Johnson, R. T., Kleinberg, R. L., Webb, R. A., and Wheatley, J. C. (1975). *J. Low Temp. Phys.* **18**, 501.
Ketterson, J. B., Roach, P. R., Abraham, B. M., and Roach, P. D. (1975). *In* "Quantum Statistics and the Many Body Problem." (S. B. Trickey, W. P. Kirk, and J. W. Dufty, eds.). Plenum Press, New York, p. 35-63.
Kojima, H., Paulson, D. N., and Wheatley, J. C. (1974). *Phys. Rev. Lett.* **32**, 141.
Landau, L. D. (1956). *Sov. Phys. JETP* **30**, 1058.
Landau, L. D. (1957). *Sov. Phys. JETP* **32**, 59.
Landau, L. D., and Lifshitz, E. M. (1959). "Fluid Mechanics." Pergamon Press, New York.
Lawson, D. T., Gully, W. J., Goldstein, S., Richardson, R. C., and Lee, D. M. (1973). *Phys. Rev. Lett.* **30**, 541.
Lawson, D. T., Gully, W. J., Goldstein, S., Richardson, R. C., and Lee, D. M. (1974). *J. Low Temp. Phys.* **15**, 169.
Lawson, D. T., Bozler, H. M., and Lee, D. M. (1975). *Phys. Rev. Lett.* **34**, 121.
Lee, D. M., and Richardson, R. C., (1978). "Superfluid ^{3}He" *in* "The Physics of Liquid and Solid Helium," Part II (K. H. Bennemann and J. B. Ketterson, eds.), John Wiley and Sons, New York, p. 287-496.
Leggett, A. J. (1975). *Rev. Mod. Phys.* **47**, 331.
Ling, R., Saunders, J., and Dobbs, E. R. (1987). *Phys. Rev. Lett.* **59**, 461.
Ling, R., Saunders, J., Wojtanowski, W., and Dobbs, E. R. (1989). *Europhys. Lett.* **1**, 323.
Ling, R., Wojtanowski, W., Saunders, J., and Dobbs, E. R. (1990). *J. Low Temp. Phys.* **78**, 187.
Lounasmaa, O. V. (1974). "Experimental Principles and Methods Below 1 K." Academic Press, London.
Mast, D. B., Sarma, B. K., Owers-Bradley, J. R., Calder, I. D., Ketterson, J. B., and Halperin, W. P. (1980). *Phys. Rev. Lett.* **45**, 266.
Masuhara, M., Candela, D., Edwards, D. O., Hoyt, R. F., Scholz, H. N. and Sherill, D. S. (1984). *Phys. Rev. Lett.* **53**, 1168.
McKenzie, R., and Sauls, J. A. (1989). *Europhys. Lett.* **9**, 459.
McKenzie, R. H. and Sauls, J. A. (1990). "Collective Modes and Nonlinear Acoustics in Superfluid ^{3}He-B" *in* "Helium Three." (W. P. Halperin and L. P. Pitaevskii, eds.) Elsevier Science Publishers, North Holland, Amsterdam, p. 255-311.

Meisel, M. W., Shivaram, B. S., Sarma, B. K., Ketterson, J. B., and Halperin, W. P. (1983a). *Phys. Rev. Lett.* **50,** 361.
Meisel, M. W., Shivaram, B. S., Sarma, B. K., Ketterson, J. B., and Halperin, W. P. (1983b). *Phys. Rev.* **B27,** 6982.
Meisel, M. W., Shivaram, B. S., Sarma, B. K., Ketterson, J. B., Halperin, W. P. (1983c). *Phys. Lett.* **98,** 437.
Movshovich, R., Varoquaux, E., Kim, N., and Lee, D. M. (1988). *Phys. Rev. Lett.* **61,** 1732.
Movshovich, R., Kim, N., and Lee, D. M. (1990). *Phys. Rev. Lett.* **64,** 431.
Mühlschlegel, B. (1959). *Z. Phys.* **155,** 313.
Osheroff, D. D., Richardson, R. C., and Lee, D. M. (1972a). *Phys. Rev. Lett.* **28,** 885.
Osheroff, D. D., Gully, W. J., Richardson, R. C., and Lee, D. M. (1972b). *Phys. Rev. Lett.* **29,** 920.
Paulson, D. N., Johnson, R. T., and Wheatley, J. C. (1973). *Phys. Rev. Lett.* **30,** 829.
Paulson, D. N., Kleinburg, R. L., and Wheatley, J. C. (1976). *J. Low Temp. Phys.* **23,** 725.
Paulson, D. N., Krusius, M., and Wheatley, J. C. (1977). *J. Low Temp. Phys.* **26,** 73.
Pekola, J. P., Simola, J. T., Hakonen, P. J., Krusius, M., Lounasmaa, O. V., Nummila, K. K., Mamniashvili, G., Packard, R. E., and Valovik, G. E. (1984). *Phys. Rev. Lett.* **53,** 584.
Osheroff, D. D., van Roosebroeck, W., Smith, H., and Brinkman, W. F. (1977). *Phys. Rev. Lett.* **38,** 134.
Pekola, J. P., Torizuka, K., Manninen, A. J., Kyynäräinen, J. M., and Valovik, G. E. (1990). *Phys. Rev. Lett.* **65,** 3293.
Pines, D., and Nozieres, P. (1966). "The Theory of Quantum Liquids." Benjamin, New York.
Polturak, E., de Vegvar, P. G. N., Zeise, E. K., and Lee, D. M. (1981). *Phys. Rev. Lett.* **46,** 1588.
Pomeranchuk, I. (1950). *JETP* **20,** 919 (in Russian).
Richardson, R. C., and Smith, E. N. (1988). "Experimental Techniques in Condensed Matter Physics at Low Temperatures." Addison-Wesley Publishing Co., New York.
Roach, P. R., Abraham, B. M., Kuchnir, M., and Ketterson, J. B. (1975a). *Phys. Rev. Lett.* **34,** 711.
Roach, P. R., Abraham, B. M., Roach, P. D., and Ketterson, J. B. (1975b). *Phys. Rev. Lett.* **34,** 715.
Roach, P. R., Meisel, M. W., and Eckstein, Y. (1982). *Phys. Rev. Lett.* **48,** 330.
Sagan, D. C., de Vegvar, P. G. N., Palturak, E., Friedman, L., Yan, S. S., Ziercher, E. L., and Lee, D. M. (1984). *Phys. Rev. Lett.* **53,** 1939.
Salmelin, R. H., Pekola, J. P., Manninen, A. J., Torizuka, K., Berglund, M. P., Kyynäräinen, J. M., Lounasmaa, O. V., Tvalashvili, G. K., Magradze, O. V., Varoquaux, E., Avenel, O., and Mineev, V. P. (1989). *Phys. Rev. Lett.* **63,** 620.
Salomaa, M. M., and Volovik, G. E. (1987). *Rev. of Mod. Phys.* **59,** 533.
Sauls, J. A., and Serene, J. W. (1984). "Proc. 17th Int. Conf. on Low Temperature Physics." **LT-17,** 775–776.
Sauls, J. A., and McKenzie, R. H. (1991). *Physica* **B169,** 170–176.
Saunders, J., Daniels, M. E., Dobbs, E. R., and Ward, P. L. (1983). "Quantum Fluids and Solids"—1983, AIP Conf. Proc. #103" (E. D. Adams and G. G. Ihas, eds.). AIP, p. 314.
Saunders, J., Ling, R., Wotanowski, W., and Dobbs, E. R. (1990). *J. Low Temp. Phys.* **79,** 75.
Schopohl, N., and Tewordt, L. (1984). *J. Low Temp. Phys.* **57,** 601.
Schopohl, N., Warnke, M., and Tewordt, L. (1983). *Phys. Rev. Lett.* **50,** 1066.
Schopohl, N., Marquardt, T., and Tewordt, L. (1985). *J. Low Temp. Phys.* **59,** 469.
Serene, J. W. (1974). Ph.D. thesis, Cornell University.
Serene, J. W. (1984). *Phys. Rev.* **B30,** 5373.
Shivaram, B. S., Meisel, M. W., Sarma, B. K., Halperin, W. P., and Ketterson, J. B. (1982). *Phys. Rev. Lett.* **49,** 1646.

Shivaram, B. S., Meisel, M. W., Sarma, B. K., Halperin, W. P., and Ketterson, J. B. (1983). *Phys. Rev. Lett.* **50,** 1070.
Shivaram, B. S., Meisel, M. W., Sarma, B. K., Halperin, W. P., and Ketterson, J. B. (1986). *J. Low Temp. Phys.* **63,** 57.
Tewordt, L., and Schopohl, N. (1979a). *J. Low Temp. Phys.* **34,** 489.
Tewordt, L., and Schopohl, N. (1979b). *J. Low Temp. Phys.* **37,** 421.
Thuneberg, E. (1986). *Phys. Rev. Lett.* **56,** 359.
Tinkham, M. (1975). "Introduction to Superconductvity." McGraw Hill-Kogakusha, Ltd.
Torizuka, K., Pekola, J. P., Manninen, A. J., Kynnäräinen, J. M., and McKenzie, R. H. (1991). *Phys. Rev. Lett.* **66,** 3152.
Varoquaux, E., Williams, G. A., and Avenel, O. (1986). *Phys. Rev.* **B34,** 7617.
Vdovin, Yu. A. (1963). *In* "Application of the Methods of Quantum Field Theory to Many-Body Problems" (A. I. Alekseyeva, ed.). Moscow: GOS ATOM ISDAT, in Russian, pg. 94.
Vollhardt, D., and Wölfle, P. (1990). "The Superfluid Phases of Helium 3." Taylor and Francis, New York.
Volovik, G. E. (1984). *Sov. Phys. JETP Lett.* **39,** 365.
Volovik, G. E. (1991). "Exotic Properties of Superfluid Helium 3." World Scientific, Singapore.
Volovik, G. E., and Khazan, M. V. (1983). *Sov. Phys. JETP* **58,** 551.
Wheatley, J. C. (1975). *Rev. Mod. Phys.* **47,** 415.
Wölfle, P. (1976). *Phys. Rev. Lett.* **37,** 1279.
Wölfle, P. (1977). *Physica* **90B,** 96.
Wölfle, P. (1978). *In* "Progress in Low Temp. Phys.," Vol. VIIa (D. F. Brewer, ed.). North Holland, Amsterdam, p. 191.
Wölfle, P., and Koch, V. E. (1978). *J. Low Temp. Phys.* **30,** 61.
Yanof, A. W., and Reppy, J. D. (1974). *Phys. Rev. Lett.* **33,** 631, 1030(E).
Zhao, Z., Adenwalla, S., Brusov, P., Ketterson, J. B., and Sarma, B. K. (1990). *Phys. Rev. Lett.* **65,** 2688.
Zhao, Z., Adenwalla, S., Sarma, B. K., and Ketterson, J. B. (1991). *Advances in Physics,* in press.

—3—
Sound Propagation in the Heavy Fermion Superconductors

BIMAL K. SARMA AND MOISES LEVY
Department of Physics, University of Wisconsin—Milwaukee, Milwaukee, Wisconsin

S. ADENWALLA AND J. B. KETTERSON
Northwestern University, Evanston, Illinois

1. Introduction 108
   1.1. Conventional Superconductivity 109
   1.2. The Ginzburg–Landau Theory 113
2. Sound Propagation in Conventional Superconductors 114
   2.1. Attenuation in the Normal State 114
   2.2. Attenuation in the Superconducting State 115
   2.3. Attenuation of a Superconductor in a Magnetic Field 116
   2.4. Velocity at the Superconducting Transition 117
3. Unconventional Superconductivity 118
   3.1. The Order Parameter of an Unconventional Superconductor 118
   3.2. Ginzburg–Landau Theory of an Unconventional Superconductor 128
   3.3. Collective Modes 137
4. Heavy Fermion Systems 140
   4.1. Normal State Properties 140
   4.2. Superconductivity 143
5. UPt_3 143
   5.1. General 143
   5.2. Sound in the Normal State 146
   5.3. Sound in the Superconducting State 150
   5.4. Phase Diagram of UPt_3 162
   5.5. Results from Other Measurements 166
   5.6. Theoretical Speculations on the Superconducting Phases 172
6. UBe_{13} 176
7. URu_2Si_2 179
8. $CeCu_6$ 181
9. Conclusions 185
   Acknowledgments 185
   References 185

1. Introduction

One of the best examples of a highly correlated condensed matter system is superfluid ^{3}He, which has been described earlier in this book. In this chapter we review the properties of the heavy-electron systems, which are intermetallic compounds. These compounds consist of one f-electron metal (usually Ce or U), with the other constituent or constituents being s, p, or d electron metals. The conduction–electron specific heats are of the order of 100 to 1,000 times larger than those found in most metals. The large electron effective masses implied by the heat capacity result from a highly correlated electronic behavior at low temperatures. These systems either remain normal, become antiferromagnetic, or become superconducting (Fisk *et al.*, 1988), as the temperature is lowered. Each of these states displays properties that are very different from the behavior typical of ordinary metals. The strong coupling between the usual s, p, d conduction electrons and the f electrons leads to the large mass and the unusual superconducting and/or magnetic properties.

These systems display a rich variety of phenomena. Their normal state behavior usually consists of a "Kondo"-like behavior at higher temperatures, followed by a gradual transition to a "Fermi-liquid" state. The superconducting behavior is unusual, and the simple phonon-mediated BCS coupling appears to be inadequate. Theoretical tools developed to explain superconductivity in the heavy fermion systems may be useful in understanding the high-T_c superconductors.

Ultrasonic attenuation measurements have proved very fruitful in the study of the conventional superconductors (for example, niobium and vanadium) where the electron–phonon interaction is the mechanism leading to the formation of Cooper pairs. We briefly review these results in Section 2. In Section 3 we introduce the definition of an unconventional superconductor and sketch the present theoretical understanding. Such superconductors can display collective modes or multiple superconducting phases. The observation of these phenomena could be taken as proof of unconventional superconductivity, if we are influenced by the observations in superfluid ^{3}He.

In Section 4 we review the properties of the heavy fermion systems, emphasizing the compounds UPt_3, UBe_{13}, URu_2Si_2, and $CeCu_6$. Extensive ultrasonic measurements have been performed on these compounds, and we devote a section to each. There are many similarities in the ultrasonic properties of the three uranium-based superconductors. UPt_3 appears to be the most interesting: It has the richest phase diagram, which was first observed, and finally established by ultrasonic measurements. Earlier experimental reviews have appeared by Stewart (1984), Rauchschwalbe (1987), and de Visser *et al.* (1987), and a theoretical review by Lee *et al.* (1986).

3. Sound Propagation in the Heavy Fermion Superconductors

1.1. CONVENTIONAL SUPERCONDUCTIVITY

Superconductors such as niobium and vanadium will be referred to as conventional superconductors. Most of their properties are well explained by the microscopic (BCS) theory proposed by Bardeen *et al.* (1957), and the phenomenological theory of Ginzburg and Landau (G–L theory) (1950), both of which are discussed in texts on superconductivity (de Gennes, 1989; Tinkham, 1985; Parks, 1969).

Ideal superconductors have two characteristics: zero electrical resistance and complete diamagnetism (the Meissner state). The superconducting transition (in zero magnetic field) is a second-order phase transition (no latent heat). In the normal state, the electronic heat capacity is linear in T. At the transition there is a jump in the heat capacity given by $\Delta C/C = 1.4$. At lower temperatures the heat capacity drops exponentially as $\exp(-\Delta/k_B T)$, where Δ is the energy gap. This energy gap appears in the electron energy spectrum at the Fermi surface and is associated with the onset of superconductivity. The gap is of the order of 10^{-4} eV for a transition temperature of $\sim 1°$K. The energy gap decreases continuously to zero as the temperature approaches the transition temperature, T_c. From the BCS theory $\Delta(T = 0) = 1.76 k_B T_c$, and $\Delta(T)$ has been tabulated by Mühlschlegel (1959). In Gorkov's derivation of the G–L theory, the energy gap is identified as the superconducting order parameter.

For conventional superconductors, the order parameter is a complex scalar function and is usually taken to be isotropic in k-space; more generally, it could be anisotropic, displaying the same symmetry as the Fermi surface. This reflects the so called $L = 0$ pairing of the BCS state. The energy gap may be directly measured by infrared absorption and by tunneling experiments. It can also be inferred from ultrasonic attenuation measurements (for suitable samples).

Associated with the Meissner effect is a length scale λ_L, the London penetration depth, entering the expression that governs the behavior of the magnetic induction near the surface of a superconductor given by

$$B(z) = B(0) \exp(-z/\lambda_L), \qquad (1)$$

where

$$\lambda_L = \left(\frac{\varepsilon_0 m c^2}{n e^2}\right)^{1/2}, \qquad (2)$$

where m is the mass of the electron and n their density. Typically, λ_L is of the order of 400 Å for a conventional superconductor such as niobium, but can be much larger for the heavy fermion ($\lambda = 7{,}000$ Å for UPt$_3$) or high-T_c superconductors.

A second characteristic length of a superconductor is the coherence length ξ; this is the length scale over which the order parameter changes in response to some perturbation. In impure metals and alloys the coherence length is reduced, because of collisions among the electrons, and is sometimes written as $\xi \rightarrow (\xi_0 \ell)^{1/2}$, where ℓ is the mean free path. The intrinsic zero temperature coherence length ξ_0 is

$$\xi_0 = \frac{2\hbar v_F}{\pi \Delta(0)}. \tag{3}$$

The coherence length can vary over a wide range, and superconductors are classified by the ratio $\kappa = \lambda/\xi$. The energy of an interface between a normal region and a superconducting region is positive if $\xi \gg \lambda$, and this results in type I superconductors. If $\xi \ll \lambda$, the interface energy is negative, and we have type II superconductors. The magnetic flux in a superconductor is quantized in integral units of the flux quantum, $\phi_0 = h/2e$. In a type I superconductor, there is a complete Meissner effect for $H < H_c$. However, in a type II superconductor, the Meissner state ($B = 0$) exists only below H_{c1}, the lower critical field. There is an upper critical field, H_{c2}, above which superconductivity is destroyed. The magnetization behavior of these two types of superconductors is shown in Fig. 1. H_{c1} and H_{c2} are related in the two characteristic lengths by

$$H_{c1} = \frac{\Phi_0}{4\pi\lambda_L^2} \ell n\, (\lambda/\xi), \tag{4}$$

$$H_{c2} = \frac{\Phi_0}{2\pi\xi^2}. \tag{5}$$

The region between H_{c1} and H_{c2} is referred to as the mixed state. In the mixed state there is flux penetration; however, because of the negative surface energy, the total flux is distributed in n individual flux lines having a core of diameter 2ξ. In the absence of any "pinning" forces (from crystal defects or impurity centers), the flux lines (or vortices) arrange themselves in a two-dimensional triangular lattice, with a lattice spacing

$$d = \left(\frac{2\Phi_0}{\sqrt{3}H}\right)^{1/2}, \tag{6}$$

where H is the applied magnetic field. These two-dimensional flux lattices have been observed by neutron diffraction (Cribier et al., 1964), and by the decoration technique (Essmann and Trauble, 1967), and most recently with a scanning tunneling microscope (Hess et al., 1989, 1990). The superconducting phase diagram in the H–T plane is shown in Fig. 2. Heavy fermion materials (as well

3. Sound Propagation in the Heavy Fermion Superconductors

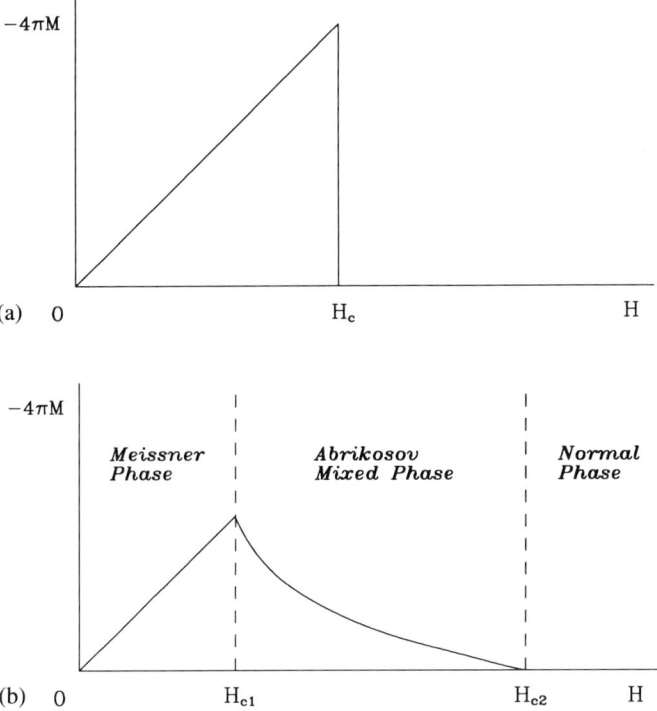

FIG. 1. Magnetization curves for (a) type I and (b) type II superconductors.

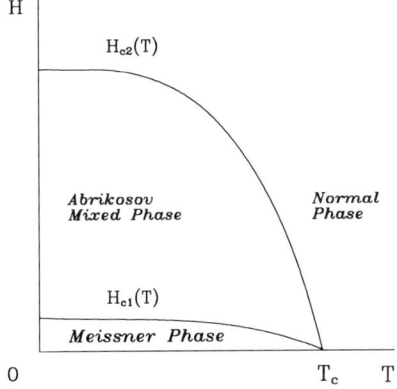

FIG. 2. H–T phase diagram for conventional superconductors.

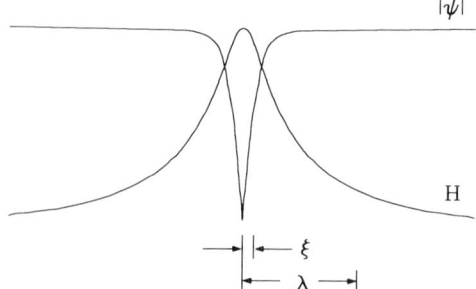

FIG. 3. Cross-section of a scalar vortex.

TABLE I

SUPERCONDUCTING PARAMETERS OF A FEW TYPICAL MATERIALS.

		T_c K	H_{c1} mT	H_{c2} T	λ_L Å	ξ_0 Å	Ref
Sn		3.722	30.9	—	340	2,300	a
Al		1.140	10.5	—	160	16,000	a
Pb		7.193	80.3	—	370	830	a
Nb		9.50	198.0	—	390	380	a
V		5.38	142.0	—	—	—	a
Nb_3Sn		18.5	—	24.0	—	—	b
Nb_3Ge		23.2	—	38.0	—	—	b
UBe_{13}		0.78	—	8.0	—	—	c
UPt_3	($\|c$)	0.52	—	2.0	7,070	200	d
	($\perp c$)	0.52	—	2.8	7,820	—	
URu_2Si_2	($\perp c$)	1.2–1.4	—	2.0	—	—	e
	($\|c$)	—	—	8.0	—	—	
$Ba_{1-x}K_xBiO_4$		30	3.5	227	2,200	36.8	f
$La_{1.85}Sr_{0.15}CuO_4$	($\|c$)	38	—	25	—	36	g
	($\perp c$)	—	—	500	—	8	
$YBa_2Cu_3O_7$	($\perp c$)	93	2.5	647	—	16	h
	($\|c$)	—	8.3	122	—	3	

[a]Kittel (1971).
[b]Narlikar and Ekbote (1983).
[c]Maple et al. (1985).
[d]Broholm et al. (1990).
[e]Kwok et al. (1990).
[f]Kwok et al. (1989).
[g]Kitazawa et al. (1989).
[h]Vandervoort et al. (1991).

3. Sound Propagation in the Heavy Fermion Superconductors

as high-T_c superconductors) are extreme type II superconductors with $H_{c2} \gg H_{c1}$. Thus, if we neglect the Meissner phase for $H < H_{c1}$, there is only one superconducting phase in the H–T plane described by a single-order parameter.

Figure 3 shows the cross-section of a (scalar) vortex. The vortex has axial symmetry. The core of the vortex is normal. The field decays exponentially as one goes out radially with a length scale given by the London penetration depth, λ_L. On the other hand, the superconducting order parameter is zero in the core of the vortex and increases radially on a length scale of the coherence length, ξ.

In Table I we list T_c, H_{c1}, H_{c2}, λ_L, and ξ for some conventional superconductors, the heavy fermion superconductors, and some high-T_c superconductors.

1.2. THE GINZBURG–LANDAU THEORY

The G–L theory is a mean field theory that accurately describes the second-order phase transition in superconductors. The order parameter, ψ, is a complex (scalar) function, and the symmetry broken is gauge symmetry (see Tinkham, 1985; de Gennes, 1989). The free energy close to T_c is expanded as a power series in ψ.

$$\delta f = f_s - f_{n0}$$

$$= \alpha |\psi|^2 + \frac{1}{2}\beta|\psi|^4 + \frac{1}{2m}\left|\left(-i\hbar\nabla - \frac{qA}{c}\right)\psi\right|^2 + \frac{h^2}{8\pi} + \text{higher order terms,}$$

where α, β are phenomenological parameters; $q = 2e$, the charge carried by a Cooper pair; h is the local magnetic field; and f_{n0} is the normal-state free energy in zero magnetic field.

Near T_c, the expansion can be truncated at the fourth power in ψ, with β being positive. α is assumed to be analytic near T_c,

$$\alpha = \alpha_0 (T - T_c), \tag{8}$$

with $\alpha_0 > 0$. The jump in the heat capacity at the transition is given by

$$\Delta C_V = T_c \frac{\alpha_0^2}{\beta}. \tag{9}$$

Inside the superconductor the magnetic field decays exponentially,

$$h(z) = H_0 \exp\left(-\frac{z}{\lambda}\right) \tag{10}$$

$$\lambda_L = \left(\frac{4\pi q^2}{mc^2}\frac{|\alpha|}{\beta}\right)^{-1/2}. \tag{11}$$

The magnitude of the order parameter, ψ, heals back to its bulk value over a length scale of the order of the coherence length ξ given by

$$\xi = \sqrt{\frac{\hbar^2}{2m|\alpha|}}. \tag{12}$$

2. Sound Propagation in Conventional Superconductors

2.1. Attenuation in the Normal State

Ultrasonic techniques have been commonly used to study normal metals and superconductors. At high temperatures, the dominant attenuation mechanism is the scattering of the sound wave from dislocations in the lattice. As the temperature is lowered and the electron mean free path becomes longer, the interaction of electrons with the lattice contributes to the attenuation.

In the limit $q\ell \ll 1$ (hydrodynamic limit), where q is the sound wavevector and ℓ is the electron mean free path, Pippard (1953) obtained expressions for the attenuation, α, of longitudinal (L) and transverse (T) sound:

$$\alpha_L = \frac{4}{15} \frac{Nmv_F}{\rho v_{sL}} q^2 \ell \tag{13}$$

and

$$\alpha_T = \frac{1}{5} \frac{Nmv_F}{\rho v_{sT}} q^2 \ell, \tag{14}$$

where N, m, v_F are the number density, effective mass, and Fermi velocity of the electrons, V_{sL} and v_{sT} are the longitudinal and transverse sound velocities, and ρ is the density. The attenuation is proportional to the mean free path and the square of the frequency ($\omega = qv_s$). As the temperature falls, the electron mean free path increases, and ultimately saturates because of scattering from imperfections in the crystal. In the collisionless limit $q\ell > 1$, i.e., for high-purity samples and/or high frequencies, the attenuation becomes independent of ℓ and is proportional to the frequency, f; the expressions are given by

$$\alpha_L = \frac{\pi}{6} \frac{Nmv_F}{\rho v_{sL}} q = \frac{\pi^2}{3} \frac{Nmv_F}{\rho v_{sL}^2} f \tag{15}$$

and

$$\alpha_T = \frac{4}{3\pi} \frac{Nmv_F}{\rho v_{sT}} q = \frac{8}{3} \frac{Nmv_F}{\rho v_{sT}^2} f. \tag{16}$$

3. Sound Propagation in the Heavy Fermion Superconductors

The limits $q\ell \ll 1$ and $q\ell \gg 1$ define regimes where the number of collisions per wavelength is respectively much greater than and much less than 1. These limits are sometimes referred to as the hydrodynamic and collisionless regimes. (Note that the number of collisions *per cycle* is *larger* by a factor v_F/v_s.)

2.2. ATTENUATION IN THE SUPERCONDUCTING STATE

In most superconducting materials, the electronic contribution to the attenuation drops sharply at T_c, approaching zero as the temperature is lowered. The superconducting state is characterized by the existence of a gap in the energy spectrum. The gap increases with decreasing temperature, approaching $\Delta = 1.764 k_B T_c$ at zero temperature. The existence of the energy gap implies that there are no unbound electrons (quasiparticles) to scatter the sound, and therefore the attenuation at $T = 0$ should go to zero. At finite T, thermally excited quasiparticles are present. The distribution of these quasiparticles is governed by the Fermi distribution, leading to an exponential temperature dependence for the ultrasonic attenuation in the superconducting state. A formal expression for the temperature-dependent attenuation is given by

$$\frac{\alpha_s}{\alpha_n} = \frac{2}{e^{\Delta(T)/k_B T} + 1}, \qquad (17)$$

where α_s and α_n are the attenuation coefficients in the superconducting and normal states, respectively, and $\Delta(T)$ is the temperature-dependent energy gap (see Fig. 4).

The above expression is valid only when the attenuation arises solely from thermally excited quasiparticles, i.e., when the phonon energy is smaller than the pair-breaking energy, 2Δ. If $\hbar\omega > 2\Delta$, acoustic phonons can break Cooper

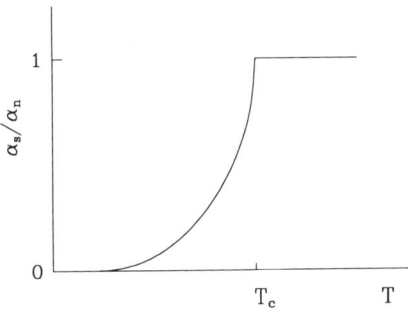

FIG. 4. Typical ultrasonic attenuation curve for a conventional superconductor.

pairs, resulting in an additional contribution to the attenuation. Typically, for a transition temperature of the order of 1K, the pair breaking energy, $2\Delta(0)$, is of the order of 77 GHz. Thus, pair breaking can be neglected, unless one is working at microwave frequencies, or very close to T_c. In the case of superfluid ^{3}He, however, because of the low T_c, pair breaking plays a significant role (see the companion article in this book).

The exponential dependence of the ultrasonic attenuation and other properties (e.g., heat capacity and NMR relaxation rate) occur when the gap is isotropic in k-space. If there are line or point zeroes (nodes) in the gap, these properties would display power-law temperature dependencies (Leggett, 1975).

2.3. ATTENUATION OF A SUPERCONDUCTOR IN A MAGNETIC FIELD

We shall restrict our discussion to type II superconductors. As discussed earlier, in a type II superconductor above H_{c1} there is an array of vortex lines. The normal cores of these vortices contribute to the sound attenuation, and thus, in the superconducting state the attenuation will increase as the field is increased. The theory of ultrasonic attenuation in the mixed state is complicated, and we shall only quote some of the theoretical results. Initially, as the field is increased, the attenuation remains constant (and very low) as long as the superconductor is in the Meissner state and no magnetic flux enters the sample. Once the magnetic field exceeds H_{c1}, the attenuation increases, at first slowly, and then more rapidly. Near the upper critical field the attenuation rises sharply, and at H_{c2} becomes equal to the normal state value (at the prevailing field). Typical attenuation curves in a magnetic field are shown in Fig. 5.

Extensive calculations for the attenuation in the mixed state have been performed (Maki, 1964, 1967, and Maki and Cyrot, 1967). The calculations were made for two different limits: the clean limit, in which $\ell \gg \xi$ (i.e., the electrons experience an order parameter averaged over many vortices) and the dirty limit, $\ell \ll \xi$ (in which the electrons experience a "local" order parameter). This is discussed in Chapter I of this book. In the clean limit ($\ell > \xi$), the attenuation close to H_{c2} is proportional to the square root of the difference in field and is given by

$$\frac{\alpha_n - \alpha_s}{\alpha_n} \simeq (H_{c2} - H)^{1/2}. \tag{19}$$

In the dirty limit ($\ell < \xi_0$), the attenuation is linearly dependent on the applied magnetic field close to H_{c2}, and we have

$$\frac{\alpha_n - \alpha_s}{\alpha_n} \simeq H_{c2} - H. \tag{20}$$

3. Sound Propagation in the Heavy Fermion Superconductors

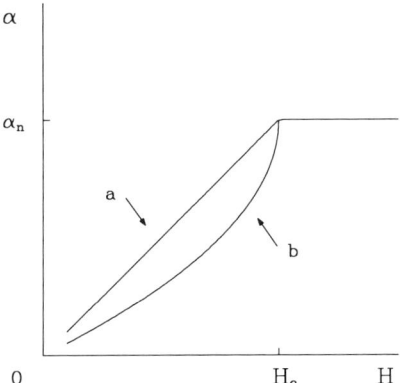

FIG. 5. Typical field dependence of ultrasonic attenuation: (a) linear (dirty limit) and (b) $(H_{c2} - H)^{1/2}$ (clean limit).

2.4. VELOCITY AT THE SUPERCONDUCTING TRANSITION

At a (mean field) second-order phase transition, the heat capacity shows a discontinuity; other second derivatives of the free energy (*viz.* elastic moduli, thermal expansion coefficient) also show an anomaly. The heat capacity, thermal expansion, and compressibility (the three second derivatives of the free energy) are related by the so-called Pippard–Buckingham–Fairbank (PBF) equation, which has been extensively applied to the study of the properties of superfluid ^{4}He (an isotropic system) at the λ-transition (Ahlers, 1978).

The elastic behavior at T_c has been examined by Testardi (1971, 1973) (for A15 superconductors), and more recently by Millis and Rabe (1988) (for the high T_c superconductors).

In the superconducting state, the free energy is lowered by

$$F_N - F_S = V_0 \frac{H_c^2(T,\varepsilon)}{8\pi} \qquad (21)$$

$$= \frac{\alpha^2(\varepsilon)}{8\pi} [T_c(\varepsilon) - T]^2.$$

Thus, the discontinuity in the heat capacity is

$$C_N - C_S = -T \frac{d^2}{dT^2} (F_N - F_S) \bigg|_{T_c} \qquad (22)$$

$$= -\frac{\alpha^2}{4\pi} T_c.$$

The second derivative of the Helmholtz free energy with respect to strain gives the isothermal elastic moduli

$$c_{ij} = \left(\frac{1}{V_0}\frac{d^2 F}{d\varepsilon_i d\varepsilon_j}\right)_T; \qquad (23)$$

(here, V_0 is the unstrained volume). Substituting the expression for the free energy, one obtains the velocity jump at the superconducting transition

$$c_{ij}^S - c_{ij}^N = -\frac{\alpha^2}{4\pi}\frac{\partial T_c}{\partial \varepsilon_i}\frac{\partial T_c}{\partial \varepsilon_j}. \qquad (24)$$

The ultrasonic velocity jump for the particular case of hexagonal symmetry is examined in greater detail in Section 5.3.3.

3. Unconventional Superconductivity

In the decade or so following the development of BCS theory, and especially after the experimental discovery of the superfluid phases of liquid ^{3}He, theorists explored the possibility of pairing with a more complicated orbital structure (e.g., p-wave or d-wave), in which the gap can vary (in both magnitude and phase) with position on the Fermi surface. In the case of odd ℓ pairing, the pairs form (as a result of the Pauli principle) in a triplet, rather than a singlet, spin state. The energy gap associated with such states may have nodes at points or lines on the Fermi surface, and consequently the number of excitations at low temperatures varies as a power of the temperature, rather than exponentially. Following a rather hectic two-year period of exploration, it was established that the pairing in liquid ^{3}He occurs in two distinct p-wave states. The A phase (also called the Anderson–Brinkman–Morel [ABM] state; Anderson and Brinkman, 1973) has point nodes on the Fermi surface, and the B phase (or Balian–Werthalmer [BW] state; Balian and Werthamer, 1963) has a gap with a constant magnitude over the Fermi surface. In these p-wave states the pairs are in triplet states, as required by the Pauli principle, and they possess magnetic properties very different from singlet paired states. For a discussion of the symmetry group classification of the superconductivity in the heavy-fermion systems, see the article by Gor'kov (1987).

3.1. The Order Parameter of an Unconventional Superconductor

Our earlier discussion of the standard Landau–Ginzburg theory of (conventional) superconductivity assumed the existence of a wave function associated with the

3. Sound Propagation in the Heavy Fermion Superconductors

superconducting electron pairs. Such a simple description is possible only when the condensate formed from the pairs has no internal degrees of freedom (other than the phase). In the BCS pairing theory of an isotropic electron liquid, the electrons are bound in a state with no orbital angular momentum (an $L = 0$ or s-wave state) and no net spin (an $S = 0$ or singlet state), and hence the pair wave function may be characterized by a single complex amplitude. One can conceive of an unconventional superconductor where the pairs form with a net orbital angular momentum, spin, or both. It is then necessary to generalize the form of the order parameter. For an isotropic liquid an appropriate form is $\psi_{\alpha\beta}(\hat{k})$ where the subscripts α and β ($= \uparrow$ or $\downarrow$) arise from the spin variables of the two electrons, and $\hat{k}$ accounts for a possible angular dependence over the Fermi surface; microscopically, the order parameter is defined as

$$\psi_{\alpha\beta}(\hat{k}) \equiv \langle c_{\mathbf{k}\alpha} c_{\mathbf{k}\beta} \rangle, \tag{25}$$

where $c_{\mathbf{k}\alpha}$ is a destruction operator for a quasiparticle with momentum $\mathbf{k}$ and spin α, and the brackets denote an ensemble average as well as an integration over the magnitude of $\mathbf{k}$. Order parameters of the above form divide naturally into two classes: so-called singlet and triplet, which are, respectively, symmetric and antisymmetric under exchange of the spin indices. By the Pauli principle the *total* wave function (orbital × spin) must be antisymmetric under interchange of the particles making up the pair. The parity of the spherical harmonics associated with pairing in a given angular momentum "channel" is $(-1)^L$; hence, singlet pairing will always be associated with even angular momentum and triplet pairing with odd angular momentum.

A firmly established example of an unconventional fermion superfluid is liquid ^{3}He, where the ^{3}He quasiparticles form in an $L = 1$ (p-wave), $S = 1$ (triplet state); it will be instructive to start with a short discussion of this system.

The spin dependence of the ^{3}He–^{3}He quasiparticle interaction, leading to the triplet pairing, has its origin in a combination of the hard-core repulsion and exchange effects (spin-orbit effects are negligible). The angular dependence of the order parameter arising from the $L = 1$ character of the pairing may be represented by the cartesian components of the unit vector $\hat{k}$ (which denotes the polar angles on the Fermi surface):

$$k_1 = \sin\theta \cos\phi, \tag{26a}$$

$$k_2 = \sin\theta \sin\phi, \tag{26b}$$

and

$$k_3 = \cos\theta; \tag{26c}$$

these forms replace the usual spherical harmonics, $Y_{1,1}$, $Y_{1,-1}$ and $Y_{1,0}$. For the spin wave functions we choose

$$\chi_1 = \frac{1}{\sqrt{2}} [-(\uparrow)(\uparrow) + (\downarrow)(\downarrow)], \tag{27a}$$

$$\chi_2 = \frac{i}{\sqrt{2}} [(\uparrow)(\uparrow) + (\downarrow)(\downarrow)], \tag{27b}$$

$$\chi_3 = \frac{1}{\sqrt{2}} [(\uparrow)(\downarrow) + (\downarrow)(\uparrow)], \tag{27c}$$

which transform as a vector under rotations. These functions have the further property that the expectation values Tr $\chi_i^\dagger S_i \chi_i$ vanish, where S_i are the components of the total spin operator $\mathbf{S} = \frac{1}{2}(\boldsymbol{\sigma}_1 + \boldsymbol{\sigma}_2)$ and $\boldsymbol{\sigma}_1$ and $\boldsymbol{\sigma}_2$ denote the vector Pauli matrices associated with the two quasiparticles. Any linear combination of (27a)–(27c) with real amplitudes or a single spin wavefunction with a complex amplitude will correspond to a superfluid with no net spin polarization. (Superfluid ^{3}He has no polarization in zero magnetic field.)

The most general order parameter for ^{3}He (which includes both polarized and unpolarized states) may be written as the superposition

$$\psi_{\alpha\beta}(\hat{k}) = \sum_{\alpha=1}^{3} \sum_{i=1}^{3} D_{\alpha i} k_i \chi_\alpha, \tag{28}$$

and the nine complex amplitudes $D_{\alpha i}$ (18 real amplitudes) define 18 different superfluid states. The two observed ground states, A and B, correspond to specific linear combinations. The equilibrium B phase can be characterized by the amplitude matrix

$$\overleftrightarrow{D} = \frac{\Delta}{\sqrt{3}} \begin{pmatrix} 1 & 0 & 0 \\ 0 & 1 & 0 \\ 0 & 0 & 1 \end{pmatrix}, \tag{29a}$$

where Δ is an arbitrary complex amplitude, and the normalization chosen is such that Tr$\overleftrightarrow{D}^\dagger \overleftrightarrow{D} = |\Delta|^2$; more generally, the unit matrix is replaced by a rotation matrix accounting for the fact that the axes defining orbital and spin space are in general different. The order parameter of the B phase corresponding to Eq. (29a) is

$$\psi(\hat{k}) = \Delta(k_1 \chi_1 + k_2 \chi_2 + k_3 \chi_3), \tag{29b}$$

where the normalization adopted here is $\int \psi^* \psi d\Omega = |\Delta|^2$.

3. Sound Propagation in the Heavy Fermion Superconductors

The A phase may be written as

$$\overleftrightarrow{D} = \frac{\Delta}{\sqrt{2}} \begin{pmatrix} 1 & i & 0 \\ 0 & 0 & 0 \\ 0 & 0 & 0 \end{pmatrix}, \tag{30a}$$

corresponding to the wave function

$$\psi = \sqrt{\frac{3}{2}} \Delta(k_1 + ik_2) \chi_1. \tag{30b}$$

In so-called weak-coupling theory, where the pairing interaction remains unaltered on entering the superfluid state, only the B phase would be stable. For the case of ^{3}He the relative stability of the possible phases is determined by the *fourth*-order terms in the G–L expansion; at the level of the second-order terms, all possible $L = 1$ superfluids would have the same transition temperature.

By applying the total angular momentum operator $J^2 = L^2 + S^2 + 2\mathbf{L} \cdot \mathbf{S}$, it is easy to verify that Eq. (29) corresponds to a $J = 0$ state. One can similarly verify that a suitable set of $J = 1$ (excited) states is

$$\frac{1}{\sqrt{2}} \begin{pmatrix} 0 & 0 & 0 \\ 0 & 0 & -1 \\ 0 & 1 & 0 \end{pmatrix}; \frac{1}{\sqrt{2}} \begin{pmatrix} 0 & 0 & 1 \\ 0 & 0 & 0 \\ -1 & 0 & 0 \end{pmatrix}; \frac{1}{\sqrt{2}} \begin{pmatrix} 0 & 1 & 0 \\ -1 & 0 & 0 \\ 0 & 0 & 0 \end{pmatrix}, \tag{31}$$
$$(a) \qquad\qquad (b) \qquad\qquad (c)$$

while a set of $J = 2$ states is

$$\frac{1}{\sqrt{2}} \begin{pmatrix} 0 & 1 & 0 \\ 1 & 0 & 0 \\ 0 & 0 & 0 \end{pmatrix}; \frac{1}{\sqrt{2}} \begin{pmatrix} 1 & 0 & 0 \\ 0 & -1 & 0 \\ 0 & 0 & 0 \end{pmatrix}; \frac{1}{\sqrt{2}} \begin{pmatrix} 0 & 0 & 0 \\ 0 & 0 & 1 \\ 0 & 1 & 0 \end{pmatrix};$$
$$(a) \qquad\qquad (b) \qquad\qquad (c)$$

$$\frac{1}{\sqrt{2}} \begin{pmatrix} 0 & 0 & 1 \\ 0 & 0 & 0 \\ 1 & 0 & 0 \end{pmatrix}; \frac{1}{\sqrt{6}} \begin{pmatrix} 1 & 0 & 0 \\ 0 & 1 & 0 \\ 0 & 0 & -2 \end{pmatrix} \tag{32}$$
$$(d) \qquad\qquad (e)$$

Excited $J = 0$ states are also possible. The amplitudes associated with (31) or (32) may be either real or imaginary; with the two $J = 0$ states, this results in a total of 18 excited states that are the collective modes (of the B phase) discussed in Chapter II.

For a singlet superfluid, the spin wave function would be

$$\chi_0 = \frac{1}{\sqrt{2}} [(\uparrow)(\downarrow) - (\downarrow)(\uparrow)]. \tag{33}$$

Thus, an s-wave superfluid has the order parameter

$$\psi = \Delta\chi_0, \tag{34}$$

where Δ is again a complex number. Singlet pairing in a higher angular momentum state involves a more complicated orbital wavefunction. For the case of singlet d-wave pairing, the order parameter can be written

$$\psi = \chi_0 \sum_{i=1}^{3} \sum_{j=1}^{3} k_i D_{ij} k_j, \tag{35}$$

where $\overleftrightarrow{D}$ is a traceless symmetric matrix. (Tr $\overleftrightarrow{D} \neq 0$ would imply an s-wave admixture). A representation of the allowed states (which also have $J = 2$) is the same as Eq. (32):

$$\frac{1}{\sqrt{2}}\begin{pmatrix} 0 & 1 & 0 \\ 1 & 0 & 0 \\ 0 & 0 & 0 \end{pmatrix}; \quad \frac{1}{\sqrt{2}}\begin{pmatrix} 1 & 0 & 0 \\ 0 & -1 & 0 \\ 0 & 0 & 0 \end{pmatrix}; \quad \frac{1}{\sqrt{2}}\begin{pmatrix} 0 & 0 & 0 \\ 0 & 0 & 1 \\ 0 & 1 & 0 \end{pmatrix};$$
$$\quad\quad (a) \quad\quad\quad\quad\quad\quad (b) \quad\quad\quad\quad\quad\quad (c)$$

$$\frac{1}{\sqrt{2}}\begin{pmatrix} 0 & 0 & 1 \\ 0 & 0 & 0 \\ 1 & 0 & 0 \end{pmatrix}; \quad \frac{1}{\sqrt{6}}\begin{pmatrix} 1 & 0 & 0 \\ 0 & 1 & 0 \\ 0 & 0 & -2 \end{pmatrix};$$
$$\quad\quad (d) \quad\quad\quad\quad\quad\quad (e) \tag{36}$$

here, also, each form can have an independent real or imaginary amplitude. These forms correspond to the usual (unnormalized) cartesian d-wave functions:

$$2k_1k_2; \quad k_1^2 - k_2^2; \quad 2k_2k_3; \quad 2k_1k_3; \quad k_1^2 + k_2^2 - 2k_3^2; \tag{37}$$
$$\;(a) \quad\quad (b) \quad\quad\;\; (c) \quad\quad\;\; (d) \quad\quad\quad (e)$$

where the k_i are defined in Eq. (26). By taking linear combinations of (a) and (b) and of (c) and (d), we may form the d-wave spherical harmonics $Y_{2,\pm 2}$ and $Y_{2,\pm 1}$; form (e) corresponds to $Y_{2,0}$. We will use the various s, p, and d wave states discussed above as a starting point for discussing some allowed electron-superconducting states.

A real metal is not an isotropic electron liquid, and the classification of various allowed electron superfluids must take account of the lowered symmetry of the host crystal lattice; to be specific, we will focus on hexagonal and tetragonal crystals (examples being UPt_3 and URu_2Si_2). A rigorous discussion would involve group theory, but most of the relevant physics can be extracted by considering the effect of various perturbations on the states of an isotropic system. Furthermore, all known heavy fermion compounds involve rare earths or actinides that, because of their high atomic numbers Z, necessarily involve spin–orbit effects. We will deal with this latter effect first.

3. Sound Propagation in the Heavy Fermion Superconductors 123

In the L–S coupling scheme, we may write the perturbing Hamiltonian arising from spin–orbit coupling as

$$V_{so} = A\mathbf{L} \cdot \mathbf{S}, \tag{38a}$$

where A is a constant. The eigenvalues of the operator $\mathbf{L} \cdot \mathbf{S}$ acting on a state of a given J, ψ_J, are given by

$$\mathbf{L} \cdot \mathbf{S}\psi_J = \tfrac{1}{2}[J(J + 1) - L(L + 1) - S(S + 1)]\psi_J. \tag{38b}$$

Hence, states with the same L and S, but differing J, will be shifted; however, the degeneracy within a multiplet of a given J is not lifted. For our p-wave superfluid, the degeneracy of the states with $J = 0$, $J = 1$, and $J = 2$ (Eqs. (29), (31) and (32)) would be lifted in the *leading order* of the GL expansion; i.e., they would each have different transition temperatures. The spin–orbit energy shifts for this case are $-2A(J = 0)$; $-A(J = 1)$; and $+A(J = 2)$.

We next consider the effect of the crystal structure. We begin by recalling the independent symmetry operations (classes) associated with the hexagonal D_{6h} (UPt$_3$) and tetragonal D_{4h} (URu$_2$Si$_2$) point groups. For D_6 we have the identity (E); two- (C_2), three- $(2C_3)$, and six- $(2C_6)$ fold rotations about the hexagonal axis; and two sets of twofold $(3U_2$ and $3U_2')$ rotations about axes in the basal plane. For D_4 we have the identity (E); two- (C_2) and four- (C_4) fold rotations about the tetragonal axis; and two sets of twofold rotations about axes in the basal plane $(2U_2$ and $U_2')$. (The number preceding the symmetry designation denotes the number of such operations.) The inclusion of inversion symmetry (which is equivalent in the present case to reflection through the basal plane) doubles the number of classes, and the corresponding point group designations are D_{6h} and D_{4h}.

The perturbing effect of the crystal potential for our cases of interest may be represented by a function $V(\hat{k})$,

$$V(\hat{k}) = \sum_{\ell,m} V_{\ell m} \cos \ell\theta \cos m\phi, \tag{39}$$

where $m = 4p$ for D_4 and $6p$ for D_6, and $\ell = 2q$; here p and q denote all positive integers. We will generally limit ourselves to the leading-order, symmetry-breaking effect: $p = q = 1$.

An important aspect of the theory of point groups is concerned with the transformation properties of (wave) functions. Wave functions (in our case, order parameters) may have two kinds of behaviors: either (i) a wavefunction transforms into itself under all symmetry operations (other than a phase factor), or (ii) a minimal set of wave functions transform into linear combinations of each other; (i) is said to be a one-dimensional (nondegenerate) representation, and the dimensionality (degeneracy) of (ii) is equal to the minimal number of required wavefunctions. As an example, consider atomic states corresponding to the total angular momentum quantum number J. Such states are $(2J + 1)$-fold degenerate; for a singlet ($L = J$), they would correspond to spherical harmonics, $Y_{L,L_z}(\theta, \phi)$, and on rotating a system in a given state L about an arbitrary axis, these functions transform into linear combinations involving the quantum number L_z. Since there is no limit on J, spherical symmetry admits very high degeneracies. Nonspherical perturbations remove much of the degeneracy.

There is an upper limit on the degeneracies of discrete point groups: For D_{6h} and D_{4h} it is two, and hence we can have only one- and two-dimensional representations. The one-dimensional representations are denoted by A and B, which are even and odd, respectively, under the minimum rotation about the principal symmetry axis (i.e., a 60° or 90° rotation for D_{6h} and D_{4h}). The two-dimensional representations are denoted by the letter E (not to be confused with the identity). If there is more than one symmetry-distinct representation, then subscripts 1, 2, etc., are included. Some of the representations have the same symmetry as the various crystal axes (x, y, or z) and the relevant axis is then included as a subscript. Finally, in the presence of inversion symmetry, a representation may be either even or odd, and the subscripts u and g, respectively, are used to denote these symmetries. Note there is no limit on the number of times a given symmetry type may occur; however, barring an accidental degeneracy, they will have different energies. The ground state corresponds to the globally lowest energy state (across all representations), which in the absence of other information could have any of the allowed symmetries (and hence degeneracies).

We now employ some simple heuristic arguments to arrive at the important representations for our chosen point groups (D_{6h} and D_{4h}), starting from the s-, p-, and d-wave pairing states of the isotropic quasiparticle superfluid. We will begin by assuming the largest energy shifts are associated with a transition from spherical to uniaxial symmetry (with an axis involving the principal four- or sixfold crystal axis). This symmetry is broken by the V_{20} term in Eq. (39); as far as the effect on symmetry breaking is concerned, it is equivalent to the application of an electric field in the (quadratic) Stark effect in atomic physics: $2J + 1$ degenerate states of a given J split into $J + 1$ levels with J doubly degenerate $\pm|J_z|$ levels and a singly degenerate $J_z = 0$ level. Note that at this elementary stage (uniaxial symmetry + spin–orbit coupling), we have already lifted most of the relevant degeneracy. Let us examine the effect of crystal field splitting on the previously enumerated s, p, and d wave states. We write the wave functions of the various states in the form $\psi = \psi^{L,S}_{J,J_z}(\hat{k})$. Clearly the singlet s wave state is unaffected and has the symmetry A_{ig}. The p-wave states split into one- and two-dimensional representation having the structure (where the quantity in parentheses is the group theoretical designation of the state):

3. Sound Propagation in the Heavy Fermion Superconductors

$J = 0$:

$$\overleftrightarrow{D} = \frac{1}{\sqrt{3}} \overleftrightarrow{1}, \tag{40}$$

$$\psi^{1;1}_{0;0}(\hat{k}) = k_1\chi_1 + k_2\chi_2 + k_3\chi_3 \quad (A_{1u}); \tag{41}$$

$J = 1, J_z = 0$:

$$\overleftrightarrow{D}_z = \frac{1}{\sqrt{2}} \begin{pmatrix} 0 & 1 & 0 \\ -1 & 0 & 0 \\ 0 & 0 & 0 \end{pmatrix}, \tag{42}$$

$$\psi^{1;1}_{1;0}(\hat{k}) = \sqrt{\frac{3}{2}}(k_2\chi_1 - k_1\chi_2) \quad (A_{2u;z}); \tag{43}$$

$J = 1, J_z = \pm 1$:

$$\overleftrightarrow{D}_x = \frac{1}{\sqrt{2}} \begin{pmatrix} 0 & 0 & 0 \\ 0 & 0 & -1 \\ 0 & 1 & 0 \end{pmatrix}, \tag{44a}$$

$$\overleftrightarrow{D}_y = \frac{1}{\sqrt{2}} \begin{pmatrix} 0 & 0 & 1 \\ 0 & 0 & 0 \\ -1 & 0 & 0 \end{pmatrix}, \tag{44b}$$

$$\psi^{1;1}_{1;\pm 1}(\hat{k}) = \sqrt{\frac{3}{2}} \begin{pmatrix} -k_3\chi_2 + k_2\chi_3 \\ k_3\chi_1 - k_1\chi_3 \end{pmatrix} \quad (E_{1u;x,y}); \tag{45}$$

$J = 2; J_z = 0$:

$$D_{r^2 - 3z^2} = \frac{1}{\sqrt{6}} \begin{pmatrix} 1 & 0 & 0 \\ 0 & 1 & 0 \\ 0 & 0 & -2 \end{pmatrix}, \tag{46}$$

$$\psi^{1;1}_{2;0}(\hat{k}) = \frac{1}{\sqrt{2}}(k_1\chi_1 + k_2\chi_2 - 2k_3\chi_3) \quad (A_{1u}); \tag{47}$$

$J = 2; J_z = \pm 1$:

$$D_{yz} = \frac{1}{\sqrt{2}} \begin{pmatrix} 0 & 0 & 0 \\ 0 & 0 & 1 \\ 0 & 1 & 0 \end{pmatrix}, \tag{48a}$$

$$D_{xz} = \frac{1}{\sqrt{2}} \begin{pmatrix} 0 & 0 & 1 \\ 0 & 0 & 0 \\ 1 & 0 & 0 \end{pmatrix}, \tag{48b}$$

$$\psi^{1,1}_{2,\pm1}(\hat{k}) = \sqrt{\frac{3}{2}}\begin{pmatrix} k_3\chi_2 + k_2\chi_3 \\ k_3\chi_1 + k_1\chi_3 \end{pmatrix} \quad (E_{1u;x,y}); \tag{49}$$

$J = 2; J_z = \pm 2$:

$$D_{xy} = \frac{1}{\sqrt{2}}\begin{pmatrix} 0 & 1 & 0 \\ 1 & 0 & 0 \\ 0 & 0 & 0 \end{pmatrix}, \tag{50a}$$

$$D_{x^2-y^2} = \frac{1}{\sqrt{2}}\begin{pmatrix} 1 & 0 & 0 \\ 0 & -1 & 0 \\ 0 & 0 & 0 \end{pmatrix}, \tag{50b}$$

$$\Psi^{1,1}_{2,\pm2} = \sqrt{\frac{3}{2}}\begin{pmatrix} k_2\chi_1 + k_1\chi_2 \\ k_1\chi_1 - k_2\chi_2 \end{pmatrix} \quad (E_{2u}). \tag{51}$$

(Note that under a 90° rotation about the z axis $k_1 \to k_2$, $k_2 \to -k_1$; $\chi_1 \to \chi_2$, $\chi_2 \to -\chi_1$, and similarly for rotations about the other two axes.) For the $L = J = 2$ (d-wave) states, we have

$L = 2; L_z = 0$:

$$\psi^{2,0}_{2,0}(\hat{k}) = \sqrt{\frac{5}{6}}(k_1^2 + k_2^2 - 2k_3^2)\chi_0 \quad (A_{1g}); \tag{52}$$

$L = 2; L_z = \pm 1$:

$$\psi^{2,0}_{2,\pm1}(\hat{k}) = \sqrt{15}\begin{pmatrix} k_2k_3 \\ k_1k_3 \end{pmatrix}\chi_0 \quad (E_{1g;x,y}); \tag{53}$$

$L = 2; L_z = \pm 2$:

$$\chi^{2,0}_{2,\pm2}(\hat{k}) = \sqrt{\frac{15}{2}}\begin{pmatrix} k_1k_2 \\ \frac{1}{2}(k_1^2 - k_2^2) \end{pmatrix}\chi_0. \quad (E_{2g}). \tag{54}$$

We next examine the effect of removing the continuous (uniaxial) rotational symmetry about the $\hat{z}$ axes and replacing it with four- or sixfold symmetry. In lowest order, this results in a perpendicular perturbing potential involving the V_{04} or V_{06} terms in Eq. (39) for the D_4 and D_6 symmetries, respectively. The two components of an E_2 representation transform as $k_1^2 - k_2^2(\sim \cos 2\phi)$ and $k_1 k_2(\sim \sin 2\phi)$. The matrix elements of the V_{04} terms of Eq. (39) (associated with the D_4 symmetry) with the components of an E_2 representation are different, and hence E_2 states split into two nondegenerate levels (the symmetry of which we will discuss shortly). For the case of D_6 symmetry (involving the V_{06} terms), the two levels making up an E_2 representation remain degenerate.

With the understanding that much of the degeneracy is lifted, it is still common

3. Sound Propagation in the Heavy Fermion Superconductors

to talk of pairing in $L = 0$, $L = 1$, $L = 2$ channels, or s-wave (singlet), p-wave (triplet) and d-wave (singlet) superconductors. We will assume that pairing effects arising from channels with $L > 2$ are negligible. Conventional s-wave superconductors correspond to a nodeless A_{1g} representation, and the gap has the full point group symmetry of the crystal.

Note that so far we have not designated any B representations. For principal axes of fourth or sixth order (D_4 or D_6), this would require wave functions containing the terms

$$\begin{pmatrix} \cos \\ \sin \end{pmatrix} m\phi,$$

where $m = 2(D_4)$ or $3(D_6)$, which change sign under a rotation of $\pi/2$ (D_4) or $\pi/3$ (D_6). (The cos $m\phi$ term peaks on the U axis and the sin $m\phi$ on the U' axis.) For D_6, such terms cannot occur for $L \leq 2$; for D_4, the $J = 2$, E_2 representation of the uniaxial symmetry case splits into a $B_1(\cos 2\phi)$ and a B_2 (sin 2ϕ) representation.

In the absence of spin–orbit coupling, the (Stark-like) potential producing the uniaxial symmetry, which may be represented by the term V_{20} in Eq. (39), has a matrix element connecting the $J = 0$, $J_z = 0$ state (Eq. 41) and the $J = 2$, $J_z = 0$ state (Eq. 47) (there are no matrix elements connecting $J = 1$, $J_z = 0$ with either of these two states). We must therefore perform a *degenerate* perturbation calculation in this subspace. The resulting wave functions and energy shifts, ΔE, are

$$\psi_{\text{polar}}^{1,1}(\hat{k}) = \sqrt{3}\, k_3 \chi_3, \qquad \Delta E = \frac{3}{5} V_{20}, \tag{55}$$

and

$$\psi_{\text{planar}}^{1,1}(k) = \sqrt{\frac{3}{2}}\, (k_1 \chi_1 + k_2 \chi_2), \qquad \Delta E = \frac{1}{5} V_{20}. \tag{56}$$

The angular dependence of the nodes of the gap follows from $\psi^\dagger(\hat{k})\,\psi(\hat{k})$. Form (55) corresponds to the polar phase of ^{3}He and has a node in the x–y (basal) plane; form (56) corresponds to the planar phase and has a node along the $\hat{z}$ axis. (Note that neither of these phases is stable in ^{3}He.) The addition of spin–orbit coupling restores forms (41) and (45) as eigenstates. Form (41) has no nodes and may be regarded as the analogue of the ^{3}He B phase; form (47) has the dependence $\sin^2 \theta$ and retains the planarlike behavior, while the polarlike state disappears.

The A or ABM phase of ^{3}He (see Eq. (30b)) may be regarded as a superposition of the $J_z = \pm 1$ states of the $J = 1$ and $J = 2$ multiplets, both of which are

$E_{1u;x,y}$ representations. However, in the presence of spin–orbit coupling, $J = 0$ and $J = 2$ have different energy (i.e., are nondegenerate), and hence a coherent superposition is not possible; thus, the A phase cannot exist for D_6 or D_4.

The antiferromagnetic ordering in the basal plane of UPt$_3$, which is suggested by neutron scattering experiments, formally lowers the symmetry from D_{6h} to D_{2h} (or D_{4h} to D_{2h} for URu$_2$Si$_2$); however, the weakness of the perturbation suggests that any degeneracies lifted involve only a very small splitting. We may impose this lowered symmetry on the representations discussed earlier by examining the effect of a term V_{02} in Eq. (39). We have already shown that E_2 is split by a V_{04} term (which is also present when the symmetry is reduced to D_2). A short calculation shows that V_{02} splits E_1. The surviving representations have the symmetries and designations: A, $B_{3;x}$, $B_{1;z}$ and $B_{2;y}$ in addition to either u or g symmetry for the triplet or singlet cases, respectively. These splittings proceed as follows:

$$E_1 \sim \begin{pmatrix} \sin \phi \\ \cos \phi \end{pmatrix} \to B_{2;y}; B_{3;x}. \tag{57a}$$

For our split E_2 representations,

$$E_2 \sim \begin{pmatrix} \sin 2\phi \\ \cos 2\phi \end{pmatrix} k_z \to A_{1;z}, A_{1;z}. \tag{57b}$$

The allowed wave function symmetries for D_2, D_4 and D_6 are summarized in Table II. The various allowed symmetry types are called "irreducible representations," because they characterize the allowed wave function symmetries using the minimum number of dimensions possible.

3.2. Ginzburg–Landau Theory of an Unconventional Superconductor

In this section we discuss the extension of the ordinary Ginzburg–Landau theory to unconventional superconductivity for some cases that may be relevant to the heavy-fermion superconductors, placing the primary emphasis on UPt$_3$. As with the general Landau theory of a second-order phase transition, it is assumed that

TABLE II

WAVEFUNCTION SYMMETRIES FOR D_2, D_4, AND D_6 CLASSES.

D_2	A	$B_{1;z}$	$B_{2;y}$	$B_{3;x}$	—	—
D_4	A_1	$A_{2;z}$	B_1	B_2	$E_{1;x,y}$	—
D_6	A_1	$A_{2;z}$	B_1	B_2	E_2	$E_{1;x,y}$

3. Sound Propagation in the Heavy Fermion Superconductors

the free energy may be expanded in terms of an order parameter, and that this order parameter can be written as a linear combination of the allowed wave function symmetries (or irreducible representations) discussed in the previous section. It will further be assumed that only one wave function symmetry (that form having the highest T_c) will be important near the transition temperature.

If the order parameter is dominated by one of the one-dimensional representations, then the Ginzburg–Landau expansion takes the same form as for an ordinary (A_{1g}) superconductor. The observation of the various phenomena discussed in Sec. 5.4, which imply the existence of more than one "transition temperature" in UPt$_3$, strongly suggests that the superconductivity is unconventional and associated with one of the two-dimensional representations, E_1 or E_2, and that the otherwise-degenerate transition temperature (associated with the two components making up a two-dimensional representation) is split by a weak symmetry breaking brought on by the antiferromagnetic ordering. It is, of course, also possible that two symmetry distinct one-dimensional representations accidentally have nearly the same transition temperatures (within a few percent for UPt$_3$); since in BCS theory the transition temperature depends exponentially on the pairing potential associated with a given representation, such an accident should be unlikely. We therefore limit further discussion to unconventional E_1 or E_2 superconductivity.

E_1 *case:* The two basis wave functions making up an E_1 representation transform as k_x and k_y. The most general E_1 state must therefore be written as a linear superposition of the two basis functions with complex coefficients, η_x and η_y:

$$\psi = \eta_x k_x + \eta_y k_y = \boldsymbol{\eta} \cdot \hat{k}_\perp. \tag{58}$$

The free energy is expanded in terms of the parameters contained in $\boldsymbol{\eta}$; these parameters must be combined such that the free energy: (i) is real, (ii) is invariant under all crystal symmetry operations, and (iii) is restricted to even powers of $|\eta_x|$ and $|\eta_y|$. At the level of the quadratic term, the only invariant function involving k_x and k_y is $(k_x^2 + k_y^2)$ with the associated invariant $(\boldsymbol{\eta} \cdot \boldsymbol{\eta}^*)$. The allowed quartic terms differ for D_4 and D_6. For D_6 the only invariant function is $(k_x^2 + k_y^2)^2$, leading to two fourth-order invariants: $(\boldsymbol{\eta} \cdot \boldsymbol{\eta}^*)^2$ and $|\boldsymbol{\eta} \cdot \boldsymbol{\eta}|^2$; for D_4 we have the additional invariant $(k_x^4 + k_y^4)$, leading to the invariant $\eta_x^2 \eta_x^{*2} + \eta_y^2 \eta_y^{*2}$. The resulting Landau–Ginzburg expansions are

$$F = F_0 + \alpha(\boldsymbol{\eta} \cdot \boldsymbol{\eta}^*) + \beta_1(\boldsymbol{\eta} \cdot \boldsymbol{\eta}^*)^2 + \beta_2|\boldsymbol{\eta} \cdot \boldsymbol{\eta}|^2 \quad (D_6) \tag{59}$$

$$F = F_0 + \alpha(\boldsymbol{\eta} \cdot \boldsymbol{\eta}^*) + \beta_1(\boldsymbol{\eta} \cdot \boldsymbol{\eta}^*)^2$$
$$+ \beta_2|\boldsymbol{\eta} \cdot \boldsymbol{\eta}|^2 + \beta_3(\eta_x^2 \eta_x^{*2} + \eta_y^2 \eta_y^{*2}) \quad (D_4). \tag{60}$$

We will restrict ourselves to the case of D_6 (UPt$_3$). Minimizing Eq. (59) with respect to $\boldsymbol{\eta}^*$ we obtain

$$\alpha\eta + 2\beta_1 (\eta \cdot \eta^*)\eta + 2\beta_2(\eta \cdot \eta)\eta^* = 0. \tag{61}$$

Near T_c we write $\alpha = \alpha_0 (T - T_c)$. Since the quantity $\eta \cdot \eta^*$ is positive definite, thermodynamic stability requires $\beta_1 > 0$; β_2 may have either sign, but thermodynamic stability requires $\beta_2 > -\beta_1$. If $\beta_2 > 0$, we achieve the lowest energy when $\eta \cdot \eta = 0$, which requires the form

$$\eta = \frac{\eta_0}{\sqrt{2}} (\hat{x} \pm i\hat{y}). \quad (\beta_2 > 0). \tag{62}$$

Inserting this in the remaining terms in (61), we obtain

$$|\eta_0|^2 = -\frac{\alpha}{2\beta_1} \quad T < T_c \quad (\beta_2 > 0), \tag{63a}$$

$$|\eta_0|^2 = 0 \quad T > T_c, \tag{63b}$$

and η_0 contains an arbitrary phase.

For the case $\beta_2 < 0$, the free energy is minimized when $\eta \cdot \eta$ is maximal; i.e., when

$$\eta = \eta_0(\cos \phi_0 \hat{x} + \sin \phi_0 \hat{y}) \quad (\beta_2 < 0), \tag{64}$$

where ϕ_0 is an arbitrary angle of rotation in the x–y plane and η_0 again contains an arbitrary phase. The value of $|\eta_0|$ minimizing Eq. (59) is

$$|\eta_0|^2 = -\frac{\alpha}{2(\beta_1 + \beta_2)} \quad (T < T_c), \tag{65a}$$

or

$$|\eta_0|^2 = 0 \quad (T > T_c). \tag{65b}$$

The two signs associated with Eq. (62), correspond to two different states that cannot be continuously deformed into each other, and hence the superfluid state can consist of a mixture of domains of each type. The state described by Eq. (64) is continuously (infinitely) degenerate with respect to the arbitrary angle, ϕ_0.

E_2 case: The E_2 representations, associated with p or d wave pairing involve the forms $k_x^2 - k_y^2$ and $2 k_x k_y$. Writing $k_x = \cos \phi$ and $k_y = \sin \phi$ and taking the appropriate linear combinations, we may schematically write the wave function of our E_2 representation as

$$\psi \sim \begin{pmatrix} e^{2i\phi} \\ e^{-2i\phi} \end{pmatrix}. \tag{66}$$

3. Sound Propagation in the Heavy Fermion Superconductors

The order parameter must be a linear combination of the two basis functions that we may write as $\eta_+ e^{2i\phi} + \eta_- e^{-2i\phi}$. If we define a vector $\begin{pmatrix} \eta_+ \\ \eta_- \end{pmatrix}$ and note that the second- and fourth-order invariants can only involve $(k_x + k_y)^2$ and $(k_x + k_y)^4$, then our free energy takes a form identical to Eq. (59):

$$F = F_0 + \alpha(\boldsymbol{\eta} \cdot \boldsymbol{\eta}^*) + \beta_1(\boldsymbol{\eta} \cdot \boldsymbol{\eta}^*)^2 + \beta_2|\boldsymbol{\eta} \cdot \boldsymbol{\eta}|^2; \tag{67}$$

the minimization of Eq. (67) is completely analogous to the E_1 case. In the ordered state we have the following two possibilities:

$$\boldsymbol{\eta} = \eta_0 \begin{pmatrix} 1 \\ \pm i \end{pmatrix} \quad (\beta_2 > 0) \tag{68a}$$

or

$$\boldsymbol{\eta} = \eta_0 \begin{pmatrix} \cos 2\phi_0 \\ \sin 2\phi_0 \end{pmatrix} \quad (\beta_2 < 0), \tag{68b}$$

with $|\eta_0|$ given by (63a) and (65a), respectively.

Inhomogenities in the order parameter: To treat the effect of boundaries or finite fields, we must incorporate the gradient terms in the G–L free energy; we will restrict ourselves to the E_1 representation of D_6. The gradient energy must be rotationally invariant in the x,y (1,2) plane. The vector $\boldsymbol{\eta}$ and the perpendicular component of the gradient, $\nabla_\perp$ (∂_x and ∂_y), each behave as $2d$ vectors, and we can form the following $2d$-rotationally invariant combinations that contribute to the free energy (Tokuyasu, 1990; Anderson and Brinkman, 1978):

$$F_1 = \kappa_1 \sum_{i,j=1}^{2} \frac{\partial \eta_i}{\partial x_j} \frac{\partial \eta_i^*}{\partial x_j}, \tag{69a}$$

$$F_2 = \kappa_2 \sum_{i,j=1}^{2} \frac{\partial \eta_i}{\partial x_i} \frac{\partial \eta_j^*}{\partial x_j}, \tag{69b}$$

and

$$F_3 = \kappa_3 \sum_{i\mu=1}^{2} \frac{\partial \eta_i}{\partial x_j} \frac{\partial \eta_j^*}{\partial x_i}. \tag{69c}$$

In addition, we must add the (single) contribution associated with variations along the z axis:

$$F_4 = \kappa_4 \sum_{i=1}^{2} \frac{\partial \eta_i}{\partial x_3} \frac{\partial \eta_i^*}{\partial x_3}. \tag{69d}$$

In the presence of a magnetic field we use the standard gauge invariant prescription

$$\frac{\partial}{\partial x_i} \rightarrow \frac{\partial}{\partial x_i} - \frac{2ie}{\hbar c} A_i, \tag{70}$$

where A_i is the vector potential and $i = 1, 2, 3$. Finally, we must add the energy density of the magnetic field:

$$F_H = \frac{1}{8\pi} H^2. \tag{71}$$

Effect of a magnetic field: The upper critical field involves the solution of the linearized Ginzburg–Landau equations in the presence of a magnetic field. Unlike the case of an ordinary superconductor, which is equivalent to the time-independent Schrödinger equation for a particle moving in a magnetic field, we obtain from Eqs. (69a)–(69c) the two coupled second-order differential equations

$$-\alpha \eta_x + \kappa_1 D_j D_j \eta_x + \kappa_2 D_x D_j \eta_j + \kappa_3 D_j D_x \eta_j + \kappa_4 D_z^2 \eta = 0, \tag{72a}$$

$$-\alpha \eta_y + \kappa_1 D_j D_j \eta_y + \kappa_2 D_y D_j \eta_j + \kappa_3 D_j D_y \eta_j + \kappa_4 D_z^2 \eta_y = 0, \tag{72b}$$

where $j = x, y$. For $H \| \hat{c}$ a convenient gauge is $\mathbf{A} = xH\hat{y}$, which, on substitution in Eqs. (72a) and (72b), yields

$$\alpha \eta_x + \left[\kappa_{123} \frac{\partial^2}{\partial x^2} + \kappa_1 \left(\frac{\partial}{\partial y} + \frac{2ie}{\hbar c} Hx \right)^2 \right] \eta_x$$
$$+ \left[\kappa_2 \frac{2ie}{\hbar c} H + \kappa_{23} \left(\frac{\partial^2}{\partial x \partial y} \right. \right.$$
$$\left. \left. + \frac{2ie}{\hbar c} Hx \frac{\partial}{\partial x} \right) \right] \eta_y = 0, \tag{73a}$$

$$-\alpha \eta_y + \left[\kappa_1 \frac{\partial^2}{\partial x^2} + \kappa_{123} \left(\frac{\partial}{\partial y} + \frac{2ieHx}{\hbar c} \right)^2 \right] \eta_y +$$
$$+ \left[\kappa_{23} \frac{2ie}{\hbar c} H + \kappa_{23} \left(\frac{\partial^2}{\partial x \partial y} + \frac{2ie}{\hbar c} Hx \frac{\partial}{\partial x} \right) \right] \eta_x = 0, \tag{73b}$$

where $\kappa_{123} \equiv \kappa_1 + \kappa_2 + \kappa_3$, etc.

We will not solve the coupled set of equations, which is a somewhat complex procedure, but rather refer the reader to Tokuyasu (1990) and Zhitomirskii (1989). The resulting E_1 phase diagram is shown in Fig. 6, of Tokuyasu there being two

3. Sound Propagation in the Heavy Fermion Superconductors

stable regions with solutions of the form $\eta \sim (\hat{x} + i\hat{y})e^{(-x/x_0)^2}$ (where $x_0^2 = \hbar c/eH$) or a linear combination of the forms $(\hat{x} \pm i\hat{y})e^{(-x/x_0)^2}$ (with unequal amplitudes). The first of these has the upper critical field

$$H_{c2} = \frac{\phi_0}{2\pi} \frac{|\alpha|}{\kappa_{13}}, \tag{74}$$

while the second is

$$H_{c2} = \frac{\phi_0}{2\pi} |\alpha| \bigg/ \left\{ \frac{3(2\kappa_1 + \kappa_{23})}{2} \pm \frac{1}{2}[(\kappa_{12} + 3\kappa_{13})^2 + 8\kappa_{23}^2]^{1/2} \right\} \tag{75}$$

(with the sign of the square root chosen so that the result is positive).

For H in the basal plane, a convenient gauge is $\mathbf{A} = yH\hat{z}$, and the resulting differential equations are

$$-|\alpha|\eta_x - \kappa_1 \frac{\partial^2 \eta_x}{\partial y^2} + \kappa_4 \left(\frac{2eHy}{\hbar c}\right)^2 \eta_x = 0, \tag{76a}$$

$$-|\alpha|\eta_y - \kappa_{123} \frac{\partial^2 \eta_y}{\partial y^2} + \kappa_4 \left(\frac{2eHy}{\hbar c}\right)^2 \eta_y = 0. \tag{76b}$$

Here η_y and η_x are uncoupled (and we have the usual harmonic oscillator form) so we may immediately write the solutions as

$$H_{c2} = \frac{\phi_0 |\alpha|}{2\pi \sqrt{\kappa_1 \kappa_4}} \quad (\eta_x \neq 0), \tag{77a}$$

or

$$H_{c2} = \frac{\phi_0 |\alpha|}{2\pi \sqrt{\kappa_{123} \kappa_4}} \quad (\eta_y \neq 0). \tag{77b}$$

The physical solution is the larger of (77a) and (77b), and hence the order parameter lies along either $\hat{x}$ or $\hat{y}$, depending on the values of κ_i.

The lower critical field depends on the energy of an isolated vortex line (de Gennes, 1989). This problem must be solved numerically due to the complex nature of E_1 vortices, and we refer the reader to Tokuyasu (1990) for details.

A very complex and somewhat controversial problem surrounds the nature of the phase transition identified as H_{FL} in the ultrasonic phase diagram. This designation implies that the transition is associated with a change in symmetry of the flux lattice, but it could equally well involve the vortex itself (or both).

Theoretically this problem is only accessible computationally, and it would be pointless for us to skim through the rather extensive calculations that have been performed; we again refer the reader to Tokuyasu (1990). (See Sec. 5.6.)

Weakly broken symmetry: An antiferromagnetic ordering with its $\hat{q}$ vector oriented in the basal plane, which has been reported in UPt$_3$, has the effect of lowering the symmetry from D_6 to D_2 (orthorhombic). As discussed earlier, this will split each E_1 and E_2 representation into two one-dimensional representations. Provided that this ordering is weak (as is suggested by the transition having a Neel temperature of about 5K), the degeneracy of the E representations is expected to be only slightly split, which can result in the presence of two closely spaced superconducting transitions. The double peak structure observed in the heat capacity and the ultrasonic velocity can be interpreted as supporting this idea. We will follow the discussion of Tokuyasu (1990).

A weak symmetry-breaking (WSB) of the above type may be modeled by including an additional second-order invariant in the G–L free energy of the form

$$F^{WSB} = \boldsymbol{\eta}^* \cdot \boldsymbol{\varepsilon} \cdot \boldsymbol{\eta}. \tag{78}$$

Here ε is a 2 x 2 matrix, referred to as the WSB field; in the orthorhombic principal axis system, ε may be written as a diagonal traceless matrix

$$\overset{\leftrightarrow}{\varepsilon} = \begin{pmatrix} \epsilon & 0 \\ 0 & -\epsilon \end{pmatrix} \tag{79}$$

(the contribution of the trace can be absorbed in the definition of the GL parameter α). (We may also think of the WSB effect as arising from a V_{02} term in the model potential $V(\hat{k})$, introduced earlier.) We write our order parameter in the form

$$\boldsymbol{\eta} = \begin{pmatrix} \eta_x e^{i\zeta} \\ \eta_y \end{pmatrix} e^{i\psi} \tag{80}$$

(where ζ is a relative phase angle between η_x and η_y and ψ is a common phase factor), and our Ginzburg–Landau free energy in the superconducting state becomes (for the $\mathbf{E}_1$ and $\mathbf{D}_6$ parent state)

$$F = F_0 + \alpha_- \eta_x^2 + \alpha_+ \eta_y^2 + \beta_1(\eta_x^2 + \eta_y^2)^2 \tag{81}$$
$$+ \beta_2(\eta_x^4 + 2\eta_x^2\eta_y^2 \cos 2\zeta + \eta_y^4),$$

where $\alpha_{\pm} = \alpha \mp \epsilon_0$. For $\beta_2 > 0$ the free energy is minimized by $\zeta = \pm \pi/2$ or $\eta_y = \pm i\eta_x$, corresponding to the state given in Eq. (76). The further min-

3. Sound Propagation in the Heavy Fermion Superconductors

imization of Eq. (81) depends on the temperature. Recalling that $\alpha \equiv a(T - T_{co})$, we see that there is an interval of temperatures where $\alpha_+ < 0$ and $\alpha_- \geq 0$ (assuming $\epsilon > 0$). On entering this interval from above, (80) is minimized by setting $\eta_x = 0$, and our order parameter becomes

$$\eta_y^2 = -\frac{\alpha_+}{2(\beta_1 + \beta_2)} \quad (T < T_{c+}), \tag{82}$$

$$\eta_x^2 = 0,$$

where we defined $T_c^{\pm} = T_{co} \pm \alpha/\epsilon_0$ (for $\epsilon < 0$ we replace T_c^+ by T_c^-, α_+ by α_-, and η_x by η_y in Eq. (82). The associated jump in the heat capacity,

$$\Delta C = -T\frac{\partial^2 F}{\partial T^2},$$

is given by

$$\Delta C_+ = \frac{T_{c+}\alpha_0^2}{2(\beta_1 + \beta_2)}. \tag{83}$$

As we continue to lower the temperature, a point T_c^* is reached where the free energy functions can be lowered by admitting a solution where the second component of the order parameter (η_x) becomes nonzero; this is a second-order transition that occurs at the point $\partial^2 F/\partial \eta_x^2 = 0$, which, due to the finite value of η_y and the fourth-order terms, does not occur at T_c^-. The resulting value for T_c^* is

$$T_c^* = T_{co}\left[1 - \frac{\beta_1}{\beta_2}\frac{\epsilon_0}{\alpha_0 T_{co}}\right], \tag{84}$$

and below this temperature the order parameter components are given by

$$\eta_x^2 = -\left[\frac{\alpha_+ + \alpha_-}{8\beta_1} - \frac{\alpha_+ - \alpha_-}{8\beta_2}\right] = -\frac{1}{4}\left(\frac{\alpha}{\beta_1} + \frac{\epsilon}{\beta_2}\right) \tag{85a}$$

and

$$\eta_y^2 = -\left[\frac{\alpha_+ + \alpha_-}{8\beta_1} - \frac{\alpha_+ - \alpha_-}{8\beta_2}\right] = -\frac{1}{4}\left(\frac{\alpha}{\beta_1} - \frac{\epsilon}{\beta_2}\right). \tag{85b}$$

The heat capacity jump at T_c^* is given by

$$\Delta C^* = T_c^* \frac{\alpha_0^2}{2\beta_1}, \tag{86}$$

and the ratio of the two discontinuities is

$$\frac{\Delta C^*}{\Delta C_+} = \frac{T_c^*}{T_c^+}\left(1 + \frac{\beta_2}{\beta_1}\right). \tag{87}$$

The WSB field is included in the linearized, field-dependent G–L equations given earlier by replacing α in (72a) and (72b) by α_- and α_+, respectively. For the basal plane, the final results along the symmetry directions are:

for $H\|x$:

$$H_{c2} = \frac{\phi_0}{2\pi}\left\{\frac{-\alpha_+}{\sqrt{\kappa_{123}\kappa_4}}\right\} \quad (\boldsymbol{\eta}\|\hat{y}), \tag{88a}$$

$$H_{c2} = \frac{\phi_0}{2\pi}\left\{\frac{-\alpha_-}{\sqrt{\kappa_1\kappa_4}}\right\} \quad (\boldsymbol{\eta}\|\hat{x}), \tag{88b}$$

and for $H\|y$:

$$H_{c2} = \frac{\phi_0}{2\pi}\left\{\frac{-\alpha_+}{\sqrt{\kappa_1\kappa_4}}\right\} \quad (\boldsymbol{\eta}\|\hat{y}), \tag{89a}$$

$$H_{c2} = \frac{\phi_0}{2\pi}\left\{\frac{-\alpha_-}{\sqrt{\kappa_{123}\kappa_4}}\right\} \quad (\boldsymbol{\eta}\|\hat{x}), \tag{89b}$$

The second result for these two field directions applies only when $T < T_c^-$; the physical H_{c2} is the greater of the two solutions (for temperatures where both could apply).

The existence of two possible solutions for H in the basal plane can lead to a transition between them at some critical field that is obtained by equating the two solutions, which yields

$$T_H = \frac{T_c^-\sqrt{\kappa_{123}} - T_c^+\sqrt{\kappa_1}}{\sqrt{\kappa_{123}} - \sqrt{\kappa_1}}. \tag{90}$$

The WSB field also affects H_{c1}, and a kink in the H_{c1} behavior is predicted at T_c^* for all field directions, with the slopes being related by

$$\frac{\left.\dfrac{dH_{c1}}{dT}\right|_{T<T^*}}{\left.\dfrac{dH_{c1}}{dT}\right|_{T>T^*}} \cong \frac{1}{4}\left(1 + \frac{\beta_2}{\beta_1}\right) \times \left\{\frac{(\kappa_1 + \kappa_{123})}{\kappa_1\kappa_{123}}\right\} \quad H\|\hat{c}; \tag{91a}$$

3. Sound Propagation in the Heavy Fermion Superconductors 137

$$\frac{\left.\frac{dH_{c1}}{dT}\right|_{T<T^*}}{\left.\frac{dH_{c1}}{dT}\right|_{T>T^*}} \cong \frac{1}{4}\left(1 + \frac{\beta_2}{\beta_1}\right) \times \left(3 + \frac{\kappa_{123}}{\kappa_1}\right) \qquad H\|\hat{y}; \qquad (91b)$$

$$\frac{\left.\frac{dH_{c1}}{dT}\right|_{T<T^*}}{\left.\frac{dH_{c1}}{dT}\right|_{T>T^*}} \cong \frac{1}{4}\left(1 + \frac{\beta_2}{\beta_1}\right) \times \left(3 + \frac{\kappa_1}{\kappa_{123}}\right) \qquad H\|\hat{x}. \qquad (91c)$$

Measurements of the lower critical field have been performed by Zhao *et al.* (1991a) and by Vincent *et al.* (1991) (See Section 5.5.).

3.3. COLLECTIVE MODES

A collective mode involves a periodic distortion of the order parameter from its equilibrium value; i.e.,

$$\boldsymbol{\eta} \to \boldsymbol{\eta} + \boldsymbol{\eta}', \qquad (92)$$

where $\boldsymbol{\eta}'$ measures the distortion. The collective mode spectrum for an unconventional superconductor of the E_1 or E_2 type has not yet been discussed. If the case of superfluid ^{3}HeB provides any insight, the problem will be rather involved. However, a phenomenological approach, where the Ginzburg–Landau free energy is used as the "potential" in a Lagrangian density, has been successful in semiquantitatively explaining most of the collective mode phenomena in ^{3}HeB (see Zhao *et al.*, 1991b); to complete the description one must add a "kinetic energy" density, $\mathcal{T}$, to the Lagrangian density; in the present case we adopt the form

$$\mathcal{T} = \Lambda \, \dot{\boldsymbol{\eta}}' \, \dot{\boldsymbol{\eta}}'^* \qquad (93)$$

To obtain the quadratic terms in the potential $\mathcal{V}^{(2)}$, we expand the free energy about the equilibrium value $\boldsymbol{\eta}$ to second order in $\boldsymbol{\eta}'$, with the result

$$\mathcal{V}^{(2)} = \alpha \boldsymbol{\eta}' \cdot \boldsymbol{\eta}'^* + \beta_1 [2(\boldsymbol{\eta} \cdot \boldsymbol{\eta}^*)(\boldsymbol{\eta}' \cdot \boldsymbol{\eta}'^*) + (\boldsymbol{\eta} \cdot \boldsymbol{\eta}'^* + \boldsymbol{\eta}^* \cdot \boldsymbol{\eta}')^2]$$
$$+ \beta_2 [(\boldsymbol{\eta} \cdot \boldsymbol{\eta})(\boldsymbol{\eta}' \cdot \boldsymbol{\eta}')^* + (\boldsymbol{\eta}^* \cdot \boldsymbol{\eta}^*)(\boldsymbol{\eta}' \cdot \boldsymbol{\eta}') \qquad (94)$$
$$+ 4(\boldsymbol{\eta} \cdot \boldsymbol{\eta}')(\boldsymbol{\eta}^* \cdot \boldsymbol{\eta}'^*)].$$

The equations of motion for the collective modes follow from Lagrange's equation

$$\frac{\partial}{\partial t}\frac{\partial \mathscr{L}}{\partial \boldsymbol{\eta}'^*} - \frac{\partial \mathscr{L}}{\partial \boldsymbol{\eta}'^*} = 0, \tag{95}$$

where

$$\mathscr{L} = \mathscr{T} - \mathscr{V}. \tag{96}$$

The resulting equation is

$$\Lambda\ddot{\boldsymbol{\eta}}' + \alpha\,\boldsymbol{\eta}' + \beta_1\{2(\boldsymbol{\eta}\cdot\boldsymbol{\eta}^*)\boldsymbol{\eta}'$$
$$+ 2\boldsymbol{\eta}[(\boldsymbol{\eta}\cdot\boldsymbol{\eta}'^*) + (\boldsymbol{\eta}^*\cdot\boldsymbol{\eta}')]\} + \beta_2\{2(\boldsymbol{\eta}\cdot\boldsymbol{\eta})\,\boldsymbol{\eta}'^* \tag{97}$$
$$+ 4(\boldsymbol{\eta}\cdot\boldsymbol{\eta}')\,\boldsymbol{\eta}^*\} = 0.$$

To proceed further we must incorporate one of the equilibrium forms discussed earlier (in general, the collective mode frequencies are a function of the form and amplitude of the equilibrium order parameter).

For the case of an E_1 superconductor with $\beta_2 < 0$, we may write the equilibrium order parameter as $\boldsymbol{\eta} = \eta_0 \hat{x}$ (case η_0 real and $\phi_0 = 0$). The distortions of the order parameter $\boldsymbol{\eta}'$ can be written as $\boldsymbol{\eta}' = \eta'_x\,\hat{x} + \eta'_y\,\hat{y}$ (here η'_x and η'_y may be complex), and our equations of motion are

$$\Lambda\ddot{\eta}'_x + \alpha\eta'_x + 2\beta_1\eta_0^2[2\eta'_x + \eta'_x{}^*] + 2\beta_2\eta_0^2[\eta'_x{}^* + 2\eta'_x] = 0 \tag{98a}$$

and

$$\Lambda\ddot{\eta}'_y + \alpha\eta'_y + 2\beta_1\eta_0^2\eta'_y + 2\beta_2\eta_0^2\eta'_y{}^* = 0. \tag{98b}$$

Note we actually have four equations of motion involving η'_x, $\eta'_x{}^*$, η'_y and $\eta'_y{}^*$; hence, there will be four independent collective modes. We define the combinations

$$\boldsymbol{\eta}^R = \frac{1}{2}\,(\boldsymbol{\eta}' + \boldsymbol{\eta}'^*) \tag{99a}$$

and

$$\boldsymbol{\eta}^I = -\frac{i}{2}\,(\boldsymbol{\eta}' - \boldsymbol{\eta}'^*). \tag{99b}$$

Taking the complex conjugate of Eqs. (98a) and (98b) and adding and subtracting, we obtain

$$\Lambda\ddot{\eta}^R_x + \alpha\eta^R_x + 6\beta_1\eta_0^2\eta^R_x + 6\beta_2\eta_0^2\eta^R_x = 0, \tag{100a}$$

$$\Lambda\ddot{\eta}^I_x + \alpha\eta^I_x + 2\beta_1\eta_0^2\eta^I_x + 2\beta_2\eta_0^2\eta^I_x = 0, \tag{100b}$$

$$\Lambda\ddot{\eta}^R_y + \alpha\eta^R_y + 2\beta_1\eta_0^2\eta^R_y + 2\beta_2\eta_0^2\eta^R_y = 0, \tag{100c}$$

3. Sound Propagation in the Heavy Fermion Superconductors

and

$$\Lambda \ddot{\eta}_y^I + \alpha \eta_y^I + 2\beta_1 \eta_0^2 \eta_y^I + 2\beta_2 \eta_0^2 \eta_y^I = 0. \tag{100d}$$

Assuming solutions of the form $\eta \sim e^{-i\omega t}$ we obtain the four frequencies

$$\omega_{xR}^2 = -\frac{2\alpha}{\Lambda}, \tag{101a}$$

$$\omega_{xI}^2 = 0, \tag{101b}$$

$$\omega_{yR}^2 = 0, \tag{101c}$$

$$\omega_{yI}^2 = \frac{2\alpha}{\Lambda} \frac{\beta_2}{\beta_1 + \beta_2}. \tag{101d}$$

Given our assumption $\beta_2 < 0$ and noting $\alpha < 0$ for $T < T_c$, we verify that (101a) and (101d) have real frequencies. On the basis of our experience with ^{3}He, we suggest that (101a) corresponds to a "gap mode" when $\hbar\omega = 2\Delta(T)$, which means we identify Δ^2 as $-\alpha\hbar^2/2\Lambda$. For small $|\beta_2|$, (101d) lies within the gap. Forms (101b) and (101c), which are called Goldstone modes, would have finite frequencies in the presence of dispersion (caused by the gradient terms in the Ginzburg–Landau expansion). The Goldstone mode (101b) associated with Eq. (100b) arises from fluctuations of the phase of the order parameter (since it involves the imaginary component and we chose our equilibrium order parameter to be real);

$$e^{i\phi(r,t)} = 1 + i\phi(\mathbf{r}, t) \quad \text{for small } \phi(\mathbf{r},t).$$

It is known that for conventional superconductors, the frequency of this mode will be lifted to the plasma frequency; this follows from the fact that time- and position-dependent phase fluctuations will have an associated velocity

$$\mathbf{v} = \frac{\hbar}{m} \nabla \phi$$

(the Josephson equation), which, through the equation of continuity, will produce charge density fluctuations, which oscillate at the plasma frequency in a metal.

The Goldstone mode (101c) is associated with a (soft) oscillation of the direction of the order parameter about its equilibrium direction $\hat{x}$, in the x–y plane. Eqs. (100d) and (101d) directly involve an infinitesimal rotation by imaginary angle, which changes the magnitude of the gap and hence has a finite frequency.

We now consider the E_1 superconductor with $\beta_2 > 0$. Equation (97) takes the form (assuming η_0 to be real)

$$\Lambda\ddot{\eta}'_x + \alpha\eta'_x + \beta_1\eta_0^2[3\eta'_x + \eta'_x{}^* - i(\eta'_y - \eta'_y{}^*)]$$
$$+ 2\beta_2\eta_0^2(\eta'_x + i\eta'_y) = 0, \quad (102a)$$

$$\Lambda\ddot{\eta}'_y + \alpha\eta'_y + \beta_1\eta_0^2[3\eta'_y - \eta'_y{}^* + i(\eta'_x + \eta'_x{}^*)]$$
$$+ 2\beta_2\eta_0^2(\eta'_y - i\eta'_x) = 0. \quad (102b)$$

We introduce the variables

$$\eta'_\pm = \eta'_x \pm i\eta'_y, \quad (103)$$

and Eqs. (98a) and (98b) become

$$\Lambda\ddot{\eta}'_+ + \alpha\eta'_+ + 2\eta_0^2(\beta_1 + 2\beta_2)\eta'_+ = 0, \quad (104a)$$

$$\Lambda\ddot{\eta}'_- + \alpha\eta'_- + 2\eta_0^2\beta_1(2\eta'_- + \eta'_-{}^*) = 0. \quad (104b)$$

Equation (104a) has the solution

$$\omega^2 = -\frac{2\alpha}{\Lambda}\frac{\beta_2}{\beta_1} \quad (105a)$$

(since $\eta'_\pm$ has the same equation of motion this mode is doubly degenerate). Solving the coupled set of equations involving (104b) and its complex conjugate, we obtain

$$\omega^2 = \begin{cases} \dfrac{-2\alpha}{\Lambda} & (105b) \\ \\ 0 & (105c) \end{cases}$$

We again identify (105b) as the gap mode. Equation (105c) is the Goldstone mode corresponding to the phase mode of an ordinary superconductor, and it will be lifted to the plasmon frequency. Since $\alpha < 0$ and $\beta_2 > 0$, Eq. (105a) has a real solution for the frequency ω.

Using perturbation theory, it is relatively straightforward to evaluate the dispersion of the collective modes from the expressions for the gradient energy given earlier.

4. Heavy Fermion Systems

4.1. NORMAL STATE PROPERTIES

As stated earlier, the heavy fermion systems are intermetallic compounds consisting of one f-electron metal and s-, p-, or d-electron metals. At room tem-

3. Sound Propagation in the Heavy Fermion Superconductors

peratures, the f-electrons behave as localized spins: the conduction electrons are the s, p or d electrons and have quite ordinary effective masses. As the material is cooled, the f-electrons begin to couple coherently to the conduction electrons, resulting in very large effective masses for the resulting hybridized carriers. These compounds have several characteristics that distinguish them from ordinary metals. The electronic heat capacities are 10^2–10^3 times larger than that observed in most metals; the Pauli susceptibility at low temperatures is also ~ 100 times larger. Both of these indicate a very large effective mass for the conduction electrons.

The resistivity and the sound attenuation both continue to change rapidly with temperature down to very low temperatures (<10 K), rather than staying constant as in most metals. At low temperatures, the resistivity is of the form $\rho = \rho_0 + AT^2$, where A is large, indicating that electron–electron scattering is important.

Figure 6 shows the resistivity of four heavy fermion compounds (normalized to their room temperature value). Most heavy fermion materials (except UPt$_3$) show a peak in the resistivity, similar to the peak seen in Kondo systems.

The susceptibility is shown in Fig. 7. At high temperatures, the susceptibility follows a Curie–Weiss Law, as would be expected for a collection of local moments (i.e., localized f-electrons). At low temperatures, the susceptibility is large (and very weakly temperature-dependent), indicating strongly interacting conduction electrons.

In addition to a large electronic specific heat (common to all heavy fermion

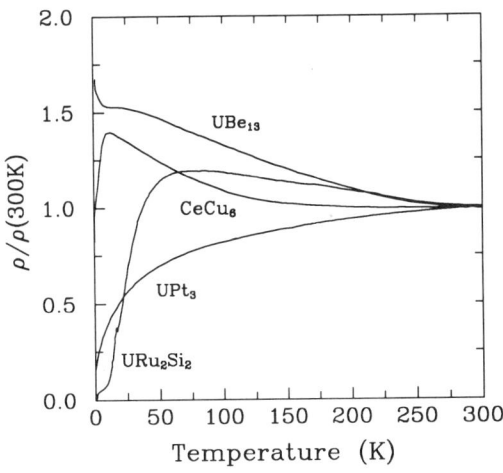

FIG. 6. Resistivity of four heavy fermion compounds. The data have been normalized to the respective room-temperature values. Adapted from Fisk et al. (1988) and Maple et al. (1986). (Copyright 1988 by the AAAS.)

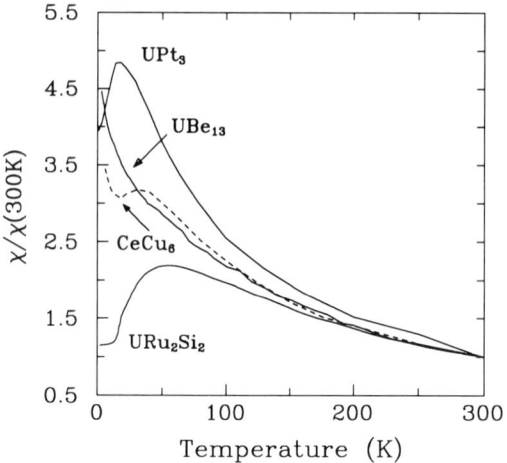

FIG. 7. Susceptibility curves for CeCu$_6$ along the a-axis (after Vrtis, 1987), UPt$_3$ in the basal plane (after Frings et al., 1983), UBe$_{13}$ (after Ott et al., 1983) and URu$_2$Si$_2$ along the c-axis (after Palstra et al., 1985). The data are normalized to their room-temperature values.

materials), UPt$_3$ has a $T^3 \log T$ term in the specific heat. A $T^3 \log T$ term in the specific heat has been observed in normal liquid ^{3}He and has been attributed to spin fluctuations. In general, a $T^3 \log T$ term is a property of an interacting Fermi liquid and can come from both spin and density fluctuations. This term in the specific heat, together with the T^2 term in the resistivity, has led to attempts to use Fermi liquid theory to account for the low-temperature properties of the

TABLE III

PROPERTIES OF SOME HEAVY FERMION COMPOUNDS.

Samples	UPt$_3$	URu$_2$Si$_2$	UBe$_{13}$	CeCu$_6$
structure	hcp	b.c. tetragonal	cubic	orthorhombic/ monoclinic ($T < 220$ K)
$a/b/c$ (Å)	5.76/4.90	4.12/9.5	10.25	5.10/8.11/10.16
ρ (10^3 kg/m^3)	19.4	10.1	4.38	8.24
ρ_{el} (300 K) ($\mu\Omega$ cm)	130	324/169	107	70/70/70
μ_{eff} (μ_B)	3.0	3.51	3.1	2.6/2.67/2.46
$\gamma(0)$ (mJ mol^{-1} K^{-2})	420	65.5	1,100	1,600
m^*/m_0	180	25	300	1,100
v_F (10^3 m/s)	6.60	47.7	3.34	2.12
T_F (K)	372	1,878	163	76.9 (?)
μ_{AFM} (μ_B)	0.02	0.03	—	—
T_N (K)	5	17.5	8.8	—
T_c (K)	0.54	1.4	0.85	—

3. Sound Propagation in the Heavy Fermion Superconductors

heavy fermion systems. However, the anistropy of these systems (due to the crystal lattice) makes this complicated.

Of the four heavy fermion compounds we will discuss, three of them (UBe_{13}, URu_2Si_2, and UPt_3) show a weak antiferromagnetic ordering followed by a subsequent superconducting transition. The Neel temperature is about 10 times higher than the superconducting transition temperature. Some of the properties of these four heavy fermion compounds are given in Table III.

4.2. Superconductivity

The heat capacity jump at the transition and the large critical field slopes dH_{c2}/dT at T_c in heavy fermion superconductors indicate that it is the heavy electrons that are responsible for the superconductivity. The properties in the superconducting state are unusual. The heat capacity and ultrasonic attenuation follow power-law temperature dependences, indicating the presence of line or point nodes in the gap. In addition, UPt_3 shows a variety of properties that make it a very strong candidate for unconventional superconductivity. We will discuss these in Section 5.

The upper critical fields of the three uranium-based heavy fermion superconductors are shown in Fig. 8. UBe_{13} is a cubic crystal with an $H_{c2}(0)$ of 8 T, and a slope of $dHc/dT = 40$ T/K at T_c. The upper critical field for UPt_3 (Shivaram et al., 1986a), a hexagonal crystal, shows a number of interesting features. The two H_{c2} curves (for $H\|c$ and $H\perp c$) *cross* at a temperature of 200 mK, implying that the anistropy of the upper critical field is not just a result of the symmetry of the Fermi surface. Recent theoretical work by Choi and Sauls (1991) shows that this behavior may be explained by odd-parity pairing with strong spin–orbit coupling (the U representations discussed in Section 3.1). Later measurements revealed a kink in the H_{c2} curve for fields in the basal plane. This kink is discussed in Section 3.

The upper critical field of URu_2Si_2 (a tetragonal crystal) is highly anisotropic, having critical fields of 2 T for $H \| c$ and 8 T for $H \| a$. There is no crossover in H_{c2} as in the case of UPt_3. The data of Kwok et al. (1990) are shown in Fig. 8c.

5. UPt_3

5.1. General

UPt_3 is a hexagonal crystal with two U atoms (six f-electrons) per unit cell. The lattice parameters are $a = 5.764$ Å and $c = 4.899$ Å, giving a density of 19.40 g/cm^3; the ratio $c/a = 0.850$ is very close to the ideal hcp value of 0.816. It

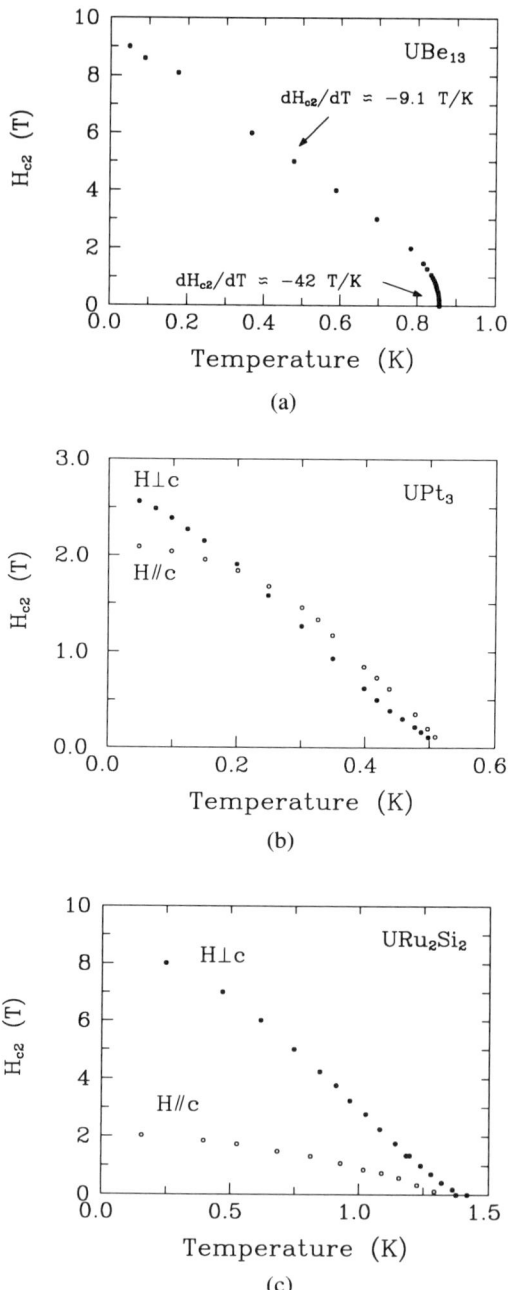

FIG. 8. H_{c2} curves of (a) UBe_{13} (after Maple *et al.*, 1985), (b) UPt_3 (after Shivaram *et al.*, 1986b), and (c) URu_2Si_2 (after Kwok *et al.*, 1990).

3. Sound Propagation in the Heavy Fermion Superconductors

has a superconducting transition temperature ranging from 0.46 to 0.56 K, depending on sample preparation technique. Experimentally, this system appears to be the most promising heavy-fermion candidate for unconventional superconductivity. High-quality single crystals have been made by float zone refining; however, there is still some ambiguity among the various groups as to what constitutes the best UPt$_3$ sample.

Neutron diffraction and muon spin resonance experiments (Aeppli et al., 1987, 1988, 1989) show an antiferromagnetic ordering at 5 K with a very weak ordered moment of 0.02 μ_B. The ordered moments lie in the basal plane. The antiferromagnetic ordering is thought to play an important role in the subsequent superconducting ordering.

The susceptibility of UPt$_3$ for temperatures in the range 4.2–300 K shows a large anisotropy between the hexagonal c-axis and the basal plane; $\chi_{a,b}$ is more than twice as large as χ_c at low temperatures. The basal plane susceptibility shows a maximum at around 17 K, which was earlier thought to be an antiferromagnetic transition, but is now believed to be due to spin fluctuations. A Curie–Weiss fit to the data in the temperature range 50–250 K yields a μ_{eff} of 2.6 ± 0.2 μ_B/U-atom for both field directions.

The electronic heat capacity coefficient is 422 mJ/K^2-mol, showing the heavy-fermion like–behavior. From the size of the unit cell (assuming a spherical Fermi surface) one obtains for the Fermi wave number $k_F = 1.08$ Å^{-1}, a renormalized mass $m/*m_e = 180$, a Fermi temperature $T_F \sim 275$ K, and a $v_F = 6,800$ m/s. (In UPt$_3$ the maximum sound velocity is in the basal plane, and has a value of ~4000 m/s. At low temperatures, there is an additional $T^3 \log T$ term in the heat capacity (which is similar to that in ^{3}He) arising from the strong electron–electron correlations; hence the heat capacity has the form

$$C = \gamma^* T + \beta_{ph} T^3 + \delta T^3 \ln\left(\frac{T}{T^*}\right). \tag{106}$$

Here T^* is a characteristic temperature, often called the spin fluctuation temperature, and is estimated to be 27 K.

In the superconducting state of UPt$_3$ the attenuation does not drop exponentially as in a conventional BCS-like superconductor (for example, niobium). The attenuation follows a power law behavior, T^n, where n may be 1, 2, or 3, depending on orientation, polarization, and sample. These temperature dependencies of the ultrasonic attenuation (Bishop et al., 1984, 1986; Qian et al., 1987; Müller et al., 1987b; Shivaram et al., 1986a) and heat capacity (Sulpice et al., 1986) in the superconducting state have been interpreted as evidence for a gap with line or point nodes, and hence to speculations of unconventional superconductivity. More convincing evidence for unconventional superconductivity came from a variety of experiments that revealed the existence of two or more superconducting phases. The earliest evidence for multiple superconducting phases came from ultrasonic attenuation measurements (Müller et al., 1986b; Qian et al., 1987; Schenstrom et al., 1989a) in the superconducting state in a magnetic field. Later experiments included torsional oscillator (Kleiman et al., 1989) and heat capacity (Fisher et al., 1989; Hasselbach et al., 1989) measurements. The most recent and complete experiments on the phase diagram of UPt$_3$ are from ultrasonic velocity measurements (Adenwalla et al., 1990; Bruls et al., 1990).

5.2. Sound in the Normal State

We will start by reviewing the ultrasonic experiments in the normal state. Velocity measurements have been performed by Yoshizawa et al. (1985) and by de Visser et al. (1985, 1987) from 600 K to 1 K. The velocities increase with decreasing temperature, down to ~50 K, for all polarizations and directions studied (involving c_{11}, c_{22}, c_{44} and c_{66}). Below 50 K there are anomalies in the velocity, with the position and size of the anomaly depending on the elastic constant involved. The elastic constants can be obtained from the velocity and the density ($\rho = 19.4$ g/cm^3). UPt$_3$ is a hexagonal crystal, and five independent measurements are required to obtain the five independent elastic constants. Usually only four of these are measured, obtained from the various velocity measurements along the principal axes. To obtain the elastic constant c_{13}, de Visser et al. (1985) propagated sound with **q** at 45° to the c-axis in the b–c plane, generating a quasi-shear mode. The elastic constants and the velocities for UPt$_3$ are listed in Table IV. The thermal expansion of UPt$_3$ is strongly anisotropic, with a maximum in the anisotropy at 14 K (de Visser, 1987), close to the temperature at which the maximum in the susceptibility is found. This is explained in terms of a correlation between the magnetic and thermal properties, with the onset of spin fluctuations confined to the basal plane. There is a relatively large dilation in the basal plane, accompanied by a contraction along the hexagonal axis.

The attenuation measurements in the normal state (Bishop et al., 1984) reveal a T^2 dependence of the attenuation down to the lowest temperatures, consistent with electron–electron scattering. The resistivity shows a similar behavior above the superconducting transition. No signature has been observed at the Neel transition, $T_N \sim 5$ K, in either the attenuation or the velocity, unlike in URu$_2$Si$_2$, even though the antiferromagnetically ordered moments are comparable (0.02 μ_B for UPt$_3$; 0.03 μ_B for URu$_2$Si$_2$).

Müller et al. (1986a) have studied the temperature dependence of the attenuation of sound in UPt$_3$; Fig. 9 shows their data for longitudinal and transverse waves. The data display a number of unusual features. Transverse waves are

3. Sound Propagation in the Heavy Fermion Superconductors

TABLE IV

SOUND VELOCITIES AND ELASTIC CONSTANTS IN UPt$_3$ (deVisser, 1987)

Type mode	Direction of propagation	Direction of particle motion	Velocity (m/s)	Elastic constant (Mbar)	
long.	c	c	$v_1 = 3860$	$c_{33} = \rho v_1^2 = 2.891$	
shear	c	in ab–plane	$v_2 = 1385$	$c_{44} = \rho v_2^2 = 0.372$	
long.	b	b	$v_3 = 3993$	$c_{11} = \rho v_3^2 = 3.093$	
shear	b	a	$v_4 = 2076$	$c_{66} = \frac{1}{2}(c_{11}-c_{12}) = \rho v_4^2 = 0.836$	
shear	b	c	$v_5 = 1388$	$c_{44} = \rho v_5^2 = 0.374$	
quasilong.	b'	b'	$v_6 = 3754$	$c_{13} = 1.732$	
quasishear	b'	c'	$v_7 = 1827$	$c_{13} = 1.695$	
shear	b'	$a' = a$	$v_8 = 1753$	$\frac{1}{4}c_{11} = \frac{1}{4}c_{12} + \frac{1}{2}c_{44} = \rho v_8^2$	

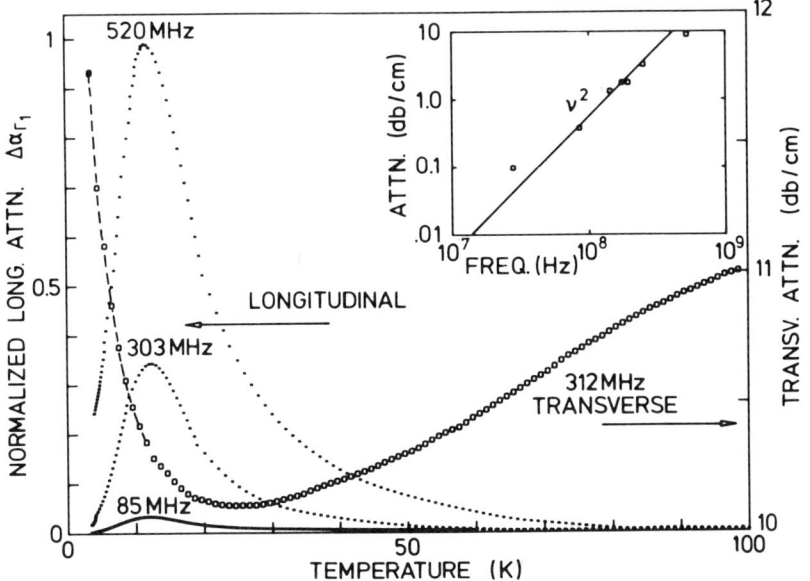

FIG. 9. Temperature dependence of the longitudinal ($q \| b$, $\varepsilon \| b$) electronic attenuation coefficient in normal UPt$_3$ at various frequencies, showing the 12 K Kondo peak. The peak is absent for transverse sound. The inset shows the ν^2 frequency dependence of the height of the peak. After Müller et al. (1986a).

shown only for the case $\hat{q}\|b$, $\hat{u} \| c$; note that at higher temperatures the attenuation falls with decreasing temperature, reaching a broad minimum in the vicinity of 25 K, after which the attenuation rises. For longitudinal sound with $\hat{q}$, $\hat{u}$, $\|c$ the attenuation at higher temperatures also falls with decreasing temperature, with the accompanying minimum near 25 K (the preceding sets of data for transverse and longitudinal waves both involve the polarization $\hat{u}\|c$). However, below the minimum the longitudinal sound passes through a relatively sharp maximum at approximately 12 K. For longitudinal sound with $\hat{q}$, $\hat{u}$, $\|b$ the attenuation initially rises with falling temperatures and passes through a similar maximum (displaced about 2 K higher), then falls to a minimum near 5 K, after which it rises.

The unusual attenuation peaks in the longitudinal sound have resulted in considerable theoretical discussion (Schotte et al., 1986; Müller et al., 1987b; Fulde et al., 1988; Yoshizawa et al., 1986). On encountering such a peak structure, the first explanation that comes to mind is that it arises from the relaxation of some internal parameter, which typically results in the attenuation displaying a behavior $\alpha \sim \omega^2\tau/(1 + \omega^2\tau^2)$ (see Eq. (107) below). Assuming τ monotonically increases with decreasing temperature, one would pass from a regime where $\alpha \sim \omega^2 \tau$ (high temperature) to another where $\alpha \sim \tau^{-1}$ (low temperature). The position of the peak occurs at $\omega\tau = 1$ and will occur at different temperatures (τ values) for different frequencies. The strength of the peak scales as ω (independent of τ). However, Müller et al. report that the peak position is essentially independent of frequency and that the attenuation at the peak scales as ω^2; hence, the peak cannot be ascribed to a conventional relaxation phenomena.

Müller et al. (1987b, 1987c) and Schotte et al. (1986) have based their analysis on a modification of the following expression for the attenuation of longitudinal sound, α_ℓ, given earlier by Akhiezer et al. (1957):

$$\alpha_\ell = \frac{\omega^2\tau}{1 + \omega^2\tau^2} \frac{g^2 N(\varepsilon_F)}{2\rho V_s^3}; \qquad (107)$$

here g^2 is a deformation-potential coupling constant and $N(\varepsilon_F)$ is the electronic density of states. In a conventional metal, $N(\varepsilon_F)$ may be taken as a temperature-independent constant. However, heavy fermion metals are characterized by a sharp peak (or resonance) in the density of states arising from a (coherent) hybridization of the narrow f levels with s–p–d electrons; the density of states arising from this effect, $N_f(\varepsilon)$, is typically written

$$N_f(\varepsilon) = n_f \frac{\Delta}{\Delta^2 + (\varepsilon - \varepsilon_0)^2}, \qquad (108)$$

3. Sound Propagation in the Heavy Fermion Superconductors

where Δ is the width of the "resonance," which is centered on the energy ε_0, and n_f is the density of f electrons. Because of this added sharp energy structure, finite temperature corrections to the density of states become important.

If we examine the derivation of Akhiezer *et al.* (1957) leading to Eq. (107), we find that the step preceeding the final result involves the integral (Schotte *et al.*, 1986)

$$\alpha = \frac{1}{2\rho V_s^3} \int (g_p - \bar{g}_p)^2 \frac{\omega^2 \tau}{1 + \omega^2 \tau} \frac{\partial f_0}{\partial \varepsilon} \frac{d^3 p}{(2\pi\hbar)^3}, \qquad (109)$$

where g_p is the momentum (and energy)-dependent deformation potential (with a global average $\bar{g}_p$) and f_0 is the Fermi function. If we approximate $\partial f_0/\partial \varepsilon$ by $-\delta(\varepsilon - \varepsilon_F)$ ($T/T_F \ll 1$), we immediately obtain Eq. (107), where g^2 is identified as the Fermi surface average of the quantity $(g_p - \bar{g}_p)^2$. If we write

$$\alpha = \alpha^{(1)} + \alpha^{(2)} \qquad (110)$$

and calculate the leading temperature-dependent correction (Kittel, 1971) to Eq. (109), we find $\alpha^{(2)}$ is proportional to T^2 and contains several terms involving energy derivatives of τ, g^2, and $N(\varepsilon)$, all of which may be large in a heavy Fermion metal, leading to a significant correction. However, the structure "smears out" on a temperature scale of order $T^* = \Delta/k_B$ (see Eq. (108) and the added contribution, which starts as T^2, would then fall off, resulting in a peak in the attenuation. Müller *et al.* (1986a) and Schotte *et al.* (1986) have proposed model expressions that qualitatively account for the observed phenomena.

Fulde *et al.* (1988) start from a general hydrodynamic description similar to that used by Martin *et al.* (1972). The electron "fluid" is introduced as a separate degree of freedom with accompanying equations of motion. Fulde *et al.* (1988) conclude that a heavy-fermion metal (or indeed, any metal possessing carriers with widely differing effective masses) will have four hydrodynamic modes: two (time-reversed) sound waves with $\omega = \pm V_s q$, an energy diffusion (heat conduction) mode with $\omega = i D_\varepsilon q^2$, and an electron (or quasiparticle) diffusion mode with $\omega = D_{q.p.} q^2$. The width of the energy-diffusion mode is narrow relative to the separation of the two sound modes, and hence the sound propagates isentropically (the same conclusion is reached by Lüthi (1985)). Fulde *et al.* conclude that the quasiparticle mode is also narrow relative to the sound mode spacing (they refer to this as the isolated regime). They obtain three separate contributions to the attenuation:

$$\alpha = \alpha_R + \alpha_\varepsilon + \alpha_{q.p.}, \qquad (111)$$

where α_R is a (direct) internal friction contribution and α_ε and $\alpha_{q.p.}$ arise from heat conduction and quasiparticle diffusion. The quasiparticle diffusion term is given by

$$\alpha_{q \cdot p} = \frac{\omega^2}{2V_s^3 \rho} NV \left(\frac{\partial \mu}{\partial V}\right)^2_{T,N_e} \frac{\sigma^2}{e^2}, \qquad (112)$$

where σ is the electrical conductivity and N is the number of ions. This expression involves the derivative of the electron chemical potential with respect to volume (at constant temperature and electron number, N_e), $(\partial \mu/\partial V)_{T,N_e}$; this quantity is not available experimentally, and Fulde et al. (1988) invoke a model calculation. The T^2 contribution to their expression is argued to be large. The theory, as cast, is then rather similar to that of Müller et al. (1987b) and Schotte et al. (1986).

The hydrodynamic theory also yields an expression for the ultrasonic velocity,

$$v^2 = v_s^2 + \frac{V}{\rho} \left(\frac{\partial \mu}{\partial V}\right)_{N_e,T} \left(\frac{\partial \langle N_e \rangle}{\partial \mu}\right)_{V,T}, \qquad (113)$$

where v_s is the adiabatic sound velocity.

Kouroundis et al. (1987) have observed a very interesting softening of the C_{11} and C_{33} elastic constants in UPt$_3$ and CeRu$_2$Si$_2$. The effect is confined to a rather narrow range of magnetic fields. The high field softening is greatly enhanced by going to lower temperatures. The data were discussed in terms of a large magnetic Gruneisen constant, Ω_B, given by

$$\Delta V = -\Omega_B^2 B^2 \chi_M, \qquad (114)$$

where χ_M is the magnetic susceptibility.

5.3. Sound in the Superconducting State

5.3.1. Attenuation in Zero Magnetic Field

The shape of the attenuation curve in the superconducting state is very different from that in a conventional superconductor. Instead of a sharp exponential drop, because of the development of the energy gap, measurements in the superconducting state reveal power-law temperature dependencies for the attenuation. Early longitudinal sound experiments (Bishop et al., 1984; Fig. 10) revealed a T^2 dependence at low temperatures (for propagation along the c-axis), which can be explained by a gap with point nodes (Rodriguez, 1985). Inferring the exact nature of the nodal structure of the gap at the Fermi surface is somewhat difficult, since the measurements have to be performed at the lowest temperatures. This is difficult for the uranium compounds, because of the small heating caused

3. Sound Propagation in the Heavy Fermion Superconductors

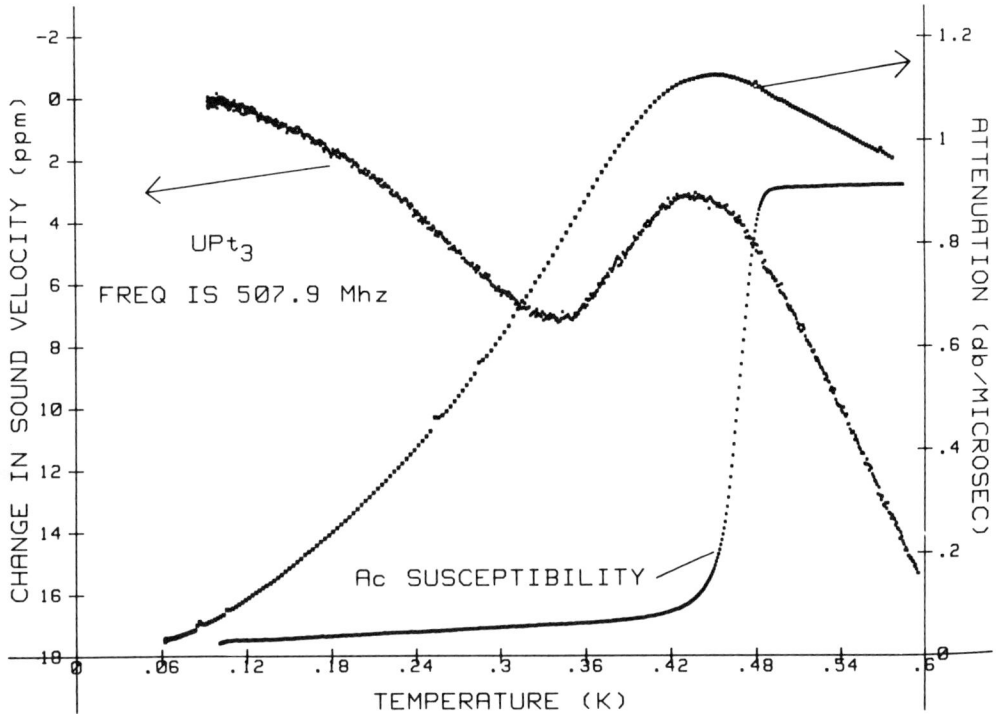

FIG. 10. Longitudinal ultrasonic attenuation, sound velocity, and ac susceptibility for UPt$_3$ at the superconducting transition in UPt$_3$. After Bishop *et al.* (1984).

by the radioactivity of the samples and the associated problems with thermal contact. Thus, it is not surprising that various groups have noted both a T^2 and a T^3 dependence for the same orientation, but on different samples. In fact, Müller *et al.* (1986b) obtained a T^3 dependence on samples obtained from the same source as those of Bishop *et al.* (1984).

Transverse sound attenuation measurements for propagation along the b-axis show different temperature dependencies depending on the direction of polarization (Shivaram *et al.*, 1986a). For polarization in the basal plane (i.e., along the a-axis), the attenuation is linear with temperature; for polarization along the c-axis, the attenuation follows a T^2 behavior (see Fig. 11). These temperature dependencies have been predicted for a gap with line nodes (Hirschfeld *et al.*, 1986; Schmitt-Rink *et al.*, 1986). A line of nodes in the basal plane is consistent with a T^2 dependence of the specific heat, but inconsistent with a T^2 dependence of the longitudinal sound attenuation. However, power-law temperature dependencies appear to be very sensitive to impurities and to the range over which

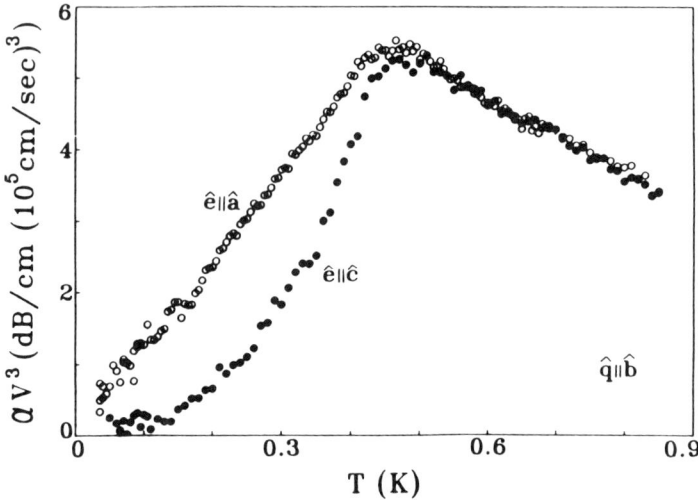

FIG. 11. Transverse sound attenuation in superconducting UPt$_3$ for $q\|b$. The attenuation is proportional to T for $\varepsilon\|a$, and to T^2 for $\varepsilon\|c$. After Shivaram et al. (1986).

the fit is made, and so they are not presently a reliable indication of the type of gap. The measurements of Shivaram et al. (1986a) are of interest, since these measurements were performed by the same group on the same sample. Thus even though one may not have confidence in the exact shape of the gap in k-space, it seems likely that the gap is highly anisotropic. There are some speculations that the gap has point nodes along the c-axis, and line nodes in the basal plane, suggesting a d-wave symmetry of the order parameter.

Müller et al. (1986b) see a very sharp peak in the attenuation just below T_c (as determined by ac susceptibility). Their data are shown in Fig. 12. This anomaly in the attenuation is seen *only* with longitudinal sound, and is referred to as the λ-peak, analogous to that observed in superfluid ^{4}He just below the λ-transition. Figure 12 also shows the T^3 dependence of the attenuation. The authors attribute their observation of the λ-peak to better sample quality: Much care was taken to remove any oxygen impurities by a long annealing at 1,200°C in a good vacuum. There was a flurry of theoretical speculation identifying the λ peak as either a collective mode or some sort of Landau–Khalatnikov damping mechanism. Subsequently, this sharp λ peak was observed in other samples (by Schenstrom et al., 1989a, and Adenwalla et al., 1990), which had not been so annealed. Evidence is accumulating that in high-quality samples, there is a very sharp velocity change at the superconducting transition, which is a characteristic of the heavy-fermion superconductors. The velocity change is accompanied by a

3. Sound Propagation in the Heavy Fermion Superconductors

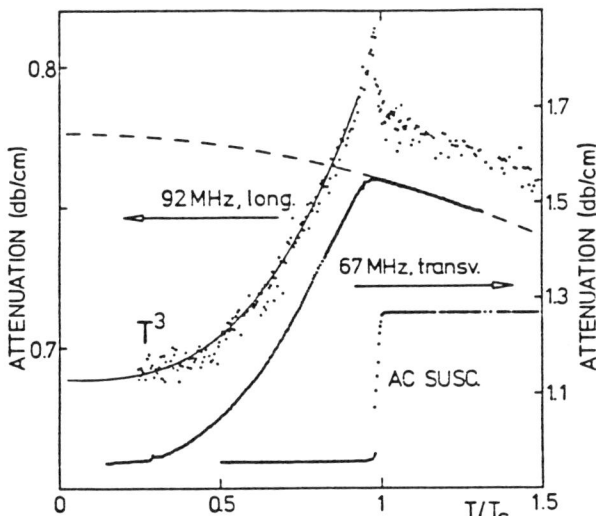

FIG. 12. Longitudinal (92 MHz) and transverse (67 MHz) ultrasonic attenuation coefficient for $q \| c$ and ac susceptibility as a function of reduced temperature in the superconducting state of UPt$_3$. The λ anomaly in attenuation is seen only for the longitudinal mode. The line is a T^3 fit to the longitudinal ultrasonic attenuation. After Müller et al. (1986b).

sharp attenuation peak just at T_c. In the first ultrasound measurements on UPt$_3$ by Bishop et al. (1985), the velocity dip at T_c was very broad because of sample quality. Adenwalla et al. (1990) have recently observed a very sharp velocity drop at T_C, with an accompanying sharp attenuation peak. The jump (of about 20 ppm) is shown to be consistent with the heat capacity jump and the strain derivative (see Section 5.3.3.).

5.3.2. Attenuation in a Magnetic Field

In UPt$_3$ the attenuation as a function of field (at a fixed T) shows some extraordinary features (Müller et al., 1987a; Qian et al., 1987). The longitudinal sound attenuation, α, at a frequency of 300 MHz and at $T = 50$ mK and along the c-axis is plotted in Fig. 13. At low fields α is linear with H, and near H_{c2} the attenuation rounds off and may be fitted to $(H_{c2} - H)^2$ rather than the Maki behavior mentioned in Section 2.3. At about 0.6 H_{c2}, there is an unusual peak in the attenuation that has not been observed in any conventional superconductor This peak has been referred to as the H_{FL} peak (FL means flux lattice) by Qian et al. (1987); they saw no frequency dependence of this peak between 75 and 300 MHz. The position of the peak has been mapped in the field–temperature plane and was found to move to lower fields with increasing temperature. The

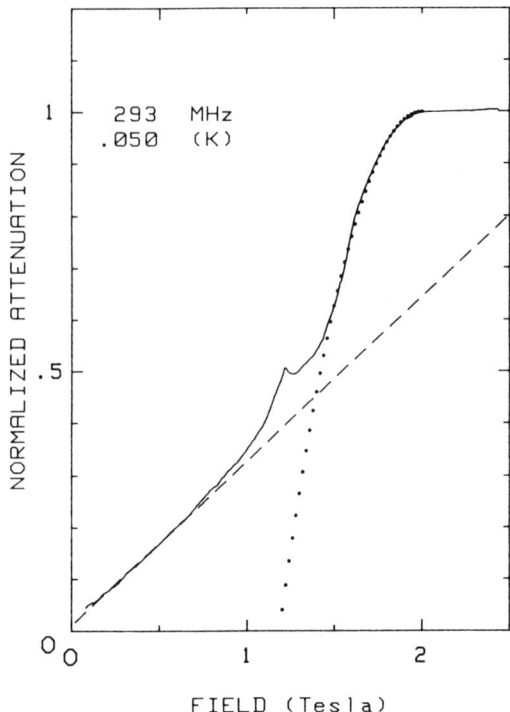

FIG. 13. Field dependence of the longitudinal ($q\|c$) ultrasonic attenuation at $T = 50$ mK for superconducting UPt$_3$. A peak in attenuation is seen at $\sim 0.6 H_{c2}$. This has been labeled the H_{FL} peak. The dashed (dotted) lines are linear (parabolic) fits to the data at low fields (close to H_{c2}) (Qian et al., 1987).

shape of the resulting curve led to speculation that the peak may be a signature of a transition between two different superconducting states. The "H_{FL}"-line separates the H–T plane into two superconducting regions, as shown in Fig. 14 (Qian et al., 1987). At $0.6H_{c2}$ the flux lattice is well-developed. For a field along the c-axis, the crystal has the same symmetry as that expected for the triangular flux lattice. This rules out a transition in the flux lattice caused by the symmetry of the underlying crystal lattice, as seen in Nb. However, if the order parameter of UPt$_3$ is unconventional, the order parameter may change as the field is increased, and this change in the order parameter could be accompanied by a change in the symmetry of the flux lattice. In the scenario proposed by Volovik (1988), the transition should disappear as the field is tilted away from the c-axis. (Note that no peak was seen for transverse sound measurements [Shivaram et al., 1987; Schenstrom et al., 1989b], implying that the phase transition coupled to density oscillations.)

3. Sound Propagation in the Heavy Fermion Superconductors

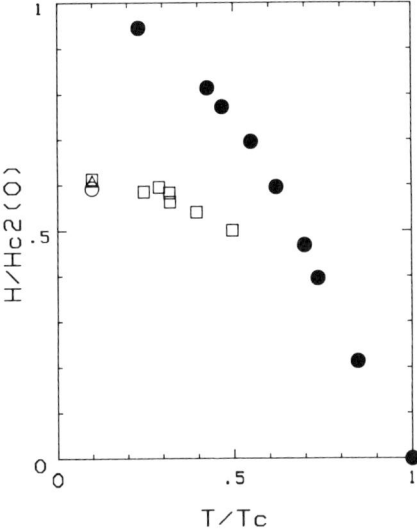

FIG. 14. Position of the H_{FL} peak in UPt$_3$ in the H–T plane. The closed circles are H_{c2} (as determined from susceptibility measurements). The H_{FL} data are denoted by the open triangles (126 MHz), open circles (249 MHz), and open squares (293 MHz). After Qian et al. (1987).

A careful study of the position of this H_{FL} peak for different orientations of field and sound propagation has been performed (Schenstrom et al., 1989a; Adenwalla, 1989) with the following results: (i) The H_{FL} peaks persist for all orientations of field as the field is tilted away from the c-axis. (ii) The position of the peak depends only on the orientation of the field with respect to the c-axis. There is no anisotropy in the basal plane. (iii) The position of the peak moves to lower fields as the field is rotated away from the c-axis into the basal plane. Figure 15 shows the anisotropy of $H_{FL}(T = 0)$, as a function of angle θ, the angle between the c-axis, and the external field. The measurements of Müller et al. (1987a) were performed with the angle θ = 20° and fall in nicely with the data of Schenstrom et al. (1989a). (iv) The overall temperature dependence of the H_{FL} transition line remains the same for all field orientations; it has a small negative slope in the H–T plane. The magnitude of the slope is largest for $H\|c$.

The H_{FL} peak becomes less defined close to the H_{c2} curve (i.e., in field sweeps at a temperature $T > 350$ mK). In the measurements of Qian et al. (1987) and Schenstrom et al. (1989a), the H_{FL} line was not extrapolated beyond their data points. It was not clear from the data whether the H$_{FL}$ line meets the $H = 0$ line just below T_c, as was suggested by Müller et al. (1987a) and Kleiman et

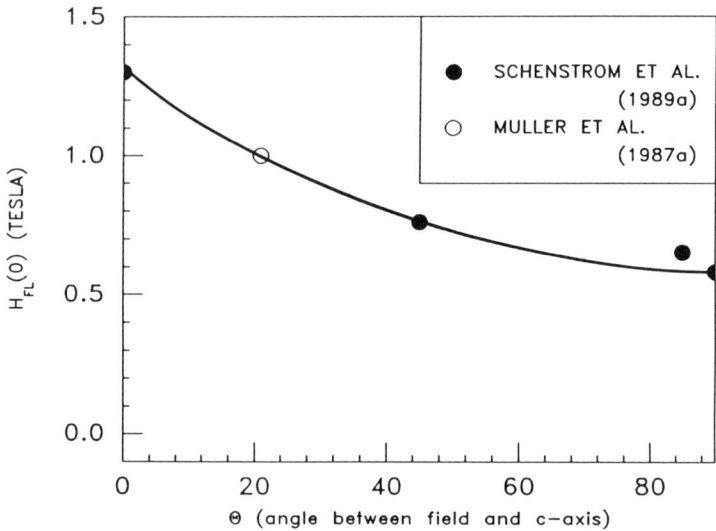

FIG. 15. Zero temperature values of the H_{FL} peak in the UPt$_3$ as a function of angle between the field and the c-axis. After Adenwalla (1989).

al. (1989), or whether it should meet the H_{c2} curve at a finite field. Immediately thereafter, heat capacity measurements in a magnetic field (Fisher et al., 1989; Hasselbach et al., 1989) showed that there was an additional low field phase close to T_c. The heat capacity (in zero field) showed *two* discrete jumps: one at T_c (their samples had a T_c of 490 mK) and the other at T_c^*, separated by a temperature interval of 50–60 mK. The heat capacity data are shown in Fig. 16 for H in the basal plane. This indicated that there were two (thermodynamic) superconducting transitions. This idea had been proposed earlier by Joynt (1988). On application of a magnetic field, both these transitions shift to a lower temperature and closer to each other and meet at a point P, at $H = 0.5T$ and $T^* = 400$ mK. For higher fields there is only one heat capacity jump. To compare with the ultrasonic data, Hasselbach et al. (1989) extrapolated the data of Schenstrom et al. (1989a) to meet the point P in a tetracritical point, thus giving three superconducting phases (Fig. 16b).

5.3.3. Velocity Measurements

At the superconducting transition, T_c, the velocity for longitudinal sound shows a sharp drop of ~20 ppm (Adenwalla et al., 1990); at the same time there is a sharp peak in the attenuation, the so-called λ peak. The transition width from velocity measurements is 14 mK and 18 mK for two different samples. The AC

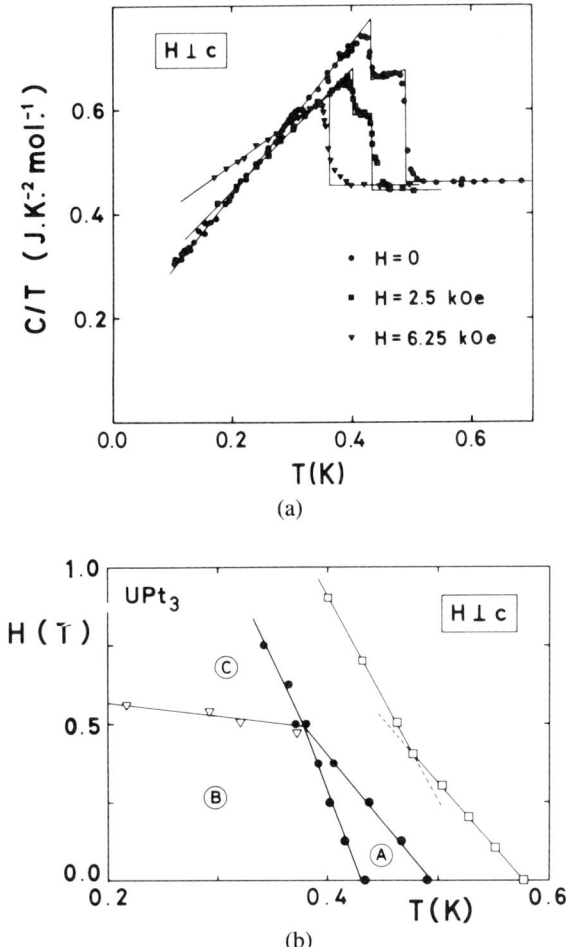

FIG. 16. Heat capacity of UPt$_3$ for different values of the magnetic field (a) for $H \perp c$. Field evolution of the double heat capacity jump in UPt$_3$, and the corresponding phase diagram (b). The triangles are the H_{FL} data from Schenstrom et al., 1989a. After Hasselbach et al. (1989).

susceptibility signature is at the same place (~10 mK higher in zero field), and this identifies the λ peak with the first superconducting transition. Earlier velocity measurements by Bishop et al. (1984) had a width in the velocity dip (100 mK) much larger than the susceptibility width. Shear wave measurements indicate only a slight change in slope at the transition.

A typical temperature sweep of the velocity is shown in Fig. 17. At T_c, there is a sharp signature that coincides with the susceptibility transition to within 10

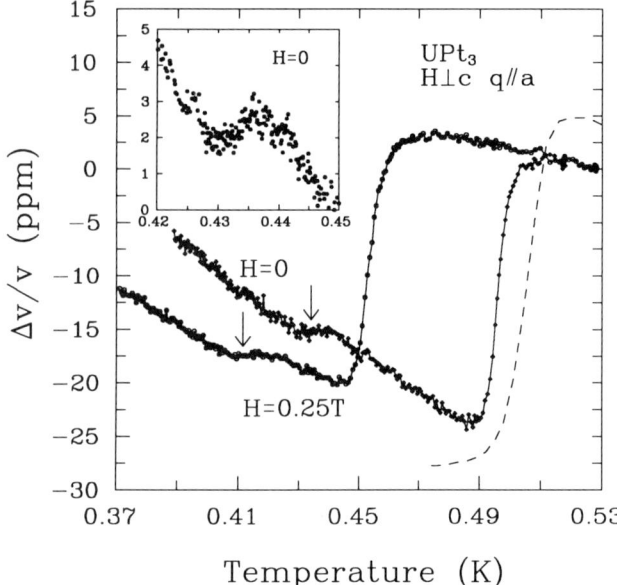

FIG. 17. Temperature dependence of the longitudinal ultrasonic velocity in UPt$_3$ at $H = 0$ and $H = 0.25$ T. The arrows denote the second transition at T_c^*. The dashed line is the susceptibility curve. The inset shows the result of a very slow temperature sweep through the anomaly at T_c^*. After Adenwalla *et al.* (1990).

mK. There is also a change in the slope of the velocity, below and above T_c. In addition, at about 60 mK below T_c, there is a small anomaly. A very slow temperature sweep through this anomaly (Adenwalla *et al.*, 1990) reveals a small velocity jump of about 3 ppm. This lower signature (denoted by T_c^*) is at the position of the lower heat capacity jump.

The jump in the velocity at the superconducting transition temperature provides valuable thermodynamic information. Here we generalize the earlier analysis of Testardi (1971) to the case of hexagonal symmetry (appropriate to UPt$_3$). To first order we may write the effect of a strain on the superconducting transition temperature as

$$T_c(\varepsilon) = T_c(0) + \Gamma \cdot \varepsilon; \qquad (114)$$

here we have written the symmetric 3×3 strain tensor ε_{ij}, and the coefficient matrix, Γ_{ij} (having the same structure), as six-component column vectors. The quantities are defined as $\varepsilon_1 = \varepsilon_{11}$, $\varepsilon_2 = \varepsilon_{22}$, $\varepsilon_3 = \varepsilon_{33}$, $\varepsilon_4 = 2\varepsilon_{23}$, $\varepsilon_5 = 2\varepsilon_{31}$, $\varepsilon_6 = 2\varepsilon_{12}$, and similarly for Γ_{ij}. Physically, the transition temperature cannot be affected by a pure shear, since changing the sign of a shear strain produces

3. Sound Propagation in the Heavy Fermion Superconductors   159

essentially the same physical state. This leaves only the diagonal components of Γ_{ij}. A second rank tensor does not distinguish symmetry axes higher than second-order; hence, the two basal plane components are identical, $\Gamma_1 = \Gamma_2 = \Gamma_a$, while the third component differs and is written $\Gamma_3 = \Gamma_c$. Hence $T_c(\varepsilon)$ becomes

$$T_c(\varepsilon) = T_c(0) + (\varepsilon_1 + \varepsilon_2) \Gamma_a + \varepsilon_3 \Gamma_c. \tag{115}$$

Thus

$$\Gamma_a = \frac{\partial T_c}{\partial \varepsilon_{1,2}}; \quad \Gamma_c = \frac{\partial T_c}{\partial \varepsilon_3}. \tag{116}$$

We now express the diagonal components ε_{ij} in terms of the diagonal stresses, σ_{ij}, as

$$\begin{pmatrix} \sigma_1 \\ \sigma_2 \\ \sigma_3 \end{pmatrix} = \begin{pmatrix} c_{11} & c_{12} & c_{13} \\ c_{12} & c_{11} & c_{13} \\ c_{13} & c_{13} & c_{33} \end{pmatrix} \begin{pmatrix} \varepsilon_1 \\ \varepsilon_2 \\ \varepsilon_3 \end{pmatrix}, \tag{117}$$

where $\sigma_1 = \sigma_{11}$, etc. Now $\boldsymbol{\sigma} = \overleftrightarrow{c} \cdot \boldsymbol{\varepsilon}$, and thus,

$$\frac{\partial}{\partial \varepsilon_i} = \frac{\partial \sigma_j}{\partial \varepsilon_i} \frac{\partial}{\partial \sigma_j}; \tag{117a}$$

we then have from Eq. (117a)

$$\frac{\partial T_c}{\partial \varepsilon_i} = c_{ij} \frac{\partial T_c}{\partial \sigma_j},$$

and, using Eq. (115),

$$\Gamma_a = (c_{11} + c_{12}) \frac{\partial T_c}{\partial \sigma_1} + c_{13} \frac{\partial T_c}{\partial \sigma_3} \tag{118a}$$

and

$$\Gamma_c = 2c_{13} \frac{\partial T_c}{\partial \sigma_1} + c_{33} \frac{\partial T_c}{\partial \sigma_3}. \tag{118b}$$

From the Ginzburg–Landau theory, the free energy is given by

$$F = F_0 + \alpha |\psi|^2 + \frac{\beta}{2} |\psi|^4. \tag{119}$$

Minimizing with respect to ψ^* yields $|\psi|^2 = -\alpha/\beta$ and

$$F = F_N - \frac{\alpha^2}{2\beta}. \tag{120}$$

Near T_c we write $\alpha = a(T - T_c)$. Noting the elastic constants are defined by

$$c_{ij} = \frac{\partial^2 F}{\partial \varepsilon_i \partial \varepsilon_j}, \quad (121)$$

we have, using Eq. (114),

$$\Delta c_{ij} = -\frac{a^2}{\beta} [\Gamma_a(\delta_{i1} + \delta_{i2}) + \Gamma_c \delta_{i3}][\Gamma_a(\delta_{j1} + \delta_{j2}) + \Gamma_c \delta_{j3}], \quad (122)$$

or

$$\Delta c_{11} = -\frac{a^2}{\beta} \Gamma_a^2, \quad (123a)$$

$$\Delta c_{12} = -\frac{a^2}{\beta} \Gamma_a^2, \quad (123b)$$

$$\Delta c_{13} = -\frac{a^2}{\beta} \Gamma_a \Gamma_c, \quad (123c)$$

and

$$\Delta c_{33} = -\frac{a^2}{\beta} \Gamma_c^2. \quad (123d)$$

We may obtain the Γ values from the changes in the ultrasonic velocity for longitudinal waves parallel to the a and c axes:

$$V_a^\ell = \sqrt{\frac{c_{11}}{\rho}} \quad (124a)$$

and

$$V_c^\ell = \sqrt{\frac{c_{33}}{\rho}}. \quad (124b)$$

This leads to

$$\Gamma_a^2 = -\frac{2\beta}{a^2} \rho V_a^\ell \Delta V_a^\ell \quad (125a)$$

and

$$\Gamma_c^2 = -\frac{2\beta}{a^2} \rho V_c^\ell \Delta V_c^\ell. \quad (125b)$$

The coefficient in the preceding expressions can be obtained from the jump in the heat capacity at T_c. From thermodynamics, $C_V = -T \partial^2 F/\partial T^2$; it then follows from Eq. (120) that

3. Sound Propagation in the Heavy Fermion Superconductors

$$\Delta C_V = -\frac{a^2}{\beta}T_c. \tag{126}$$

Using Eq. (118), we may compare the ultrasonic results with those obtained directly from the stress dependence of T_c. Note the change in T_c with hydrostatic pressure is given by

$$\frac{\partial T_c}{\partial p} = 2\frac{\partial T_c}{\partial \sigma_{1,2}} + \frac{\partial T_c}{\partial \sigma_3}. \tag{127}$$

The changes in the T_c of UPt$_3$ with hydrostatic pressure and uniaxial stress have been measured by Greiter *et al.* (1990). These measurements reveal the following: Under hydrostatic pressure, $\partial T_c/\partial P = -24$ mK/kbar, for *uniaxial* stress along the *c*-axis $\partial T_c/\partial \sigma_3 = -26$ mK/kbar, and for *uniaxial* stress in the basal plane $\partial T_c/\partial \sigma_{1,2} = \sim 0$. Using Eq. (117), we can obtain (Taillefer, 1990)

$$\Gamma_a = 43.5 \text{ K}, \qquad \Gamma_c = 74 \text{ K}.$$

Using Eqs. (124)–(126) we get

$$\frac{\Delta V_a^\ell}{V_a^\ell} = \frac{1}{2}\frac{\Delta C}{T_c}\frac{\Gamma_a^2}{C_{11}}, \tag{128}$$

$$\frac{\Delta V_c^\ell}{V_c^\ell} = \frac{1}{2}\frac{\Delta C}{T_c}\frac{\Gamma_c^2}{C_{33}}, \tag{129}$$

Using previously measured values of ΔC (Fisher *et al.*, 1989), c_{11} and c_{33} (de Visser *et al.*, 1987), we obtain

$$\Delta \frac{V_a^\ell}{V_a^\ell} \simeq 18 \text{ ppm},$$

$$\Delta \frac{V_c^\ell}{V_c^\ell} \simeq 55 \text{ ppm}.$$

The experiments of Adenwalla *et al.* (1990) and Bruls *et al.* (1990) measure $\Delta V_a^\ell/V_a^\ell$ to be 20 ppm. However, the experimental values do not show a marked difference in the velocity jump for different directions of sound propagation (Bruls *et al.*, 1990). The smaller velocity signature at T_c^* is related to the lower of the two heat capacity jumps seen in UPt$_3$. The same relationship given above can be used to calculate the strain derivative of T_c^*. Depending on the size of the heat capacity jump used in the calculation (which is sample dependent), the strain derivative could be the same or a factor of three larger than that at T_c.

Anomalies in the velocity are also observed by sweeping the field. As the field is increased, the velocity decreases monotonically. This seems to be a universal phenomenon with all the heavy fermion compounds studied (including

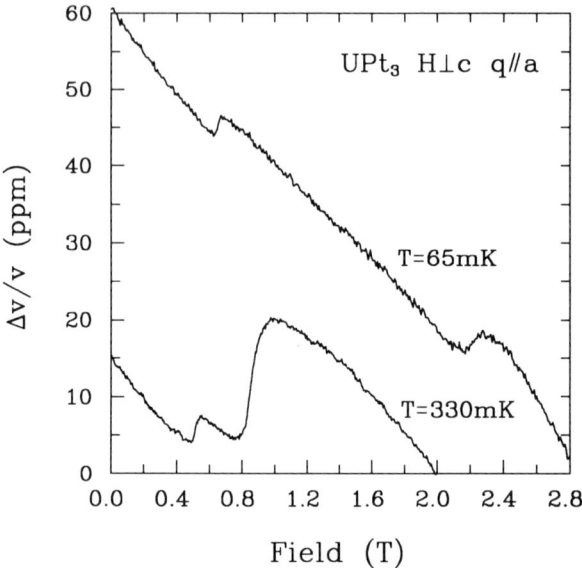

FIG. 18. Magnetic field dependence of the longitudinal ultrasonic velocity in UPt$_3$ at two temperatures. The signatures at H_{FL} and H_{c2} are both clearly visible. After Adenwalla et al. (1990).

CeCu$_6$) and may be simply a magnetostriction effect. A field sweep at $T = 50$ mK is shown in Fig. 18. At the position of the H_{FL} attenuation peak, there is a dip in the velocity of about 3 ppm; at H_{c2} there is a distinct signature in the velocity as the sample becomes normal. No heat capacity anomaly (latent heat or heat capacity jump) has been seen at the position of the H_{FL} signature. Field sweeps at higher temperatures show a shift of the H_{FL} to lower fields. The H_{FL} signature in the velocity was also observed in temperature sweeps at higher temperatures where dH_{FL}/dT is steepest. None of the anomalies was observed in transverse sound experiments.

The three signatures (T_c, H_{FL} and T_c^*) were tracked as functions of field and temperature, and a phase diagram was obtained. The velocity signatures remain sharp and unambiguous to within the width of the superconducting transition. This new phase diagram unifies all of the known signatures that have previously been identified as possible phase transitions.

5.4. Phase Diagram of UPt$_3$

There are several experiments that indicate UPt$_3$ has several superconducting phases in the H–T plane. The existence of multiple superconducting phases is a clear indication that UPt$_3$ is an unconventional superconductor. Each super-

3. Sound Propagation in the Heavy Fermion Superconductors

conducting phase is described by a separate (nonvanishing) order parameter. Multiple superconducting phases also exist in the other well known unconventional superconductor, superfluid ^{3}He, discussed in an earlier chapter.

The first evidence of multiple superconducting phases came from longitudinal attenuation measurements (Qian et al., 1987; Müller et al., 1987a) in a magnetic field. For $H\|c$, this occurs at $0.6H_{c2}$ at the lowest temperature, shifting to slightly lower fields as the temperature is increased. This peak was called the H_{FL} peak and was thought to be the signature of a phase transition either in the flux lattice or in the vortex core (Qian et al., 1987). This transition was present for all field orientations (Schenstrom et al., 1989a). The H_{FL} line separated the H–T plane into two superconducting regions, I and II. However, as one approached the H_{c2} curve it was difficult to resolve the H_{FL} peak from the background attenuation, and it was not certain whether the H_{FL} line met the H_{c2} curve at a finite field (~0.57 Tesla) or met (on extrapolation) the $H = 0$ line just below T_c. This was conclusively resolved in the velocity measurements by Adenwalla et al. (1990).

The velocity measurements revealed all the known signatures that had previously been identified as possible phase transitions, the H_{FL}, T_c^* (at the position of the lower heat capacity jump), and H_{c2}. Figures 19 and 20 show the phase

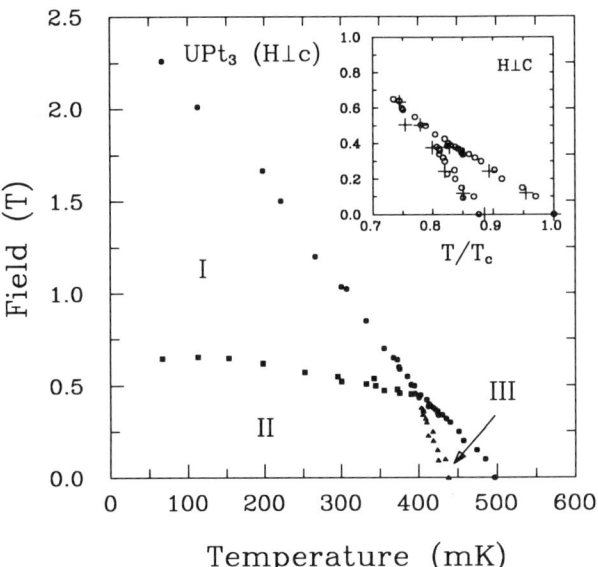

FIG. 19. Phase diagram of superconducting UPt$_3$ obtained from velocity measurements for $H \perp c$. The inset shows the comparison of the velocity data (○) with the heat capacity data (+). After Adenwalla et al. (1990).

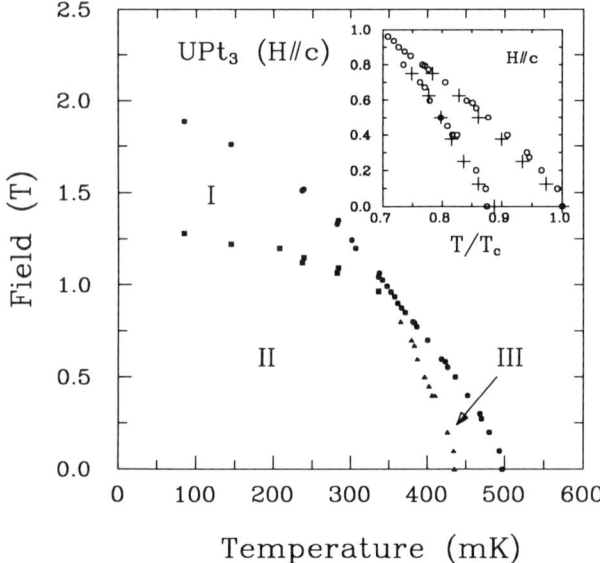

FIG. 20. Phase diagram of superconducting UPt$_3$ obtained from velocity measurements for $H\|c$. The inset shows the comparison of the velocity data (○) in comparison with the heat capacity data (+). After Adenwalla et al. (1990).

diagram obtained by Adenwalla et al. (1990) from velocity measurements. The insets show the T_c^* and T_c from velocity measurements compared with the heat capacity data of Hasselbach et al. (1989). The velocity signatures remain sharp and unambiguous to within the width of the transition. This phase diagram (the first to be obtained by a single measuring technique on the same sample) indicated that UPt$_3$ has three distinct superconducting phases. These three phases and the normal phase intersect at a tetracritical point on the H_{c2} curve, for fields parallel and perpendicular to the c-axis.

Figure 21 shows the upper critical field curves from velocity measurements. There is a crossover at 200 mK (as seen earlier in the data of Shivaram et al., 1986b). For H in the basal plane, there is a kink in H_{c2} at the tetracritical point. Similar results have been obtained by Bruls et al. (1990). However their interpretation of the data is different from that of Adenwalla et al. (1990).

Prior to these velocity measurements, Blount et al. (1990) had proposed a phase diagram for the two possible orientations $H\|c$ and $H\perp c$. For $H\|c$, there were only two phases, with the phase boundary running parallel to the H_{c2} curve, which was similar to the phase diagram proposed (based on extrapolation) by Müller et al. (1987a, attenuation), Kleiman et al. (1989, torsional oscillator), Aeppli et al. (1989, neutron diffraction) and Ellman et al. (1990, heat capacity—

3. Sound Propagation in the Heavy Fermion Superconductors

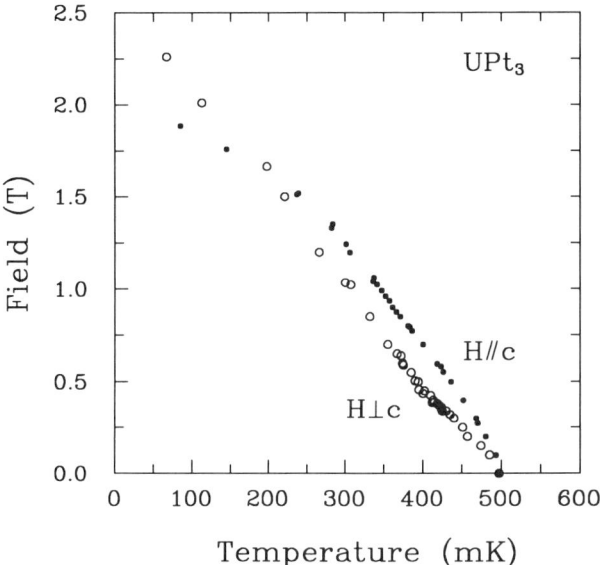

FIG. 21. H_{c2} of UPt$_3$ from velocity measurements for $H\|c$ and $H\perp c$. Note the crossover at 200 mK and the kink for $H\perp c$. Adapted from Adenwalla et al. (1990).

in magnetic field sweeps). For H in the basal plane they predicted three superconducting phases: the H_{FL} line and the T_c^* line meet the H_{c2} curve at two tricritical points P_1 and P_2. They ruled out the possibility of a tetracritical point.

The limit of resolution of the data of Adenwalla et al. (1990) is the width of the normal-to-superconducting transition. If the H_{FL} and T_c^* are different segments of the same transition line (as suggested by Joynt, 1988 and Blount et al., 1990), then this transition line would have to approach the H_{c2} curve to within 12–15 mK without intersection. If the H_{FL} and T_c^* intersect H_{c2} at two tricritical points (as suggested by Blount et al., 1990, for $H \perp c$), then their separation is less than $\delta H = 0.07T$ and $\delta T = 18$ mK.

A recent analysis by Yip et al. (1991) shows that certain phase diagrams are not allowed by thermodynamics. In particular, three second-order phase transitions cannot meet at a point unless an additional phase transition line (either first or second order) emerges from that point. If T_c^* and H_{c2} are both second-order lines (which is likely in view of the heat capacity measurements), then a tricritical point is not allowed.

From their analysis, the slopes of the four phase transition lines (close to the tetracritical point) can be related to the ratios of the heat capacity jumps. Using the slopes obtained from the velocity phase diagram, Adenwalla et al. (1991)

have checked for consistency with the observed heat capacity jumps and find that the agreement is good. Furthermore, using a thermodynamic analysis, they obtain a relationship between the slopes of the four transition lines at the tetracritical point and the heat capacity jumps. Using the data from the new velocity phase diagram to obtain the slopes, Adenwalla et al. (1991) have checked for consistency with the observed heat capacity jumps and find that the values agree to within 10%. They also predict that for $H \perp c$, the H_{FL} line is a first-order phase transition line and calculate the latent heat to be ~500 erg/mol (at a temperature 10 mK below the tetracritical point).

5.5. Results from Other Measurements

In this section we briefly mention some of the other measurements that have been performed on the superconducting state of UPt_3. These measurements either directly or indirectly confirm the unconventional nature of the superconductivity in UPt_3.

5.5.1. Thermal Expansion Coefficient

The thermal expansion coefficient, α, is also a second-order derivative of the Gibbs potential and hence related to the heat capacity jumps at T_c through the Ehrenfest relation. This is given by

$$\left.\frac{dT_c}{dP}\right|_{P=0} = \frac{V_m T_c \Delta \alpha_v}{\Delta C_P,} \tag{130}$$

where α_v is the volume thermal expansion coefficient,

$$\alpha_v = \alpha_\| + 2\alpha_\perp. \tag{131}$$

Both the heat capacity, C, and the thermal expansion coefficient, α, have been measured on two samples from the same source (Hasselbach et al., 1990). The coefficient of linear thermal expansion $\alpha = L^{-1} dL/dT$ was measured along the a- and c-axes of a single crystal bar of dimensions $2 \times 1 \times 3$ mm^3. (This is the same sample on which the velocity measurements were made by Adenwalla et al., 1990.) A three-dimensional capacitance method was employed, making use of a low-temperature pre-amplifier and a low temperature reference capacitor with a detection limit of $\Delta L/L \sim 2 \times 10^{-10}$. The thermal expansion is anisotropic: (i) the basal plane expands on heating ($\alpha_\perp > 0$), while the c-axis contracts ($\alpha_\| < 0$). (ii) At the superconducting transition (both at T_c and T_c^*) the response along the c-axis is much greater than that along the a-axis. An abrupt increase in $\alpha_\|$ is seen, whereas in the basal plane, there is maybe only a small change

3. Sound Propagation in the Heavy Fermion Superconductors 167

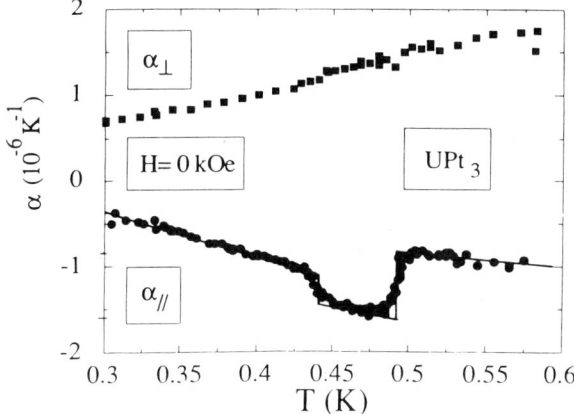

FIG. 22. Thermal expansion coefficient of UPt₃ parallel and perpendicular to the c-axis. There are two discontinuities, one at T_c, the other 60 mK below. This was the sample used in the velocity measurements of Adenwalla et al. (1990). After Hasselbach et al. (1990).

in slope. This anisotropy in the discontinuity at T_c would imply that a uniaxial pressure along the c-axis strongly depresses T_c, whereas a stress along the a-axis would hardly have any effect on T_c. This is consistent with the experimental results of Greiter et al. (1990), who observe a depression of $dT_c/d\sigma = -26$ mK/kbar for a uniaxial stress along the c-axis, and a negligible effect when the stress is along the a- or b-axis.

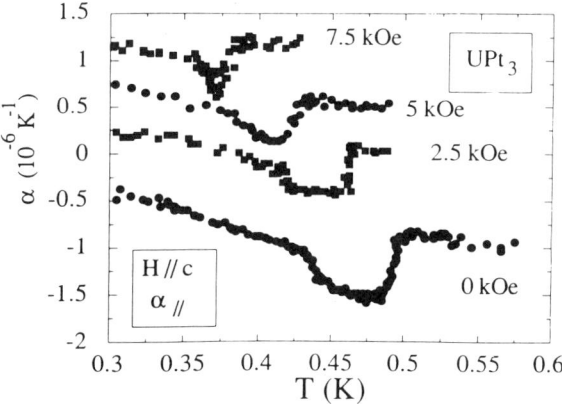

FIG. 23. Thermal expansion coefficient of UPt₃ for different values of magnetic field. After Hasselbach et al. (1990).

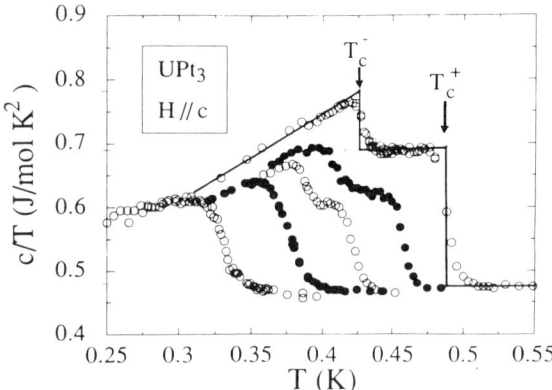

FIG. 24. Heat capacity of UPt$_3$ for different values of magnetic field for $H\|c$. After Hasselbach et al. (1990).

For $\alpha_\|$, clear evidence is seen for the low-temperature transition at T_c^*; this is not so clear for $\alpha_\perp$. (See Fig. 22.) It is to be noted that the two discontinuities at T_c and T_c^* are of the opposite signs, the discontinuity at T_c being more prominent. As the field is increased, the two $\Delta\alpha$ values decrease in amplitude similar to the decrease in ΔC_P values: They shift to a lower temperature and move close to each other. The transitions merge at about 8 kG and match the data obtained from heat capacity measurements. The data are shown in Figs. 23 and 24.

5.5.2. Neutron Scattering Measurements

Extensive neutron measurements have been done on the heavy fermion superconductor UPt$_3$ by Aeppli and co-workers (Aeppli et al., 1987, 1988, 1989). Strong antiferromagnetic fluctuations have been observed in neutron scattering experiments. Typically, the local magnetization (mostly f-electrons) fluctuates randomly with an energy of the order of 10 meV. At a lower energy level (i.e., on a long time scale) of 0.3 meV, there is a weaker antiferromagnetic alignment with a characteristic wavevector $Q = (\pm\frac{1}{2},0,1)$. Below 5 K, this ordering becomes static, leading to a staggered magnetization in the basal plane. Only a small fraction of the moments is effectively ordered on a long time scale ($\mu_{\text{eff}} \sim 0.02\ \mu_B$). This ordering is very weak and not very long-ranged—the antiferromagnetic correlation length is less than 300 Å. The staggered (antiferromagnetic) magnetization, M_s, as measured by the area under the $(\frac{1}{2},0,1)$ Bragg peak, increases slowly as the temperature is lowered below T_N (as shown in Fig.

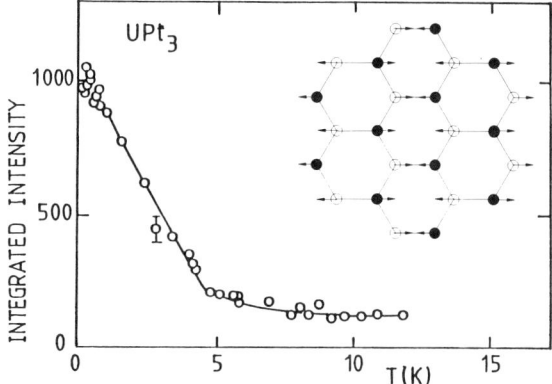

FIG. 25. The intensity of the AFM Bragg peak in UPt$_3$ as a function of temperature. After Aeppli et al. (1990).

25). This increase flattens at the superconducting transition T_c. From measurements made in a magnetic field, in the superconducting state Aeppli et al. (1989) obtain a phase diagram in general agreement with the torsional oscillator measurements (Kleiman et al., 1989).

Hydrostatic pressure studies have been performed of the magnetic Bragg scattering peak (Taillefer et al., 1990) and the two peaks in the specific heat (Trappmann et al., 1991), and both show a similar behavior: (i) The antiferromagnetic magnetization (the integrated intensity of the $(\frac{1}{2},0,1)$ Bragg peak) falls by a factor of 2.5 on application of a pressure of 2.5 kbar. This extrapolates to a zero intensity (disappearing antiferromagnetism) at 3.0 kbar. The heat capacity measurements (see Fig. 26) yield a shift to lower temperatures in both T_c and T_c^* on application of (hydrostatic) pressure; the two peaks merge into one at a pressure of 3.5–4 kbar. From the similar behavior under pressure of the antiferromagnetic order parameter and the splitting of the superconducting transition (from heat capacity measurements), one is led to conclude that the two types of ordering are intimately connected. The coupling to the AFM order parameter lifts the degeneracy of the superconducting state, which is expected to have a multicomponent) vector order parameter (see Sections 3 and 5.6).

Unfortunately, no other bulk measurements (ultrasonic velocity/attenuation, heat capacity, thermal expansion, magnetization or resistivity) have observed the AFM ordering at 5 K, and a systematic study of the nature of the coupling of the AFM order with the superconducting order parameter has yet to be performed.

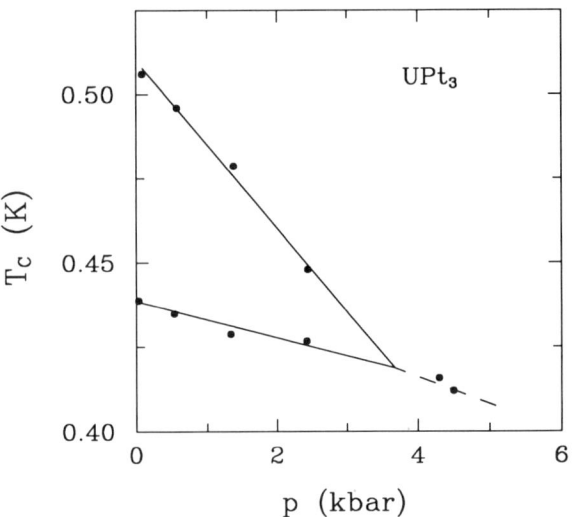

FIG. 26. Temperature–pressure phase diagram of UPt$_3$ derived from heat capacity measurements. The two heat capacity jumps merge into one at a pressure of about 3.5 kbar. After Trappmann et al., (1991).

5.5.3. H_{c1}, λ_L

The London penetration depth, λ_L, in UPt$_3$ has been measured by Shivaram et al. (1989) in rf absorption measurements using a lower-temperature tunnel diode: A sensitivity of 1 in 10^7 was obtained in measuring the frequency shifts. From these measurements, they obtain an anisotropic λ_L. From measurements in a magnetic field, they infer a value of the lower critical field, H_{c1}, ($\sim$ 100 g) and obtain the temperature dependence $H_{c1}(T)$. These measurements were performed on long, thin samples. Dc magnetization measurements have been performed by Zhao et al. (1991a) on a sample in the shape of a sphere (3 mm diameter). A spherical geometry is needed for proper magnetization measurements, as otherwise the demagnetization effects would make it difficult to estimate H$_{c1}$. A kink in the H_{c1} is seen for all orientations (of the magnetic field with respect to the c-axis), as predicted by Hess et al. (1989). The sphere was obtained from the same piece of crystal from which the sample for the ultrasonics experiments was obtained.

The best H_{c1} data on UPt$_3$ are by Vincent et al. (1991) who have used a SQUID magnetometer. Their sample had an electron mean free path of 1,300 Å. Their data are shown in Fig. 27. A change in slope of the H_{c1} curve is seen for both field orientations occurring at $T = T_0$, which is the same as T_c^*, (the temperature of the lower heat capacity jump), as expected. The increase in slope

3. Sound Propagation in the Heavy Fermion Superconductors

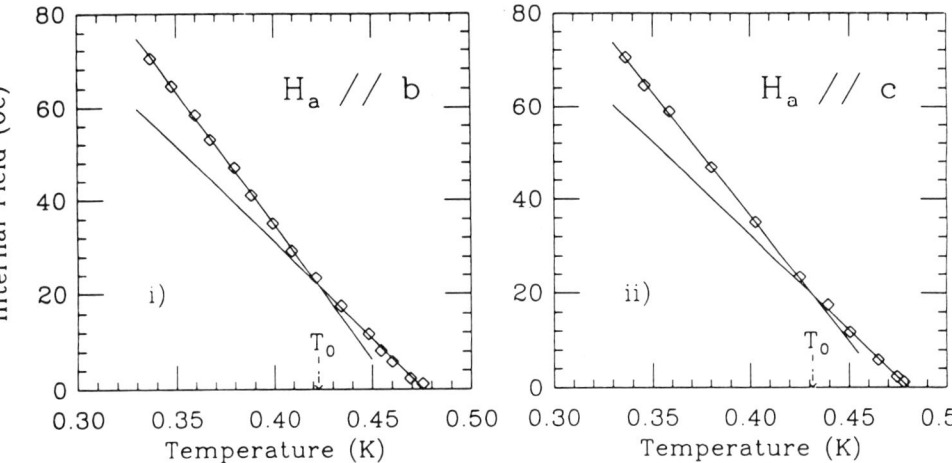

FIG. 27. Lower critical field data for UPt$_3$ for $H\|c$ and $H\perp c$. There is a kink in the critical field curve for both orientations. After Vincent et al. (1991).

of H_{c1} at T_0 is due to the appearance of a new superconducting order parameter associated with the lower transition.

λ_L has also been determined by muon spin relaxation (μ^+SR) experiments by Broholm et al. (1990). μ^+SR is an excellent probe of the field inhomogeneities of the vortices in type II superconductors. The samples had an electronic mean free path of $\ell_\| = 2{,}700$ Å and $\ell_\perp = 660$ Å (relative to the c-axis). μ^+SR gives both $\lambda_\perp$ and $\lambda_\|$. From an analysis of their data they obtain the following information: (i) the superconducting gap $\dfrac{\Delta(0)}{T_c} = 2.0$; (ii) the extrapolated zero temperature penetration depths are $\lambda_\|(0) = 7{,}070$ Å and $\lambda_\perp(0) = 7{,}820$ Å. λ_L is related to the effective mass, m_{eff}, through the London equation

$$\frac{1}{\lambda^2(0)} = \frac{4\pi n e^2}{c^2 m_{\text{eff}}}. \tag{130}$$

This yields an effective mass of 260–280 m_e, in good agreement with the heat capacity results.

From their $\lambda_L(0)$ values, Broholm et al. (1990) infer a H_{c1} value of less than 30 G, in contrast with the near 100 G value obtained from magnetization measurements.

The temperature variation of λ_L has been obtained. The value of $\lambda_L(0)$ is nearly isotropic; however, their temperature variation is very different for the two orientations. $1/\lambda_\perp^2(T)$ decreases linearly with T, and $1/\lambda_\|^2(T)$ decreases with a higher power.

5.6. THEORETICAL SPECULATIONS ON THE SUPERCONDUCTING PHASES

From the experiments discussed so far, a convincing case can be made that the superconducting state in UPt_3 is of an unconventional nature. The main features are (i) zeroes (lines or nodes) in the energy gap resulting in the order parameter having a lower symmetry, (ii) the existence of at least three superconducting phases in the mixed state, and (iii) the unusual anistropy of H_{c2}. The nature of the order parameter for an unconventional superconductor has been discussed in Section 3. The identification of the three superconducting phases has not yet been agreed upon. There have been many theoretical proposals (Volovik, 1988; Joynt, 1988; Hess *et al.*, 1989; Machida and Ozaki, 1989; Blount *et al.*, 1990) as to the possible nature of these phases, with considerable disagreement. Since it is not possible to discuss all these theories in this text, we shall only describe the results of the Northwestern group.

Neutron scattering experiments have shown strong antiferromagnetic tendencies, and it is believed that the exchange of spin fluctuations may play a dominant role in the pairing mechanism instead of the conventional electron–phonon interaction. Since there is yet no well understood microscopic mechanism for the pairing, several authors have argued in favor of both even- (d-wave) or odd-parity (p-wave) representations. In both cases the order parameter is a two-component array (η_1, η_2) which transforms like a vector in the x-y plane. The G-L expansion is carried out to the sixth order in the order parameter, η, to take care of the degeneracies of the representation and introduce a coupling to the AFM order parameter (in the a–b plane) as a symmetry breaking field (SBF). This results in a small splitting of the superconducting transition, which is seen in the heat capacity measurements. They (Tokuyasu *et al.*, 1990) also show the axial symmetry of $H_{c2}(T)$ in the basal plane may be broken if the SBF is locked to the crystal lattice. A kink appears in $H_{c2}(T)$ at a temperature T_H, which depends on the strength of the SBF coupling and the stiffness coefficients appearing in the G–L free energy. The lower critical field will also display a kink for all field orientations at the temperature of the zero-field transition. From analyzing the heat capacity data, Hess *et al.* (1989) predicted the ratio of the lower critical field slopes at $T = T_c^*$ to be ~1.46. Kinks in the H_{c1} curves have been seen by various groups; the experimentally determined ratio of the slopes is much smaller, ~1.2.

Using the time-dependent relaxation method of Thuneberg (1987), Tokuyasu *et al.* (1990) have solved the G–L equations over a wide range of the G–L parameters, with boundary conditions such that far away from the vortex core, the vector potential is given by its London limit form. For various values of the G–L parameters, they obtain different classes of vortex solutions that are en-

3. Sound Propagation in the Heavy Fermion Superconductors

ergetically stable. Axial vortices are obtained for relatively large values of β, which determines the difference in condensation energy between the different possible ground states. For higher stiffness ratios κ, they find non-axial vortices to be stable: The two cores split and lower the energy (see Fig. 28), resulting in vortices with broken hexagonal symmetry (triangular or crescent-shaped vortices).

In either case, the vortex–vortex interaction will be anisotropic, having a tendency to line up the neighboring vortices with the edges parallel. As the field is increased, the vortices move closer, and an orientational frustration develops, resulting in a transition. There are two possible purely-structural vortex–lattice transitions at intermediate fields. If the vortex cores are rigid at these fields, the favored lattice will be a honeycomb lattice, with a reduced number of nearest neighbors. Alternatively, the lattice remains hexagonal, but one-third of the vortices have axially symmetric cores. A hexagonal lattice, with six triangular vortices, all satisfying the edge constraint, and a single axial vortex in the center, eliminates the orientational frustration. Thus, one would expect a structural transition in the flux lattice (F-L) to occur when the vortex lattice spacing is approximately equal to the core size. (See Fig. 29.)

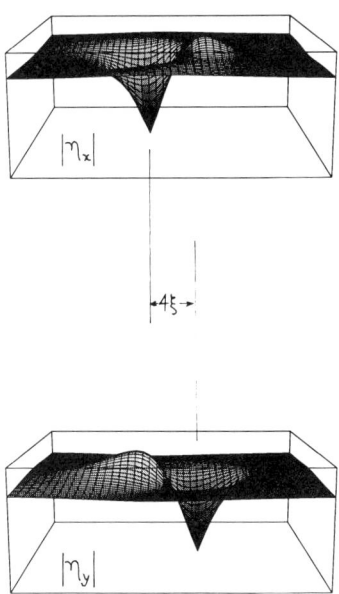

FIG. 28. Plots of η_x and η_y, the two components of the older parameter, for a triangular vortex. The vortex core is split. After Tokuyasu *et al.* (1990).

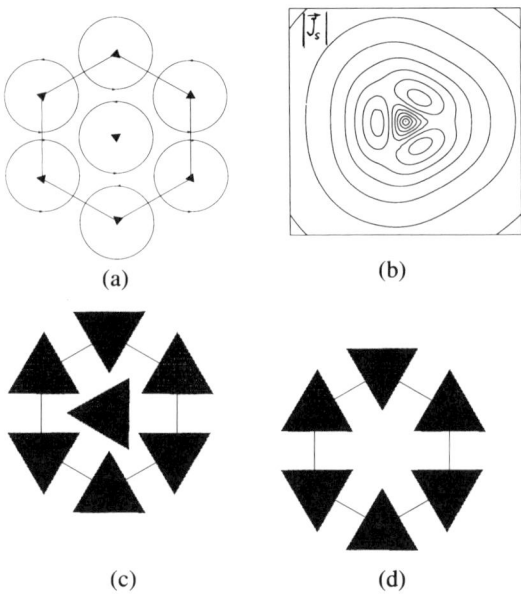

FIG. 29. "Frustration" for triangular vortices. At low fields (a) the vortex–vortex interaction is isotropic and the lattice is hexagonal. As the field is raised and the vortices move closer together, the interaction between vortices is anisotropic and the triangular vortices prefer to align with their edges parallel (b). This is not possible for a hexagonal lattice (c); however, a honeycomb lattice (d) will allow this. After Tokuyasu et al. (1990).

Tokuyasu and Sauls (1990) also find other novel solutions in the intermediate field region: Often there are doubly quantized vortices (containing two quanta of flux). Figure 30a,b,c shows these doubly quantized vortices. Doubly quantized vortices are axially symmetric, but have an intricate internal structure. The direction of current flow inside and outside of the vortex core ($\sim 10\xi$ in diameter) is reversed (Fig. 30b), and the magnetic field distribution has a minimum at the center of the vortex (Fig. 30c). In this model a lattice of doubly quantized vortices is stable below "H_{FL}," while at high fields, $H_{FL} < H < H_{c2}$, the stable lattice is a more conventional lattice of singly quantized vortices. Numerical solutions of the G–L equations show a dissociation transition at $H \sim 0.5 \, H_{c2}$ for $H\|c$ (Tokuyasu and Sauls, 1990), which is close to the experimentally observed value of $H_{FL} = 0.6 H_{c2}$ ($H\|c$).

Information on flux lattices in conventional superconductors has been obtained from neutron diffraction (Cribier et al., 1964) and decoration techniques (Bitter patterns, Essmann and Trauble, 1967), both of which seem formidable in the case of UPt_3. The London penetration depth λ_L ($\sim 7,000$ Å), is about 20 times larger than the coherence length and the flux-lattice spacing (distance between

3. Sound Propagation in the Heavy Fermion Superconductors

(a)

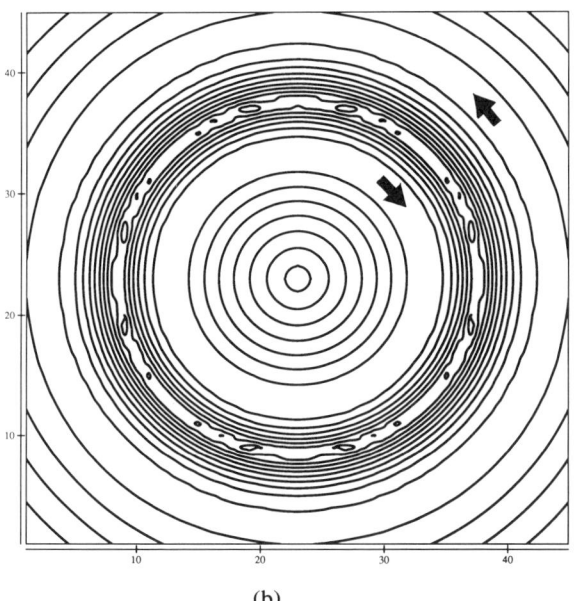

(b)

FIG. 30. (a) Double (oval) and singly (circular) quantized vortices. (b) Current density for a doubly quantized vortex. (c) Field distribution for a two-quantum vortex. After Tokuyasu (1990).

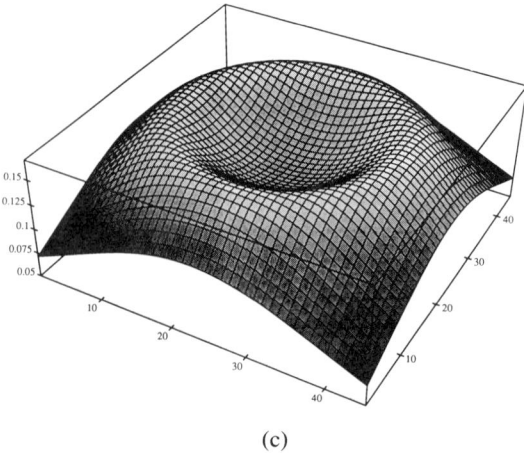

(c)

FIG. 30. Continued

the vortices) at H_{FL}. This would make the observation of the flux lattice close to H_{FL} rather difficult (the neutrons see the magnetic field distribution around the vortex cores), and it would be virtually impossible to resolve a difference in the vortex structure or lattice at H_{FL}. Recently, Kleiman *et al.* (1991) have seen a flux lattice in UPt$_3$ at 50 mK by neutron diffraction using counting times of the order of 24 hours; but they have yet to resolve any change at H_{FL}.

The best promise of being able to tell experimentally what the various superconducting phases are would be to do a low temperature STM (scanning tunneling microscope) study. Hess *et al.* (1989, 1990) have been very successful in developing a low-temperature STM to study the layered superconductor NbSe$_2$, and it may be possible to obtain results on UPt$_3$.

6. UBe$_{13}$

Superconductivity in UBe$_{13}$ was discovered by Ott *et al.* (1983), with a superconducting transition temperature $T_c = 0.85$ K. UBe$_{13}$ is a cubic crystal with $a = 5.13$ Å, giving a mass density of 5.5 g cm^{-3}. The sound velocities in UBe$_{13}$ are given in Table V. In the normal state, the low-temperature specific heat (and hence the effective mass) are among the largest for the heavy fermion superconductors, with $\gamma = 1,100$ mJ/mole K^2. There is a peak in the resistivity at 2 K, below which the resistivity drops rapidly. Recent magnetostriction measurements (Kleiman *et al.*, 1990) show evidence for an antiferromagnetic transition at ~9 K, confirming a general feature seen in the other heavy fermion superconductors, i.e., a magnetic ordering at $T_N \sim 10T_c$. To our knowledge, no

3. Sound Propagation in the Heavy Fermion Superconductors

TABLE V

SOUND VELOCITIES IN UBe_{13} AND URu_2Si_2.

	q along	longitudinal	shear
UBe_{13}	a	6,500	5,292
URu_2Si_2 (tetragonal)	c	5,394	3,210
	a	5,241	3,613 ($\varepsilon \| b$)
			3,127 ($\varepsilon \| c$)

ultrasonic anomaly at the AFM transition has been reported. In the superconducting state, the low-temperature specific heat is nonexponential, following a T^3 power-law dependence, indicating the presence of nodes in the gap. Ultrasonic measurements on superconducting UBe_{13} were performed by Golding et al. (1985) over a temperature range from 1 K to 0.1 K. Longitudinal sound measurements at frequencies ranging from 0.9 GHz to 2.4 GHz were performed along the [001] direction, using thin-film ZnO transducers sputtered directly on the sample. The ultrasonic attenuation and velocity and the ac susceptibility were simultaneously monitored. (See Fig. 31.) The transition temperature and the width of the transition (as measured by susceptibility) were 0.86 K and 0.040 K, respectively. The attenuation in the superconducting state was very different from the predictions of BCS theory. At low temperatures ($T < 0.5T_c$), the attenuation follows a T^2 behavior, suggesting the presence of nodes in the gap. There is a sharp sound absorption peak at T_c. Both these features are very different from the predictions of BCS theory, which predicts an exponential temperature dependence of the attenuation, with a sharp drop at T_c. Measurements were performed at six frequencies between 0.98 and 2.4 GHz, and the temperature of the attenuation peak remained unchanged, showing no frequency dependence. Both the height of the attenuation peak and the absorption relative to the normal state, $\alpha_N - \alpha_s$, scaled as the square of the frequency, showing that these measurements were all performed in the hydrodynamic limit. The overall magnitude of the electronic absorption, $\alpha_N - \alpha_s$, was very small. The low density of UBe_{13} results in a large sound velocity (6,500 m/s). Consequently, the change in the attenuation is much smaller when compared to UPt_3, all other things being equal.

The velocity also shows a sharp signature at T_c, a discontinuous drop of 27 ppm, exactly where the ac susceptibility shows the superconducting transition. The velocity change at T_c can be related to the heat capacity jump and the strain derivative of T_c, as discussed for UPt_3. Using previously measured values for these quantities, Golding et al. (1985) found good agreement.

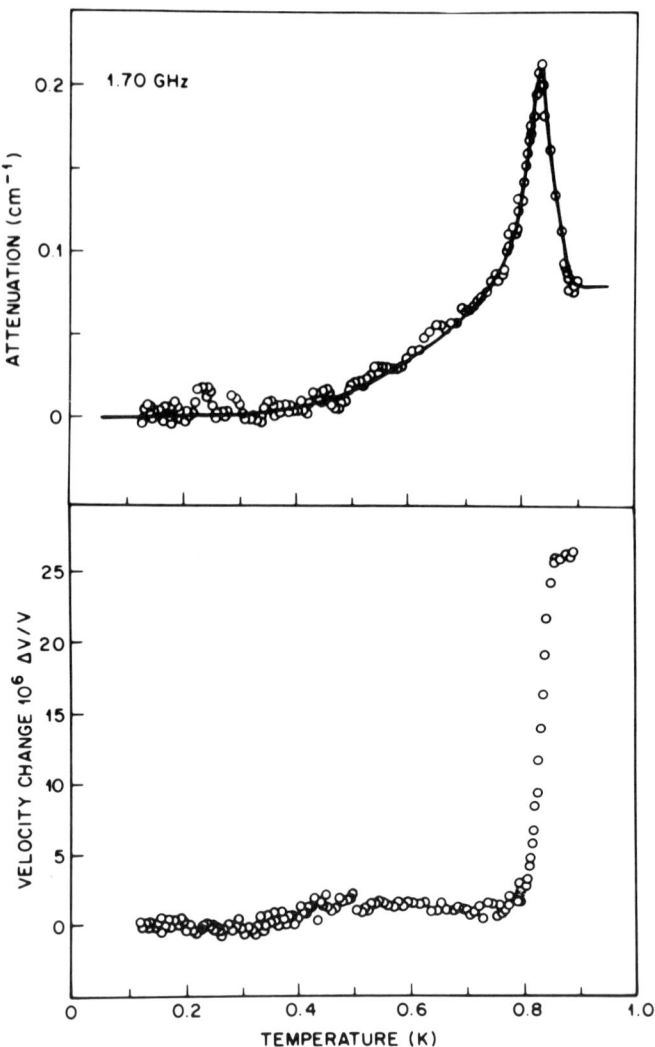

FIG. 31. Acoustic attenuation and velocity changes in UBe_{13} below the superconducting transition $T_c = 0.9$ K. After Golding et al. (1985).

Golding et al. (1985) have studied the effect of magnetic fields up to 2T (note: $H_{c2} = 8T$) on the absorption. The velocity discontinuity remains the same, both in magnitude and in sharpness. The peak moves towards a lower temperature, in agreement with T_c, but surprisingly the amplitude increases by about 20%.

A peak in the ultrasonic attenuation at T_c appears to be a common feature to most heavy fermion superconductors. The existence of an attenuation peak at or just below T_c is not expected for a conventional superconductor. Golding *et al.* proposed that the peak in the attenuation is caused by absorption into a collective mode. However, there is no frequency dependence of the position of the peak, as would be the case for an order parameter collective mode. The large velocity jump at T_c (due to the large heat capacity jump associated with the heavy mass of these materials) suggests a possible explanation of this anomalous attenuation peak.

7. URu_2Si_2

The heavy fermion compound URu_2Si_2 is a tetragonal crystal with $a = 4.12$ Å and $c = 9.5$ Å, and a density of 15 g cm^{-3}; it exhibits two phase transitions at low temperatures, a superconducting and an antiferromagnetic transition, with critical temperatures T_c and T_N of 1.28 K and 17.5 K, respectively. Magnetism and superconductivity may coexist in URu_2Si_2. Neutron diffraction measurements (Broholm *et al.*, 1987) of URu_2Si_2 reveal antiferromagnetic ordering along the c-axis with a Neel temperature $T_N = 17.5$ K. The magnitude of the ordered moment is small, $0.03\mu_B$. The sublattice magnetization is along the c-axis. The resistivity shows a small peak at T_N (Maple *et al.*, 1986), and the susceptibility shows a sharp drop in slope at T_N (Onuki *et al.*, 1987). Specific heat measurements indicate that a charge- or spin-density-wave transition opens an energy gap of 11 meV over part of the Fermi surface below T_N (Maple *et al.*, 1986).

Longitudinal ultrasonic attenuation measurements have been performed by several groups (Fukase *et al.*, 1987; Sun *et al.*, 1989; Bullock *et al.*, 1990; Lin *et al.*, 1990a; Ran *et al.*, 1990) near both the Neel temperature T_N and the superconducting transition temperature T_c of URu_2Si_2. The longitudinal and transverse velocities are 5.2×10^5 cm/s and 3.1×10^5 cm/s at room temperature, respectively, for sound propagation along the a-axis. The sound velocities are listed in Table V.

In Fig. 32 is a plot of the shear wave velocity, V_T, with polarization along the c-axis, and the longitudinal sound velocity, V_L, as a function of temperature, for sound propagating along the a-axis. In both cases there is a sharp signature at $T_N = 17.5$ K. Above T_N the velocity increases linearly with decreasing temperature. Below T_N there is a sharp increase in the velocity, showing a cusplike behavior. It is surprising that both the longitudinal and shear velocity show approximately the same behavior at T_N. Since the reported velocities have not been corrected for any length changes, this suggests that there may be a corresponding dilation effect (a magnetostriction due to the onset of antiferromagnetic ordering) at T_N. The transverse attenuation as a function of temperature for two frequencies, 166 MHz and 346.5 MHz, is shown in Fig. 33. Above T_N

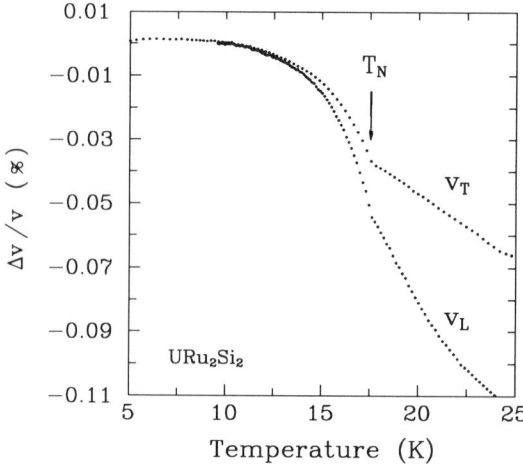

FIG. 32. Ultrasonic velocity measurements (longitudinal and transverse) at the antiferromagnetic transition in URu$_2$Si$_2$ (Ran *et al.*, 1990). (© 1990 IEEE.)

the attenuation is flat, and below T_N it increases monotonically with decreasing temperature. Fukase *et al.* (1987) have also measured T_N as a function of applied magnetic field: T_N is constant along the *a*-axis; however, it decreases along the *c*-axis by 0.5 K for a field increase of 6 Tesla.

Figure 34 shows a plot of the longitudinal ultrasonic attenuation, velocity,

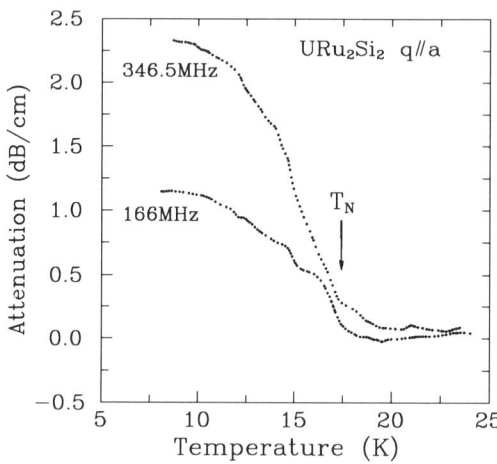

FIG. 33. Transverse ultrasonic attenuation vs. temperature for URu$_2$Si$_2$ for two frequencies at the AFM transition. After Ran *et al.* (1990). (© 1990 IEEE.)

3. Sound Propagation in the Heavy Fermion Superconductors 181

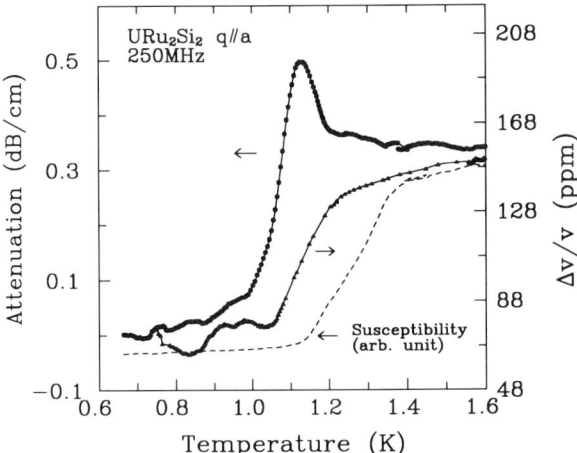

FIG. 34. Longitudinal velocity, attenuation, and ac susceptibility at the superconducting transition in URu$_2$Si$_2$ (Ran et al., 1990). (© 1990 IEEE.)

and ac susceptibility of URu$_2$Si$_2$ as a function of temperature near T_c. The onset of superconductivity is determined (from the ac susceptibility) to be at 1.4 K, and the superconducting transition temperature $T_c = 1.28$ K is placed at the midpoint of the susceptibility change. The transition width, ΔT_c, corresponding to the 10% and 90% change of the susceptibility transition, is about 150 mK, which is relatively large. Above T_c, the attenuation is flat. Just below (or at) T_c there is a peak in the attenuation, similar to (but not as sharp as) the peak seen in UBe$_{13}$ and UPt$_3$. As the temperature is lowered, the attenuation drops. The peak disappears on application of a small magnetic field (Sun et al., 1989).

The relative attenuation changes $\alpha_N(1.5K)-\alpha_0(0.8K)$ at 85 MHz and 250 MHz are 0.039 dB/cm (Sun et al., 1989) and 0.35 dB/cm (Ran et al., 1990) respectively. There have been some recent heat capacity measurements (Maple et al., 1991) that seem to indicate the possibility of a double heat capacity jump in URu$_2$Si$_2$, but it was not clear whether this is from two different structural phases or from two superconducting phases in a single structure.

8. CeCu$_6$

CeCu$_6$ is another heavy fermion compound on which extensive ultrasonic measurements have been performed. CeCu$_6$ does not undergo any superconducting or magnetic (antiferromagnetic) transition down to the lowest temperature studied (20 mK). Clearly this would be an ideal system to study, since one could follow the development of heavy fermion–like behavior down to the lowest temperature.

As noted earlier, the heavy fermion systems show Fermi liquid behavior; however, this is suppressed in UPt_3, UBe_{13}, and URu_2Si_2 because of the superconducting transition. Also, $CeCu_6$ has the largest effective mass of these compounds (~1,100), resulting in a Fermi velocity $v_F = 2,120$ m/s, which is less than the sound velocity.

$CeCu_6$ has an orthorhombic structure at room temperature: There are four Ce atoms per unit cell, and the cell dimensions are $a = 8.11$ Å, $b = 5.10$ Å and $c = 10.16$ Å. It has a large susceptibility of 0.027 emu/moleG at 1.5 K and a paramagnetic moment of $2.58\mu_B$ at higher temperatures. The electrical resistivity

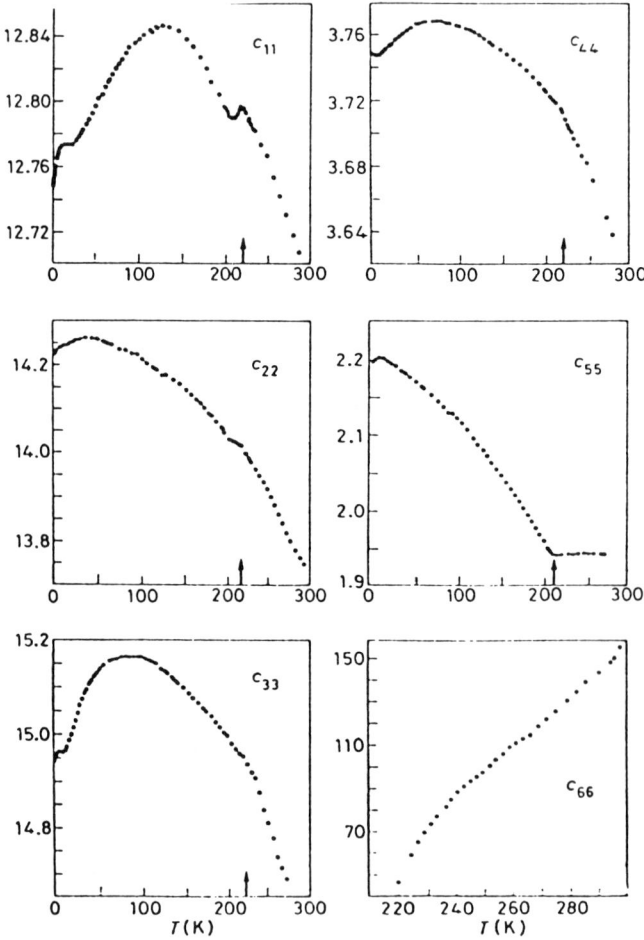

FIG. 35. Elastic constants of $CeCu_6$ from ultrasonic measurements in units of 10^{11} erg/cm^3, except for c_{66} (10^8 erg/cm^3). From Weber et al. (1987).

3. Sound Propagation in the Heavy Fermion Superconductors

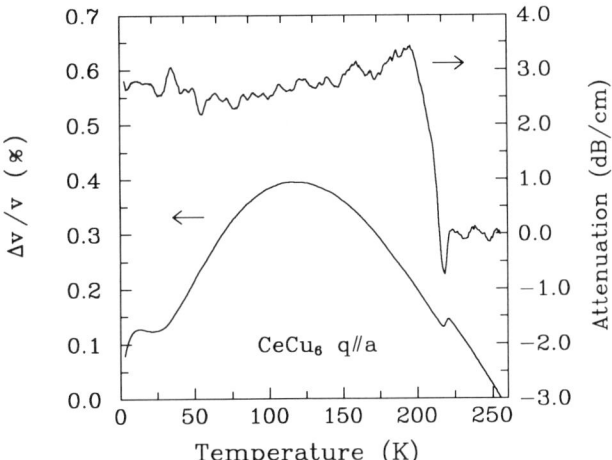

FIG. 36. Temperature dependence of the longitudinal attenuation and velocity for CeCu$_6$ along the a-axis at 130 MHz. The structural transition occurs at $T_s \simeq 220$K. After Lin et al. (1990b). (© 1990 IEEE.)

shows the classic Kondo behavior: The electrical resistivity increases as the temperature is lowered, passes through a maximum around 10 K, and decreases linearly below 1 K and quadratically below 100 mK (Amato et al., 1985). The linear decrease (below 1 K) has been attributed to a coherent transition in the Kondo system (Zemirli and Barbara, 1985), and the quadratic temperature dependence (below 100 mK) is characteristic of a Fermi liquid system.

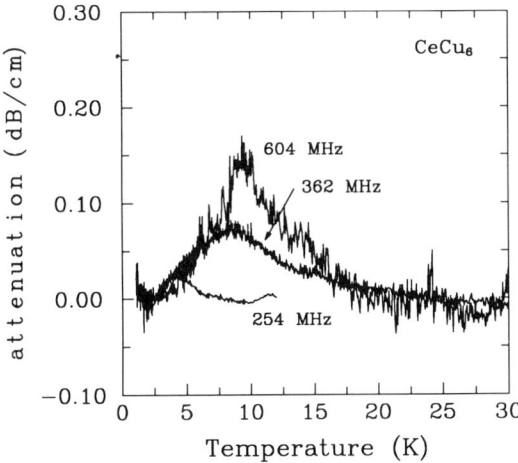

FIG. 37. Frequency dependence of the longitudinal sound attenuation (along the a-axis in CeCu$_6$ in the vicinity of the Kondo temperature.

$CeCu_6$ undergoes a structural phase transition to a monoclinic phase at a temperature, T_s, in the vicinity of 220 K. A lower transition temperature is obtained if the crystal is strained (which may arise just from the pulling speed during crystal growth). Weber et al. (1987) have measured the ultrasonic velocity in $CeCu_6$ and find a strong signature in all the elastic constants at T_s (see Fig. 35). There is also a large discontinuous change (a step) in the attenuation at T_s (Lin et al., 1990b, Fig. 36) of 3 dB/cm at a frequency of 130 MHz. As the temperature is lowered there is an attenuation peak at 2.5 K, which may be similar to the Kondo peak seen in UPt_3 by Müller et al. (1986a). The frequency dependence of this low-temperature peak has been measured (Lin et al., 1991). The attenuation peak shifts to higher temperatures with higher frequencies: At 254 MHz the peak is at 4 K, and this shifts to near 10 K at a frequency of 604 MHz. The data are shown in Fig. 37. At lower temperatures the attenuation increases, along with a monotonic decrease in the resistivity, while the velocity decreases. In a magnetic field, there is a large decrease in the velocity, which is consistent with measurements on other heavy fermion system (Weber et al., 1987; Lin et al., 1990b). This is due to the large magnetic Grüneisen parameters of the heavy fermion compounds. Figure 38 shows the velocity change in $CeCu_6$ at two different temperatures, 1 K and 70 mK, when the field is increased from 0 to 8 T.

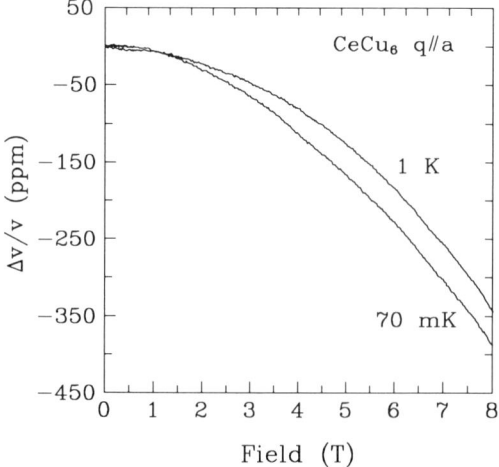

FIG. 38. Variation of longitudinal velocity in $CeCu_6$ with magnetic field, at 70 mK and 1 K. The field was applied along the a-axis. After Lin et al. (1990b). (© 1990 IEEE.)

9. Conclusions

The heavy fermion superconductors are intriguing systems to study. UPt_3 clearly shows unconventional behavior, with a nonphonon mechanism for the pairing. Ultrasonic techniques have proved very useful in discovering the unconventional characteristics: (i) zeroes in the gap from power-law temperature dependencies in the attenuation, and (ii) existence of multiple superconducting phases from sound attenuation. The first complete phase diagram has been obtained from velocity measurements, showing the existence of a tetracritical point on the H_{c2} curve, for all field orientations. Other thermodynamic measurements agree very well with those from ultrasonic measurements, when performed on the same sample.

The information on the other two heavy fermion superconductors is rather weak: Good samples (as compared to UPt_3) are not available. However, there are indications that URu_2Si_2 may show a double heat capacity jump and multiple superconducting phases. This system has the advantage that the AFM ordering at 17.5 K is readily observed in sound velocity, and the coupling of AFM ordering with the superconducting ordering could be studied.

$CeCu_6$ does not show any ordering to the lowest temperature, and the transition to the various coherent states can be studied. $CeCu_6$ is especially interesting, being probably one of those few metallic compounds where the sound velocity is larger than the Fermi velocity. As a consequence, interesting possibilities arise if one could get samples with large mean free paths. Recently, there have been some preliminary results indicating that $CeCu_6$ may also show some ordering, possibly antiferromagnetic, if cooled to very low temperatures (< 6 mK) (Jin et al., 1991). Because of the low ordering temperature, available sound frequencies are now of the order of the AFM energy.

ACKNOWLEDGMENTS

The work at UW—Milwaukee was supported by the Office of Naval Research, and at Northwestern by the National Science Foundation under grant DMR-89-07396. We thank Jim Sauls for many useful discussions and for also doing the calculations of $\Delta v/v$ (pg 159) and S. W. Lin for help in preparing this manuscript.

References

Adenwalla, S. (1989). Ph.D. thesis, Northwestern University.
Adenwalla, S., Lin, S. W., Ran, Q. Z., Zhao, Z., Ketterson, J. B., Sauls, J. A., Taillefer, L., Hinks, D., Levy, M., and Sarma, B. K. (1990). *Phys. Rev. Lett.* **65**, 2298.

Adenwalla, S., Zhao, Z., Ketterson, J. B., Lin, S. W., Ran, Q. Z., Levy, M., and Sarma, B. K. (1991). *Bull. Am. Phys. Soc.* **36**, 606.
Aeppli, G., Goldman, A., Shirane, G., Bucher, E., and Lux-Steiner, M. Ch. (1987). *Phys. Rev. Lett.* **58**, 808.
Aeppli, G., Bucher, E., Broholm, C., Kjems, J. K., Baumann, J., and Hufnagl, J. (1988). *Phys. Rev. Lett.* **60**, 615.
Aeppli, G., Bishop, D., Broholm, C., Bucher, E., Siemensmeyer, K., Steiner, M., and Stüsser, N. (1989). *Phys. Rev. Lett.* **63**, 676.
Akhiezer, A. I., Kaganov, M. I., and Liubarskii, G. Ia. (1957). *Sov. Phys. JETP* **5**, 685.
Ahlers, G. (1978). In "The Physics of Liquid and Solid Helium" (K. H. Bennemann and J. B. Ketterson, eds.), Wiley, New York, 1978, Part I. pp. 85–206.
Amato, A., Jaccard, D., Walker, E., and Flouquet, J. (1985). *Solid State Commun.* **55**, 1131.
Anderson, P. W. & Brinkman, W. F. (1973). *Phys. Rev. Lett.* **30**, 1108.
Anderson, P. W., and Brinkman, W. F. (1978). In "The Physics of Liquid and Solid Helium" (K. H. Bennemann and J. B. Ketterson, eds.), Wiley, New York, 1978, Part II. pp. 177–286.
Balian, R., and Werthamer, N. R. (1963). *Phys. Rev.* **131**, 1553.
Bardeen, J., Cooper, L. N., and Schrieffer, J. R. (1957). *Phys. Rev.* **108**, 1175.
Bishop, D. J., Varma, C. M., Batlogg, B., Bucher, E., Fisk, Z., and Smith, J. L. (1984). *Phys. Rev. Lett.* **53**, 1009.
Bishop, D. J., Batlogg, B., Varma, C. M., Bucher, E., Fisk, Z., and Smith, J. L. (1984). *Physica* **126B**, 455–456.
Blount, E. I., Varma, C. M., and Aeppli, G. (1990). *Phys. Rev. Lett.* **64**, 3074.
Broholm, C., Kjems, J. K., Buyers, W. J. L., Matthews, P., Palstra, T. T. M., Menovsky, A. A., and Mydosh, J. A. (1987). *Phys. Rev. Lett.* **58**, 1467.
Broholm, C., Aeppli, G., Kleiman, R. N., Harshman, D. R., Bishop, D. J., Bucher, E., Williams, D. L., Ansaldo, E. J., and Heffner, R. H. (1990). *Phys. Rev. Lett.* **65**, 2062.
Bruls, G., Weber, D., Wolf, B., Thalmeier, P., Lüthi, B., deVisser, A., and Menovsky, A. (1990). *Phys. Rev. Lett.* **65**, 2294.
Bullock, G. L., Shivaram, B. S., and Hinks, D. G. (1990). *Physica* **C169**,. 497.
Choi, C. H., and Sauls, J. A. (1991). *Phys. Rev. Lett.* **66**, 484–487.
Cribier, D., Jacrot, B., Rao, L. M., and Farnoux, B. (1964). *Phys. Lett.* **9**, 106.
de Gennes, P. (1989). "Superconductivity of Metals and Alloys," Reprint Edition, Addison Wesley.
de Visser, A., Franse, J. J. M., Menovsky, A., and Palstra, T. T. M. (1985). *J. Phys.* **F15**, L53.
de Visser, A., Menovsky, A., and Franse, J. J. M. (1987). *Physica* **147B**, 81–160.
Ellman, B., Yang, J., Rosenbaum, T. F., and Bucher, E. (1990). *Phys. Rev. Lett.* **64**, 1569.
Essman, V., and Trauble, H. (1967). *Phys. Lett.* **24A**, 526.
Fisher, R. A., Kim, S., Woodfield, B. W., Phillips, N. E., Taillefer, L., Hasselbach, K., Flouquet, J., Giorgi, A. L., and Smith, J. L. (1989). *Phys. Rev. Lett.* **62**, 1411.
Fisk, Z., Hess, D. W., Pethick, C, J., Pines, D., Smith, J. L., Thompson, J., and Willis, J. O. (1988). *Science* **239**, 33.
Frings, P. H., Fanse, J. J. M., de Boer, F. R., and Menovsky, A. (1983). *J. Magn. Magn. Mat.* **31–34**, 240.
Fukase, T., Kotke, Y., Nakanomyo, T., Shiokawa, Y., Menovsky, A. A., Mydosh, J. A., and Kes, P. H. (1987). *Jpn. J. Appl. Phys.* **26**, 1249.
Fulde, P., Keller, J., and Zwicknagl, G. (1988). *Solid State Phys.* **41**, 1.
Ginzburg, V. L., and Landau, L. D. (1950). *Zh. Eksp. Teor. Fiz.* **20**, 1064.
Golding, B., Bishop, D. J., Batlogg, B., Haemmerle, W. H., Fisk, Z., Smith, J. L., and Ott, H. R. (1985). *Phys. Rev. Lett.* **55**, 2479.
Gor'kov, L. P. (1987). *Sov. Sci. Rev.* **A9**, 1.
Greiter, M., Taillefer, L., Austin, N., and Lonzarich, C. G. (1990). In "The Proceedings of the International Conference on Valence Fluctuations," Rio de Janeiro (to be published).

3. Sound Propagation in the Heavy Fermion Superconductors

Hasselbach, K., Taillefer, L., and Flouquet, J. (1989). *Phys. Rev. Lett.* **63**, 93.
Hasselbach, K., Lacerda, A., Behnia, K., Taillefer, L., Flouquet, J., and de Visser, A. (1990). *J. Low Temp. Phys.* **81**, 299.
Hess, D. W., Tokuyasu, T. A., and Sauls, J. A. (1989). *J. Phys. Condens. Matter* **1**, 8135.
Hess, H. F., Robinson, R. B., and Waszczak, J. V. (1990). *Phys. Rev. Lett.* **64**, 2711.
Hess, H. F., Robinson, R. B., and Waszczak, J. V. (1991). *Physica* **B169**, 422–431.
Hirschfeld, P., Volhardt, D., and Wolfle, P. (1986). *Solid State Commun.* **59**, 111–115.
Jin, C., Pollack, L., Smith, E. N., and Lee, D. M. (1991). *Bull. Am. Phys. Soc.* **36**, 717.
Joynt, R. (1988). *Sup. Sci. Tech.* **1**, 210.
Kitazawa, K., Kambe, S., and Naito, M. (1989). preprint.
Kittel, C. (1971). "Introduction to Solid State Physics," Fourth Edition, J. Wiley & Sons, New York.
Kleiman, R. N., Gammel, P. L., Bucher, E., and Bishop, D. J. (1989). *Phys. Rev. Lett.* **62**, 328.
Kleiman, R. N., Bishop, D. J., Ott, H. R., Fisk, Z., and Smith, J. L. (1990). *Phys. Rev. Lett.* **64**, 1975.
Kleiman, R. N., Broholm, C., Aeppli, G., Bucher, E., Stücheli, E., Bishop, D. J., Clausen, K. N., Howard, B., Mortensen, K., and Petersen, J. S. (1991). *Bull. Am. Phys. Soc.* **36**, 608.
Kouroundis, I., Weber, D., Yoshizawa, M., Luthi, B., Puech, L., Haen, P., Flouquet, J., Bruls, G., Welp, U., Franse, J. M., Menovsky, A., Bucher, E., and Hufnagl, J. (1987). *Phys. Rev. Lett.* **58**, 820.
Kwok, W. K., Welp, U., Crabtree, G. W., Vandervoort, K. G., Hulscher, R., Zheng, Y., Dabrowski, B., and Hinks, D. G. (1989). *Phys. Rev.* **B40**, 9400.
Kwok, W. K., DeLong, L. E., Crabtree, G. W., Hinks, D. G., and Joynt, R. (1990). *Phys. Rev.* **B41**, 11649–52.
Lee, P. A., Rice, T. M., Serene, J. W., Sham, L. J., and Wilkins, J. W. (1986). *Comments Condensed Matter Phys.* **12**, 99.
Lin, S. W., Adenwalla, S., Ran, Q. Z., Sun, K. J., Hinks, D. G., Ketterson, J. B., Levy, M., and Sarma, B. K. (1990a). *Physica* **B165 & 166**, 423.
Lin, S. W., Ran, Q.-Z., Adenwalla, S., Zhao, Z., Edelstein, A., Das, B. N., Ketterson, J. B., Sarma, B. K., and Levy, M. (1990b). *In* "IEEE 1990 Ultrasonics Symposium Proceedings." (90 CH2938-9, B. R. McAvoy, ed). IEEE, New York, pp. 1313–1316.
Lin, S. W., Ran, Q.-Z., Adenwalla, S., Ketterson, J. B., Hinks, D. G., Levy, M., and Sarma, B. K. (1991). *IEEE-Symposium, Orlando, FL (to be published)*.
Lüthi, B. (1985). *J. Mag. and Magnetic Materials* **52**, 70.
Machida, K., and Ozaki, M. (1989). *J. Phys. Soc. Jpn.* **58**, 2244.
Maki, K. (1964). *Physica* **1**, 127.
Maki, K. (1967). *Phys. Rev.* **156**, 437.
Maki, K., and Cyrot, M. (1967). *Phys. Rev.* **156**, 433.
Maple, M. B., Chen, J. W., Lambert, S. E., Fisk, Z., Smith, J. L., Ott, H. R., Brooks, J. S., and Naughton, M. J. (1985). *Phys. Rev. Lett.* **54**, 477.
Maple, M. B., Chen, J. W., Dalichaouch, Y., Kohara, T., Rossel, C., Torikachvili, M. S., McElfresh, M. W., and Thompson, J. D. (1986). *Phys. Rev. Lett.* **56**, 185.
Maple, M. B., Dalichaouch, Y., Lea, B. W., Seaman, C. L., Tsai, P. K., Armstrong, P. E., Fisk, J., Rossel, C., and Torikachvili, M. S. (1991). *Physica* **B171**, 219.
Martin, P. C., Parodi, O., and Pershan, P. S. (1972). *Phys. Rev.* **A6**, 167.
Millis, A. J., and Rabe, K. M. (1988). *Phys. Rev.* **B38**, 8908.
Mühlschlegel, B. (1959). *Z. Phys.* **155**, 313.
Müller, V., Maurer, D., de Groot, K., Bucher, E., and Bömmel, H. E. (1986a). *Phys. Rev. Lett.* **56**, 248.
Müller, V., Maurer, D., Scheidt, E. W., Roth, Ch., Luders, K., Bucher, E., and Bommel, H. E. (1986b). *Sol. State Comm.* **57**, 319.

Müller, V., Roth, Ch., Maurer, D., Scheidt, E. W., Luders, K., Bucher, E., and Bommel, H. E. (1987a). *Phys. Rev. Lett.* **58,** 1224.
Müller, V., Maurer, D., Roth, C., Scheidt, E. W., Luders, K., Bucher, E., and Bömmel, H. E. (1987b). *Proc. 18th Int. Conf. on Low Temp. Phys., J. Jnl. Appl. Phys.* **26,** 1223–24.
Müller, V., Maurer, D., and de Grott, K. (1987c). *Physica* **148B,** 73.
Narlikar, A. V., and Ekbote, S. N. (1983). "Superconductivity and Superconducting Materials." South Asia Publishers, New Delhi.
Onuki, Y., Yamazaki, T., Ukon, I., Omi, T., Shibutani, K., Komatsubara, T., Sakamoto, I., Sugiyama, Y., Onodera, R., Yonemitsu, K., Umezawa, A., Kwok, W. K., Crabtree, G. W., and Hinks, D. G. (1987). *Physica* **148B,** 29–32.
Ott, H. R., Rudigier, H., Fisk, Z., and Smith, J. L. (1983). *Phys. Rev. Lett.* **50,** 1595.
Palstra, T. T. M., Menovsky, A. A., van den Berg, J., Dirkmaat, A. J., Kes, P. H., Nieuwenhuys, G. J., and Mydosh, J. A. (1985). *Phys. Rev. Lett.* **55,** 2727.
Parks, R. D. (1969). "Superconductivity." Marcel Dekker, Inc., New York.
Pippard, A. B. (1953). *Proc. Roy. Soc.* **A216,** 547.
Qian, Y. J., Xu, M.-F., Schenstrom, A., Baum, H.-P., Ketterson, J. B., Hinks, D., Levy, M., and Sarma, B. K. (1987). *Solid State Commun.* **63,** 599.
Ran, Q.-Z., Lin, S. W., Adenwalla, S., Hinks, D. G., Zhao, Z., Ketterson, J. B., Levy, M., and Sarma, B. K. (1990). *In* "IEEE 1990 Ultrasonics Symposium Proceedings;" (90 CH2938-9, B. R. McAvoy, ed.) IEEE, New York, pp. 1225–1228.
Rauchschwalbe, U. (1987). *Physica* **B147,** 1.
Rodriguez, J. P. (1985). *Phys. Rev. Lett.* **55,** 250.
Schenstrom, A., Xu, M.-F., Hong, Y., Bein, D., Levy, M., Sarma, B. K., Adenwalla, S., Zhao, Z., Tokuyasu, T., Hess, D. W., Ketterson, J. B., Sauls, J. A., and Hinks, D. G. (1989a). *Phys. Rev. Lett.* **62,** 332.
Schenstrom, A., Xu, M.-F., Hong, Y., Levy, M., Sarma, B. K., Adenwalla, S., Zhao, Z., Ketterson, J. B., and Hinks, D. (1989b). *J. Less Common Metals* **149,** 353.
Schmitt-Rink, S., Miyake, K., and Varma, C. M. (1986), *Phys. Rev. Lett.* **57,** 2575.
Schotte, K. D., Förster, D., and Schotte, U. Z. (1986). *Phys. B (Cond. Matt.)* **64,** 165.
Shivaram, B. S., Jeong, Y. H., Rosenbaum, T. F., and Hinks, D. G. (1986a). *Phys. Rev. Lett.* **56,** 1078.
Shivaram, B. S., Rosenbaum, T. F., and Hinks, D. G. (1986b). *Phys. Rev. Lett.* **57,** 1259.
Shivaram, B. S., Jeong, Y. H., Rosenbaum, T. F., Hinks, D. G., and Schmitt-Rink, S. (1987). *Phys. Rev.* **B35,** 5372.
Shivaram, B. S., Gannon, J. J., Jr., and Hinks, D. G. (1989). *Phys. Rev. Lett.* **63,** 1723.
Stewart, G. R. (1984). *Rev. Mod. Phys.* **56,** 755.
Sulpice, A., Gandit, P., Chaussy, J., Flouquet, J., Jaccard, D., Lejay, P., and Tholence, J. L. (1986). *J. Low. Temp. Phys.* **62,** 39.
Sun, K. J., Schenstrom, A., Sarma, B. K., Levy, M. and Hinks, D. J. (1989). *Phys. Rev.* **B40,** 11284.
Taillefer, L. (1990). *Physica* **B163,** 278–284.
Taillefer, L., Flouquet, J., Hayden, S. M., and Vettier, C. (1990). *Proc. of YAMADA Conf. (Osaka),* April.
Testardi, L. (1971). *Phys. Rev.* **B3,** 95.
Testardi, L. R. (1973). *In* "Elastic Behavior and Structural Instability of High Temperature A-15 Structure Superconductors," Physical Acoustics, Vol. X (W. P. Mason and R. N. Thurston, eds). Academic Press, New York, pp. 193–296.
Thuneberg, E. V. (1987). *Phys. Rev.* **B36,** 3583.
Tinkham, M. (1985). "Introduction to Superconductivity." Robert E. Krieger Publishing Co., Malabar, Florida.
Tokuyasu, T. A. (1990). "Vortex States in Unconventional Superconductors." Ph.D. thesis, Princeton University.

Tokuyasu, T. A., and Sauls, J. A. (1990). *Physica* **B165–166,** 347.
Tokuyasu, T. A., Hess, D. W., and Sauls, J. A. (1990). *Phys. Rev.* **B41,** 8891.
Trappmann, T., Löhneysen, H. V., and Taillefer, L. (1991). *Phys. Rev.* **B43,** 13714–16.
Vandervoort, K. G., Welp, U., Kessler, J. E., Claus, H., Crabtree, G. W., Kwok, W. K., Umezawa, A., Veal, B. W., Downey, J. W., Paulikas, A. P., and Lin, J. Z. (1991). *Phys. Rev.* **B43,** 13042.
Vincent, E., Hammann, J., Taillefer, L. Behnia, K., Keller, N., and Flouquet, J. (1991). preprint.
Volovik, G. E. (1988). *J. Phys.* **C21,** L221.
Vrtis, M. (1987). Ph.D. thesis, Northwestern University.
Weber, D., Yoshizawa, M., Kouroudis, I., Lüthi, B., and Walker, E. (1987). *Europhys. Lett.* **3,** 827.
Yip, S., Li, T. C., and Kumar, P. (1991). *Phys. Rev.* **B43,** 2742.
Yoshizawa, M., Lüthi, B., Goto, T., Suzuki, T., Render, B., de Visser, A., Frings, P., and Franse, J. J. M. (1985). *J. Magn. Mag. Mat.* **52,** 413.
Yoshizawa, M., Lüthi, B., and Schotte, K. D. (1986). *Z. Phys. B. (Cond. Matt.)* **64,** 169.
Zemirli, S., and Barbara, B. (1985). *Solid State Commun.* **56,** 385.
Zhao, Z., Behroozi, F., Adenwalla, S., Guan, Y., Ketterson, J. B., Sarma, B. K., and Hinks, D. G. (1991a). *Phys. Rev. B.* **43,** 13720.
Zhao, Z., Adenwalla, S., Sarma, B. K., and Ketterson, J. B. (1991b). Advances in Physics (in press).
Zhitomirskii, M., (1989). *Zh. Eksp. Teor. Fiz. Picma* **49,** 333.

—4—
Ultrasonic Attenuation in the Magnetic Superconducting System $Er_{1-x}Ho_xRh_4B_4$

KEUN JENN SUN
Department of Physics, The College of William and Mary, Williamsburg, Virginia

MOISES LEVY
Department of Physics, University of Wisconsin—Milwaukee, Milwaukee, Wisconsin

1. Introduction 191
2. Experimental Details and Physical Properties 194
3. Common Attenuation Behavior 196
4. Special Attenuation Behavior 198
   4.1. Enhanced Ultrasonic Attenuation in the Superconducting State and
        Spin–Phonon Interaction 198
   4.2. Ultrasonic Attenuation of $Er_{0.705}Ho_{0.295}Rh_4B_4$ and $ErRh_4B_4$ 207
   4.3. Relaxation Mechanism of Ultrasonic Attenuation in Ho-Rich $Er_{1-x}Ho_xRh_4B_4$ 218
   4.4. Sound Absorption of $Er_{0.088}Ho_{0.912}Rh_4B_4$ at High Magnetic Fields 227
5. Summary 230
   Acknowledgments 232
   References 233

1. Introduction

The effects of magnetic impurities on the properties of superconductors have been of continuing interest, because magnetic order can not easily coexist with superconductivity (Ginsburg, 1957; Herring, 1958; Suhl and Matthias 1959). Early experimental results (Matthias *et al.*, 1958a, 1958b; Lynn *et al.*, 1980) have shown that the addition of magnetic impurities to superconductors reduced the superconducting transition temperature T_c to lower values, or even quenched superconductivity completely. It is generally true that a magnetic force, either internal or external to a superconductor, tends to break the superconducting-electron pairs and quench the superconducting state (Ihikawa, 1982).

Nevertheless, from the time when it became possible to synthesize rare-earth (RE) ternary compounds such as $RERh_4B_4$ (Matthias *et al.*, 1977; Vandenberg and Matthias, 1977), $REMo_6S_8$ (Fischer *et al.*, 1975), and $REMo_6Se_8$ (Shelton *et al.*, 1976), magnetic superconductors have been studied intensively. It was found experimentally that a single-crystal sample of $ErRh_4B_4$ did exhibit evidence of coexistence of a long-range magnetic order and superconductivity in a small temperature interval (Sinha et al., 1982a, 1982b); $ErRh_4B_4$ underwent a superconducting transition at 8.7 K. However, the onset of ferromagnetic order at a lower temperature, 0.71 K (Fertig et al., 1977), terminated its superconducting behavior and turned the sample back to the normal state (so-called reentrant superconductor). A sinusoidally modulated magnetic order with wavelength ~ 100 Å (Moncton et al., 1980) has been detected in the superconducting state at temperatures between 0.71 K and 1.2 K. Theoretically (Blount and Varma, 1979; Matsumoto et al., 1979; Kuper et al., 1980), it is possible to have this coexistence state if the electromagnetic interaction between superconducting currents and periodically placed magnetic ions plays a predominant role over the exchange interaction between the magnetic ions in these materials. However, not all the $(RE)Rh_4B_4$ type compounds have this reentrant (to the normal state) behavior. Some of them, such as $NdRh_4B_4$, $SmRh_4B_4$, and $TmRh_4B_4$, are antiferromagnetic superconductors whose superconductivity persists and coexists with magnetic order below Neel's temperature. Others, such as $LuRh_4B_4$, are superconductors with no magnetic transitions. And the rest are ferromagnetic compounds. $HoRh_4B_4$ is one of the last group. The dominance of a strong exchange interaction between rare earth ions in these ferromagnets prevents the occurrence of superconductivity (Ihikawa, 1982). Detailed information on the properties of $(RE)Rh_4B_4$ can be obtained in "Ternary Superconductors," edited by Shenoy et al. (1981), and in "Superconductivity in Ternary Compounds," edited by Fischer and Maple (1982). In this chapter, we will concentrate on the systems with RE = Er and Ho.

Holmium (Ho) and erbium (Er) are neighbors in the periodic table; the former has 10 $4f$ electrons with total angular momentum $J = 8$, and the latter has 11 $4f$ electrons with $J = 15/2$. Both of them have a well-localized f-electron shell and the same configuration for the rest of the electrons. However, $HoRh_4B_4$ and $ErRh_4B_4$ have quite different characteristics at low temperatures, in spite of their similar crystal structure (Woolf et al., 1979; Maple 1983). For example, magnetization and specific-heat measurements have shown that $HoRh_4B_4$ has mean-field behavior (Ott et al., 1980, 1982) at its magnetic transition, while $ErRh_4B_4$ shows relatively large spin fluctuations at its Curie temperature (Woolf et al., 1979). In $Er_{1-x}Ho_xRh_4B_4$ samples, the Er ions tend to retain the superconductivity caused by the $4d$ electrons of the rhodium (Rh) atoms, while the Ho ions tend to destroy it. When x, which represents the Ho concentration, is less than 0.9, the materials have a superconducting state above their respective Curie temperatures. On the other hand, when x is larger than 0.9, samples display only ferromagnetic phase transitions (Johnston et al., 1978). Therefore, for a study of the interplay between long-range ferromagnetic order and superconductivity,

$Er_{1-x}Ho_xRh_4B_4$ should be a system that could yield interesting information about the evolution from mean-field ferromagnetic to reentrant superconducting behavior. Besides, because the rare earth ions are located in an ordered sublattice, these compounds also exhibit unusual magnetic and superconducting characteristics (Crabtree *et al.*, 1982; Behroozi *et al.*, 1983), which have not been seen in conventional superconductors and are worth investigating.

The ultrasonic technique has been a useful tool for studying the temperature variation of superconducting energy gap of conventional superconductors, such as Nb, V, Pb, Sn, Al, etc. It is natural to apply this measurement technique to the magnetic superconductors in order to see how the electron–phonon interaction is affected by the presence of magnetic ions. Experiments have been conducted on several samples of the $Er_{1-x}Ho_xRh_4B_4$ system. However, the typical ultrasonic attenuation behavior described in the BCS theory has not been observed. These results do not indicate that the phonon-mediated superconductivity mechanism fails to apply to magnetic superconductors. As a matter of fact, tunneling experiments (Pobell, 1981) on $Cu_{1.8}Mo_6S_8$ and $PbMo_6S_8$ suggest a strong electron–phonon interaction for the superconductivity of these systems. The major reason for the absence of the expected temperature-dependent attenuation variations was due to the complex nature of $Er_{1-x}Ho_xRh_4B_4$. Several additional interactions stemming from crystalline electric fields (CEF), critical spin fluctuations, as well as superconducting currents (or electrons), compete with each other at low temperatures. Also, the predominance of each interaction in different temperature ranges results in a corresponding additional sound absorption, which may obscure the attenuation due to electron–phonon interaction. Although these unexpected sources of energy dissipation of sound have diminished the opportunity for exploring, ultrasonically, how the magnetic ions affect the temperature variation of the superconducting energy gap of $Er_{1-x}Ho_xRh_4B_4$, the mechanisms developed to analyze the ultrasonic attenuation data of $Er_{1-x}Ho_xRh_4B_4$ are interesting and informative—especially when attenuation anomalies have also been observed at the superconducting transitions of high-temperature superconductors $YBa_2Cu_3O_7$ (Xu *et al.*, 1988; Sun *et al.*, 1988), Tl–Ca–Ba–Cu–O (Sun *et al.*, 1989d), and Bi–Sr–Ca–Cu–O (Sun *et al.*, 1990), and the models for explaining these anomalies are still being developed.

Since the discovery of high-temperature superconductors, the association of superconductivity with antiferromagnetism has been studied intensively. A magnetic mechanism (Baskaran *et al.*, 1987) that is responsible for the electron pairing has been postulated. The coupling of spins of the Cu ions with charge carriers (Warren *et al.*, 1989; Kurihara, 1989), as well as the exchange-type (Kurihara, 1989) magnetic interactions between spins of the Cu ions in adjacent Cu–O planes, was proposed to explain several experimental observations. It was predicted (Kurihara, 1989) that a possible lattice distortion that involves the displacement of oxygen ions along the c-axis would occur below T_c if the superexchange interaction between Cu–O planes plays a significant role for the pairing mechanism. In this case, an ultrasonic attenuation peak may occur and may be closely associated with this distortion. Another possible model of a magnetic mechanism also concerns the exchange interaction, but is more directly related to sound propagation (Tachiki and Maekawa, 1974). Since the exchange constant between spins is spatially dependent, lattice deformation caused by the traveling sound waves will vary this constant in an oscillatory manner. It is possible that through this mechanism, an anomalous sound energy dissipation will be observed at temperatures near a magnetic or a superconducting phase transition if critical spin fluctuations occur. In this respect, the analysis of ultrasonic attenuation data of the magnetic superconductors should provide fruitful information concerning the ultrasonic properties of high-T_c superconductors.

Ultrasonic attenuation measurements as a function of temperature at constant magnetic fields and as a function of magnetic field at constant temperatures have been performed on samples with various values of x in the $Er_{1-x}Ho_xRh_4B_4$ system. Frequency-dependent attenuation data were also obtained for two of the samples. These samples not only exhibited common behavior, such as the unusual relaxation-type attenuation maxima whose source was the split $4f$ electron energy levels of magnetic ions, but also displayed individually characteristic properties that resulted from the existence or the nonexistence of the superconducting state. The enhanced ultrasonic attenuation observed in the superconducting state of the reentrant superconductors, and the anisotropy of the ultrasonic attenuation in magnetic fields observed in the samples with $x = 0$ and $x = 0.295$, are two examples of the latter case. In the following, a general survey of the experimental results is given in order to ascertain the common behavior in this system and to examine the characteristic trends in ultrasonic attenuation that could possibly elucidate the nature of the system. A few models are suggested to interpret the data as well.

2. Experimental Details and Physical Properties

The samples of $Er_{1-x}Ho_xRh_4B_4$ were synthesized by arc melting the high-purity elements under argon gas, followed by annealing in sealed tantalum tubes at 1,200°C for one week and then at 900°C for three weeks (Johnston et al., 1978). Magnetic-susceptibility measurements (Johnston et al., 1978), specific-heat measurements (Mackay et al., 1979), and neutron diffraction measurements (Mook et al., 1982a and 1982b) revealed that samples with $x > 0.9$ showed mean-field

4. Magnetic Superconducting System

behavior and were ferromagnets with Curie temperature T_m. Those with $x < 0.9$ were reentrant superconductors; the onset of long-range ferromagnetic order quenched superconductivity at a second critical temperature T_{c2} that was lower than the temperature T_{c1} at which the sample became superconducting and which was of the order of T_m. In addition, for the samples with $x < 0.3$, there was a sinusoidally modulated magnetic order that coexisted with superconductivity at temperatures between T_m and T_{c2}. Here, T_m is the temperature at which the sinusoidally modulated magnetic order occurs and $T_m > T_{c2}$. For those with $0.3 < x < 0.9$, there is no coexistence state and then the value of T_m is the same as that of T_{c2}.

Temperature-dependent attenuation in zero magnetic field was measured on the samples with $x = 0$, 0.295, 0.6, 0.813, 0.912, and 1. A 15-MHz quartz transducer was epoxy-bonded to one of the parallel faces of each sample. The pulse echo technique described elsewhere (Sun, 1986) was employed for the attenuation measurements. Both double-echo and single-echo measurements have been conducted, especially on the sample with $x = 0.6$, and the measured attenuation showed the same results. Longitudinal as well as shear wave attenuation was also measured for the sample with $x = 0.6$, and both wave modes displayed the same characteristic attenuation features. Therefore, the following discussions will be mainly based on single-echo measurements of longitudinal waves. The dimensions, longitudinal sound velocities, and phase transition temperatures of the samples are listed in Table I. Values of critical temperatures were determined by ac magnetic-susceptibility measurements that were done simultaneously with temperature-dependent attenuation measurements and were

TABLE I

SOUND VELOCITIES[a] AND CRITICAL TEMPERATURES OF POLYCRYSTALLINE SAMPLES OF $Er_{1-x}Ho_xRh_4B_4$

Sample[b]	Dimensions (cm)	Sound Velocity (m/s)[a]	T_{c1}/T_{c2} (K)	T_m (K)
$x = 1$	1.27 × 0.66 × 0.87	5.4 × 10³ ± 7% (L)	—	6.4
$x = 0.912$	0.74 × 0.54 × 0.43	6.3 × 10³ ± 1% (L)	—	6.2
$x = 0.813$	0.70 × 0.53 × 0.43	5.9 × 10³ ± 1% (L)	6.0/4.95	4.95
$x = 0.6$	0.71 × 0.51 × 0.45	6.0 × 10³ ± 1% (L)	6.7/3.45	3.45
		2.8 × 10³ ± 2% (S)		
$x = 0.295$	1.30 × 0.68 × 0.48	7.0 × 10³ ± 7% (L)	7.6/0.99	1.31[c]
				0.89[d]
$x = 0$	0.42 × 0.35 × 0.31	6.0 × 10³ ± 5% (L)	8.7/0.96[b]	0.96[b]

[a] (L) = longitudinal wave, (S) = shear wave.
[b] Polycrystalline sample.
[c] Ho.
[d] Er.

complemented by magnetic-field–dependence attenuation measurements. The sample of $Er_{0.705}Ho_{0.295}Rh_4B_4$ has two ferromagnetic transition temperatures, 0.99 K and 0.89 K (Curie temperatures), which correspond to the alignments of the Ho^{3+} and Er^{3+} ions, respectively. The polycrystalline samples of $Er_{0.4}Ho_{0.6}Rh_4B_4$ and $Er_{0.187}Ho_{0.813}Rh_4B_4$ did not exhibit a coexistence region; therefore, their T_ms are the same as their respective T_{c2}s.

3. Common Attenuation Behavior

The temperature-dependent ultrasonic attenuation curves at 15 MHz in zero magnetic field for samples with $x = 1, 0.912, 0.813$, and 0.6 are shown in Fig. 1. Except in the regime where the ultrasonic behavior was affected by the superconducting properties, the low-temperature attenuation data of all the samples but $ErRh_4B_4$ displayed a broad maximum at temperatures close to 10 K, and the temperature position (T_p) of this maximum depends on the Ho concentration (x) ($ErRh_4B_4$ has this type of maximum at 5 K). The attenuation curves

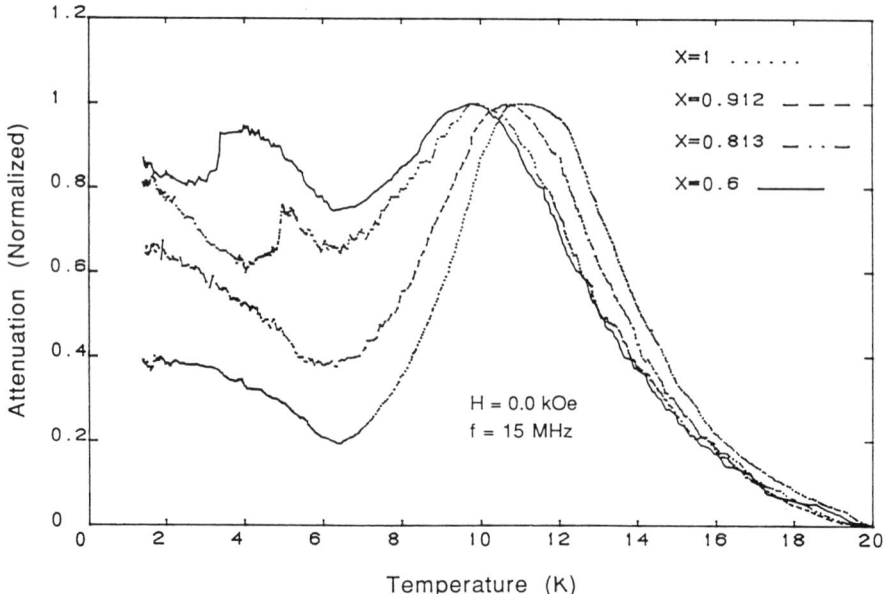

FIG. 1. Temperature-dependent ultrasonic attenuation of $Er_{1-x}Ho_xRh_4B_4$ with $x = 1, 0.912, 0.813$, and 0.6 at 15 MHz and zero external magnetic field.

4. Magnetic Superconducting System

are normalized in the following way: Taking the attenuation at 20 K as the reference of zero attenuation, the attenuation value on each curve at any temperature below 20 K is determined relative to this reference attenuation, and then normalized to the maximum attenuation value of the respective curve. Several features of the attenuation curves can be observed from the figure: (1) The position T_p of the bell-shaped relaxation maximum on each curve shifts to a lower temperature as the concentration x of the Ho atoms decreases. (2) The ratio of the attenuation at the magnetic phase transition temperature T_m to the maximum attenuation on each curve increases as x decreases. (3) For the samples with $x = 1$, and $x = 0.912$, the valleylike relative minimum on the low-temperature side of the bell-shaped maximum of each curve is located at the Curie temperature of each sample, which is 6.4 K and 6.2 K, respectively. However, for the samples with $x = 0.813$ and $x = 0.6$, there is a steplike change in attenuation at each Curie temperature, which is 4.95 K and 3.45 K, respectively, and there is a relative minimum at each superconducting phase transition temperature T_{c1}, which is 6.0 K and 6.7 K, respectively. (4) For the reentrant superconductors, the attenuation increases when the temperature is lowered from T_{c1} to T_m. (5) In the ferromagnetic state, the attenuation of all the samples increases as the temperature is lowered.

Figure 2 displays data of similar measurements but for the samples with $x = 0.295$ and $x = 0$. The curve for $Er_{0.705}Ho_{0.295}Rh_4B_4$ was taken at 79 MHz. If we take the same normalization procedure as before, it may appear that the curves in this figure behave in quite a different manner when compared with those in Fig. 1. Several features are worth mentioning in Fig. 2. First, there are a maximum at 3.45 K and a shoulder centered at 12 K for the sample with $x = 0.295$ at 79 MHz, and a big hump centered at 5 K in the attenuation curve of $ErRh_4B_4$ at 15 MHz. If the same mechanism that caused the broad maxima described in feature (1) of Fig. 1 is applicable, and if we assume that the shoulder is associated with Ho^{3+} ions, and that the maximum as well as the hump are associated with Er^{3+} ions, then the broad attenuation maxima of the attenuation curves in the two figures show a trend consistent with that stated in feature (1) when the upward shift in temperature due to frequency is taken into account for the 79 MHz curve. It was possible to perform frequency-dependent measurements on the samples with $x = 1$ as well as $x = 0.813$, and it was found that the bell-shaped maximum moved to higher temperatures as the frequency was increased. This is indicative of a relaxation-type attenuation maximum. A model that attributes attenuation to the electron-population changes of 4f-electron energy levels induced by the propagation of sound in the sample, and a derived equation that relates the attenuation to the Schottky specific heat, are presented later in this chapter. Second, there is a sharp and large peak at 0.96 K for $ErRh_4B_4$ that is

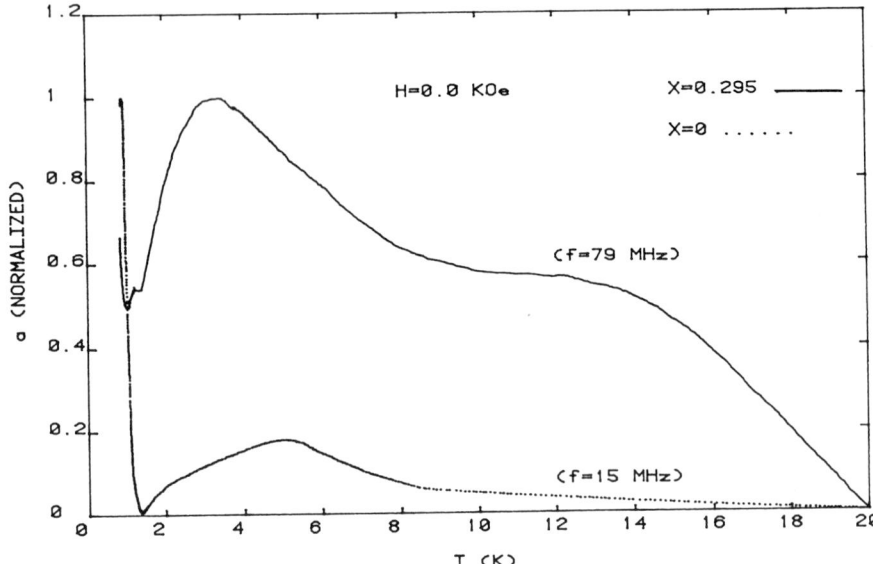

FIG. 2. Temperature-dependent ultrasonic attenuation of $Er_{1-x}Ho_xRh_4B_4$ with $x = 0.295$ and 0 at 79 MHz and 15 MHz, respectively, and at zero external magnetic field.

not seen on the attenuation curves of Ho-rich samples. Because the Er^{3+} ions have much stronger spin fluctuations at their magnetic phase transition, we believe that this attenuation anomaly results from the strong coupling between spins and phonons. This spin–phonon interaction is also involved in the increasing attenuation in the superconducting state. We will discuss these facts in detail in the next section.

4. Special Attenuation Behavior

4.1. Enhanced Ultrasonic Attenuation in the Superconducting State and Spin–Phonon Interaction

The ultrasonic attenuation in the superconducting state of $Er_{1-x}Ho_xRh_4B_4$ reentrant superconductors does not decrease with temperature as described in the BCS theory. In fact, the attenuation in the superconducting state of samples with $x = 0.6$ and $x = 0.813$ increases when the temperature decreases from T_{c1} to T_m. Temperature-dependent attenuation and ac susceptibility curves for $Er_{0.4}Ho_{0.6}Rh_4B_4$ are given in Fig. 3. As can be seen, the attenuation increased

4. Magnetic Superconducting System

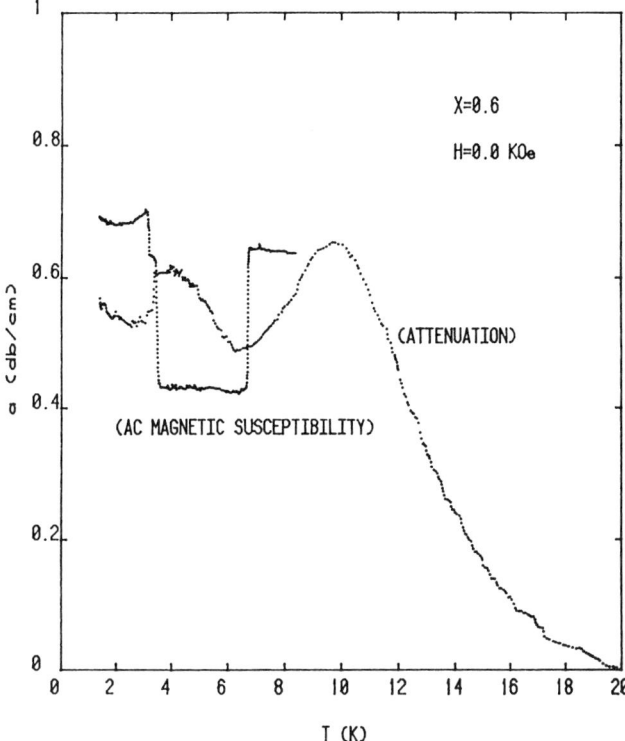

FIG. 3. Temperature-dependent ultrasonic attenuation and ac magnetic susceptibility of $Er_{0.4}Ho_{0.6}Rh_4B_4$ in the absence of a magnetic field. The two measurements are performed simultaneously.

immediately below 6.7 K till 3.45 K, which were identified by the ac susceptibility curve shown in Fig. 3 as T_{c1} and T_m, respectively. To confirm that the sample was in the superconducting state in the range between 3.45 K and 6.7 K, an external magnetic field was applied. Figure 4 displays the results of the same measurements as in Fig. 3, but at constant fields. An applied magnetic field did decrease the attenuation in this temperature range as long as the magnitude of the field was below H_{c2}. However, the attenuation outside this temperature range remained the same up to 3.0 kOe. Similar features have also been observed for the sample with $x = 0.813$. Also shown in Fig. 5 are the ac magnetic susceptibility curves of $Er_{0.187}Ho_{0.813}Rh_4B_4$ at constant fields. The diminishing of the U-shaped portion in the curves indicates the shrinkage of superconducting fraction in the sample. Therefore, the rise in attenuation in the superconducting state is truly associated with superconductivity. Furthermore, the results obtained from attenuation as a function of magnetic field at constant

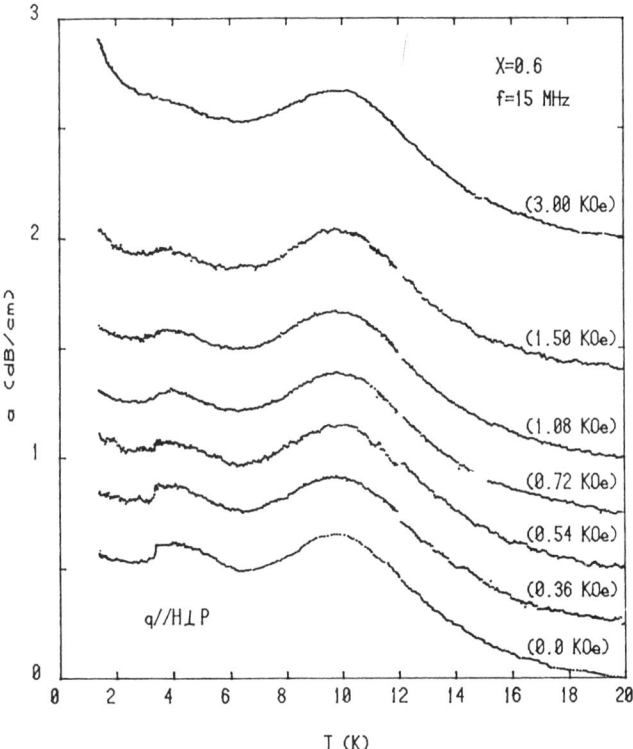

FIG. 4. Temperature-dependent ultrasonic attenuation of $Er_{0.4}Ho_{0.6}Rh_4B_4$ at constant external magnetic fields.

temperatures (Fig. 6) were also consistent with this interpretation. In Fig. 6, at the temperatures where the sample was superconducting, the attenuation decreased down to a certain field strength before it went up at higher fields. The field at which the minimum in attenuation located was identified as H_{c2}. Again, this behavior is different from what is observed in either type-I or type-II superconductors at their critical magnetic fields. Before proposing different interpretations for this enhanced attenuation in the superconducting state, let us look into the attenuation data of $ErRh_4B_4$ and $HoRh_4B_4$ separately.

$ErRh_4B_4$ is a well-investigated reentrant superconductor. The magnetic ordering of Er^{3+} ions that develops with decreasing temperature in the superconducting state eventually destroys superconductivity and returns the compound to the normal state at T_{c2} ($T_{c2} = 0.96$ K for our polycrystalline sample). Specific heat measurements (Moncton et al., 1977) on $ErRh_4B_4$ observed a spike-shaped feature associated with the long-range ferromagnetic ordering of Er^{3+} magnetic

4. Magnetic Superconducting System

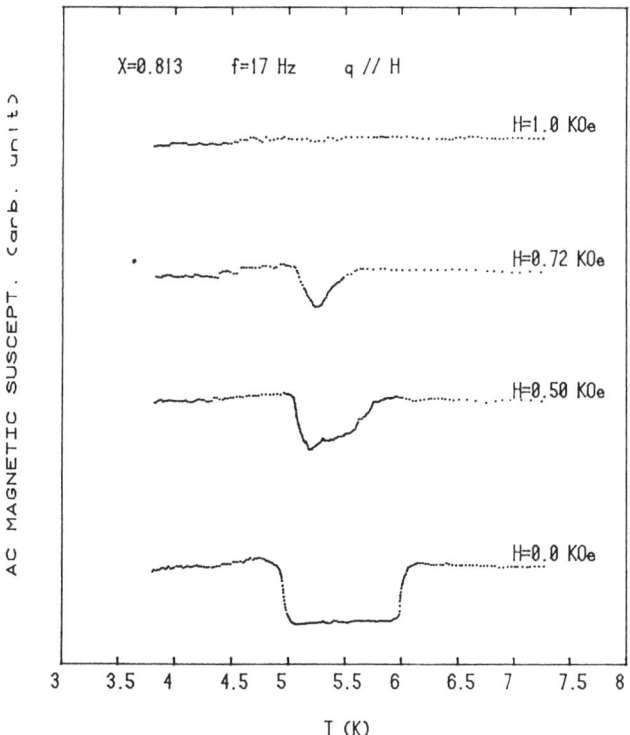

FIG. 5. Temperature-dependent ac susceptibility curves of $Er_{0.187}Ho_{0.813}Rh_4B_4$ at constant external magnetic fields. The diminishing of the U-shaped portion on the curves indicates the shrinking of superconducting state in the sample.

moments in the vicinity of T_{c2}. Ultrasonically, a relatively sharp attenuation peak was observed (Sun, 1986) at T_{c2}. An applied magnetic field shifts this anomaly to a higher temperature, broadens its width, and dwarfs its height (Fig. 7). Similarly but less obviously, $Er_{0.705}Ho_{0.295}Rh_4B_4$ shows a rapid attenuation change at the ordering temperature of Er^{3+}, as shown in Fig. 8 (Sun, 1986). Comparing these results with those of some rare-earth magnetic materials (Maekawa et al., 1976; Treder and Levy, 1977; Treder et al., 1979; Tachiki et al., 1974, 1975), such as Ho, Tb, Dy, and Er, one could infer that the anomaly could be attributed to spin–phonon interaction.

Spin–phonon interaction has been well described theoretically (Tachiki and Maekawa, 1974) and agrees with the observed experimental results (Maekawa et al., 1976; Treder and Levy, 1977; Treder et al., 1979; Tachiki et al., 1974, 1975). In principle, this interaction arises from the deformation of the magnetic ion lattice that is induced by the propagation of sound waves. Since the

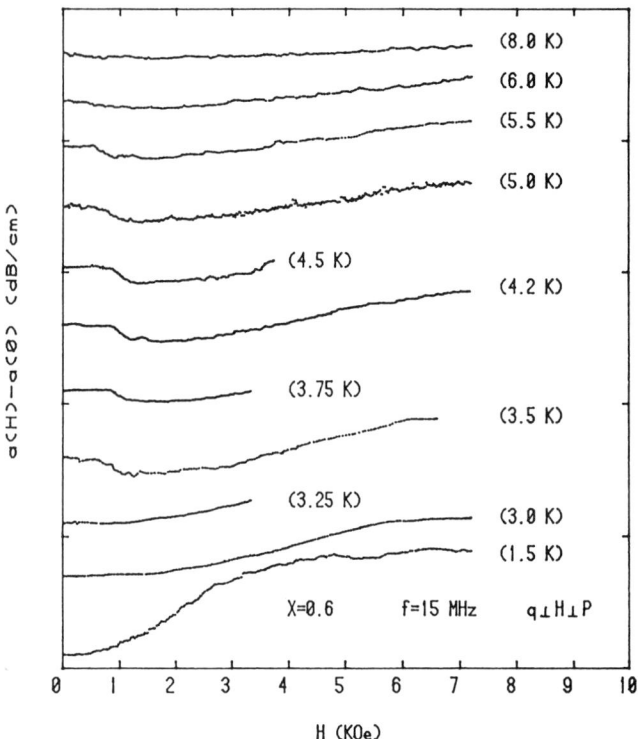

FIG. 6. Magnetic-field–dependent ultrasonic attenuation of $Er_{0.4}Ho_{0.6}Rh_4B_4$ at constant temperatures. Between 3.45 K and 6.7 K, the sample is in the superconducting state when the field is zero. The curves have been shifted with respect to each other for clarity.

Heisenberg-type exchange interaction between the spins (or magnetic moments) is spatially dependent, traveling sound waves in these materials vary the exchange constants in an oscillatory manner. Energy dissipation of sound occurs when the random forces that are caused by the thermal spin fluctuations act on the acoustic normal modes via this interaction. According to Mori's theory for irreversible processes (Mori, 1965), the sound attenuation coefficient is expressed as a function of the time correlation of the random force acting on the phonons. If a relationship is introduced between random forces and boson operators for phonons and their time derivative, a final result showing the attenuation to be proportional to the spin correlation function is obtained (Tachiki and Maekawa, 1974). In a magnetic field, the spin correlation function is divided into two parts. One is the cross product of the static spin polarization induced by the magnetic field and the two-spin correlation function, and the other is the correlation function of four spins. The latter part is considered to be responsible for the critical

4. Magnetic Superconducting System

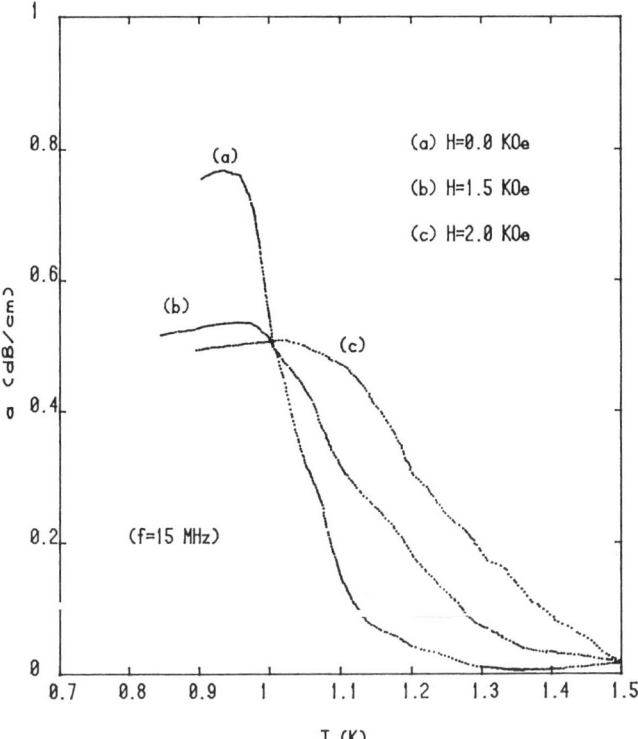

FIG. 7. Temperature-dependent ultrasonic attenuation of ErRh$_4$B$_4$ betewen 0.8 K and 1.5 K at constant magnetic fields. The attenuation zero at 1.5 K is arbitrarily chosen.

attenuation anomaly in the absence of a magnetic field, However, if a magnetic field is applied, the spin fluctuations are suppressed, and the magnitude of the spin correlation function is decreased. Then, the former part becomes predominant and displays a quadratic field dependence for the attenuation. The attenuation coefficient can be written as (Tachiki and Maekawa, 1974)

$$\alpha_q = g_1(T,q)M^2[\chi_q^z(0)]^2 + \Sigma_{i,k}\, g_2(T,q)[\chi_q^i(k)]^3, \qquad i = x, y, \qquad (1)$$

when the magnetization M is induced by an external magnetic field applied in the z direction above the phase transition temperature, where $g_1(T,q)$ and $g_2(T,q)$ are temperature-dependent functions related to the spin–phonon coupling. q is the wavevector of sound, and $\chi_q^i(k)$ is the staggered susceptibility in the i direction with wave number k. Usually, the second term of Eq. (1) is much smaller than the first term. Through this mechanism, an anomalous sound energy dissipation may be observed at temperatures near a magnetic phase transition where critical

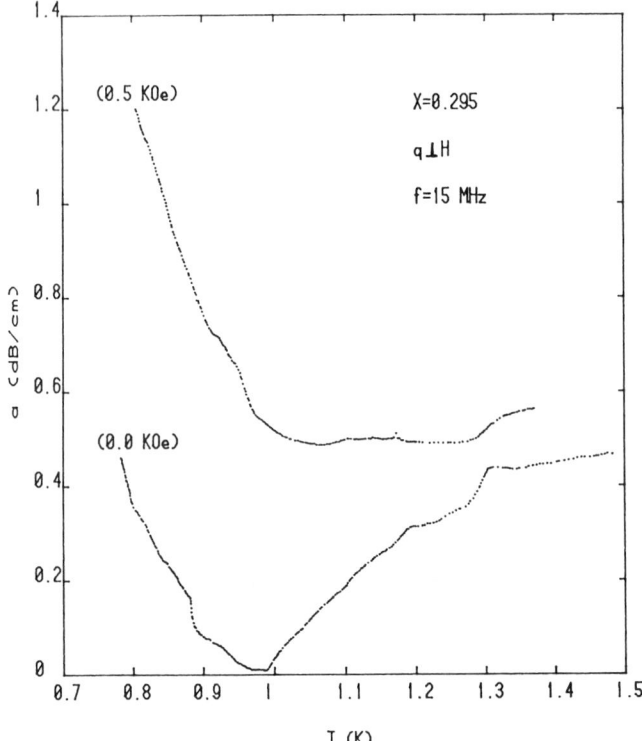

FIG. 8. $\alpha(T)$ of $Er_{0.705}Ho_{0.295}Rh_4B_4$ at 15 MHz in magnetic fields of zero and 0.5 kOe.

spin fluctuations occur. For $ErRh_4B_4$, the spike-shaped feature in specific heat and the peak in attenuation are believed to provide evidence for the existence of spin fluctuations.

As can be seen in Fig. 1, the valleylike minima of the samples with $x = 1$ and 0.912 are located at the magnetic phase transition temperatures. Based on the quadratic dependence on the magnetic moment, as mentioned in the previous paragraph, the attenuation of $HoRh_4B_4$ and $Er_{0.088}Ho_{0.912}Rh_4B_4$ below their Curie temperature should increase with decreasing temperature. This increase superimposed on a decreasing attenuation background forms a local minimum as is displayed in the figure. There is no superconducting state for these two compounds (Johnston *et al.*, 1978). A dominant exchange interaction (Shenoy *et al.*, 1981) between Ho^{3+} magnetic moments is responsible for the absence of superconductivity. $HoRh_4B_4$ (Ott *et al.*, 1980, 1982) exhibits mean-field behavior in its magnetic, thermal, and transport properties. Therefore, the lack of an anomalous attenuation peak at the Curie temperature is a result of the fact that

4. Magnetic Superconducting System

the transition has no spin fluctuations associated with it, which is unlike $ErRh_4B_4$. Similarly, $Er_{0.088}Ho_{0.912}Rh_4B_4$ exhibits the same absence of spin fluctuations (Mackay et al., 1979) as well. The reason for the disappearance of critical spin fluctuations is not clear yet. However, we may be able to deduce the mechanism after comparing the attenuation results of the four Ho-rich $Er_{1-x}Ho_xRh_4B_4$ samples.

As opposed to $HoRh_4B_4$ and $Er_{0.088}Ho_{0.912}Rh_4B_4$, the valleylike attenuation minima of the reentrant superconductors $Er_{0.187}Ho_{0.813}Rh_4B_4$ and $Er_{0.4}Ho_{0.6}Rh_4B_4$ are located at their superconducting transition temperatures T_{c1}, and there is a steplike change in attenuation at T_m for each sample. The relative minimum at T_{c1} is caused by the increasing attenuation in the superconducting state. However, it is the onset of superconductivity, instead of magnetization, that induces this increase. It is very unusual to observe that the energy dissipation of sound is enhanced in the superconducting state instead of being decreased as described by the BCS theory. In order to interpret this experimental result, it may be necessary to suggest that the increase in attenuation of sound waves in the superconducting state is caused by the existence of critical spin fluctuations. Thus, we believe that there exist critical spin fluctuations of both of the magnetic ions Er^{3+} and Ho^{3+} in $Er_{1-x}Ho_xRh_4B_4$, although they may be suppressed in the samples with $x > 0.9$. When the sample temperature is near T_m, the attenuation should exhibit an increase at T_m as a result of spin–phonon interaction. Since the T_ms of $Er_{0.187}Ho_{0.813}Rh_4B_4$ and $Er_{0.4}Ho_{0.6}Rh_4B_4$ are lower than respective T_{c1}s, the attenuation would show an increase from T_{c1} to T_m as is observed. Furthermore, the observed steplike change at T_m is a consequence of the truncation of this attenuation increase. That means that the low-temperature half (below T_m) of this peak is deleted completely by an internal field, which is present in the ferromagnetic state. And, for some reason, this field is suppressed in the superconducting state, letting the other half of the attenuation curve (above T_m) remain as shown. Since the spin correlation length is short-ranged as compared to the superconducting coherence length of this system, the superconductivity will not be quenced immediately after the appearance of spin fluctuations at T_{c1}, but after the long-ranged ferromagnetic order is established at T_{c2}.

Now, we ought to point out which magnetic ions, Ho^{3+} or Er^{3+}, contribute to the fluctuations before continuing the discussion about the possible models. The phase diagram of T_{c1} and T_m vs. x for this system (Johnston et al., 1978) reveals that the T_m for the Er^{3+} ions in all the $Er_{1-x}Ho_xRh_4B_4$ samples is below 0.96 K (T_{c2} for polycrystalline $ErRh_4B_4$) and depends on x. As shown in Figs. 2 and 7, the temperature range within which the spins of the Er^{3+} ions begin to fluctuate is quite narrow and fairly near the T_{c2} of $ErRh_4B_4$. Thus, the increase in attenuation of the Ho-rich reentrant superconductors of $Er_{1-x}Ho_xRh_4B_4$ may not due to the fluctuations associated with the Er^{3+} ions. If the steplike changes of attenuation that are located at 3.45 K and 4.95 K for the samples with $x = 0.6$ and 0.813, respectively, are the results of critical spin fluctuations, it is more plausible to assume that they are the spin fluctuations of the Ho^{3+} ions. Since the exchange interaction between conduction electrons and the well-localized 4f-electrons of the Er^{3+} ions in $ErRh_4B_4$ is so weak as to make the existence of a superconducting state in this material possible, the substitution of Er^{3+} by Ho^{3+} ions in $HoRh_4B_4$ should weaken the indirect exchange interaction of the Ho^{3+} ions and may help indirectly in the appearance of spin fluctuations of the Ho^{3+} ions.

Studies of magnetic superconductors have established that crystalline electric fields (CEF) affect the fundamental properties of these materials in many respects (Woolf et al., 1979; Johnston et al., 1978; Moncton et al., 1977). Here, we assume that CEF also play an important role in determining the appearance of spin fluctuations of the magnetic ions.

To qualitatively explain the observed attenuation enhancement in the superconducting state and the difference in ultrasonic attenuation behavior between the reentrant superconductors and the mean-field ferromagnets in the $Er_{1-x}Ho_xRh_4B_4$ system, we suggest the following two models (Sun et al., 1989a). One model proposes that the crystalline electric fields (CEF) in $Er_{1-x}Ho_xRh_4B_4$ align the magnetic moments of Ho^{3+} ions along the tetragonal c-axis (Lander et al., 1979) and suppress completely their spin fluctuations at temperatures close to the T_m of each sample with $x > 0.9$. As the concentration of Er^{3+} ions becomes large enough to allow the superconducting state to exist, the suppressed (by CEF) spin fluctuations would be recovered in the superconducting state at temperatures close to the T_m of the Ho^{3+} ions. However, these critical spin fluctuations do not show their full strength because they are also screened by the superconducting currents to some extent. When an external field is applied, it depresses superconductivity, and consequently spin fluctuations as well (Sun et al., 1983). As its magnitude reaches H_{c2}, the magnetic field destroys superconductivity, magnetizes the sample, and quenches spin fluctuations. In fact, field-dependent attenuation curves of samples with $x = 0.6$ (Fig. 6) and $x = 0.813$ (Sun, 1986) show a minimum at H_{c2}. However, this model has a flaw since it may imply that the superconducting electrons shield the CEF, and since the CEF are very short-ranged, while superconducting-electrons screening is usually only effective for long range. Therefore, a self-consistent model that includes the effect of fluctuations on superconducting-electron screening has to be developed in order to properly assess the validity of this phenomenological model.

Another model that might account for the increase in attenuation in the superconducting state could be that the BCS coherence factor, which cancels the singularity in the density of states at the gap edge for electron–phonon interaction, does not do so in a reentrant superconductor because the electron spins are flipped for a magnetically mediated electron–phonon interaction, and therefore the coherence factor has the opposite sign. The attenuation is proportional to the real part of the conductivity in the superconducting state σ_1, which for the latter case and a BCS value for the energy gap, will exhibit a maximum immediately below T_c and an exponential decay at low temperatures (Tinkham, 1975). However, recent calculations have shown that for energy gaps that are smaller than the BCS value of $2\Delta = 3.52 k_B T_c$, it is possible to obtain a broad maximum in σ_1 that extends below $T_c/2$ (Baum et al., 1989). Since T_m is above $T_c/2$ in the samples shown in Fig. 1, the interaction is quenched before the attenuation starts to decrease in the superconducting state.

4.2. Ultrasonic Attenuation of $Er_{0.705}Ho_{0.295}Rh_4B_4$ and $ErRh_4B_4$

$Er_{0.705}Ho_{0.295}Rh_4B_4$ is an interesting sample not only because of the existence of a magnetic competition between the Er^{3+} ions, which are ferromagnetically ordered in the tetragonal basal plane, and the Ho^{3+} ions, which are ferromagnetically ordered along the c-axis (Lander et al., 1979), but also because of the interplay between these magnetic ions and the superconducting currents associated with the rhodium (Rh) atoms.

As illustrated by ac susceptibility measurements on $Er_{0.705}Ho_{0.295}Rh_4B_4$ (Johnston et al., 1978; Sun et al., 1989b), Ho^{3+} ions undergo a magnetic phase transition at 1.31 K and are ferromagnetically ordered at 0.99 K. In addition, Er^{3+} ions are ferromagnetically ordered at 0.89 K. Ultrasonic attenuation data as a function of temperature at 15 MHz for this material in magnetic fields of zero and 0.5 KOe are shown in Fig. 8. The zero field attenuation curve has a relatively sharp change at 1.31 K and a relative minimum located at 0.99 K, as well as a steeper variation at 0.89 K. The last might be associated with the activation of critical spin fluctuations of the Er^{3+} ions. However, instead of having a peak at 0.89 K, as would be expected for a phase transition and as we observed for polycrystalline $ErRh_4B_4$ at its T_m (Figs. 2 and 7), the attenuation keeps increasing as the temperature is lowered from 0.89 K. This difference could be due to the growing magnetization of the Ho^{3+} ions with decreasing temperature, as well as the quadratic dependence of the attenuation on magnetization (Tachiki and Maekawa, 1974). At constant fields up to 1.5 kOe (Sun et al., 1989b), the fact that this steeper variation moves to higher temperatures and becomes less obvious may provide additional evidence for this interpretation, since it appears that this signature is correlated with the changes of the phase transition temperature produced by the constant magnetic fields. The relative minimum located at 0.99 K disappeared when an external field of 0.5 kOe was applied. Similar experimental results have also been observed at 79 MHz (Sun, 1986). A tentative conclusion that may be drawn from this disappearance is that 0.99 K is the reentrant temperature (T_{c2}) of the sample. The onset of magnetic order of the Ho^{3+} ions is the main reason for the pair-breaking process in this superconducting sample. However, instead of quenching the superconducting state immediately after ordering magnetically at 1.31 K as for the Ho-rich reentrant superconductors (samples with $x = 0.6$ and $x = 0.813$), the Ho^{3+} ions in $Er_{0.705}Ho_{0.295}Rh_4B_4$ may undergo a transition into a sinusoidally modulated ordered state as predicted (Tachiki, 1981 and 1982), which coexists with the superconducting state till 0.99 K. In what follows we will address this point in more detail.

Energy band calculations of the electronic structure for magnetic superconductors show that the exchange-type interaction between magnetic moments is extremely weak in these compounds, and therefore the electromagnetic interaction between the persistent currents and the rare-earth magnetic moments becomes important and may predominate. The spin–spin interaction, which mainly originates from magnetic dipolar interaction, is mediated by an internal magnetic field. This internal field induces persistent currents and is in turn screened by these persistent currents (Tachiki, 1981 and 1982). As a consequence, a spin-spiral or sinusoidally modulated magnetic order (Blount and Varma, 1979; Matsumoto et al., 1979; Kuper et al., 1980; Tachiki et al., 1980b) in the superconducting state of the ferromagnetic superconductors, an anisotropy of the ultrasonic attenuation (Tachiki et al., 1980a) associated with spin fluctuations, and a screening of the forward magnetic scattering of neutrons are expected to be observed in these systems.

Therefore, the observed decrease in attenuation that ranges from 1.31 K to 0.99 K may indicate that the sample is still in the superconducting state. However, whether there is a long-range sinusoidally modulated magnetic order, and whether ferromagnetism and superconductivity coexist throughout the whole volume of the material (Kuper et al., 1980; Tachiki, 1982) or stay in separate domains with the superconducting domains containing only sinusoidally modulated magnetic order (Freeman and Jarlborg, 1979; Tachiki, 1979b) remains to be resolved, perhaps by other measurement techniques.

As stated in Eq. (1), ultrasonic attenuation is a function of the staggered susceptibility. In the superconducting state of a magnetic superconductor, the screening effect of persistent currents on the spin–spin interaction will affect the behavior of the staggered susceptibility, and thus can be observed by the measurement of ultrasonic attenuation (Tachiki et al., 1980a, 1981). Furthermore, since the persistent current is a transverse current and screens only the transverse component of spin–spin interaction, the attenuation that arises from the transverse

4. Magnetic Superconducting System

component of spin fluctuations through spin–phonon interaction turns out to be very much suppressed in the superconducting state, while the longitudinal component is free from this effect. When the propagation direction of sound is perpendicular to the magnetization, the transverse susceptibility will be (Tachiki, 1981)

$$\chi^t(q) = C/\{T - C[r_q - 4\pi c(q)/\lambda^2 q^2 + c(q)]\}, \tag{2}$$

Here, $\chi^t(q) = \chi^t_q(0)$ in Eq. (1), C is the Curie constant, T is the temperature, $c(q)$ is the boson characteristic function whose order of magnitude is about 1, λ is the penetration depth, and r_q is the Fourier transform of the exchange constant $r(x)$. In the superconducting state, the wave number of an impressed sound q is usually much smaller than λ, so that $\chi(q)$ can be reduced to

$$\chi^t_s(q) = C/\{T - C[r_q - 4\pi]\}, \tag{3}$$

and in the normal state λ is taken to be infinite; then

$$\chi^t_n(q) = C/(T - Cr_q), \tag{4}$$

By looking at Eqs. (3) and (4), it can be seen that $\chi^t_s(q)$ is smaller than $\chi^t_n(q)$. Thus, the attenuation in the superconducting state will be smaller than it is in the normal state. In the longitudinal case where the magnetic field is parallel to q, the susceptibility is expressed as

$$\chi^l(q) = C/\{T - C[r_q - 4\pi]\}, \tag{5}$$

which is the same as $\chi^t_s(q)$ and stays at the same value both in the superconducting state and in the normal state because there are no superconducting properties involved in this expression.

Figure 9A shows the attenuation as a function of magnetic field at 1.5 K for the sample with $x = 0.295$. The peculiar anisotropic dependence of the attenuation on the propagation direction of the sound waves agrees at least qualitatively with the theory mentioned in the previous paragraph. At fields larger than H_{c2}, we can see that the attenuation not only increases with magnetic field because of the growing magnetization, but also depends on the relative orientation of the field with respect to the sound direction. Similar results were also observed for $ErRh_4B_4$. The magnetic field dependence of the attenuation coefficient (Schneider et al., 1981) in $ErRh_4B_4$ for the case when the wavevector q is parallel to the magnetic field is shown in Fig. 9b, while the transverse case $q \perp H$ is shown in Fig. 9c. A discontinuous increase in the attenuation curve occurred at 4.8 kOe in Fig. 9c, but is not seen in Fig. 9b. The presence of an increase at 4.8 kOe in Fig. 9c can be understood using the above model if we assume that 4.8 kOe is H_{c2} for this sample. Thus, below H_{c2}, the sample is superconducting, and the attenuation

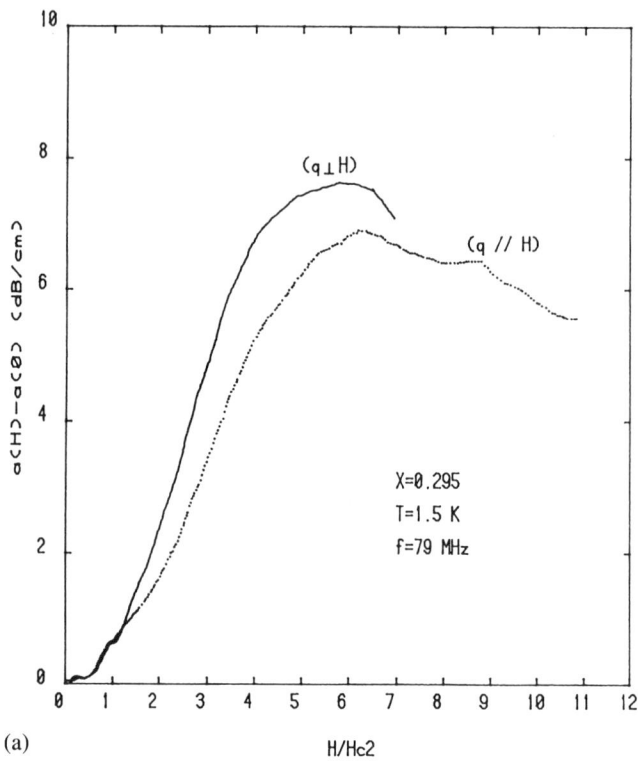

(a)

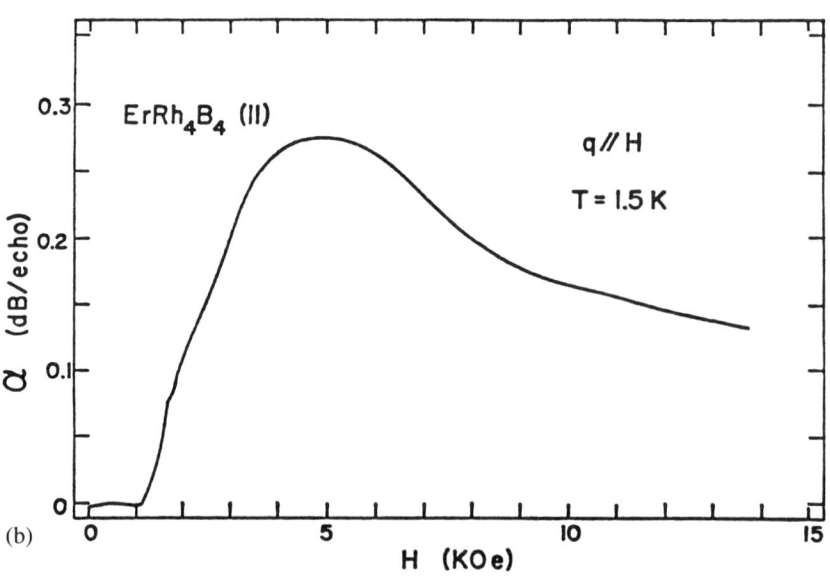

(b)

4. Magnetic Superconducting System

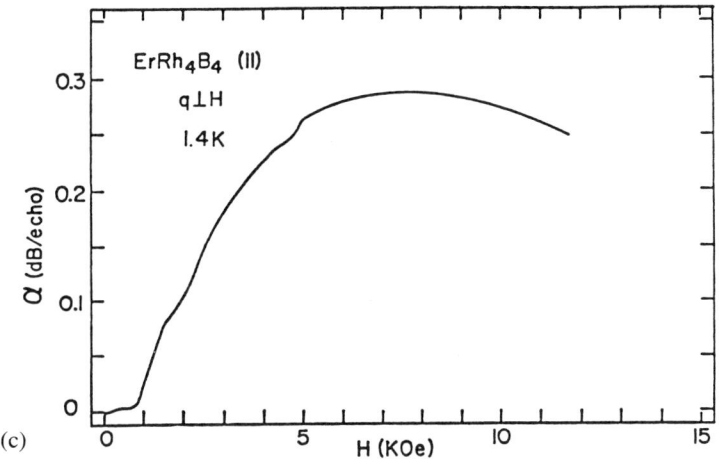

FIG. 9. Continued

is suppressed by superconducting screening according to Eq. (3). Above H_{c2}, the sample is in the normal state, and the attenuation is proportional to $\chi_n^t(q)$ as given in Eq. (4). Therefore, there should be an increase in attenuation upon going from the superconducting state to the normal state across H_{c2}. When the magnetic field is parallel to the propagation direction, there is no screening effect due to superconducting currents, and the transition across H_{c2} should be smooth, as is observed in Fig. 9b. In fact, the measurements shown in Figs. 9b and 9c were some of the first to confirm the electromagnetic interaction model proposed for the magnetic superconductors by Matsumoto et al. (1979) and Tachiki et al. (1980a).

In addition to the anisotropic effect, several common features are evident in Figs. 9b and 9c. The attenuation remains constant up to a field that we name H_1 (1.1 kOe for 9b and 0.85 kOe for 9c); it increases up to a small plateau at 1.65 kOe and 1.75 kOe, respectively, which we will identify as H_2; then both curves have a maximum at 4.8 kOe and 7.0 kOe, respectively. We identify H_1 with H_{c1}, the lower critical field, once the respective demagnetization factor in each orientation is taken into account.

The appearance of the maxima in the curves in Figs. 9a, 9b, and 9c is also the result of phonon interaction with spin fluctuations and can be explained by Eq. (1). In a magnetic field, the second term in Eq. (1) becomes negligible, and thus the expression for α_q becomes proportional to $M^2[\chi_q^z(0)]^2$. As shown in Fig. 10, M^2 is an increasing function of field, while $\chi_q^z(0)$ is a decreasing function,

---

FIG. 9. (a) $\alpha(H)$ of $Er_{0.705}Ho_{0.295}Rh_4B_4$ shows anisotropic behavior at 1.5 K. (b) $\alpha(H)$ at $T = 1.5$ K for $ErRh_4B_4$ ($q\|H$). (c) $\alpha(H)$ at $T = 1.4$ K for $ErRh_4B_4$ ($q \perp H$).

$$\alpha_q \simeq \langle S_0 \rangle^2 \langle S_q S_{-q} \rangle^2$$

$$\simeq \langle S_0 \rangle^2 \chi_q^2$$

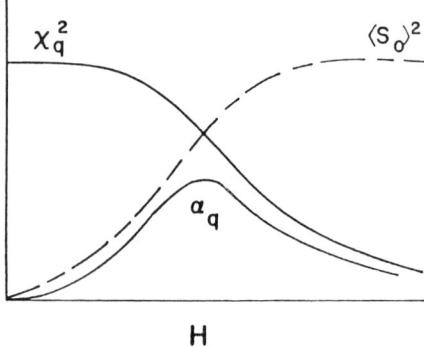

FIG. 10. Magnetic-field–dependent attenuation (α_q) behavior of a spin–phonon interaction. α_q is proportional to the square of the product of static magnetization $\langle S_0 \rangle$ and staggered susceptibility χ_q.

since as the magnetic moments are aligned by an external field the Fourier components of the fluctuations with wavevector q are suppressed. The product of these two functions yields a maximum in the attenuation as a function of H, which is what we observed in Figs. 9a, 9b, and 9c.

Moreover, the model of Tachiki et al. (1979a) also suggests that there is a possibility that a type II superconductor will become a type II-I superconductor or even a type I superconductor when its temperature is close to the Curie temperature. The major difference in magnetization among these types of superconductors is their behavior at H_{c1}. As is known, the magnetization (M) of a type I superconductor has a steep change from $-4\pi M$ to zero at H_c; that of a type II increases relatively gradually from $-4\pi M$ at H_{c1} to zero at H_{c2}, with a large change in slope from negative value to positive at H_{c1}. For a type II-I, its total magnetization increases steeply from $-4\pi M$ to some value at H_{c1}, and then gradually to a positive value, which is the ferromagnetization of the magnetic system at H_{c2}. The occurrence of this intermediate type in magnetic superconductors results from the flux quantization of the magnetic induction $b(x)$ of a vortex, which is the sum of its magnetization $m(x)$ and in-vortex magnetic field $h(x)$ and can be written as

4. Magnetic Superconducting System

$$\int b(x)d^2x = \int h(x)d^2x + 4\pi \int m(x)d^2x = hc/2e, \qquad (6)$$

where x is the distance from the center of a vortex. As the temperature approaches the magnetic phase transition temperature, the increasing $\int m(x)d^2x$ ($\equiv I_2$) in the material will lessen the $\int h(x)d^2x$ ($\equiv I_1$) as required to meet the flux quantization restriction. Let us elaborate a little bit more on this point. In the purely paramagnetic state of a magnetic superconductor, $m(x)$ in I_2 is just proportional to the $h(x)$ in I_1, and therefore I_2 is proportional to I_1, with no pathological effects emerging, and it is straightforwardly possible to obtain monotonically decreasing positive values of $h(x)$ with increasing x that satisfy Eq. (6). However, for a ferromagnetic superconductor close to the Curie temperature, the exchange interaction becomes significant, such that $m(x)$ is not only a function of $h(x)$, but also of the adjoining magnetization $m(x \pm \Delta x)$. Thus, even though $h(x)$ could be zero, $m(x)$ might still have a finite value, which would have to gradually decrease as a function of x, and consequently would contribute to the magnitude of I_2. And if, consequently, I_2 becomes too large, then I_1 will have to compensate for this by having portion of itself becomes negative. This is accomplished by making the tail of $h(x)$ at a large value of x, negative (Tachiki et al., 1979a). Since the mutual interaction between two vortices is proportional to $h(x)$, the negative value of a part of $h(x)$ can result in the existence of an attractive force between two vortices and thus in a jump of the magnetization at H_{c1}. Therefore, depending on the value of $\int m(x)d^2x$, a ferromagnetic superconductor in the superconducting state can evolve from a type II superconductor to a type II-I to a type I as the temperature of the sample is moved towards the Curie temperature. For a type II-I superconductor, part of the magnetization has a first-order transition at H_{c1}, while the rest of it decreases gradually up to H_{c2}, where it has a second-order phase transition. A schematic diagram of the magnetization curve of such a type II-I superconductor is shown in Fig. 11 (Schneider, 1981). Experimentally, the magnetization of a single crystal of ErRh$_4$B$_4$ (Behroozi et al., 1983) did exhibit the predicted behavior at H_{c1} and H_{c2} at various temperatures. However, due to the demagnetization factor of the sample, the field where the sharp change in magnetization occurred was broadened. Therefore, the determination of H_{c1} by the magnetization measurements becomes less precise. Ultrasonically, as is described in Eq. (1), a jump at H_{c1} in magnetization would result in a quick enhancement in attenuation. Again, the geometric shape and the quality of a sample decide how obvious this predicted result will be because of the involvement of demagnetization factors and anisotropic crystalline effects.

A magnetic-field–dependent attenuation curve at 15 MHz, along with an ac magnetic susceptibility curve at 1.2 K, is shown in Fig. 12 for the sample with $x = 0.295$. Important information about the phase transitions in this sample

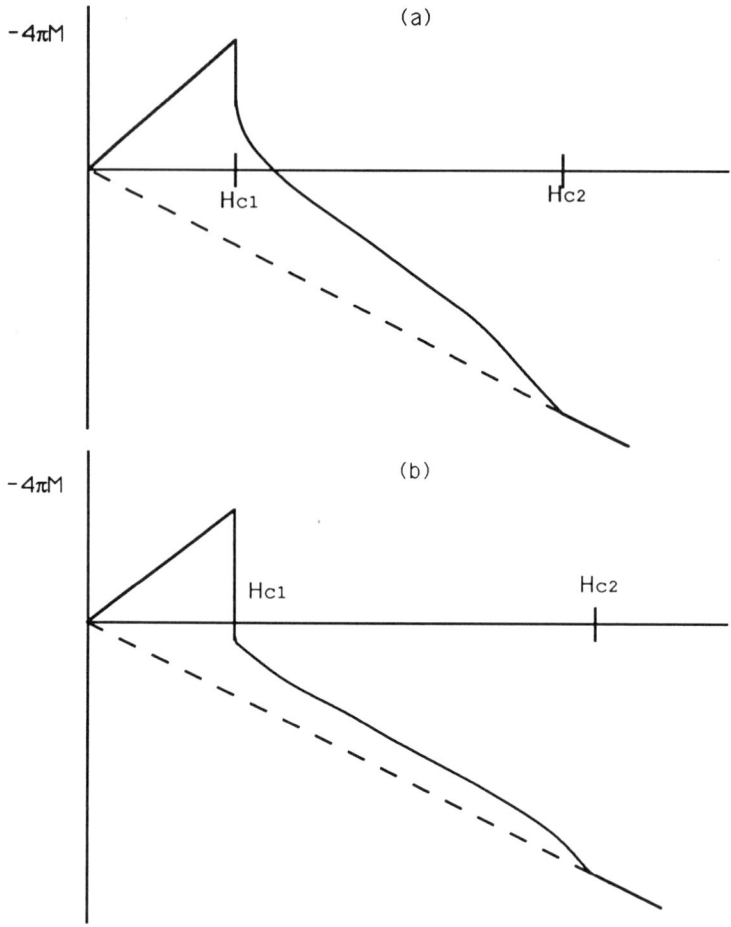

FIG. 11. Schematic diagram for the magnetization curve of a type II-I superconductor: (a) drop to diamagnetic state at H_{c1}, (b) drop to paramagnetic state at H_{c1}.

may be uncovered by looking into both curves in the small field range. When the effective field (taking the demagnetization factor into account, which is 0.148) is below 0.27 kOe and above 0.43 kOe, the slope of the ac susceptibility curve is relatively small and close to zero. In between, it is quite steep and bends down a little bit at 0.32 kOe. The attenuation is constant below 0.27 kOe and increases almost monotonically from 0.27 kOe to at least 0.7 kOe. These data can be interpreted as follows: (1) 0.27 kOe is the lower critical magnetic field H_{c1} at 1.2 K, and 0.43 kOe is the upper critical field H_{c2}. (2) The relative large changes of susceptibility in the narrow magnetic field range (from H_{c1} to H_{c2})

4. Magnetic Superconducting System

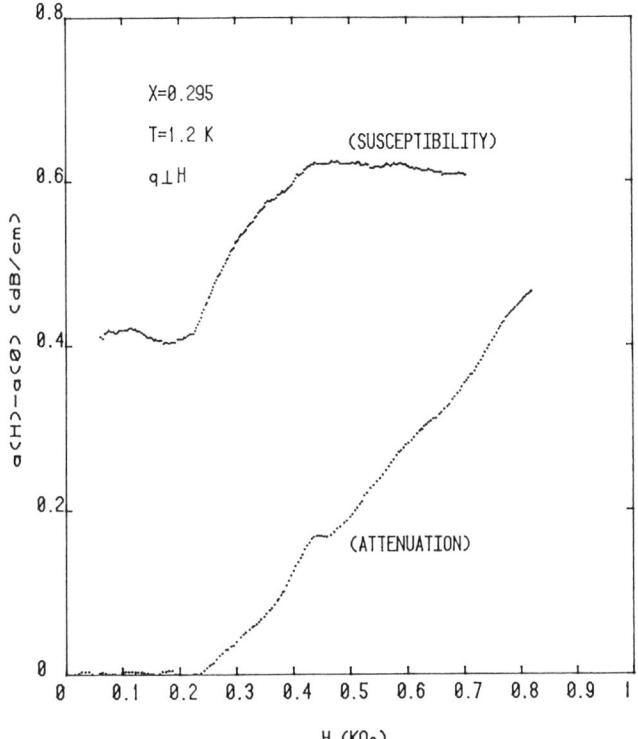

FIG. 12. Magnetic-field–dependent ultrasonic attenuation and ac magnetic susceptibility of $Er_{0.705}Ho_{0.295}Rh_4B_4$ at 1.2 K. The variations in attenuation reflect the changes in ac susceptibility.

imply sharp changes of magnetization. The attenuation curve displays corresponding variations between the two critical fields. Therefore, it is quite possible that the sample is a type II-I superconductor at this temperature, and 0.32 kOe is the field where the sharp change in magnetization at H_{c1} ends and the diamagnetism of the superconducting subsystem is replaced by the paramagnetism of the magnetic ions and the magnetic flux lines in the sample. Similarly, for $ErRh_4B_4$, we identify the attenuation changes between H_1 and H_2 in Figs. 9b and 9c as corresponding to the steep changes in magnetization at H_{c1}, for a type II-I transition. As mentioned above, the slightly broadening field between H_1 and H_2 may be due to the demagnetization factors. (3) The increase of attenuation at fields above 0.32 kOe may be due to the fact that magnetic flux lines penetrate the sample, which then is in the mixed state, and the increase above 0.32 kOe could be ascribed to the increase of the magnetization in the normal state.

To show how the attenuation varies with field at other temperatures below

1.5 K, Fig. 13 is plotted by taking the data from the attenuation curves of temperature dependence at several constant external fields and those of field dependence at 1.5 K. The inflection points at 0.21 kOe for 1.1 K and 0.34 kOe for 1.3 K were interpolated and could possibly correspond to the lower critical fields at these temperatures. Together with the 0.27 kOe point at 1.2 K, these curves show the possibility for the existence of a Meissner state between 0.99 K and 1.31 K, and they might be used to confirm the existence of a superconducting state down to 0.99 K, even with the Ho^{3+} ions magnetically ordered at 1.31 K and below.

As displayed in Fig. 2, the attenuation data of $Er_{0.705}Ho_{0.295}Rh_4B_4$ did not exhibit a special feature at T_{c1}, which is 7.6 K. This was the case even when external magnetic fields were applied (Schneider, 1981; Sun, 1986). When compared with the attenuation results shown in Fig. 1 and those for $ErRh_4B_4$ in the

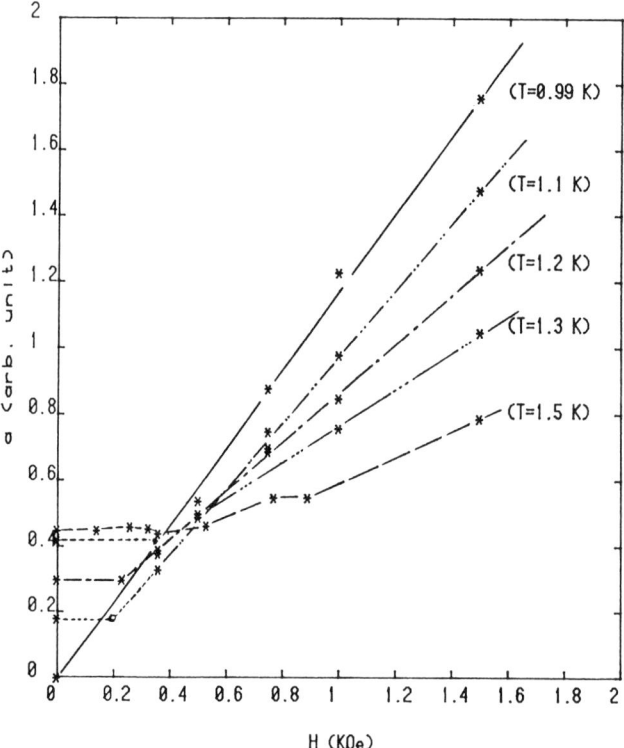

FIG. 13. 15 MHz $\alpha(H)$ of $Er_{0.705}Ho_{0.295}Rh_4B_4$ at constant temperatures. Dotted lines shown at low magnetic fields are interpolated lines. Constant attenuation at low fields implies the sample is in the Meissner state.

4. Magnetic Superconducting System

absence of magnetic field shown in Fig. 2, the bell-shaped maximum centered at 3.5 K and the shoulder around 12 K in Fig. 2 for $x = 0.295$ are believed to be relaxation-type attenuation maxima, which are associated with Er^{3+} and Ho^{3+} ions, respectively. The effect of the maximum corresponding to the Er^{3+} ions at 79 MHz is so pronounced that the maximum at 12 K belonging to the Ho^{3+} ions is merged and appears as a shoulder. That this shoulder is at 12 K instead of being located at a temperature below 10 K can be attributed to the frequency dependence of the temperature location of a relaxation attenuation maximum (Sun *et al.*, 1989c).

It is also possible to deduce a phase diagram of critical fields against superconducting transition temperatures for $Er_{0.705}Ho_{0.295}Rh_4B_4$ obtained from the ultrasonic attenuation and ac susceptibility measurements. Such a phase diagram is shown in Fig. 14. It is a typical H_c vs. T_c diagram for reentrant superconductors. However, it is quite different from those of conventional BCS superconductors.

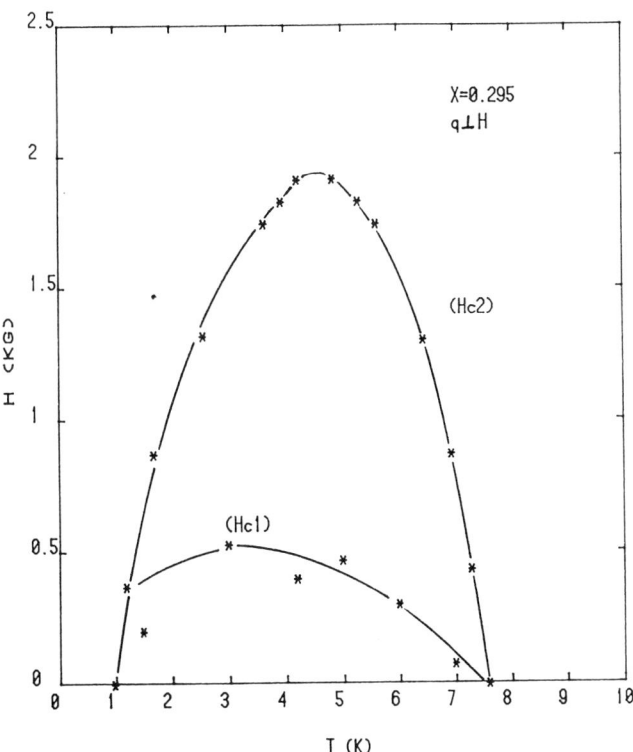

FIG. 14. The phase diagram of critical fields vs. superconducting transition temperature for $Er_{0.705}Ho_{0.295}Rh_4B_4$.

In summary, the experimental results of ultrasonic attenuation for $Er_{0.705}Ho_{0.295}Rh_4B_4$ at 15 and 79 MHz provide indirect evidence for the coexistence of superconductivity and long-range magnetic order at temperatures between 0.99 K and 1.31 K. By analyzing the attenuation and ac magnetic susceptibility data, a phase diagram of critical magnetic fields vs. superconducting transition temperatures could also be deduced.

The attenuation behavior at temperatures close to 0.99 K implies that $Er_{0.705}Ho_{0.295}Rh_4B_4$ could be a type II-I superconductor in this temperature range, as discussed in the theory by Tachiki et al. (1979a, 1980a). Furthermore, the anisotropy in the attenuation—as shown in Fig. 9, which is the result of varying the orientation of the magnetic field with respect to the ultrasound propagation direction—further confirms the assumptions of this theory: namely, that the electromagnetic interaction between the superconducting currents and the magnetic ions and that between the magnetic ions are the main mechanisms that cause the interesting and unique behavior of magnetic superconductors.

4.3. Relaxation Mechanism of Ultrasonic Attenuation in Ho-Rich $Er_{1-x}Ho_xRh_4B_4$

A characteristic attenuation behavior of the $Er_{1-x}Ho_xRh_4B_4$ system is the broad maximum present on the temperature-dependent attenuation curves as displayed in Fig. 1 and Fig. 2. Since the bell-shaped attenuation maxima are located at temperatures close to 10 K for all the measured samples of the $Er_{1-x}Ho_xRh_4B_4$ system except for $ErRh_4B_4$, it is reasonable to assume that this maximum is associated with Ho^{3+} ions instead of Er^{3+} ions. However, when x is decreased, the effect of the Er^{3+} ions becomes stronger, and the temperature-dependent attenuation curves exhibit a second maximum associated with Er^{3+} ions at lower temperatures, such as those shown on the attenuation curves of the samples with $x = 0.295$ and $x = 0$ (Fig. 2). Similar measurements on $HoRh_4B_4$, but at higher sound frequencies, show that T_p (temperature position of the broad maximum) moves to higher temperatures when the frequency is increased. Moreover, experimental results revealed that the T_ps of all the bell-shaped maxima stayed unchanged when a constant magnetic field of up to 6 kOe was applied (Sun, 1986). Such a dependence of T_p on frequency can be the result of an acoustic relaxation process (Lindsay, 1960) occurring at low temperatures. We believe that this relaxation process is associated with the population changes of split 4f-electron energy levels of magnetic ions caused by the propagation of sound waves in the sample (Sun et al., 1989c). A detailed description of the experimental results, and a derivation of an equation that relates the Schottky specific heat and an effective relaxation time to the attenuation, will be presented. Because

4. Magnetic Superconducting System

the same mechanism could be applied to the maxima contributed by Ho^{3+} ions and Er^{3+} ions, respectively, for convenience let us concentrate on the following discussion on the Ho-rich samples.

Studies of the interaction between long-range magnetic order and superconductivity of $RERh_4B_4$ have revealed that crystalline electric fields (Shenoy et al., 1982; Dunlap and Niarchos, 1982; Radousky et al., 1983) play an important role in the physical properties of these magnetic superconductors. The CEF remove the degeneracy of the Hund's rule ground state of the rare-earth ions in $RERh_4B_4$ at low temperatures, and therefore result in the appearance of Schottky anomalies in specific heat (Woolf et al., 1979; Ott et al., 1980, 1982), strong magnetic anisotropy (Crabtree et al., 1982; Behroozi et al., 1983; Lander et al., 1979), and deviation of the magnetic transition temperatures from the trends normally expected (Johnston et al., 1978; Maekawa et al., 1980). The split of the rare-earth ions' ground states may also affect the propagation of sound waves in these materials.

Figure 15 displays the attenuation as a function of temperature for $HoRh_4B_4$ at three different frequencies. As is shown, T_ps are 11.1 K, 13.6 K, and 15.2 K for 15 MHz, 50 MHz, and 81.3 MHz, respectively. That is, the temperature location of the bell-shaped maximum is frequency-dependent and moves to higher temperatures as the frequency is increased. The same behavior was also exhibited by $Er_{0.187}Ho_{0.813}Rh_4B_4$, with T_ps of 9.9 K and 13 K for 15 MHz and 52.5 MHz, respectively (Sun et al., 1986).

Basically, this frequency-dependent behavior can be qualitatively described by a relaxation attenuation equation (Lindsay, 1960):

$$\alpha(T)/\alpha_{max} = 2\omega\tau(T)/[1 + \omega^2\tau^2(T)], \tag{7}$$

where $\alpha(T)$ is the temperature-dependent ultrasonic attenuation; α_{max}, whose material-dependent parameters will be discussed in detail later, is the maximum of attenuation at angular frequency ω; and $\tau(T)$ is the temperature-dependent relaxation time, which is an intrinsic property of the material. As described in the Eq. (7), the temperature at which a relaxation type attenuation maximum occurs will be determined by the product of ω and τ. Whenever $\omega\tau = 1$, α will reach its maximum value at an angular frequency equal to $1/\tau$. However, τ is a temperature-dependent quantity, implying that for one ω, there is one corresponding τ which makes $\omega\tau$ equal to 1, and thereby τ determines the temperature at which $\alpha = \alpha_{max}$. In fact, we found that τ is a monotonically decreasing function of increasing temperature. Hence, at higher frequencies, α_{max} is located at higher temperatures.

The energy dissipation of sound waves due to the relaxation process occurring in $HoRh_4B_4$ can be qualitatively explained as follows. The 17-fold degeneracy

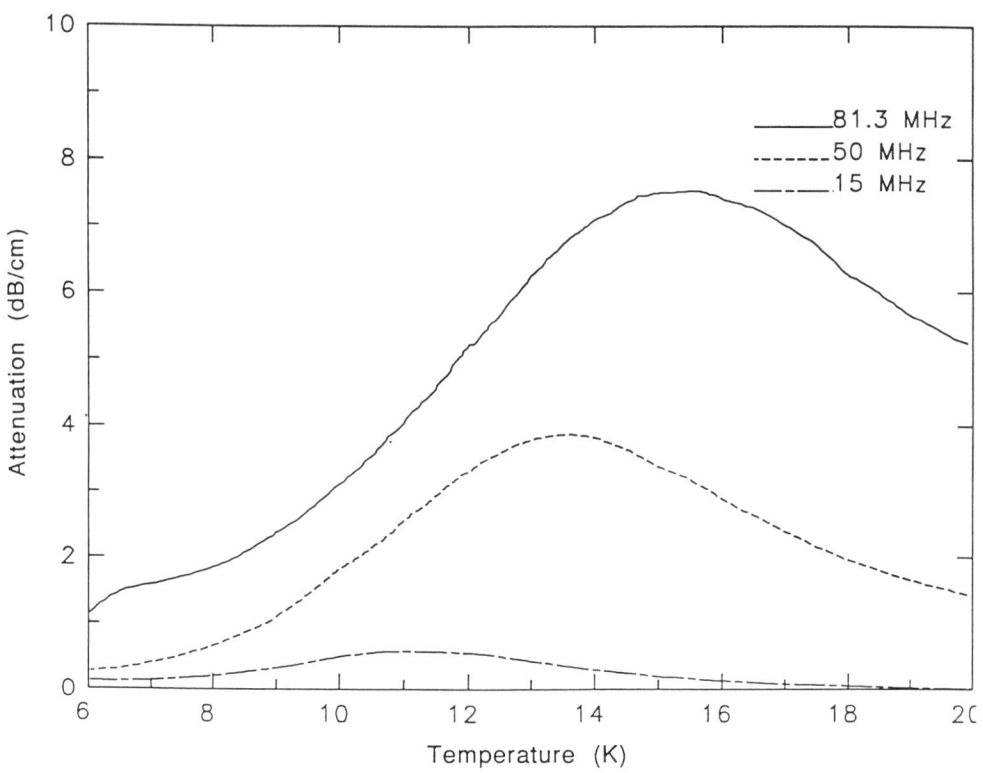

FIG. 15. Temperature-dependent ultrasonic attenuation of HoRh$_4$B$_4$ at 15, 50, and 81.3 MHz.

of the Hund's rule multiplet of the Ho^{3+} ions is lifted by the CEF. The strength of the CEF is determined by lattice symmetry, the lattice charge distribution, Steven factors, and radial integrals (Elliott, 1972). Therefore, when a stress wave is traveling in the sample, the deformation of the lattice will alter the CEF strength and result in a periodic variation of the energy difference between the energy levels. Consequently, the electron populations of each energy level will change as well. The lack of the instantaneous return of the overpopulated and the underpopulated electron levels to their instantaneous equilibrium states causes the electrons to relax out of phase with respect to the traveling stress wave and results in energy dissipation of the waves.

To quantitatively describe this relaxation attenuation, we begin with a two-level model. Let the energy difference between the two energy levels be ε, and let the population densities for the higher energy level (level 2) and the lower energy level (level 1) in the equilibrium state be n_{20} and n_{10}, respectively. The transition rates of electrons from level 2 to level 1 and from level 1 to level 2

4. Magnetic Superconducting System

in the equilibrium state will be denoted by k_{21}^0 and k_{12}^0, respectively. The presence of a sound wave alters the relative numbers of electrons in the two states, as well as the transition rates. We denote the number densities and the transition rates of electrons in the nonequilibrium state for level 1 and level 2 by n_1, k_{12}, and n_2, k_{21}, respectively. Then the rate of change of the population in the excited state can be written as

$$dn_2/dt = n_1 k_{12} - n_2 k_{21}. \qquad (8)$$

By introducing $\Delta n = n_2 - n_{20}$ and $\Delta k_{12} = k_{12} - k_{12}^0$, and using the Boltzmann relation $n_{20}/n_{10} = \exp(-\varepsilon/k_B T)$, we find that

$$\Delta n = [(n_{20} k_{21}^0 \tau \varepsilon)/(1 + i\omega\tau)] [(\Delta T/T) - (\Delta\varepsilon/\varepsilon)] [1/k_B T], \qquad (9)$$

where $\Delta\varepsilon$ and ΔT represent deviations from their equilibrium values due to the passing sound wave, and we have used $(d\Delta n/dt) = i\omega\Delta n$ with ω being the angular frequency of the sound wave. Here τ is defined as $1/(k_{12}^0 + k_{21}^0)$ (Lindsay, 1960).

To simplify our notation, we introduce the quantity $C_0 \equiv (n_{20} k_{21}^0 \tau \varepsilon^2 V)/(k_B T^2)$, which is the Schottky-type heat capacity of a two-level magnetic system at constant volume V and at zero frequency ($\omega = 0$), and we define $C(\omega) \equiv C_0/(1 + i\omega\tau)$. Equation (9) now becomes

$$\varepsilon V \Delta n = C(\omega) T (\Delta T/T - \Delta\varepsilon/\varepsilon). \qquad (10)$$

We consider HoRh$_4$B$_4$ to be composed of two subsystems: the internal (rare-earth 4f-electrons) subsystem and the external (background thermal phonons) subsystem. The variations of the electron-occupation number of the magnetic sublattice energy levels cause sound-wave energy loss, and the variations of the occupation numbers of the background phonons will also contribute to the dissipation.

For simplicity we adopt Barett's model (1969) for the effect of sound waves on thermal phonons, i.e., we regard the external subsystem of thermal phonons as being lumped into a single effective mode. Let the frequency of this mode be ω_2, with corresponding energy level $\varepsilon_2 = \hbar\omega_2/2\pi$, and let Δn_2 be the variation of the phonon-occupation number for that mode due to the presence of the sound wave. By extending our analysis for the electron subsystem to the case of the phonon subsystem (see also Barett, 1969), we can obtain an expression like Eq. (10). Thus, for each subsystem, we can write

$$\varepsilon_i V \Delta n_i = C_i(\omega) T (\Delta T/T - \Delta\varepsilon_i/\varepsilon_i), \qquad (11)$$

where $i = 1$ refers to quantities for the electron subsystem and $i = 2$ refers to quantities for the phonon subsystem. In particular, for the phonon subsystem

$C_2(\omega) = C_{20}/(1 + i\omega\tau_2)$, where C_{20} is the lattice heat capacity (at $\omega = 0$) and τ_2 is the phonon relaxation time.

The heat (dQ) produced in the material as a result of the electron and phonon population variations can be expressed as

$$dQ = TdS = \sum_{i=1,2} TdS_i, \qquad (12)$$

where (Zemansky, 1968)

$$TdS_i = \varepsilon_i V\Delta n_i = C_i(\omega)T(\Delta T/T - \Delta\varepsilon_i/\varepsilon_i), \quad i = 1,2. \qquad (13)$$

Introducing the Gruneisen constant

$$\Gamma_i = -\partial \ln(\varepsilon_i)/\partial \ln(V) = -(V/\varepsilon_i)(\partial\varepsilon_i/\partial V),$$

we have

$$TdS_i = C_i(\omega)T(\Delta T/T + \Gamma_i \Delta V/V), \quad i = 1,2. \qquad (14)$$

Since sound-wave propagation is basically an adiabatic process, it follows that $\sum_{i=1,2} TdS_i = 0$, and Eq. (14) then gives

$$\Delta T/T = -\Delta V/V \left[(C_1\Gamma_1 + C_2\Gamma_2)/(C_1 + C_2)\right], \qquad (15)$$

where, for notational simplicity, we have written C_i for $C_i(\omega)$. Substitution of expression (15) into Eq. (13) yields,

$$\varepsilon_i \Delta n_i = C_i T(\Delta V/V) \left[\Gamma_i - (C_1\Gamma_1 + C_2\Gamma_2)/(C_1 + C_2)\right]. \qquad (16)$$

The mean energy loss of the sound waves in one oscillation can be expressed as $\langle \sum_i \Delta n_i (d\Delta\varepsilon_i/dt)\rangle$, which implies that

$$\langle dQ/dt \rangle = (-1/2)\sum_i \omega \text{Im}(\Delta n_i \Delta\varepsilon_i^*)$$
$$= (\omega/2)|\Delta V/V|^2 T(\Gamma_1 - \Gamma_2)^2 \text{Im}[C_1 C_2/(C_1 + C_2)], \qquad (17)$$

where we have used the relation $\Delta\varepsilon_i = -\varepsilon_i \Gamma_i \Delta V/V$. Here $\text{Im}(X)$ means the imaginary part of X, and $\Delta\varepsilon_i^*$ is the complex conjugate of $\Delta\varepsilon_i$.

The relationship between the ultrasonic attenuation coefficient per centimeter α and the energy loss is $\alpha = \langle dQ/dt\rangle [\frac{1}{2}|\Delta V/V|^2 \rho v_s^3]^{-1}$, where ρ is the mass density and v_s is the speed of sound. It follows that

$$\alpha(T) = A \omega T \text{Im}[C_1 C_2/(C_1 + C_2)], \qquad (18)$$

where $A = (\Gamma_1 - \Gamma_2)^2/\rho v_s^3$. Explicit evaluation of the factor $\text{Im}[C_1 C_2/(C_1 + C_2)]$ leads to the final expression

$$\alpha(T) = A(TC_{10}C_{20}/C_t)\left[\omega^2\tau^*/(1 + \omega^2\tau^{*2})\right], \qquad (19)$$

4. Magnetic Superconducting System

where $C_{i0} = C_i(0)$, and we have defined $C_t = C_{10} + C_{20}$, which is the total heat capacity of the system, and $\tau^* = (C_{10}\tau_2 + C_{20}\tau_1)/C_t$, which is an effective relaxation time.

Note that $\alpha(T)$ has the characteristic relaxation-attenuation form and that it is a function of temperature, sound frequency, and the specific heat, as well as the relaxation time of each subsystem.

Using Eq. (19), there are two ways to find the temperature dependence of τ^* from our data. The first is to take the experimental attenuation results of HoRh$_4$B$_4$ and the heat capacity data of HoRh$_4$B$_4$ and LuRh$_4$B$_4$ (Ott *et al.*, 1980) to calculate the relaxation time for HoRh$_4$B$_4$ at temperatures between 7 K and 20 K (at 6.7 K, HoRh$_4$B$_4$ undergoes a magnetic phase transition). The other approach is to take the ratio of the attenuation at different frequencies. In this way, the term $f(T)$ $(= ATC_{10}C_{20}/C_t)$ will be cancelled, and τ^* can be determined without employing empirical specific heat data. Before calculating τ^* by following either of these two ways, we assume that the background attenuation linearly depends on temperature for each frequency, and the attenuation (α) that arises only from the relaxation process is determined by the expression $\alpha = \alpha_0 - ST + \alpha_c$ (where α_0 is the attenuation shown in Fig. 15; S is the slope of the straight line for this linear background attenuation-temperature relationship and is proportional to the frequency of the sound wave; T is temperature; and α_c is a constant background attenuation whose value does not depend on frequency). Figure 16 shows three curves of the τ^*s obtained by using three attenuation ratios. These curves are quite close to each other, especially at temperatures below 15 K, which could be an indication that Eq. (19) is applicable for interpreting the observed relaxation attenuation. A single function $\tau^*(T)$ can, in fact, be chosen to give a reasonably consistent description of the three measured attenuation curves. The following simple function accomplishes this:

$$\tau^*(T) = 2.27 \times 10^{-8} \exp[-0.26(T - 6)]. \qquad (20)$$

This function is a rather good straight-line fit to the τ^*_{13} curve in Fig. 16. If the relaxation time of the background phonon subsystem is relatively small or tends to zero, then $\tau^* \sim C_{20}\tau_1/C_t$ and would be expected to decrease with increasing temperature. Qualitatively, this is consistent with expression (20).

Inserting expression (20) for τ^* into Eq. (19), and using the measured $\alpha(T)$, three sets of functions of $f(T)$ can be calculated. By taking the average of $f(T)$ at each temperature and using expression (20) for τ^* in Eq. (19), we can calculate the ultrasonic attenuation and compare it with the experimental data. The results are shown in Fig. 17. If a 5% experimental error is assumed, the calculated values agree with the experimental data quite well. Thus, Fig. 17 shows that

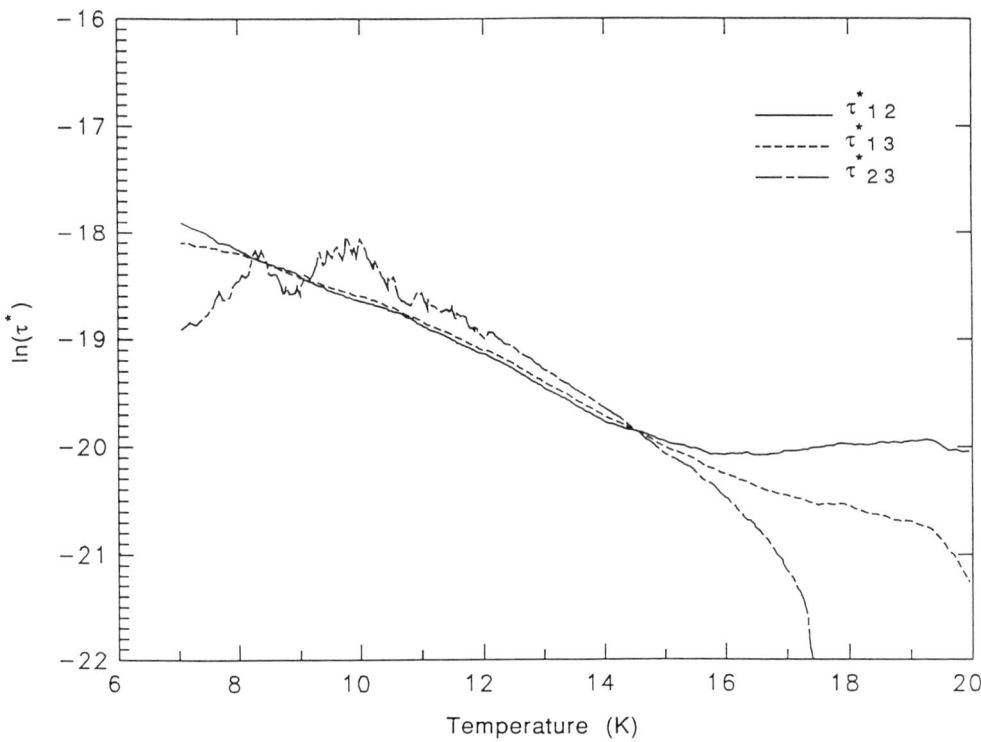

FIG. 16. Relaxation time (τ^*) of HoRh$_4$B$_4$ as a function of temperature. τ^*_{12} represents the curve obtained by using the ratio of attenuation at 15 MHz to that at 50 MHz. τ^*_{23} is that for 50 MHz and 81.3 MHz. τ^*_{13} is that for 15 MHz and 81.3 MHz.

the data can be fitted with a single $\tau^*(T)$ and a single $f(T)$ in the relaxation-attenuation expression (19).

As displayed in Fig. 18, the values of $f(T)$ that are obtained by the procedure stated in the previous paragraph do not match well with those calculated by using the experimental values of specific heat (Ott et al., 1980), C_{10}, C_{20}, and C_t, and an appropriately assigned value of A. This disagreement may be attributed to our simplified model. For the Schottky specific heat, $C_0 \sim (\varepsilon/k_BT)^2 e^{\varepsilon/k_BT}/(e^{\varepsilon/k_BT} + 1)^2$. If ε is 40 K and 60 K (Ott et al., 1980; Radousky et al., 1983), which are the energy differences of the first and second excited states from the ground state of Ho^{3+} ions, respectively, and T is between 10 K and 20 K, the magnitudes of C_{10} from both states will be of the same order, which implies that the 60 K energy level contribution can be a large fraction of C_{10}. In contrast, for attenuation, $\alpha \sim C_{10}\tau^*/(1 + \omega^2\tau^{*2})$, and τ^* for the higher energy level is expected to be much smaller than that of the lower energy level. It follows that

4. Magnetic Superconducting System

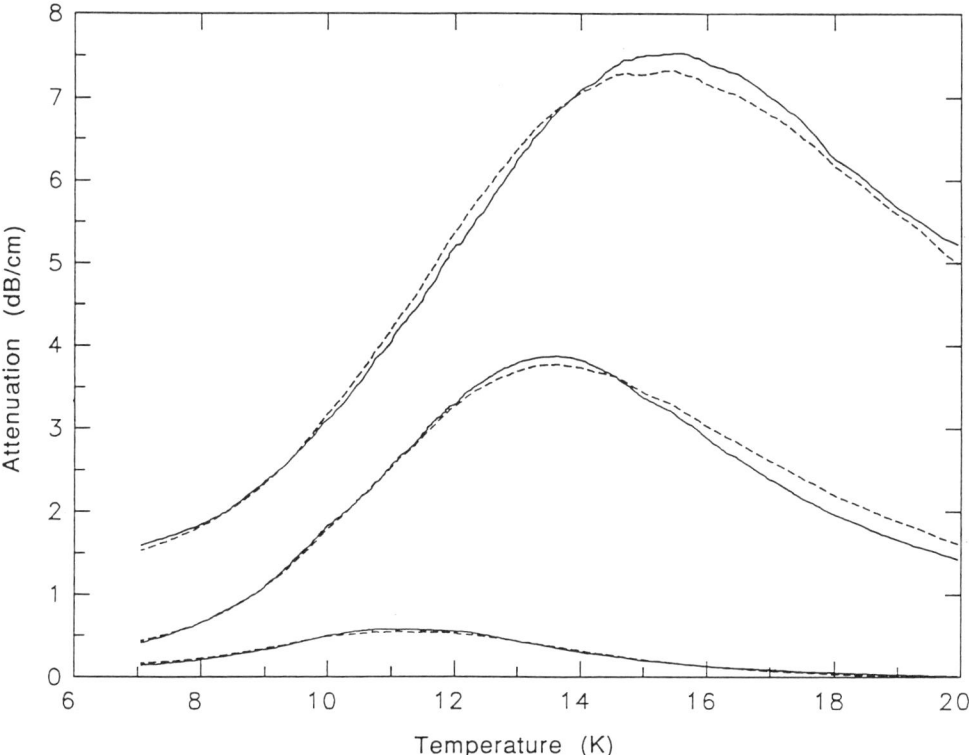

FIG. 17. Comparison of experimental attenuation curves (solid lines) with those of calculated values (broken lines).

the 60 K level would not contribute significantly to α at low temperatures. Thus, while the two-level energy model describes the behavior of relaxation ultrasonic attenuation rather well, the temperature dependence of specific heat involves other excited states with higher values of energy.

At this point, we should make a few remarks concerning the first two features displayed in Fig. 1. First, the attenuation maxima move to lower temperatures as the Ho concentration (x) is decreased. We can deduce, from the relaxation expression Eq. (19), that this result implies that the relaxation time decreases with decreasing concentration of Ho. A possible reason is that as Er^{3+} ions are added to the samples of $Er_{1-x}Ho_xRh_4B_4$ to replace Ho^{3+} ions, additional channels for relaxation are provided. That means τ^* is shorter at the same temperature because of the addition of Er^{3+} ions, and thus shifting the maximum position down to a lower temperature. Secondly, the increase of normalized attenuation at magnetic phase transition as x decreases can be attributed to the increasing

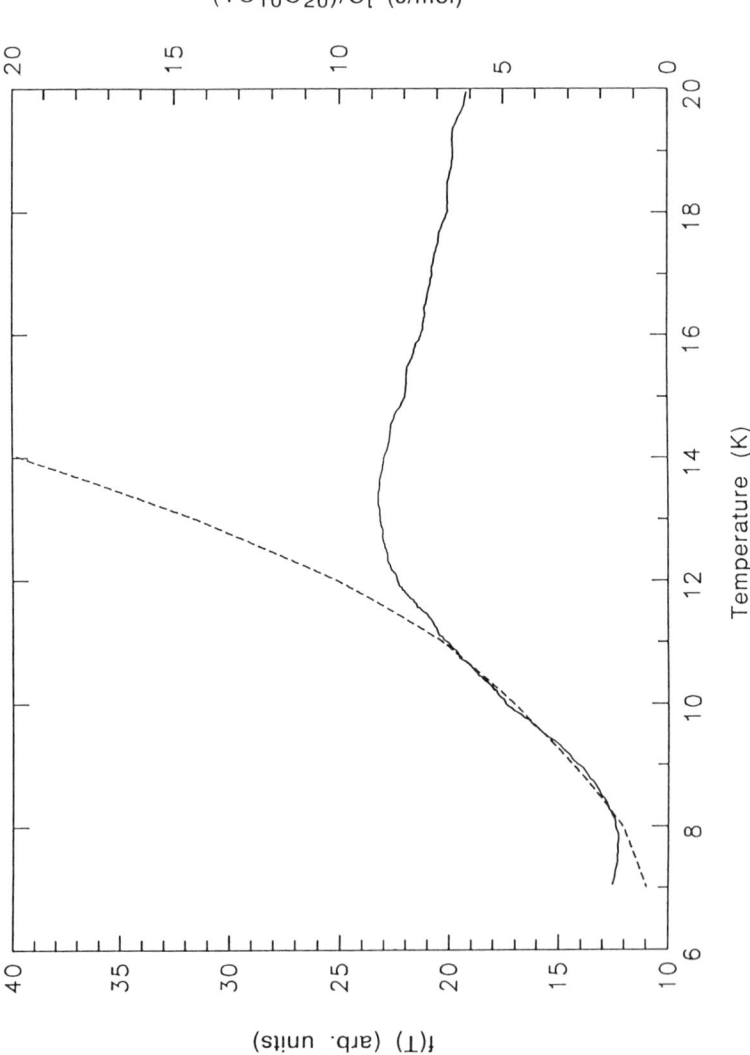

FIG. 18. Comparison of the average of $f(T)$ (solid line) with $TC_{10}C_{20}/C_t$, obtained by employing the experimental data of specific heat (broken line).

4. Magnetic Superconducting System

Er^{3+} concentration also. As can be seen in Fig. 2, the relaxation attenuation maximum associated with Er^{3+} of $ErRh_4B_4$ locates at 5 K. This relaxation-type sound-energy dissipation contributes to the background attenuation in the neighborhood of 5 K for all the $Er_{1-x}Ho_xRh_4B_4$ samples where the influence of Ho^{3+} is still dominant. Furthermore, the degree of this contribution is an increasing function of Er^{3+} concentration, as implied in Eq. (19). Therefore, the background attenuation at low temperatures should increase with Er^{3+} concentration as shown in Fig. 1. Because of the dependence of attenuation on specific heat as well as relaxation time of both magnetic ions, a quantitative explanation for this experimental result is quite involved, especially when it is necessary to consider the effects of superconductivity and spin fluctuations.

In brief, the CEF split ground states of the magnetic sublattice in $Er_{1-x}Ho_xRh_4B_4$ contribute to the relaxation-type ultrasonic attenuation behavior at low temperatures. A theoretical model that considers the oscillating energy levels as a result of the propagation of sound waves in the samples yield a relaxation-type ultrasonic attenuation that is related to the Schottky specific heat. The effective relaxation time that is obtained from our model decreases monotonically with increasing temperature, and this is consistent with the experimental attenuation results.

4.4. Sound Absorption of $Er_{0.088}Ho_{0.912}Rh_4B_4$ at High Magnetic Fields

As was mentioned previously, crystalline electric fields in a magnetic superconductor produce strong effects on many of its properties at low temperatures. The lifting of the degeneracy of the ground state of the magnetic ions by CEF has resulted in a relaxation-type ultrasonic atteuation mechanism in $Er_{1-x}Ho_xRh_4B_4$ through the phonon–phonon interaction, as discussed in the last section. There is one more interaction, which associates energy dissipation of sound waves to the Zeeman effect and to the crossing of CEF-split energy levels, that may have been observed when $Er_{0.705}Ho_{0.295}Rh_4B_4$ was placed in high magnetic fields.

Figure 19 shows magnetic-field–dependent attenuation curves obtained with the double-echo measurement technique at 5 K, 6 K, 7 K, and 8.5 K with fields up to 60 kOe. The zero field attenuation point (relative value) at different temperatures has been shifted for clarity. As can be seen, every curve shows a broad maximum at lower magnetic fields (<25 kOe). At higher temperatures, the variations of sound energy loss with magnetic field are not as dramatic as those observed at low temperatures. In fact, the results of temperature-dependent attenuation measurements at constant magnetic fields reveal that the dissipation at temperatures above 14 K is insensitive to an applied magnetic field (Sun, 1986).

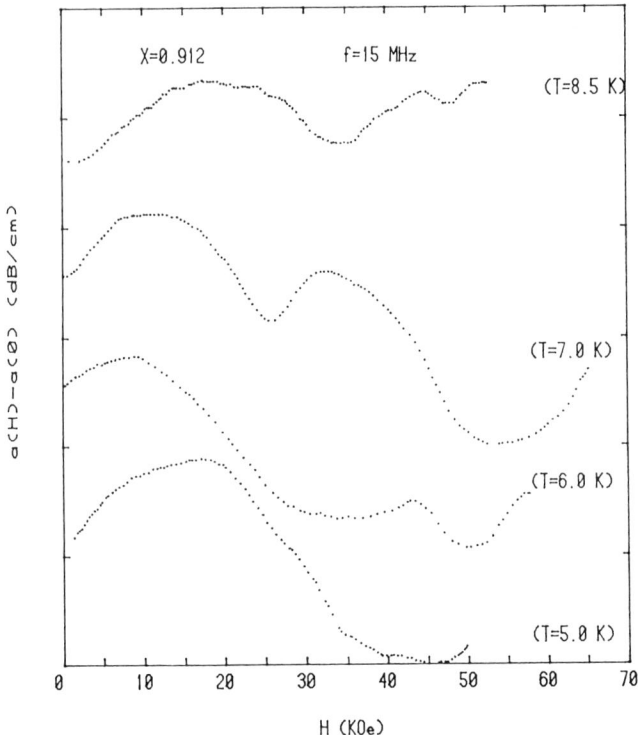

FIG. 19. $\alpha(H)$ of $Er_{0.088}Ho_{0.912}Rh_4B_4$ at 15 MHz at several constant temperatures. The relative positions of curves are shifted for clarity. (Unit scale shown in y axis is 0.05 dB/cm.)

Even if the increase of attenuation at low magnetic fields can be ascribed to the growth of magnetic moments until they are saturated at the low fields, it is still necessary to explain the decrease and the subsequent increase to the other relative maxima in attenuation at higher fields.

For polycrystalline samples of metallic systems containing rare-earth ions, conduction electron–phonon interaction, spin–phonon interaction, and magnetoelastic interactions are among the absorption mechanisms that can produce distinctive attenuation behavior. Since the magnitudes of the attenuation shown in Fig. 19 are too large to be the result of electron–phonon interaction (which would be of the order of 10^{-4} dB/cm; Pippard, 1955), and the appearance of the relatively smaller maxima at higher fields can not be explained by the simple saturation of magnetization, it seems that an additional mechanism, which involves the effect of high magnetic fields, will be necessary for interpreting the data. Becker et al. (1978) have calculated the sound-energy dissipation originating from magnetoelastic interactions. They consider the effects of electron population variations on the sound propagation when the scheme of CEF-induced 4f-electron energy levels of rare-earth ions is changed by external magnetic fields. Depending on the lattice symmetry and the energy-level scheme, the characteristic attenuation behavior due to the magnetoelastic coupling can be observed at relatively high magnetic fields and low temperatures. While the crystal structure of the isomorphic RERh$_4$B$_4$ is relatively complex, the magnetic ions are seated in a face-centered cubic sublattice. Even so, the field-dependent energy-level schemes are still too complex to predict in the case of total angular momentum quantum number $J = 8$ for Ho^{3+} and $J = 15/2$ for Er^{3+}, especially when the zero field wavefunctions are significantly affected by the applied fields (Dunlap et al., 1989). Nevertheless, the experimental results shown in Fig. 19 may still be able to be interpreted qualitatively by the magnetoelastic interaction. When $\mu H \ll k_B T$, the electrons mainly move between energy levels of separation μH owing to the thermal energy. As the magnetic field increases, the net magnetization increases, which enhances the attenuation. When the field increases to values such that $\mu H \cong k_B T$, the magnetic moments become aligned with field, the magnetization starts to saturate, and the susceptibility begins to decrease. Consequently, the attenuation will reach a maximum value, as shown in Fig. 19, and start to decrease. The widening energy gap, with increasing field, decreases the thermal excitations between energy levels. Therefore, sound attenuation will remain at low values despite the increasing field unless a crossing of energy sublevels occurs and yields an anomaly (Becker et al., 1978). When energy levels cross, thermal excitations can again occur between them, causing, in effect, a repetition of the process that was described earlier, which would then produce another maximum in attenuation as the levels are separated by a further increase of magnetic field. Thus, we believe that the local attenuation maxima at 33 kOe when $T = 7$ K and at 43 kOe when $T = 6$ K imply the occurrence of such energy-level crossovers.

The magnetic-field–dependent attenuation of HoRh$_4$B$_4$ has also been measured up to 8 kOe. The experimental results (Sun, 1986) display similar behavior. There is a broad attenuation maximum that occurs at low magnetic fields at each preset temperature. However, the slope of the increasing attenuation beginning at zero field is much smaller than that for the Er$_{0.088}$Ho$_{0.912}$Rh$_4$B$_4$. In addition, the broad attenuation maxima of HoRh$_4$B$_4$ are located at much smaller fields (around 3.5 kOe at 6 K and 4.2 K), and their magnitudes are about 20 percent of the magnitude of the maximum of Er$_{0.088}$Ho$_{0.912}$Rh$_4$B$_4$. The ground state (Ott et al., 1980) of the Ho^{3+} ion is a doublet and is separated from its first excited state by about 40 K (3.45 meV). The degeneracy (Radousky et al., 1983) of the low-lying state of Er^{3+} is four (which includes a doublet ground state and a doublet at 1.4 K), and the next doublet is at 21 K. Therefore, if the attenuation maxima are associated with the changes of electron populations in the different energy levels, we may conclude that the electrons associated with the two doublets at Er^{3+} sites dominantly contribute to the field-dependent sound-energy loss in $Er_{0.088}Ho_{0.912}Rh_4B_4$.

In summary, magnetic-field–dependent ultrasonic absorption data at constant temperatures exhibit consistent features. Magnetoelastic interaction that is associated with the Zeeman effect and the crossing of 4f-electron energy levels is proposed as the mechanism responsible for the observed sound attenuation behavior of $Er_{0.088}Ho_{0.912}Rh_4B_4$ at high magnetic fields. Further calculations would be necessary for ascertaining the validity of this model.

5. Summary

The system $Er_{1-x}Ho_xRh_4B_4$, which includes reentrant superconductors when $x < 0.9$, exhibits new phenomena and provides interesting interactions in the superconducting state. Its ultrasonic attenuation behavior displays the effects of the presence of magnetic ions in the superconducting materials. As a matter of fact, a couple of unique phonon-related interactions have been observed in the measurements.

All of the samples with $x > 0$ exhibit a broad maximum in the temperature-dependent attenuation curves at around 10 K, which moves to higher temperatures as x is increased, and also as the frequency is increased for the same value of x. The frequency dependence of this bell-shaped maximum identifies it as a possible relaxation maximum. A theory is presented that associates the relaxation processes with the electron transitions between the ground and the first excited state of Ho^{3+} ions, and also with the phonon–thermal phonon interaction. A typical relaxation relation is obtained, $\alpha \sim \omega^2\tau^{*2}/(1 + \omega^2\tau^{*2})$. However, in our case τ^* is an effective relaxation time that contains both of the above relaxation times modified by the ratio of their specific heats to the total specific heat.

Samples with $x > 0.9$ undergo a ferromagnetic transition that can be described by mean field theory. The principal consequence of this fact is that there is no sound attenuation peak at the phase transition, since there are no fluctuations present. Obviously, below the temperature where the relaxation maximum is located, the attenuation decreases. It continues to do so until the Curie temperature T_m is reached, whereupon the attenuation starts to increase. Thus, there is a minimum in attenuation located at T_m. The enhancement in attenuation is due to the fact that the magnetization increases in the ferromagnetic state, and the attenuation should be proportional to the square of magnetization. Another interesting feature was found when large magnetic fields ~6 T were applied to

4. Magnetic Superconducting System

the sample with $x = 0.912$. Here, probably for the first time ever, an effect of magnetoelastic interaction was detected when the electron energy levels crossed each other in a large magnetic field. This interaction produced a relative maximum in attenuation. Presumably, what is happening at an energy-level crossing is that the energy levels get close enough to each other that the impressed phonons have sufficient energy to effect electron transitions between the two levels and produce a maximum. When the increasing field continues to separate the energy levels, the phonon energy will not be large enough to produce transitions, and the attenuation decreases.

For all the samples with $x < 0.9$ reentrant superconductors, several new effects that are intimately associated with the superconductivity of the samples, and the interplay between superconductivity and magnetism and their electromagnetic interaction, were discovered and identified. The most perplexing phenomenon has been the increase in attenuation found in the superconducting state of the samples with $x = 0.6$ and $x = 0.813$. When the measurements were done in magnetic fields, it was verified that the effect disappeared when superconductivity was quenched by a magnetic field. Two models are offered to explain this increase. In both models, the increase in energy dissipation of sound in the superconducting state is due to spin–phonon interaction with the Ho^{3+} ions. In one model, it is proposed that there is an enhancement of spin fluctuations in the superconducting state that may be due to the superconducting screening currents. Spin–phonon interaction would then result in an increase in attenuation. The other model suggests that spin–phonon interaction has a BCS coherence factor that is associated with electromagnetic absorption, just like in nuclear magnetic resonance measurements. This coherence factor does not cancel the singularity in the density states at the energy gap edge, as it would for electron–phonon interaction, and therefore the absorption increases in the superconducting state. A resolution of this phenomenon awaits a quantitative theoretical calculation.

When x is decreased to 0.295, the superconducting region extends from 7.6 K to 0.99 K, and the effects due to the increased concentration of Er^{3+} ions become more predominant. In the first place, there is a relaxation maximum at 3.5 K that moves to 5 K when $x = 0$. For both of these cases, there is a peak in attenuation at the ferromagnetic transition that is associated with spin fluctuations of the Er^{3+} ions. For the $x = 0$ case, this peak is shifted to a higher temperature and broadened by a magnetic field, as would be expected since a magnetic field will increase the magnetic transition temperature and decrease the sharpness of the transition. Although no special features of attenuation are observed at the superconducting transition of these two samples, some very interesting results are uncovered in the superconducting state at temperatures close to the magnetic transition, and in a magnetic field throughout the whole superconducting temperature range. First, for $x = 0.295$, from the change in attenuation in zero magnetic field and in 0.5 kOe, it is possible to deduce that there is a coexistence region between ferromagnetism and superconductivity in the temperature range from 0.99 K to 1.3 K. Also, by measuring attenuation as a function of magnetic field and identifying the field where the attenuation starts to increase as H_{c1}, since this is where the magnetic flux lines begin to penetrate the sample, it is possible to plot a phase diagram for $x = 0.295$ in the H–T plane. Second, measuring the attenuation as a function of magnetic field at the lowest temperatures, but still in the superconducting state, it is possible to verify the models that used electromagnetic interaction to explain the superconducting interaction in these reentrant superconductors. It turns out that in the normal state when the applied magnetic field is perpendicular to the propagation direction, spin–phonon interaction is larger than when the field is parallel to the propagation direction. However, in the superconducting state, the two orientations yield the same magnitude for the interaction, which is in fact equal to that in the parallel case in the normal state. Therefore, by performing the transverse measurements on $x = 0$ and $x = 0.295$ at 1.5 K, it was possible to observe an increase in attenuation at H_{c2} that confirmed the theoretical predictions. It was also shown experimentally that in general, the attenuation for the transverse case in the normal state was larger than for the parallel case. And finally, at these lower temperatures the theory predicts that the samples should experience a type II-I transition at H_{c1}. The changes in attenuation observed in this magnetic field range exhibited the signature that would be expected for such a transition.

It may be possible that the effects that were first discovered in the $Er_{1-x}Ho_xRh_4B_4$ system will find their counterparts in the high-T_c superconductors, since at least some of them do contain magnetic ions and undergo antiferromagnetic transitions at low temperatures.

Acknowledgments

We would like to acknowledge that all the samples of $Er_{1-x}Ho_xRh_4B_4$ we have measured were made and provided by M. B. Maple and his group, Department of Physics, University of California, San Diego. We would also like to thank Professors Masashi Tachiki, Richard Sorbello, and Susan Schneider for numerous enlightening discussions. The work at CWM was supported by NASA-Langley Research Center, and the research at UWM was supported by the Office of Naval Research.

4. Magnetic Superconducting System

References

Barett, H. H. (1969). *Phys. Rev.* **178,** 743–762.
Baskaran, G., Zou, Z., and Anderson, P. W. (1987). *Solid State Commun.* **63,** 973–976.
Baum, H. P., Schentrom, A., Zheng, Y., Sarma, Bimal, K., and Levy, M. (1989). *IEEE Tran. Magnetics,* 987–989.
Becker, K., Thalmeier, P., and Fulde, P. (1978). *Z. Phys. B.* **31,** 257–267.
Behroozi, F., Crabtree, G. W., Campbell, S. A., and Hinks, D. G. (1983). *Phys. Rev. B* **27,** 6849–6852.
Blount, E. I., and Varma, C. M. (1979). *Phys. Rev. Lett.* **42,** 1079–1082.
Crabtree, G. W., Behroozi, F., Campbell, S. A., and Hinks, D. G. (1982). *Phys. Rev. Lett.* **49,** 1342–1345.
Dunlap, B. D., and Niarchos, D. G. (1982). *Solid State Commun.* **44,** 1577–1581.
Elliott, R. J. (1972). "Magnetic Properties of Rare Earth Metals." Plenum Press, New York.
Fertig, W. A., Johnston, D. C., Delong, L. E., McCallum, R. W., Maple, M. B., and Matthias, B. T. (1977). *Phys. Rev. Lett.* **38,** 987–990.
Fischer, D., and Maple, M. B. (1982). "Superconductivity in Ternary Compounds," I and II, Topics in Current Physics, Vol. 32 and Vol. 34. Springer, Berlin, Heidelberg, New York.
Fischer, D., Treyvaud, A., Chevrel, R., and Sergent, M. (1975). *Solid State Commun.* **17,** 721–724.
Freeman, A. J., and Jarlborg, T. (1979). *J. Appl. Phys.* **50,** 1876–1879.
Ginsburg, V. L. (1957). *Sov. Phys. JEPT* **4,** 153.
Herring, C. (1958). *Physica* **24,** S184.
Ihikawa, M. (1982). *Contemp. Phys.* **23,** 443–468.
Johnston, D. C., Fertig, W. A., Maple, M. B., and Matthias, B. T. (1978). *Solid State Commun.* **26,** 141–144.
Kuper, C. G., Revzen, M., and Ron, A. (1980). *Phys. Rev. Lett.* **44,** 1545–1548.
Kurihara, S. (1989). *Phys. Rev. B* **39,** 6600–6606.
Lander, G. H., Sinha, S. K., and Fradin, F. Y. (1979). *J. Appl. Phys.* **50,** 1990–1990.
Lindsay, R. B. (1960). "Mechanical Radiation." McGraw-Hill, New York.
Lynn, J. W., Moncton, D. E., Passell, L., and Thomlinson, W. (1980). *Phys. Rev. B* **21,** 70–78.
Mackay, H. B., Woolf, L. D., Maple, M. B., and Johnston, D. C. (1979). *Phys. Rev. Lett.* **42,** 918–921.
Mackay, H. B., Woolf, L. D., Maple, M. B., and Johnston, D. C. (1980). *J. Low Temp. Phys.* **41,** 639.
Maekawa, S., Smith, J. L., and Huang, C. Y. (1980). *Phys. Rev. B* **22,** 164–167.
Maple, M. B. (1983). *In* "Advances in Superconductivity" (M. B. Maple, B. Deaver, and J. Ruvalds, eds.). Plenum Press, New York, pp. 279–346.
Matsumoto, H., Umezawa, H., and Tachiki, M. (1979). *Solid State Commun.* **31,** 157–161.
Matthias, B. T., Suhl, H., and Corenzwit, E. (1958a). *Phys. Rev. Lett.* **1,** 92–94.
Matthias, B. T., Suhl, H., and Corenzwit, E. (1958b). *Phys. Rev. Lett.* **1,** 449–450.
Matthias, B. T., Corenzwit, E. J., Vandenberg, M., and Barz, H. (1977). *Proc. Natl. Acad. Sci. USA* **74,** 1334–1335.
Moncton, D. E., Mowhan, D. B., Eckert, J., Shirane, G., and Thomlinson, W. (1977). *Phys. Rev. Lett.* **39,** 1164–1166.
Moncton, D. E., McWhan, D. B., Schmidt, P. H., Shirane, G., Thomlinson, W., Maple, M. B., Mackay, H. B., Woolf, L. D., Fisk, Z., and Johnston, D. C. (1980). *Phys. Rev. Lett.* **45,** 2060–2063.
Mook, H. A., Koehler, W. C., Maple, M. B., Fisk, Z., Johnston, D. C., and Woolf, L. D. (1982a). *Phys. Rev. B* **25,** 372–380.
Mook, H. A., Koehler, W. C., Sinha, S. K., Crabtree, G. W., Hinks, D. G., Maple, M. B., Fisk, Z., Johnston, D. C., Woolf, L. D., and Hamaker, H. C. (1982b). *J. Appl. Phys.* **53,** 2614–2618.

Mori, H. (1965). *Prog. Theo. Phys. (Japan)* **33**, 423–455.
Ott, H. R., Keller, G., Odoni, W., Woolf, L. D., Maple, M. B., Johnston, D. C., and Mook, H. A. (1982). *Phys. Rev. B* **25**, 477–480.
Ott, H. R., Woolf, L. D., Maple, M. B., and Johnston, D. C. (1980). *J. Low Temp. Phys.* **39**, 383–396.
Pippard, A. B. (1955). *Phil. Mag.* **46**, 1104–1114.
Pobell, F. (1981). In "Ternary Superconductors" (G. K. Shenoy, B. D. Dunlop, and F. Y. Fradkin, eds.). North Holland, Amsterdam, pp. 35–41.
Radousky, H. B., Dunlap, B. D., Knapp, G. S., and Niarchos, D. G. (1983). *Phys. Rev. B* **27**, 5526–5529.
Schneider, S. C. (1981). Ph.D. thesis, University of Wisconsin—Milwaukee.
Schneider, S. C., Chen, R., Tachiki, M., Levy, M., Johnston, D. C., and Matthias, B. T. (1981). *Solid State Commun.* **40**, 61–64.
Shelton, R. N., McCallum, R. W., and Adrian, H. (1976). *Phys. Lett.* **56A**, 213–214.
Shenoy, G. K., Dunlop, B. D., and Fradkin, F. Y. (1981). "Ternary Superconductors." North Holland, Amsterdam.
Shenoy, G. K., Noakes, D. R., and Hinks, D. G. (1982). *Solid State Commun.* **42**, 411–414.
Sinha, S. K., Crabtree, G. W., Hinks, D. G., and Mook, H. (1982a). *Phys. Rev. Lett.* **48**, 950–953.
Sinha, S. K., Crabtree, G. W., and Hinks, D. G. (1982b). *Physica* **109 & 110B**, 1693–1698.
Suhl, H., and Matthias, B. T. (1959). *Phys. Rev.* **114**, 977–988.
Sun, K. J. (1986). Ph.D. thesis, University of Wisconsin—Milwaukee.
Sun, K. J., Levy, M., Maple, M. B., and Torikachvilli, M. S. (1983). *IEEE Ultrasonics Symposium Proc.*, pp. 1087–1090.
Sun, K. J., Sorbello, R. S., Levy, M., Maple, M. B., and Torikachvili, M. S. (1986). *IEEE Ultrasonics Symposium Proc.*, pp. 1123–1126.
Sun, K. J., Winfree, W. P., Xu, M.-F., Sarma, Bimal K., Levy, M., Caton, R., and Selim, R. (1988). *Phys. Rev. B* **38**, 11988–11991.
Sun, K. J., Levy, M., Maple, M. B., and Torikachvalli, M. S. (1989a). *Phys. Rev. Lett.* **63**, 453–456.
Sun, K. J., Levy, M., Maple, M. B., and Torikachvilli, M. S. (1989b). *Phys. Rev. B* **39**, 2159–2164.
Sun, K. J., Sorbello, R. S., and Levy, M. (1989c). *Phys. Rev. B* **40**, 2133–2137.
Sun, K. J., Winfree, W. P., Xu, M.-F., Levy, M., Sarma, Bimal K., Singh, A. K., Osofsky, M. S., and Le Tourneau, V. M. (1989d). *Phys. Rev. B* **42**, 2569–2572.
Sun, K. J., Parker, F. R., and Winfree, W. P., Syed, H. I., Meng, R. L., Sun, Y. Y., Hor, P. H., and Chu, C. W. (1990). *IEEE Ultrasonics Symposium Proc.*, pp. 1293–1296.
Tachiki, M. (1981). In "Ternary Superconductors" (G. K. Shenoy, B. D. Dunlop, and F. Y. Fradkin, eds.). North Holland, Amsterdam, pp. 267–273.
Tachiki, M. (1982). *Physica* **109 & 110B**, 1699–1709.
Tachiki, M., and Maekawa, S. (1974). *Prog. Theo. Phys.* **51**, 1–25.
Tachiki, M., Lee, M. C., Treder, R. A., and Levy, M. (1974). *Solid State. Commun.* **5**, 1071–1074.
Tachiki, M., Maekawa, S., Treder, R. A., and Levy, M. (1975). *Phys. Rev. Lett.* **34**, 1579–1582.
Tachiki, M., Matsumoto, H., and Umezawa, H. (1979a). *Phys. Rev. B* **20**, 1915–1927.
Tachiki, M., Kotoni, A., and Umezawa, H. (1979b). *Solid State Commun.* **32**, 599–602.
Tachiki, M., Koyama, T., Matsumoto, H., and Umezawa, H. (1980a). *Solid State Commun.* **34**, 269–273.
Tachiki, M., Matsumoto, H., Koyama, T., and Umezawa, H. (1980b). *Solid State Commun.* **34**, 19–23.

4. Magnetic Superconducting System

Tinkham, M. (1975). "Introduction to Superconductivity." McGraw-Hill, New York, pp. 50–59.
Treder, R. A., and Levy, M. (1977). *J. Magn. Magn. Mater.* **5,** 9–19.
Treder, R. A., Tachiki, M., and Levy, M. (1979). *J. Magn. Magn. Mater.* **12,** 167–175.
Vandenberg, J. M., and Matthias, B. T. (1977). *Proc. Natl. Acad. Sci. USA* **74,** 1336.
Warren, W. W., Jr., Walstedt, R. E., Brennert, G. F., Cava, R. J., Tycko, R., Bell, R. F., and Dabbagh, G. (1989). *Phys. Rev. Lett.* **62,** 1193–1196.
Woolf, L. D., Johnston, D. C., Mackay. H. B., McCallum, R. W., and Maple, M. B. (1979). *J. Low Temp. Phys.* **35,** 651.
Xu, M.-F., Baum, H-P, Schenstrom, A., Sarma, B. K., Levy, M., Sun, K. J., Loth, L. E., Wolf, S. A., and Gubser, D. U. (1988). *Phys. Rev. B* **37,** 3675–3677.
Zemansky, M. W. (1968). "Heat and Thermodynamics," 5th edition. McGraw-Hill, New York, pp. 269–269.

—5—
Ultrasonic Propagation in Sintered High-T_c Superconductors

MOISES LEVY, MIN-FENG XU AND BIMAL K. SARMA
Department of Physics, University of Wisconsin — Milwaukee, Milwaukee, Wisconsin

KEUN JENN SUN
Department of Physics, The College of William and Mary, Williamsburg, Virginia

1. Introduction 237
2. Attenuation and Velocity in $La_{2-x}Sr_xCuO_4$ 238
3. Ordinary Sintered $YBa_2Cu_3O_{7-\delta}$ 243
   3.1. Attenuation 243
   3.2. Velocity 252
4. Oriented $YBa_2Cu_3O_{7-\delta}$ 254
   4.1. Attenuation 254
   4.2. Velocity 264
5. Sound Propagation in $GdBa_2Cu_3O_{7-\delta}$ and $ErBa_2Cu_3O_{7-\delta}$ ... 271
6. BiSrCaCuO and $T\ell BaCaCuO$ Superconducting Compounds 274
7. Sound Propagation in $Ba_{1-x}K_xBiO_3$ 280
8. Summary 289
   Acknowledgment 292
   Appendix A Crystal Structure of High-T_c Superconductors 292
   Appendix B Selected Velocities and Elastic Constants for High-T_c Superconductors 295
   Appendix C Temperature Position of Attenuation Peaks for Sinter-Forged
      $YBa_2Cu_3O_7$ Samples 297
   Appendix D Activation Energies for Relaxation Times Associated with Relaxation
      Attenuation Peaks 297
   References 298

1. Introduction

High-T_c superconductivity was discovered in $Y_1Ba_2Cu_3O_7$ (Wu *et al.* 1987; Zhao *et al.* 1987; Hikami *et al.* 1987; Takagi *et al.* 1987). However, to date, single crystals of $YBa_2Cu_3O_{7-\alpha}$ still have thicknesses of the order of one millimeter to one-tenth of a millimeter. Therefore, ultrasonic propagation measurements in such thin materials are generally not easy to perform. They usually require very fast spectometers to produce nanosecond-wide pulses at very high frequencies, or they may need buffer rods to separate the rf pulses. Nevertheless, the authors in Chapters VII and VIII discuss in detail ultrasonic measurements on single crystals. They use either very short pulses at GHz frequencies (Ch. VII), or a new and very promising resonance technique (Ch. VIII).

In this chapter, we discuss the results of ultrasonic attenuation and velocity measurements in polycrystalline high-T_c materials, because these samples are easier to obtain and usually have suitable sizes for ultrasonic measurements. However, discrepancies in different samples due to differences in grain size, packing ratio, and impurities are inevitable. In addition, there could be differences due to variations in oxygen content in different samples, which could even be the case for single crystals. In the latter case, there could also be oxygen content in homogeneity. These complications are addressed in Chapter IX. Nevertheless, despite small discrepancies, ultrasonic measurements in polycrystalline samples are generally in agreement with each other.

We start our discussion with a brief survey of the $La_{2-x}Sr_xCuO_4$ system (Section 2). Then, in Sections 3 and 4, detailed discussions are given on the $YBa_2Cu_3O_{7-\delta}$ system. In Section 5, results on other 1–2–3 compounds such as $ErBa_2Cu_3O_{7-\delta}$ and $GdBa_2Cu_3O_{7-\delta}$ are given. Section 6 covers the BiSrCaCuO system the TlBaCaCuO system. These two systems have nearly the same structure as $YBa_2Cu_3O_{7-\delta}$, but some of the compounds have higher T_c's than $YBa_2Cu_3O_{7-\delta}$. In Section 7, a different superconducting system, $Ba_{1-x}K_xBiO_3$, which has a simple cubic structure and does not have Cu–O planes, is discussed in detail. And finally, Section 8 summarizes the results presented in this chapter. There are four appendices: The first one shows the crystal structures of the known high-T_c superconductors; the second, the velocities or elastic constants at room temperature; the third, the temperature position of the attenuation peaks for the sinter-forged $Y_1Ba_2Cu_3O_7$ samples, and the fourth, the activation energies for the relaxation times associated with these and other relaxation peaks.

2. Attenuation and Velocity in $La_{2-x}Sr_xCuO_4$

High-T_c superconductors were first discovered in the La pseudoternary compounds by Bednorz and Müller (1986). The crystal structure of these compounds is shown in Appendix A. Sound velocity measurements in $La_{2-x}Sr_xCuO_4$ by Bourne et al. (1987) and Bishop et al. (1987a) are among the earliest works performed acoustically in the ceramic high-T_c superconductors. Both ultrasonic attenuation or low-frequency internal friction and sound velocity measurements in the La compounds were also reported by Horie et al. (1987a), Esquinazi et al. (1987), Fossheim et al. (1987), Makarov et al. (1987), and Bhattacharya et al. (1988a, b).

It was found that the undoped La_2CuO_4 is a semiconductor, while the Ba- or Sr-doped $La_{2-x}Sr_xCuO_4$ or $La_{2-x}Ba_xCuO_4$ could be a superconductor, depending

5. Sintered High-T_c Superconductors

on the value of x. The divalent metal concentration x plays an important role in the phase diagrams of these materials. For $x = 0.15$, the superconductors of $La_{2-x}Sr_xCuO_4$ have their maximum T_c, about 35 K. It is very useful to study the properties of the $La_{2-x}Sr_xCuO_4$ superconductors with different concentrations x, for this may provide additional clues toward the understanding of these materials. Sound measurements in samples with different concentrations have been performed by Bourne et al. (1987), Horie et al. (1987a), and Klochko et al. (1989). The velocity and elastic constants are tabulated in Appendix B.

Discrepancies in sound measurements in $La_{2-x}Sr_xCuO_4$ appeared mainly because of the ceramic nature of the samples used. Difficulties in controlling parameters during sample preparation resulted in samples with different characteristics. Different sound frequencies could also cause differences in observations, since wave scattering at grain boundaries would depend both on the size of the grains and the wavelength of the sound.

Difficulties in characterizing the samples also contributed to the discrepancies in the measurements. But despite these discrepancies, general features could be extracted from the measurements.

Figure 1 shows the Young's modulus of a $La_{1.85}Sr_{0.15}CuO_4$ sample as a function of temperature (Bourne et al. (1987). It can be clearly seen that there appears a huge softening on cooling between 200 K and 100 K. This huge softening at high temperatures is a common feature in velocity measurements in this system,

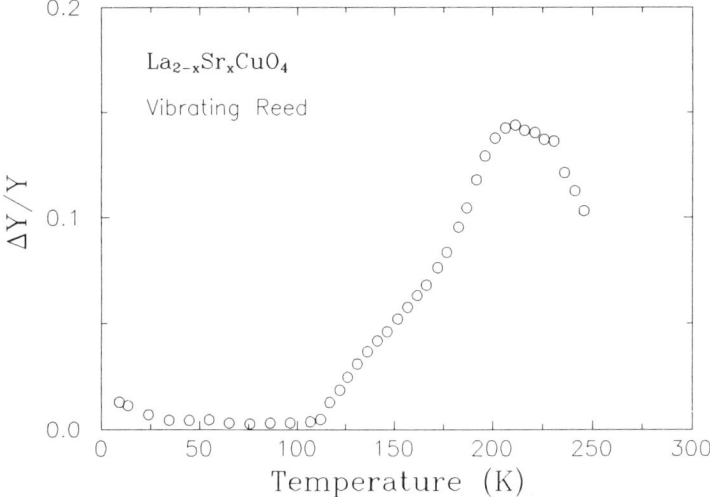

FIG. 1. Relative Young's modulus change in $La_{1.85}Sr_{0.15}CuO_4$. After Bourne et al. (1987).

since it has also been reported by other groups — for example, Bishop et al. (1987a,b), Horie et al. (1987a), Esquinazi et al. (1987), Fossheim et al. (1987), and Bhattacharya et al. (1988a). The temperature range for the softening was different for the various groups. It is believed that different annealing processes would affect the softening temperature range. Bourne et al. (1987) reported that the lower bound of the range changed from 100 K to 70 K when measured in a sample with different annealing conditions, and hence different T_c's. The range was also reported to be between 250 K and 120 K (Bishop et al. 1987b) and between 300 K and 150 K (Fossheim et al. 1987). The softening range also depended on the value of x. For $x = 0.14$, Horie et al. (1987a) observed a softening between 240 K and 150 K. Bhattacharya et al. (1988a) reported a gigantic softening between 220 K and 190 K in their sample with $x = 0.2$. For $x = 0.3$, the compound was hardly a superconductor, and the softening even disappeared (Bourne et al. 1987), becoming harder at lower temperatures like a normal metal.

The softening at high temperatures is typically around 10%, as reported by several groups. This is in contrast to that observed in the undoped La_2CuO_4. Luthi et al. (1987) compared the doped and the undoped compounds and showed that for both longitudinal and transverse waves, the strontium-doped compound exhibited large softening, while the undoped compound had little velocity change compared to the doped compound. They also showed that the softening was frequency-independent, as they obtained the same softening behavior for soundwaves both at 5 MHz and at 15 MHz.

For the $La_{2-x}Sr_xCuO_4$ system, there is a tetragonal-to-orthorhombic phase transition near 200 K when $x = 0.15$ (Jérome and Kang 1988). The huge softening in velocity with decreasing temperature may be directly related to the temperature-dependent transition into orthorhombicity (Crawford et al. 1990). The frequency independence of the softening reported by Luthi et al. (1987) suggests that the softening may be associated with a phase transition.

A lattice instability preceding a structural phase transition and then followed by superconductivity in A15 compounds has been reported and studied earlier (Testardi, 1973, 1975a). The behavior of the anomalous velocity appearing at high temperatures in the $La_{2-x}Sr_xCuO_4$ system is very similar to that of the A15 compounds V_3Si and Nb_3Sn. It is reasonable to conjecture that the instability of the lattice in the $La_{2-x}Sr_xCuO_4$ compound is the signature of the onset of superconductivity, as in the A15 compounds. However, the increase in magnitude of the lattice parameters with decreasing temperature below the structural phase transition in the A15 compounds was locked by the onset of the superconducting state, which was not observed for $La_{2-x}Sr_xCuO_4$ (Fleming et al. 1987). In addition, Klochko et al. (1989) observed similar behavior in sound velocity in both

5. Sintered High-T_c Superconductors

single-crystal La_2CuO_4 and ceramic $La_{2-x}Sr_xCuO_4$ with $x = 0.2$. A strong hardening instead of softening was observed for both of the systems in the high-temperature range between 200 K and 100 K.

At low temperatures, the velocity of $La_{2-x}Sr_xCuO_4$ was observed to increase with decreasing temperature. This was consistently reported by almost all of the research groups who measured velocity in the system. The starting temperature for the increase, though, is sample-dependent and may depend on the concentration x. For $x = 0.15$, the upturn in velocity was seen to be around 100 K, after the saturation of the huge softening, by Bourne et al. (1987), at about 100 K by Fossheim et al. (1987), and at about T_c by Bishop et al. (1987a). For a different concentration $x = 0.2$, Esquinazi et al. (1987) observed the increase of velocity at 20 K, Bhattacharya et al. (1988a) observed the change starting at around 170 K, and Klochko et al. (1989) reported the upturning at around 50 K. For $x = 0.14$, Horie et al. (1987a) observed the change around 115 K. Large discrepancies in the starting temperature of the upturn in velocity were evident, but the velocity did increase at low temperatures. Structural phase transformations in $La_{2-x}Ba_xCuO_4$ and $La_{2-x}Sr_xCuO_4$ reported by Crawford et al. (1990) and J. D. Axe et al. (1989) indicate that there is an orthorhombic-to-tetragonal phase transition at low temperatures. The stoppage of the orthorhombicity in $La_{2-x}Sr_xCuO_4$ may result in the upturn of sound velocity, and the increase could then be just regular stiffening due to cooling.

A velocity discontinuity at T_c was observed by several groups (Bourne et al. 1987, and Bishop et al. 1987b), and a distinct discontinuity in the slope of velocity with respect to temperature was observed by Bhattacharya et al. (1988b). These are thermodynamically related to the specific-heat jump at T_c, which was also experimentally observed in the $La_{2-x}Sr_xCuO_4$ system. A detailed discussion of the relationship is given in the next section, since a similar feature was observed in the $YBa_2Cu_3O_{7-\delta}$ system as well.

In ultrasonic attenuation or low-frequency internal friction measurements, peaks or shoulders in attenuation or internal friction were observed near 100 K and 200 K. Bourne et al. (1987) had a shoulder in their internal friction at about 90 K in the low kilohertz frequency range for a concentration $x = 0.15$. Fossheim et al. (1987) observed a hump in longitudinal attenuation below 100 K and a shoulder at about 180 K at a frequency of 24 MHz in a sample with $x = 0.15$. Horie et al. (1987a) reported a pronounced peak and a small maximum at 100 K and 200 K, respectively, in longitudinal attenuation at 10 MHz in their sample with $x = 0.14$. For $x = 0.2$, Bhattacharya et al. (1988a) observed shoulders near 100 K and 200 K in longitudinal attenuation at 15 MHz, and Klochko et al. (1989), Fig. 2, reported shoulders at 100 K and 180 K at 50 MHz for longitudinal modes.

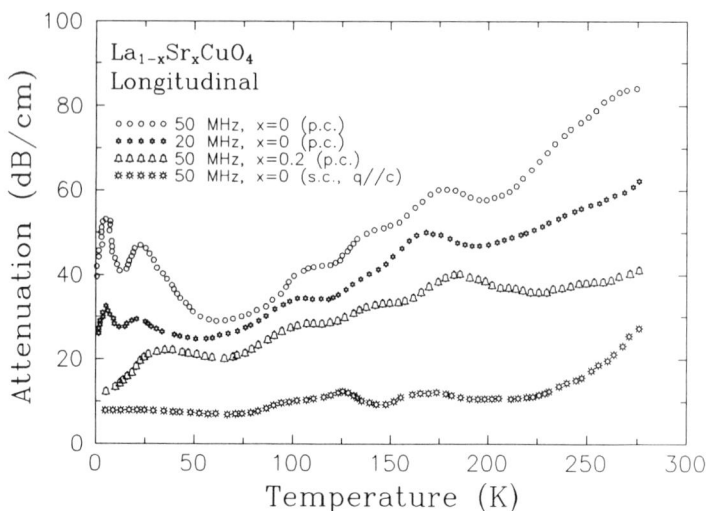

FIG. 2. Temperature dependence of sound absorption in LaCuO$_4$ (top two), La$_{1.8}$Sr$_{0.2}$CuO$_4$ (third), and single-crystal LaCuO$_4$. After Klochko et al. (1989).

It seems that the attenuation maxima near 100 K and 200 K are common for the La$_{2-x}$Sr$_x$CuO$_4$ system and are probably not strongly dependent on the concentration x. It is very possible that both maxima are associated with the occurrence of structural phase transformations at low and high temperatures, respectively. The one at 200 K may be due to the tetragonal-to-orthorhombic phase transition near this temperature, and the other may be caused by the proximity to the low-temperature orthorhombic-to-tetragonal transformations (Crawford et al. 1990).

In some of the measurements, a peak or a maximum in attenuation was observed below T_c (Bhattacharya et al. 1988a; Klochko et al. 1989). This is very similar to what is observed in the YBa$_2$Cu$_3$O$_{7-\delta}$ system, which is discussed in more detail in Section 3. The peak may not be directly related to superconductivity. Instead, it might be associated with relaxation processes involving excitations that may mediate the interactions that produce superconductivity in these systems. The fact that similar peaks at approximately the same temperature had been observed in the undoped La$_2$CuO$_4$ system (Klochko et al. 1989) shows that the peak itself may have nothing to do with the superconducting transition. In addition, the peak at low temperatures in La$_2$CuO$_4$ seems to move with frequency. For 50 MHz sound waves, the peak shifted to a slightly higher temperature than that for 20 MHz waves. This is the characteristic behavior of a thermally activated relaxation process. If the peak for the La$_{2-x}$Sr$_x$CuO$_4$ system

5. Sintered High-T_c Superconductors

behaves similarly to that in La_2CuO_4, then a relaxation process may most likely be responsible for it.

In summary, for the $La_{2-x}Sr_xCuO_4$ system, velocity measurements have shown a large softening from around 200 K to around 100 K, which may be associated with the tetragonal-to-orthorhombic phase transition occurring in this temperature range. Analogies have been drawn with the A15 compounds such as V_3Si and Nb_3Sn, which also show an instability before a structural phase transition from cubic to tetragonal; their phase transitions are followed by a superconducting transition, a sequence that is common to all A15 compounds. Therefore, the instability was said to be a precursor of superconductivity in the A15 compounds. In this feature, the $La_{2-x}Sr_xCuO_4$ system may resemble the A15 compounds. A discontinuity in velocity at T_c was observed by several groups. This may be thermodynamically related to the specific-heat jump at T_c that is associated with the second-order phase transition. The fact that the velocity increases below T_c with decreasing temperature in all the reported measurements by different groups proves that this is an intrinsic effect; however, the large increase in velocity, which is several orders of magnitude higher than expected, has not yet been explained.

Attenuation measurements generally show peaks at around 200 K and 100 K. The 200 K peak may be related to a structural phase transition from a high-temperature tetragonal phase to an orthorhombic phase displaying a large softening in velocity. The 100 K peak may be due to a structural phase transition from the orthorhombic phase to a low-temperature tetragonal phase. In addition, a peak below the superconducting transition has often been observed. This peak may not be directly related to the superconducting state. Instead, relaxation processes are more likely to be responsible for it, as evidenced by the fact that similar peaks were also observed in the semiconducting compound La_2CuO_4, and the latter peaks shifted with frequency. Of course, frequency-dependent measurements have to be performed in the superconducting state of $La_{2-x}Sr_xCuO_4$ in order to fully justify this deduction. Similar features both in attenuation and in velocity have been found in the $YBa_2Cu_3O_{7-\delta}$ compounds and chemically isomorphic compounds with yttrium replaced by rare-earth elements. We will cover those in the following sections.

3. Ordinary Sintered $YBa_2Cu_3O_{7-\delta}$

3.1. ATTENUATION

Ultrasonic attenuation measurements in $YBa_2Cu_3O_{7-\delta}$ materials generally yield three attenuation maxima in the temperature range from 4 K to 300 K in zero magnetic field. One of the three maxima is around the superconducting transition temperature, and the other two are located approximately at 250 K and 180 K. The positions of these maxima display considerable frequency and sample dependence.

Figure 3 shows ultrasonic attenuation measurements in a single-phase $YBa_2Cu_3O_{7-\delta}$ sample for 15 MHz longitudinal waves (Xu et al. 1988). The sample had a relative density of 78% compared to that of a single crystal, which would signify a porosity of 22%, and a transition temperature of 91 K. Two maxima were observed in this experiment, with one located at 84 K and a very broad one around 250 K. The broad maximum at 250 K in Fig. 3 might comprise features around 180 K that could not be resolved in these measurements. Figure 4 shows a different measurement with longitudinal waves at 5 MHz (Bhattacharya et al. 1988a). The measurement was concentrated in the temperature range between 4 K and 150 K. Two maxima were observed, with one at 70 K and one at 130 K. These might correspond to the 84 K and 180 K maxima, respectively. Figure 5 displays the attenuation data for 10 MHz longitudinal waves in a single-phase sample with a T_c of 92.1 K (Lagreid and Fossheim, 1988). The data shown are in the high temperature range. Again two maxima were observed, with one at 265 K and one around 180 K. Other researchers also have observed attenuation maxima in these temperature ranges. He et al. (1987) found peaks in their attenuation measurements around 160 K and 250 K for longitudinal

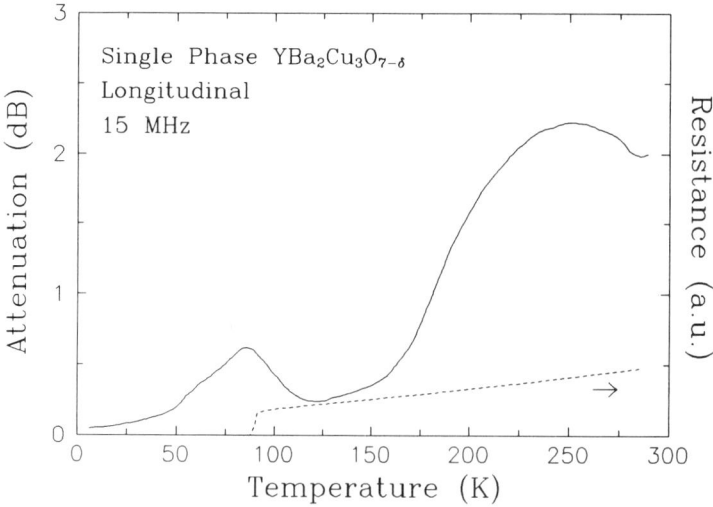

FIG. 3. Ultrasonic attenuation of 15 MHz longitudinal waves in a single-phase sample of $YBa_2Cu_3O_{7-\delta}$. The lower curve shows the resistive transition at 91 K. After Xu et al. (1988)

5. Sintered High-T_c Superconductors

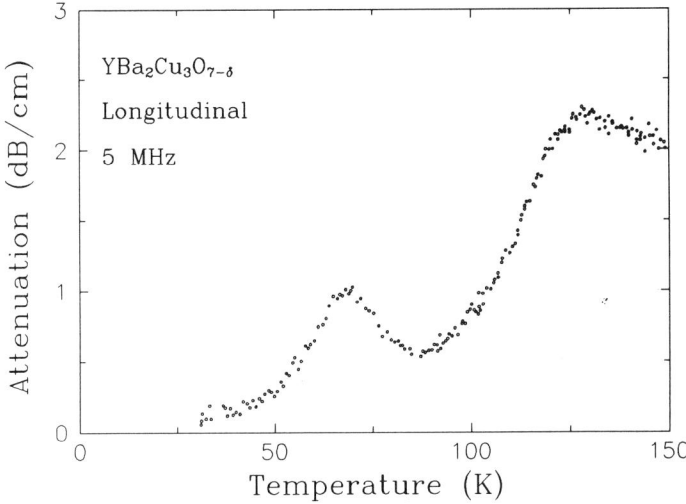

FIG. 4. Temperature dependence of attenuation in $YBa_2Cu_3O_{7-\delta}$ at 5 MHz. Residual attenuation at 4.2 K has been subtracted. After Bhattacharya *et al.* (1988a).

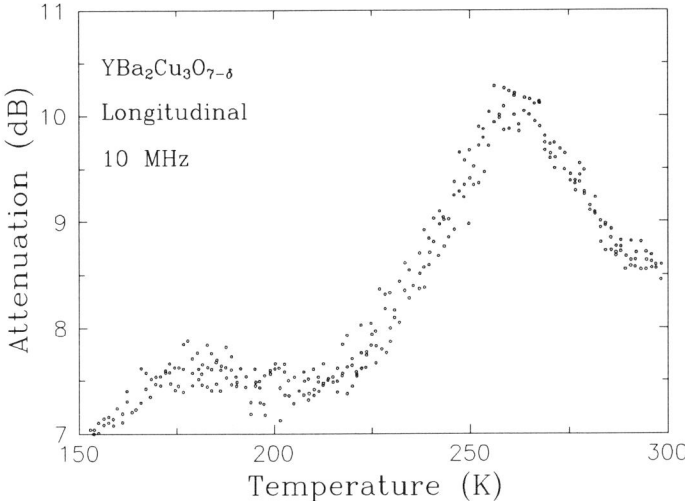

FIG. 5. Attenuation of longitudinal ultrasound in $Y_1Ba_2Cu_3O_{7-\delta}$ at 10 MHz, showing clear peaks about 265 K and 180 K. After Lagreid and Fossheim (1988).

sound waves at 10 MHz in samples with T_c = 80 K and 85 K. Wang et al. (1987) observed peaks at around 230 K and 95 K for 5 MHz longitudinal sound waves in a sample with T_c = 90 K. Horie et al. (1987b) reported attenuation peaks at about 245 K and 160 K for longitudinal waves at 10 MHz in a sample with T_c = 93 K.

The peak below T_c in Fig. 3 was very intriguing. Ultrasonic attenuation peaks in heavy fermion systems such as UPt$_3$ were observed just below T_c (Qian et al. 1987). It had been suggested that gapless regions on the Fermi surface of these heavy fermion superconductors could be responsible for these peaks. Would high-T_c superconductors resemble heavy fermion superconductors? If one determines the total change of the attenuation below T_c in the data for high-T_c superconductors, one would find that the change is typically on the order of 1 dB/cm. This change in attenuation is too large to be produced by electron–phonon interaction in the high-T_c superconductors.

For a normal metal, using a free electron model, the total attenuation of longitudinal waves due to electron phonon interaction for $ql < 1$ is given by (see Chapter I for a review)

$$\alpha = \frac{4}{15} \frac{Nmv_F^2}{\rho v_s^3} \omega^2 \tau, \qquad (1)$$

where q is the sound wave vector, $l = v_F \tau$ is the electron mean free path, N is the electron density, m is the electron mass, v_F is the Fermi velocity of the electrons, ρ is the mass density of the material, v_s is the sound velocity for longitudinal waves, ω is the angular frequency, and τ is the electron relaxation time. To evaluate this relation we need to know τ, which could be obtained from the electrical conductivity expression, if we assume the τ's in these two processes are the same. The electrical conductivity σ is related to the relaxation time τ by

$$\sigma = \frac{Ne^2 \tau}{m}, \qquad (2)$$

where e is the electron charge. The Fermi velocity for a free electron is

$$v_F = \frac{\hbar}{m} (3\pi^2 N)^{1/3} \qquad (3)$$

By substituting (2) into (1), one obtains

$$\alpha = \frac{4}{15} \frac{m^2 v_F^2 \omega^2}{\rho e^2 v_s^3} \sigma, \qquad (4)$$

which shows that the attenuation in a metal due to electron–phonon interaction is proportional to the electrical conductivity. Substituting the values, for 10 MHz

5. Sintered High-T_c Superconductors

longitudinal waves in $YBa_2Cu_3O_{7-\delta}$, $N = 5.8 \times 10^{21}$, $\omega = 2\pi \times 10^7$ rad/s, $\sigma = (1/200[\mu\Omega\text{ cm}])$ (optimistic), $\rho = 6.37$ g/cm^3, and $v_s = 4.4 \times 10^5$ cm/s, one obtains for the attenuation due to electron–phonon interaction $\alpha = 10^{-6}$ dB/cm. This value, due to the low electrical conductivity and low carrier density, is surprisingly small compared to the attenuation change observed in $YBa_2Cu_3O_{7-\delta}$ compounds. Recent surface acoustic wave measurements (Baum 1990 and Baum et al. 1991) allow the deduction that the intrinsic resistivity along the ab planes of crystallites composing an $Y_1Ba_2Cu_3O_7$ film is 12.5 $\mu\Omega$ cm, or eight-times better than what is measured in the best sintered samples. Since electron–phonon interaction is directly proportional to the electron mean free path, then it should sample mean free paths in parallel, and therefore should be determined by the longest mean free path in the system — as opposed to the resistivity, which samples them in series, and therefore should be determined by the shortest mean free path in the system. Following this argument, we could estimate that the attenuation could be an order of magnitude larger, 10^{-5} dB/cm.

The attenuation value calculated above was based on a free electron model. The actual attenuation in real metals is usually larger than the one calculated above because of anisotropies of the Fermi surface. Compared to the case for a vanadium sample with a resistivity ratio of 20 (Waynert et al. 1974), one would expect a total attenuation change for $YBa_2Cu_3O_{7-\delta}$ to be of the order of 10^{-2} dB/cm, after taking into consideration the new resistivity ratio, the carrier density, the mass density, and the sound velocity of the high-T_c samples (Xu et al. 1988). The experimental result of a total attenuation change of 1 dB/cm is still too big if it were due to electron–phonon interaction only.

To verify experimentally whether the peak in attenuation below T_c is due to electron–phonon interaction, one can apply a magnetic field strong enough to make the material normal and at the same time measure the attenuation to see if any change happens to the peak. But for the $YBa_2Cu_3O_{7-\delta}$ material, H_{c2} at zero temperature is on the order of 100 Tesla, which is not easily achieved in most laboratories. An alternate way would be to change the frequency of the sound waves to see if the peak position changes with frequency or remains at the same position in temperature.

Figure 6 shows the frequency dependence of the attenuation peaks for longitudinal sound waves at 10 MHz, 27 MHz, and 32 MHz (Sun et al. 1988). The sample had a T_c of 89 K and a relative density of 75%. Clearly, as shown in the figure, at 10 MHz the attenuation peak was observed below T_c at about 81 K, but the peak was shifted to 100 K, which is above T_c when the sound frequency was changed to 27 MHz. Increasing the frequency further caused the peak position to shift to an even higher temperature. For 32 MHz, the peak was shifted to about 120 K. This proves that the peak below T_c is not directly

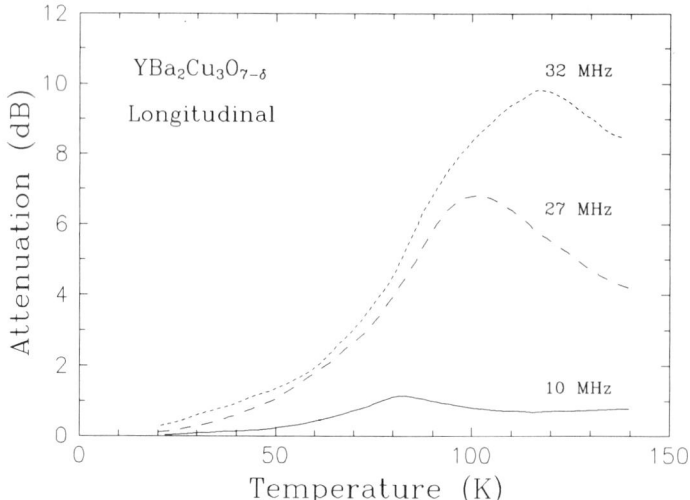

FIG. 6. Temperature-dependent attenuation measurements at three different frequencies, 10 MHz, 27 MHz, and 32 MHz in ordinary sintered YBa$_2$Cu$_3$O$_{7-\delta}$. After Sun et al. (1988).

associated with the superconducting transition. Rather, the behavior is characteristic of a relaxation process for sound wave attenuation. It was also observed that the attenuation maximum approximately had a quadratic frequency dependence.

The relaxation process could be either intrinsic or extrinsic. Extrinsic causes include Bordini relaxation processes, relaxation of grain boundary movement, and relaxation due to the mobility of twinning boundaries, etc. Intrinsic relaxation may be explained with the aid of two-energy-level systems that may be associated with plasmons, excitons, splitting of magnetic moments by crystalline electric fields, or oxygen tunneling due to oxygen deficiency in the Cu–O chains and planes, etc.

For a two-energy-level system with an energy separation ε, because the population of the energy levels cannot instantly follow the sound waves interacting with them, it can be shown that the attenuation of the sound waves owing to the relaxation from the upper energy level to the lower level is given by (Sun and Levy, Chapter IV of this book; Sun et al. 1989a)

$$\frac{\alpha}{\alpha_{\max}} = 2 \frac{\omega\tau(T)}{1 + \omega^2\tau^2(T)}, \tag{5}$$

where $\alpha_{\max}$ is the maximum attenuation at angular frequency ω, and $\tau(T)$ is the characteristic relaxation time, which is usually temperature-dependent. The right-

5. Sintered High-T_c Superconductors

hand side of the relation has a maximum when $\omega\tau = 1$. If we assume that the relaxation time τ is a decreasing function of temperature T, then we would expect that the maximum in attenuation moves to higher temperatures when the frequency of the sound waves is raised, as is seen in Fig. 6.

More information could be extracted from the relaxation process. Sun et al. (1988) calculated the temperature dependence of the relaxation time τ from the attenuation data and found that τ has a nearly exponential temperature dependence, as indicated by the nearly linear relation of a semilog plot of τ versus T. The straight-line fit of the high temperature data yields $\tau \sim e^{450/T}$, while the low temperature data yields $\tau \sim e^{50/T}$. Although these three semilog plots for the different frequencies do not fall on top of each other, especially at low temperatures (below 70 K), the authors believe that an appropriate background attenuation subtraction would improve the agreement among the three curves. However, the preceding expressions may mean that there exist relaxation processes with activation energies of 38 meV (450 K) and 4.3 meV (50 K). The former value is roughly equal to the Debye temperature obtained by specific heat measurements on $YBa_2Cu_3O_{7-\delta}$. Although this energy is less than the electron pairing energy required by some theoretical models, it is possible that these phonons are still important for superconductivity, since it has been proposed that both phonon and nonphonon mechanisms are necessary for providing high superconducting transition temperatures (Kresin, 1987a), especially when a multigap or anisotropic energy gap model is considered (Kresin, 1987b).

There are various models that can be assumed for the appearance of a relaxation-type ultrasonic attenuation curve. How the magnitude of the attenuation maximum depends on the frequency, and sometimes how the properties of the material, such as the specific heat, are incorporated into the relaxation process, may help to determine the possible mechanism involved. A report by Horn et al. (1987) about the temperature variations of the lattice constants of $YBa_2Cu_3O_{7-\delta}$ indicated that an orthorhombic distortion occurs at temperatures between 60 K and 140 K and showed that a maximum difference in the lattice constants b and a appeared around the superconducting transition. This distortion does not result in changes of the volume of the unit cell or the area of unit basal plane. It may be possible that the softening of the sound velocity occurring at T_c, which we cover later, reflects this structural instability. It also has been found (Bhattacharya et al. 1988b) that softening due to the change of the shear modulus at T_c is predominant over that of the bulk modulus. This further illustrates that the distortion is shear in nature.

It is possible that this distortion also enhances the energy loss of sound at temperatures around T_c. The difference between the temperature variation rates of the lattice constants a and b may produce anisotropic grain boundary expansions or contractions with respect to the propagation direction of the sound waves. These grain boundary motions (dc motions) together with their vibrations (ac motions) induced by the traveling sound waves result in a relaxation attenuation. Usually, for a relaxation process, the α_{max} varies linearly with ω. The quadratic dependence of α_{max} on ω in the data may also be interpreted as being caused by this broad temperature range structural distortion. Some of the A15 structure superconductors have a similar frequency-dependent attenuation behavior resulting from their structural transformations at low temperatures.

However, the possibility that an attenuation anomaly results directly from the intrinsic properties of $YBa_2Cu_3O_{7-\delta}$ cannot be totally excluded (Sun et al. 1988) — such as (1) a perturbed tunneling effect when the sound wave deforms the lattice potential, which may be related to the motion of the copper electrons in a multiwell potential set by the surrounding oxygens in the Cu-O planes; or (2) an acoustoelectric effect (Mason, 1958), which arises from the simultaneous bunching of electrons and holes in materials, caused by the passing of sound waves. The return of these carriers to their instantaneous equilibrium state in a multiwell potential shows a relaxation time, which exponentially depends on the inverse of the temperature and has a similar mathematical expression to

$$\frac{1}{\tau(T)} = \frac{1.3 \times 10^7}{(e^{50/T} - 1) + \frac{9.95 \times 10^9}{(e^{450/T} - 1)}} \tag{6}$$

which gives a reasonable fit to the semilog plot of the attenuation in Fig. 6. The attenuation coefficient of this effect is proportional to the square of the frequency of the sound waves. The fact that the electric current carriers of $YBa_2Cu_3O_{7-\delta}$ are holelike in the a–b plane and electronlike along the c-axis (or a–c plane) could provide an environment for the occurrence of this type of relaxation process.

There is some evidence that, at least at low temperatures, these materials behave similarly to glasslike solids. Specific heat measurements in the $La_{2-x}Sr_xCuO_4$ and the La_2CuO_4 systems reveal that there is a linear term in the temperature dependence of their specific heat at low temperatures (Fischer et al. 1988). These two systems are superconducting and semiconducting, respectively, at low temperatures, but both have linear terms. One explanation could be that the linear contribution is from the nonsuperconducting electrons with heavy effective masses, but this is less likely for the La_2CuO_4 system, which is semiconducting. Alternatively, it could also be explained by analogy to the linear specific heat term observed in most amorphous solids, which was believed to be produced by low-energy two-level systems. Thus, the linear term contribution in the $La_{2-x}Sr_xCuO_4$

system may come from the glasslike properties of the oxide superconductors. Similarly, the $YBa_2Cu_3O_{7-\delta}$ system may have the same property as the $La_{2-x}Sr_xCuO_4$ system, although no linear specific heat term has been observed at low temperatures, but it might be masked by a large background in the specific heat of $YBa_2Cu_3O_{7-\delta}$. The observation of a logarithmic temperature dependence of the sound velocity change at low temperatures may be proof that low-energy two-level systems do exist in $YBa_2Cu_3O_{7-\delta}$ materials (Golding *et al.* 1987, and Chapter VII). Thermal conductivity measurements showed power-law temperature behavior at low temperatures. This is reminiscent of observations in amorphous materials and glasslike solids, which are explained within the framework of two-level-system tunneling models (Fischer *et al.* 1988).

It has been proposed that the low-temperature specific heat, which is in excess of the lattice contribution in $YBa_2Cu_3O_{7-\delta}$, may be associated with isolated paramagnetic Cu^{2+} ions, whose presence may be due to impurities, lattice defects, and oxygen content in homogeneities. The magnetic moments of the ions may have an energy splitting distribution due to interactions with the local crystalline electric fields (Fischer *et al.* 1988).

It should be pointed out that different oxygen content in the $YBa_2Cu_3O_{7-\delta}$ compounds may result in different magnitudes of the attenuation peaks, and also may result in different peak positions in temperature. The density of states of the two-level systems at low temperatures could be a function of the oxygen deficiency by a mechanism similar to that reported by Hikata *et al.* (1989), who performed relative velocity measurements at very low temperatures for various oxygen deficiencies δ and found changes in the density of states with changes in δ. The change of the density of states of the two-level system would, in turn, affect the magnitude of the attenuation peak. Specific-heat measurements in single crystal $YBa_2Cu_3O_{7-\delta}$ showed that oxygen deficiency may affect the specific heat contributed by the lattice (von Molnar *et al.* 1988). This would change the total effective relaxation time τ (Sun *et al.* 1989a; Sun and Levy, Chapter IV) because τ is related to the specific heat of the phonon subsystems. Different τ's would then result in changing the positions of the attenuation peaks in temperature.

In summary, we have shown that the attenuation exhibits peaks that move up in temperature as the frequency is increased. This behavior is typical of a relaxation process that may be associated with a two energy-level system produced by an excitation that could be intrinsically related to the mechanisms that produce superconductivity in this system. Furthermore, there is evidence that at low temperatures there are relaxation peaks that are associated with tunnelling between two low-lying energy levels.

3.2. VELOCITY

Velocity measurements in $YBa_2Cu_3O_{7-\delta}$ generally give consistent results over a wide temperature range for both sintered polycrystalline ceramics and single crystals when the porosity of the materials is taken into consideration. Figure 7 displays the temperature dependence of the velocity in a polycrystalline sample (Xu, 1990). Since the thermal expansion of $YBa_2Cu_3O_{7-\delta}$ is very small, typically on the order of a hundred ppm over the whole temperature range, corrections due to thermal expansion were not taken into account. The velocity generally increases with decreasing temperature. At the vicinity of T_c, there is a change of the derivative of the velocity with respect to temperature. The velocity increases below T_c, maybe even more rapidly than above T_c. Similar results have been obtained by Bishop et al. (1987b). Figure 8 shows the velocity as a function of temperature in the vicinity of T_c with higher resolution (Golding et al. 1988). A jump in velocity at T_c was seen, with the total change being about 30 ppm, by taking the difference of the extrapolated velocity data from both above and below T_c. Similar phenomena were observed by other research groups — for example, Horie and Mase (1989) and Saint-Paul et al. (1989) on single crystals. Recently, Breazeale and Jiang (1990) have reported that ultrasonic nonlinearity of the elastic constants, as measured by the generation of the second harmonics, appears to vanish at T_c.

The discontinuity of both the velocity and its temperature derivative at T_c

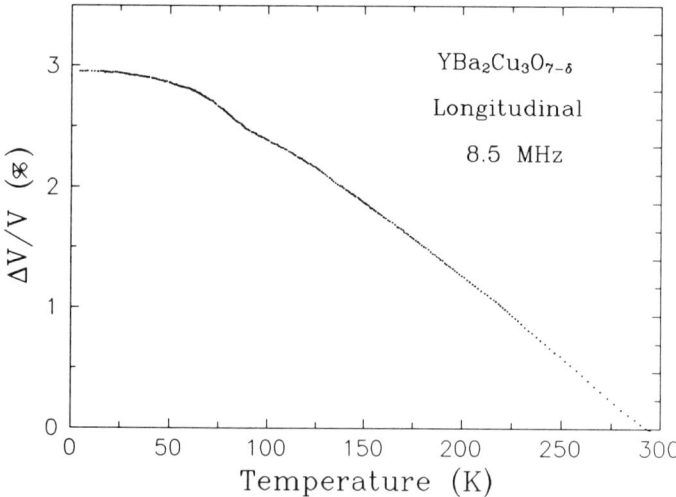

FIG. 7. Relative velocity change of longitudinal waves as a function of temperature in ordinary sintered $YBa_2Cu_3O_{7-\delta}$. After Xu (1990).

5. Sintered High-T_c Superconductors

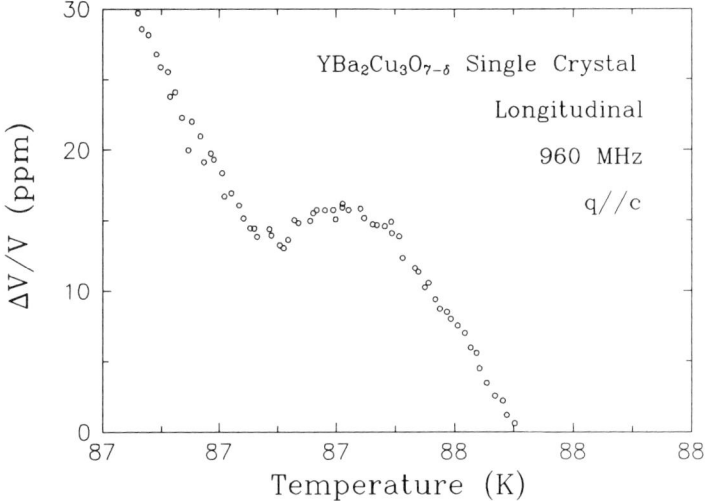

FIG. 8. Longitudinal sound along the c-axis. The discontinuity is estimated to be $\Delta v_c/v_c = 30$ ppm at T_c. After Golding *et al.* (1988). © 1988 IEEE.

could be explained thermodynamically by the jump in specific heat observed at T_c and by the pressure dependence of T_c. At zero magnetic field, the superconducting transition is a second-order phase transition. Starting from one of the TdS equations $TdS = C_P dT - TV\beta dP$ and a differential equation for $V(P, T)$,

$$dV = \frac{\partial V}{\partial T} dT + \frac{\partial V}{\partial P} dP, \tag{7}$$

one obtains Ehrenfest's equations for a second-order phase transition, by using the fact that the entropy and the specific volume are continuous across a second-order phase transition,

$$\frac{dP}{dT} = \frac{\Delta C_P}{TV\Delta\beta}, \tag{8}$$

$$\frac{dP}{dT} = \frac{\Delta\beta}{\Delta\kappa}, \tag{9}$$

where we have defined the thermal expansion coefficient β and the compressibility coefficient κ as follows:

$$\beta = \frac{1}{V} \frac{\partial V}{\partial T}, \tag{10}$$

$$\kappa = -\frac{1}{V} \frac{\partial V}{\partial P}, \tag{11}$$

Substituting Eq. (9) into (8), in order to eliminate $\Delta\beta$, one obtains

$$\Delta\kappa = \frac{1}{T_c V}\left(\frac{dT_c}{dP}\right)^2 \Delta C_P, \qquad (12)$$

Since the compressibility κ is the reciprocal of the bulk modulus B, it follows that

$$\Delta B = -\frac{\Delta\kappa}{\kappa^2} = -B^2\,\Delta\kappa. \qquad (13)$$

If we neglect the contribution from the shear modulus, then

$$\frac{\Delta V_\ell}{V_\ell} = \frac{1}{2}\frac{\Delta B}{B}, \qquad (14)$$

and the longitudinal velocity change ΔV_ℓ can then be written as

$$\frac{\Delta V_\ell}{V_\ell} = -\frac{1}{2}B\frac{\Delta C_P}{T_c V}\left(\frac{dT_c}{dP}\right)^2. \qquad (15)$$

Using the values for $B = 64 \times 10^{10}$ dyn/cm^2 (Fischer et al. 1988; Zhao et al. 1989), $\Delta C_p/T_c V = 0.415$ mJ/cm^{3}K^2 (Fischer et al. 1988, for average of values given), and $dT_c/dP = 0.07$ K/kbar (Schirber et al. 1987), one obtains $\Delta V_\ell/V_\ell = 6.5 \times 10^{-6}$, which is generally within an order of magnitude of the velocity changes determined experimentally.

The change of the velocity derivative with respect to temperature is related to the specific-heat jump and the second derivative of T_c with respect to stress. Testardi (1975b) derived a general relation between them. For longitudinal waves along a principal axis, the change is directly proportional to the specific heat jump at T_c and the corresponding elastic constant. It is also proportional to the sum of three terms involving the derivative of T_c with respect to stress, the square of the derivative, and the second derivative of T_c with respect to stress.

4. Oriented YBa$_2$Cu$_3$O$_{7-\delta}$

4.1. Attenuation

So far we have discussed ultrasonic attenuation and velocity measurements that were mainly performed on ordinary sintered isotropic samples. Information from single crystals is discussed in Chapters VII and VIII. An alternative way to obtain information on anisotropic properties of high-T_c materials is to measure samples with preferentially aligned crystallites. The advantages of measuring such oriented samples are many: Large-size samples are easier to obtain as compared to

5. Sintered High-T_c Superconductors

single crystals; pulsed signals are easier to separate due to the proper size of the samples; oxygen content is usually more homogeneous because of the small size of the crystallites; and anisotropic effects are observable because of the preferential orientation of the crystallites in the samples. Oriented samples may be obtained from processes using strong magnetic field alignment or sinter-forging. In this section, we shall discuss measurements on sinter-forged materials.

Sinter-forged $YBa_2Cu_3O_{7-\delta}$ samples (Robinson et al. 1987) are made by repetitive calcining, soaking, and grinding of the initially stoichiometrically mixed powders. After the preceding process is repeated five times, pellets are made from the powder. The last process is carried out in a resistance furnace, by forging the pellets under constant pressure at high temperature in a flowing oxygen atmosphere. The resulting pellets show preferential orientation, with 80% of the c-axis of the crystallites aligned within 20° of the forging axis, as determined by x-ray diffraction and optical micrographs. The individual crystallites are disklike, with an average size of 11 μm in thickness and 37 μm in diameter. For our measurements, samples were cut from the pellets, and the density of the samples was determined to be 95% of that of a single crystal. T_c was found to be 91 K as determined by the midpoint of the resistive transition.

Previously, anisotropy was observed in the resistivity, the magnetization (Song et al. 1987), and the critical current density. The lower critical field was determined to be 700 G and 1,300 G at 5 K when the magnetic field was applied perpendicular and parallel to the forging axis, respectively. Critical currents were determined to have values of 9.4×10^4 and 5.9×10^4 A/cm^2, respectively, when currents were sent perpendicular and parallel to the forging axis at 5 K and zero magnetic field.

Anisotropy was also observed in ultrasonic attenuation and velocity measurements in the sinter-forged samples when propagating different modes of sound waves along different directions with respect to the forging axis. Since in the direction perpendicular to the forging axis, the a and b axes of the crystallites are randomly oriented, the samples show rotational symmetry. There are five different propagation configurations for sound waves: longitudinal waves propagating perpendicular or parallel to the forging axis, transverse waves propagating parallel to the forging axis, and transverse waves propagating perpendicular to the forging axis with the polarization direction perpendicular or parallel to the forging axis.

Ultrasonic attenuation measurements were performed in the sinter-forged samples by Xu et al. (1989) in three different configurations. These were longitudinal sound waves propagating perpendicular and parallel to the forging axis, and transverse waves travelling parallel to the forging axis. Later, they also performed measurements in a fourth configuration, transverse waves with the propagation and polarization perpendicular to the forging axis (Xu, 1990). Results are generally consistent. Measurements in the fifth configuration were difficult to perform because the samples were thin (typically 1.6 mm in thickness), and transverse waves propagating in this material with the polarization perpendicular to the broad surfaces appeared to exhibit considerable interference effects.

Figure 9a shows the attenuation as a function of temperature for 12 MHz longitudinal waves propagating perpendicular to the forging axis. Three peaks in attenuation were observed in the temperature range from room temperature to 4 K. The peak located at 250 K is very similar in position to the one shown in Fig. 3 and Fig. 5. The peak at around 180 K was not present in the single phase sample (Fig. 3), probably because of the high background in the latter. Lagreid and Fossheim (1988) observed a similar peak at almost the same temperature as shown in Fig. 5. The peak below T_c at about 70 K is similar to the ones observed in Figs. 3 and 4, and, as shown in Fig. 6, it is believed to be produced by a relaxation process.

For a different configuration, longitudinal waves propagating along the forging axis at 12 MHz, a different attenuation curve was obtained (Fig. 9b). Surprisingly, no peaks were seen around 250 K and below T_c. Only one peak was found at about 180 K, similar to the previous configuration. This peak is very pronounced.

For transverse waves, propagation along the forging axis was performed at two different frequencies, 14.5 MHz and 41.0 MHz. Figure 10a shows the attenuation at 14.5 MHz. The attenuation generally increases upon cooling till it saturates at the lowest temperature. This is different from what is observed for longitudinal waves propagating perpendicular to the forging axis, as shown in Fig. 9a and 9b. There is a shoulder in attenuation around 180 K. If we subtract a background, which is obtained by interpolating the attenuation from above 250 K and below 120 K indicated by the dashed line in the figure, a maximum in attenuation is clearly seen, as shown in the bottom curve. The maximum is located at about 180 K, similar to the location of the maxima located in both Figures 9a and 9b.

Figure 10b exhibits the attenuation for transverse waves at the higher frequency, 41 MHz. Generally, the curve looks very similar to that shown in Fig. 10a, except that the total attenuation change is approximately 4.2 times that in Fig. 10a, and there appears to be a slight rounding down at the lowest temperature. As in the procedure utilized in Fig. 10a, a background (indicated by the dashed line) is subtracted from the attenuation, and the resultant bottom curve clearly shows a maximum at about 195 K. Thus, the maximum moved to a higher temperature when the frequency was increased from 14.5 MHz to 41 MHz. Again, a relaxation process is inferred.

5. Sintered High-T_c Superconductors

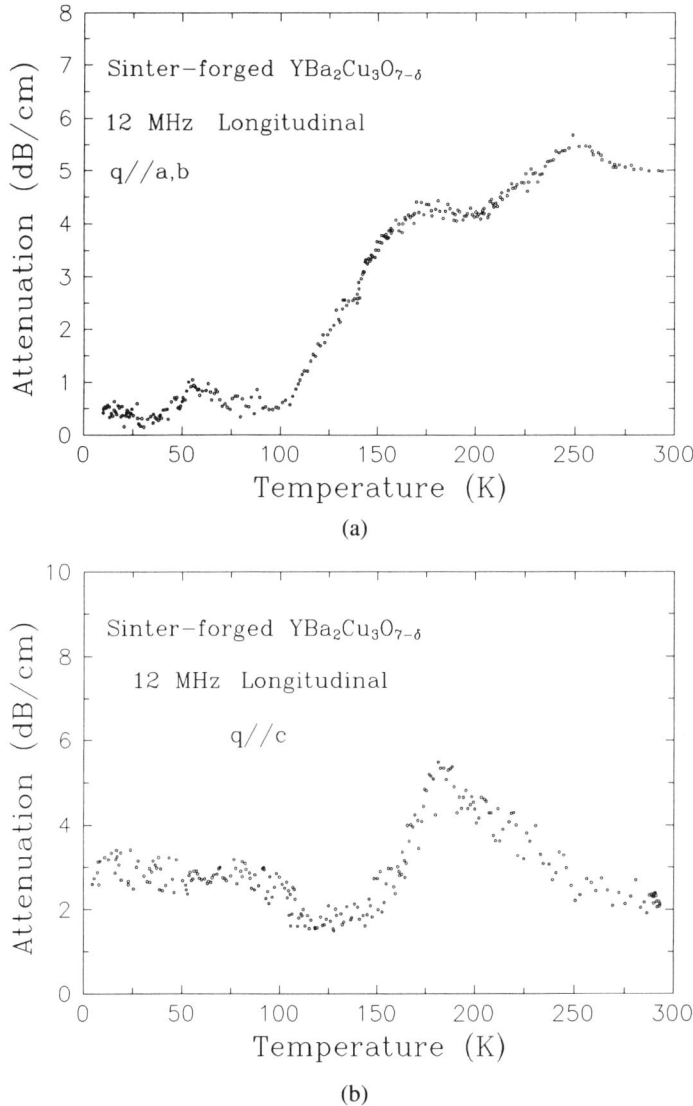

FIG. 9. (a) Attenuation of 12 MHz longitudinal waves propagating perpendicular to the forging axis. (b) Attenuation of 12 MHz longitudinal waves propagating parallel to the forging axis. After Xu *et al.* (1989).

Anisotropy in attenuation is clearly exhibited in these oriented materials. When longitudinal sound waves were propagated perpendicular to the forging axis, three peaks were seen at 70 K, 180 K, and 250 K, but when both longitudinal and transverse waves were propagated parallel to the forging axis, only one peak

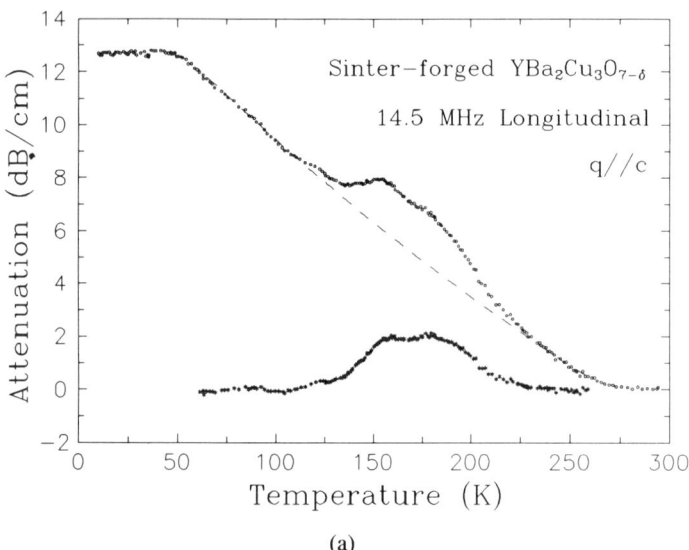

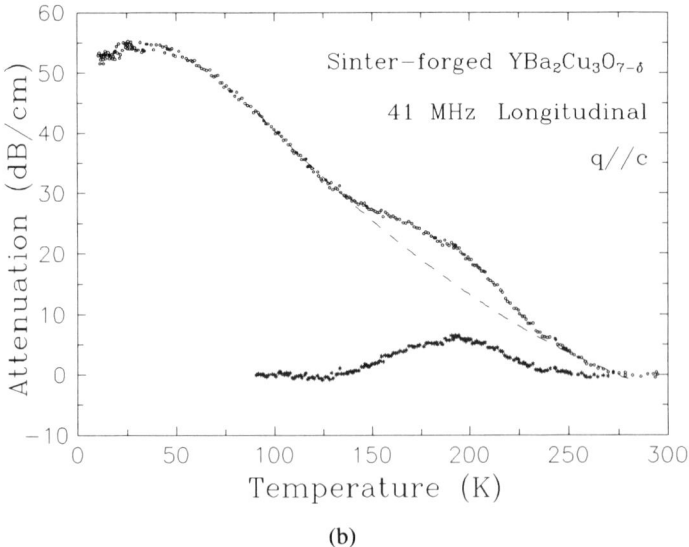

FIG. 10. Attenuation of 14.5 MHz transverse waves propagating parallel to the forging axis. The lower curve is the resultant attenuation after a background, indicated by the dashed line, is subtracted. (b) Attenuation of 41 MHz transverse waves propagating parallel to the forging axis. The lower curve is the resultant attenuation after background, indicated by the dashed line, is subtracted. After Xu et al. (1988).

5. Sintered High-T_c Superconductors

was seen around 180 K. Since the c-axis of more than 80% of the crystallites is aligned parallel to the forging axis, it is reasonable to say that the direction of the forging axis is the same as the direction of the c-axis. Consequently, when sound waves travel parallel to the forging axis, it is equivalent to sound waves travelling parallel to the c-axis of the crystallites.

In an orthorhombic $YBa_2Cu_3O_{7-\delta}$ unit cell, there are two Cu–O planes that are separated by one yttrium atom that is situated at the center of the cell. The planes are perpendicular to the c-axis. There are also two Cu–O chains that are parallel to each other and to the b-axis. The chains are also perpendicular to the c-axis, and thus parallel to the Cu–O planes. The crystal structure is shown in Appendix A.

Therefore, when longitudinal waves are propagated perpendicular to the c-axis or parallel to the a–b planes, the sound waves generate distortions in the Cu–O planes and the Cu–O chains of the crystals. This seems to produce three peaks in attenuation over the whole temperature range. But when longitudinal waves are propagated parallel to the c-axis or perpendicular to the a–b planes, only the separation between the planes is changed, i.e., there are no distortions of the Cu–O planes or of the Cu–O chains. This, in turn, seems to produce only one peak in the attenuation. Similarly, when transverse waves were propagated along the c-axis, the planes undergo shear motion, and there are no distortions of the planes. This again produced only one peak in attenuation. Hence, it has been proposed (Xu et al. 1989) that the peak in attenuation at 250 K and the one below T_c are associated with the interaction of sound waves with the excitations in the Cu–O planes or in the Cu–O chains, while the peak at 180 K seems to be an isotropic effect in the $YBa_2Cu_3O_{7-\delta}$ material.

The attenuation of transverse waves along the forging axis increased with decreasing temperature. This is contrary to what is observed in normal cases, where the attenuation generally decreases with decreasing temperature because the effects associated with dislocations and with thermal phonons interacting with sound waves are reduced when the temperature is decreased. In the sinter-forged samples, the disklike crystallites are oriented so that the flat faces of the disks are perpendicular to the forging axis. When transverse waves are propagated parallel to the forging axis, it is easy for the disklike crystallites to experience shear motion in the background sintered matrix. Therefore, we suggest that the increase of attenuation with decreasing temperature is not intrinsic, but rather is due to the relaxation process involving the shear motion of the disklike crystallites in the surrounding matrix. Actually, as pointed out earlier, the attenuation of the higher-frequency waves (41 MHz) had a slight local maximum at around 25 K. This may be a relaxation peak due to the crystallites' motion, where the product of the relaxation time associated with it and the sound wave angular frequency approaches unity. If this were the case, it would be easy to explain why there was no such peak at low temperatures at 14.5 MHz, simply because the τ in this case was too short, so that $\omega\tau$ was still less than unity even at the lowest temperature of the experiment. In Fig. 11 we have plotted the temperature dependence of τ obtained by analyzing the data of Fig. 10b using Eq. (5). By extrapolating to low temperatures, we see that at 14.5 MHz a relaxation maximum does not occur even at the lowest measured temperature.

In the work by Xu et al. (1989), it was predicted that the attenuation for transverse waves propagating perpendicular to the forging axis with the polarization also perpendicular to the forging axis should exhibit similar features to those observed for the configuration of transverse waves traveling parallel to the forging axis. Figure 12a shows that attenuation in the aforementioned configuration. The measurements were performed at a frequency of 6.5 MHz. Only one broad maximum in attenuation was seen at 165 K. In this configuration, the sound waves shear the planes perpendicular to the c-axis, but generate no area changes in the planes. This should produce only one attenuation peak, as deduced from the arguments presented above. The background attenuation did not increase as contrasted to the case for transverse waves parallel to the forging axis when cooling to the lowest temperature, probably because only the edges of the crystallites are sheared with respect to the surrounding matrix.

Results of attenuation at a higher frequency, 10.1 MHz, in the same configuration are shown in Fig. 12b. The curve looks very similar to that shown in

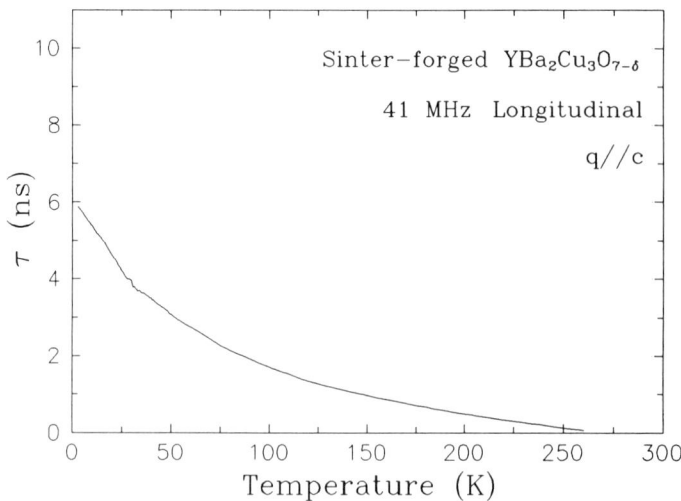

FIG. 11. Relaxation time τ versus temperature for the attenuation in Fig. 10b. After Xu (1990).

5. Sintered High-T_c Superconductors

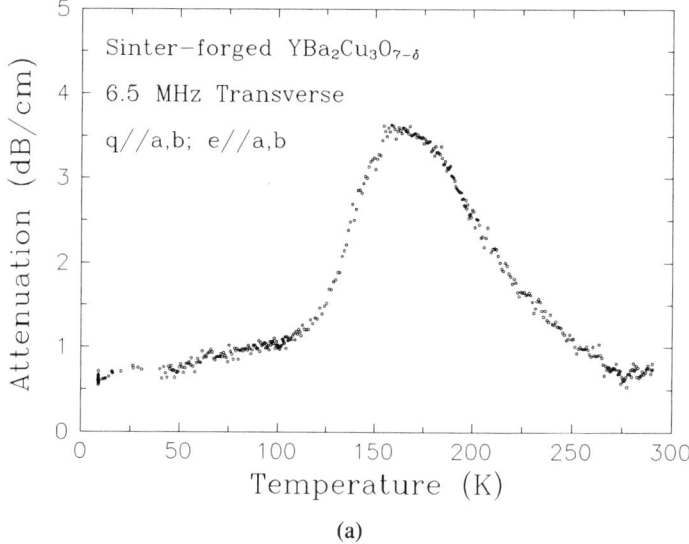

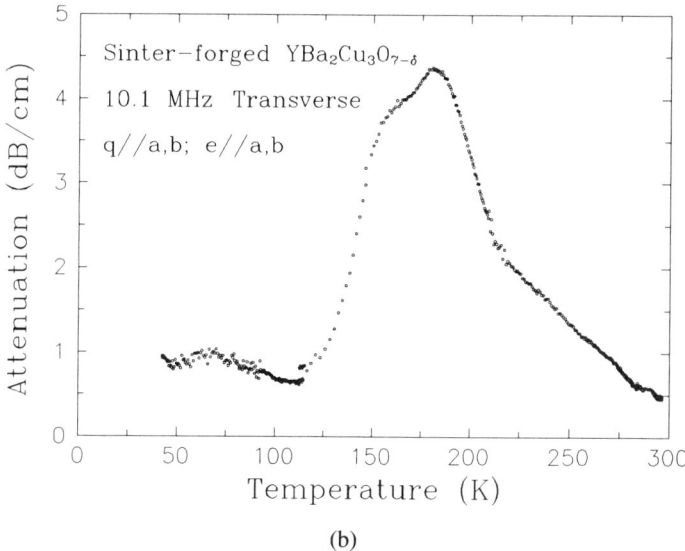

FIG. 12. (a) Attenuation of 6.5 MHz transverse sound waves with propagation and polarization direction perpendicular to the forging axis. (b) Attenuation of transverse sound waves at 10.1 MHz with both propagation and polarization direction perpendicular to the forging axis. After Xu (1990).

Fig. 12a, exhibiting only one maximum, except that its position shifted to a higher temperature 170 K. Once again, this is indicative of a relaxation process.

We have already discussed the attenuation peak below T_c that may be attributed to a relaxation process associated with the excitations of a two-energy-level system. The attenuation in the neighborhood of 180 K may have a similar mechanism, because the positions of the peaks are frequency-dependent. But for the peak around 250 K, frequency-dependent attenuation measurements have not been performed on the same sample. Therefore, whether or not this peak is associated with a relaxation process is still not definitely established. However, if it were, the temperature dependence of the relaxation time associated with such a process is given in Fig. 13a. An activation energy of 750 K may be obtained from the slope of this curve. The data are reproduced on an expanded scale in Fig. 13b. The solid line is plotted using Eq. (5) with a thermally activated relaxation time with an activation energy equal to 750 K. The positions of the three peaks are tabulated in Appendix C. The activation energies of the excitations that may be associated with these peaks are tabulated in Appendix D. In this Appendix it may be seen that the 180 K peak is associated with excitations with an activation energy of 1100 K. This is the energy of the excitations that several theorists have postulated for the production of Cooper pairs in these high T_c superconductors. So indirectly we may have extracted evidence for the presence of excitations that may replace the virtual phonons in a BCS-like interaction in the formation of Cooper pairs. And our selection rules evolved from the anisotropies in the attenuation seem to indicate that these excitations are associated with distortions of the reservoir of electrons between the Cu-O planes, since they do not require distortions of the Cu-O planes to contribute to the attenuation. The temperature positions of the three peaks are tabulated in Appendix C.

Attenuation peaks in the vicinity of 250 K were observed by several other research groups. He *et al.* (1987) reported attenuation peaks at 250 K and 270 K in two different samples for longitudinal sound waves at 10 MHz with T_cs of 80 K and 85.5 K, respectively. They also observed a peak at 235 K for longitudinal waves at 10 MHz in a different compound, $Y_3Ba_5Cu_4NbO_{18-y}$. Horie *et al.* (1987b) observed a broad peak at 245 K for longitudinal waves at 10 MHz in a sample with $T_c = 91.4$ K (midpoint of resistive transition). Suzuki *et al.* (1988) measured 10 MHz longitudinal sound attenuation in a sample of relative density of 83% and T_c of 90 K and found a peak near 240 K. Wang *et al.* (1987) performed both internal friction and ultrasonic attenuation measurements. They found peaks at 220 K and 240 K in internal friction measurements at 5 kHz and 120 kHz, and 230 K and 260 K in longitudinal ultrasonic attenuation measurements at 5 MHz. Cannelli *et al.* (1987) reported a peak at 230 K from their internal friction measurements at 6.5 kHz in a sample with $T_c = 91$ K. Shi *et*

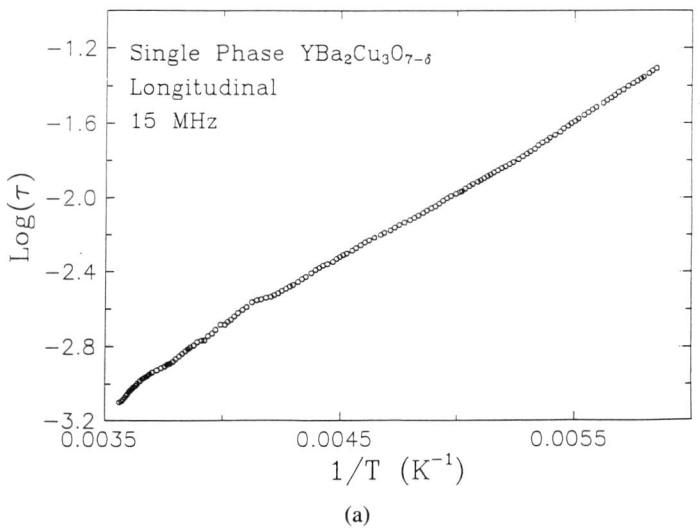

(a)

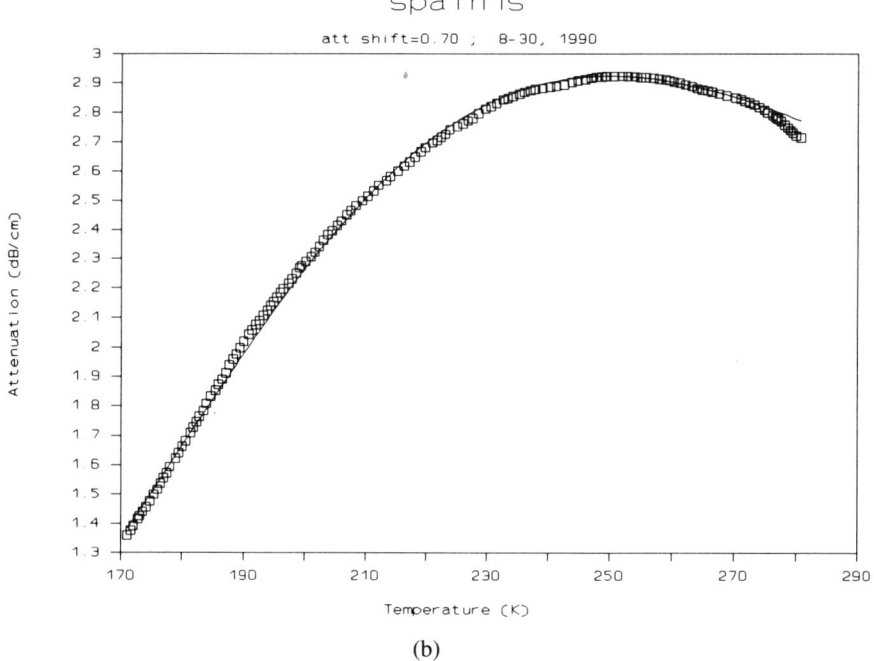

(b)

FIG. 13. (a) Plot of relaxation time τ versus $1/T$ for 250 K peak in Fig. 3. Equation (5) was used to obtain τ. The slope of this curve yields an activation energy of 750 K. (b) High-temperature attenuation peak from Fig. 3 uses an exponential fit to the relaxation time with an activation energy $\Delta = 750$ K, as obtained from Fig. 13a. After Xu (1990).

al. (1989) measured internal friction in a single crystal and had a peak at 245 K in their curve.

In addition to the model presented above by Xu et al. (1989), who suggested from their measurements in oriented samples that the peak around 250 K may be related to the interaction of sound waves with excitations in the Cu–O planes, other groups have proposed different models, the principal model of which is that the maximum is produced by a first-order phase transition. Specific heat measurements have shown an anomalous peak around 220 K (Lagreid et al. 1987) when warming up from below 200 K. Chen et al. (1987) observed a step in their resistivity measurement. Calemczuk et al. (1988) found anomalies between 220 K and 230 K in their specific-heat and resistivity measurements. In lattice parameter measurements, Srinivasan et al. (1988) observed a change in the c lattice parameter in the temperature range from 220 K to 240 K using x-ray diffraction, and Francois et al. (1988) found changes in $(a - b)/(a + b)$ near 240 K in their neutron powder diffraction measurements. They claim a first-order phase transition in this temperature range. He et al. (1987) also suggest a lattice instability in the same temperature range. Lagreid and Fossheim (1988) proposed a glasslike phase, after a lattice instability related to the ordering of the oxygen vacancies, based on their analysis of a plot of $\log(\tau^{-1})$ versus $\log(T - T_0)$, where τ is the relaxation time and T_0 is the freezing temperature. Some argue (Chapter IX) that an antiferromagnetic phase transition may contribute to the anomalies at this temperature range, since impurities like CuO may have an antiferromagnetic transition near 230 K.

4.2. Velocity

Velocity measurements in the sinter-forged samples generally show as strong an anisotropy as was observed in the attenuation measurements. Large hysteresis was observed in these materials while cycling from warm-up to cool-down. Similar hysteric phenomena have been reported by other groups on ordinary sintered samples of single phase or multiphase materials composed of large grains.

Figure 14a shows the velocity versus temperature for 12 MHz longitudinal sound waves propagating perpendicular to the forging axis. This curve looks different from what was measured in ordinary sintered materials, where the velocity generally increases on cooling down to liquid-helium temperature. The velocity in Fig. 14a initially increased upon cooling from room temperature, reaching a maximum around 130 K. From there it started to decrease until well below T_c. Then it increased below 30 K. On the warming-up curve, hysteresis was observed starting from around 50 K. The warming-up curve merges with the cooling-down curve before reaching room temperature.

5. Sintered High-T_c Superconductors

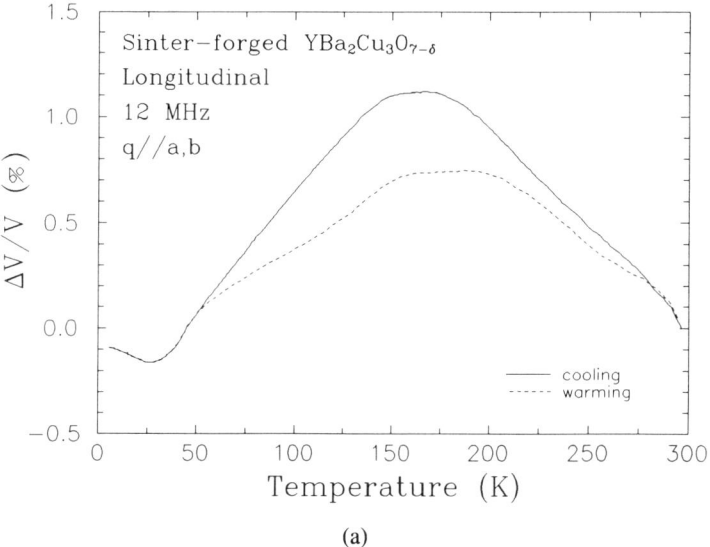

(a)

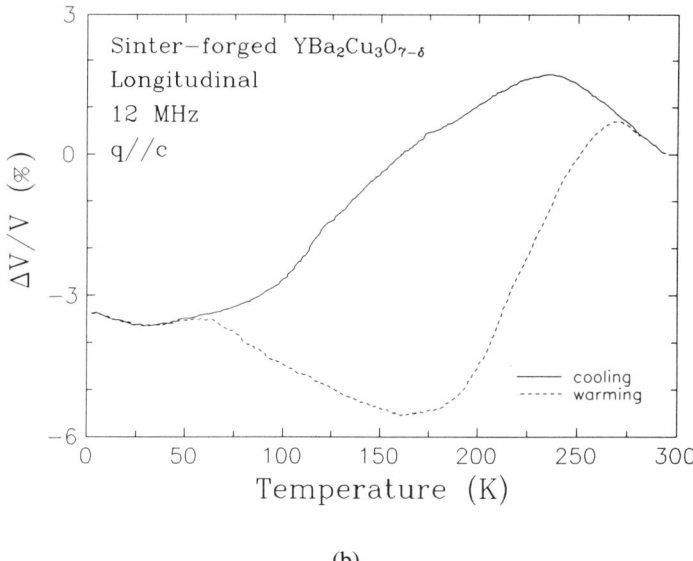

(b)

FIG. 14. (a) Temperature-dependent velocity change when propagating longitudinal sound waves perpendicular to the forging axis. (b) Velocity change of longitudinal sound waves when propagating parallel to the forging axis. (c) Velocity change of transverse sound waves propagating parallel to the forging axis. (d) Velocity change of transverse waves with both propagation and polarization direction perpendicular to the forging axis. [(a-c) After Zhao et al. (1989); (d) after Xu et al. (1990).]

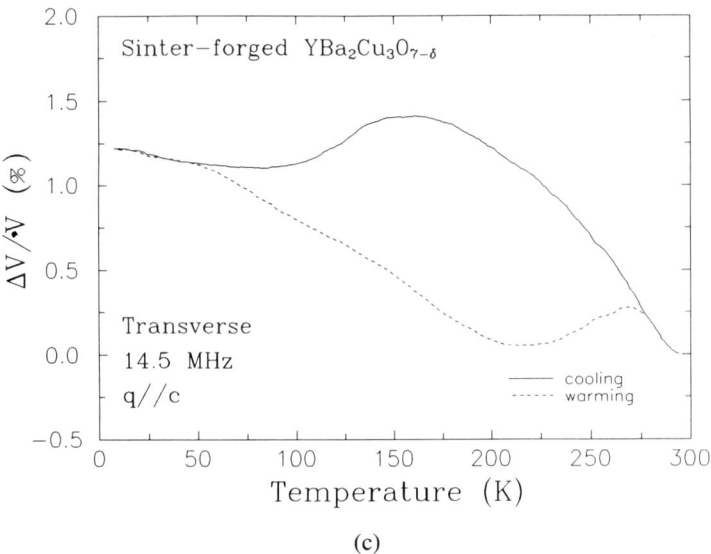

(c)

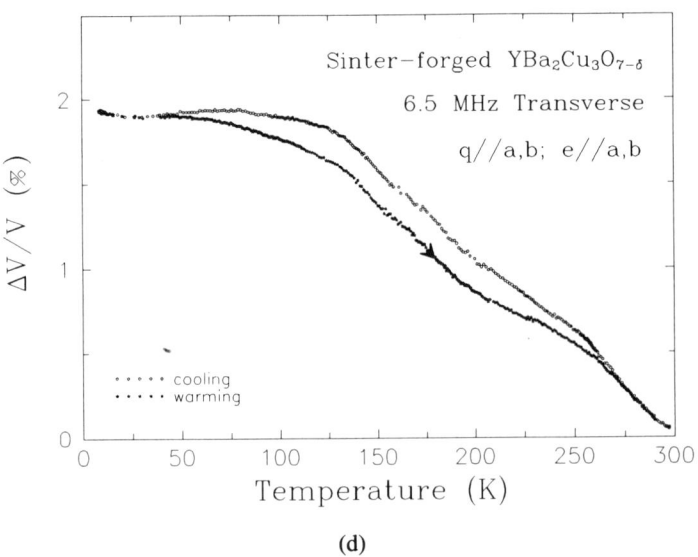

(d)

FIG. 14. Continued

5. Sintered High-T_c Superconductors

In a different propagation configuration for longitudinal waves traveling parallel to the forging axis at 12 MHz, a different curve was obtained, as shown in Fig. 14b. The velocity increased upon cooling till around 200 K, where a maximum is located; it decreased thereafter to the lowest temperature in the measurements. On warming up, there was a sudden softening at 65 K. The velocity stayed at lower values till around 170 K, where it started hardening and merged with the cooling curve at 270 K, yielding a distinctive hysteresis curve.

When 14.5 MHz transverse waves were propagated parallel to the forging axis, the velocity behaved differently, as shown in Fig. 14c. A large hysteresis was again observed in velocity, with a maximum near 160 K on cooling and a minimum at about 200 K on warming. The hysteresis vanished below 50 K and above 270 K. A small velocity slope change was also observed at T_c when warming up, similar to that observed in ordinary isotropic samples as discussed above.

In the fourth propagation configuration, 11.2 MHz transverse waves were sent perpendicular to the forging axis, with the polarization also perpendicular to the forging axis. The results are shown in Fig. 14d. The velocity generally had higher values at low temperatures than at higher temperatures, both on cooling and on warming, which is similar to the velocities measured in ordinary sintered isotropic samples, except that a small hysteresis was present in the temperature range between 50 K and 270 K, as seen in Fig. 14d.

In summary, common features as well as differences were observed for different propagation configurations. Anisotropy in the behavior of the velocity was evident. The elastic constants and velocities are tabulated in Appendix B. One of the features in common is the hysteresis observed in all four different propagation configurations, although the magnitude of the effect varied. The temperature range for the hysteresis seemed to be within 50 K and 270 K, and the velocity generally increases upon cooling near room temperature. Differences in the velocity curves are: first, the velocity values at room temperature are different for different configurations, indicating anisotropy in the elastic constants of the lattice; second, the difference between the room-temperature value and the value at the lowest temperature is different depending on the configuration, implying that the change of elastic constants is also anisotropic; third, the positions of the maxima when cooling are different; and, last, the magnitude of the hysteretic effect is not the same for all configurations.

Hysteresis in velocity was also observed in ordinary sintered $YBa_2Cu_3O_{7-\delta}$ samples by other researchers. Müller et al. (1987) have reported velocity hysteresis in their sample with 80% density and 91.5 K T_c for 6 MHz longitudinal waves, Fig. 15. The velocity that they measured was the group velocity obtained from measuring the time of flight between successive echoes. They found that

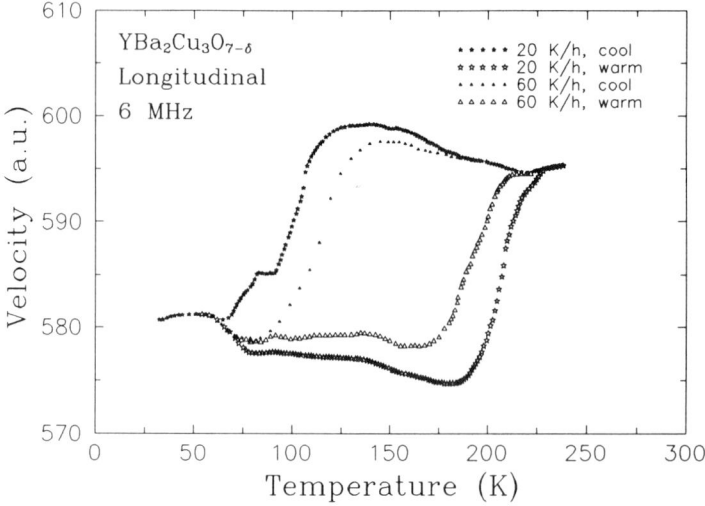

FIG. 15. Temperature dependence of the group velocity in $YBa_2Cu_3O_{7-\delta}$ for longitudinal sound at different cooling and heating rates. After Müller et al. (1987).

the starting temperature of the large softening on cooling was dependent on the cooling rate, and therefore the hysteresis loop depended on the cooling and warming rate. Ewert et al. (1987) measured velocity in two different grain-size samples: coarse-grained and fine-grained samples. The coarse-grained sample had grains approximately 50 μm in diameter and 10 μm in thickness, an overall relative density of 78%, and a T_c of 87 K. From their relatively fast cooling and warming rate measurements (5 K/min), they reported hysteresis in their coarse-grained sample but not in their fine-grained sample. Ledbetter and Kim (1988) measured both longitudinal and transverse wave velocities in a sample with 94% relative density and a T_c of 91.4 K at 5 MHz, and calculated shear modulus, bulk modulus, and Poisson ratio. All three of the quantities showed hysteresis, with the bulk modulus having the largest and the shear-mode–related elastic constants having the smallest. They also observed hysteresis in $EuBa_2Cu_3O_{7-\delta}$ and $HoBa_2Cu_3O_{7-\delta}$, with the former exhibiting a larger effect, and the latter, a smaller effect. Velocity hysteresis was also observed by different research groups on different samples: Lemmens et al. (1988), Lang et al. (1988), Almond et al. (1987a,b), Lagreid and Fossheim (1988), and Cannelli et al. (1988).

A tentative phenomenological explanation for the hysteresis is presented here based on the facts quoted above. For large-size crystallites, there exist more likely two different phases: a high-temperature phase that is stiffer, and a low-temperature phase that is softer. The second phase may be attributed to the twinning boundaries that are believed to be more numerous in large-size crystallites. Hysteresis in velocity may result from the hysteretic effect of the two

5. Sintered High-T_c Superconductors

thermally different phases. At high temperatures near room temperature, there is only one phase, the high-temperature stiffer phase, in the grain. As the temperature is dropped below a certain value, the low-temperature phase sets in, causing a decrease in the velocity because the low-temperature phase is softer. The position of the maximum in velocity below which the velocity starts to drop depends on many factors. A different sintering process would result in a different number and nature of twinning boundaries, which may cause different onset temperatures for the second phase. Different cooling rates seem to vary the position of the maximum as well, because the transition to the second phase is a slow thermal process.

When warming up from low temperatures, the low-temperature softer phase would initially dominate, resulting in lower velocity values. The reason for this may be that the system stays in a state of lower elastic energy by keeping the softer phase. This was observed in all the measurements when hysteresis was present where the velocity was always in the lower branch of the hysteresis loop when warming up from low temperatures. When a high enough temperature is reached where the high-temperature phase eventually breaks in, there is a turnover in the velocity. Near room temperature, the whole material becomes single-phase, and we see that the cooling-down and the warming-up curves merge together.

In an anisotropic sample such as the sinter-forged samples, the anisotropy of the hysteretic effect may be due to the different propagation directions of the sound waves relative to the sandwiched bulk phase and the twinning boundary phase. The twinning boundaries are parallel to the c-axis of the crystallites and, as mentioned, may act as nucleation sites for the softer low-temperature phase. Therefore, when both longitudinal and transverse sound waves are propagated perpendicular to the c-axis, the softer phase in the two phases would dominate. But when both longitudinal and transverse waves are propagated parallel to the c-axis, the stiffer phase would dominate. The former would produce a smaller hysteresis loop, as shown in Figs. 14a and 14d, while the latter appears to give a larger hysteresis effect, as shown in Figs. 14b and 14c. Thus, since a small fraction of the stiffer phase has a larger effect when the sound waves are propagating parallel to the forging axis, hysteresis in its density produces a larger hysteretic velocity effect in this orientation.

It was once reported (Bhargava *et al.*, 1987) that T_c's were enhanced by thermally cycling the samples below a certain temperature (236 K). In order to see if the reported enhancement had anything to do with the softer phase identified from the hysteresis loops in the velocity measurements, Xu *et al.* (1989) performed multicycle velocity measurements on a sinter-forged sample with 15 MHz longitudinal waves propagating parallel to the forging axis. The results are shown in Figs. 16a and 16b. The multiple hysteresis loops are a verification of

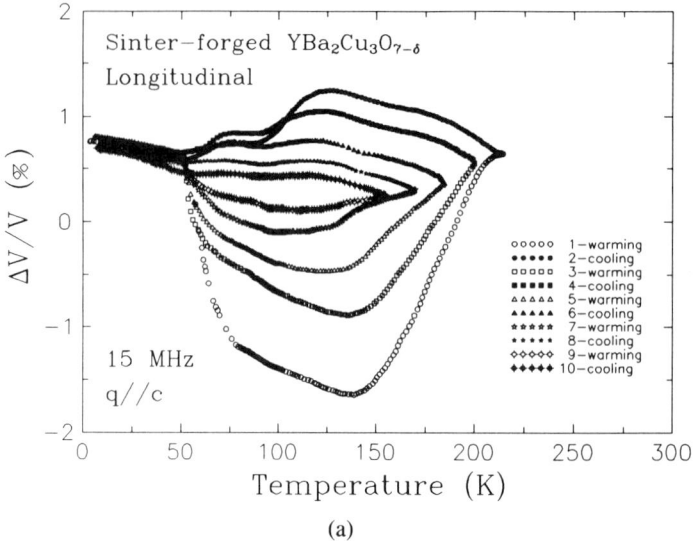

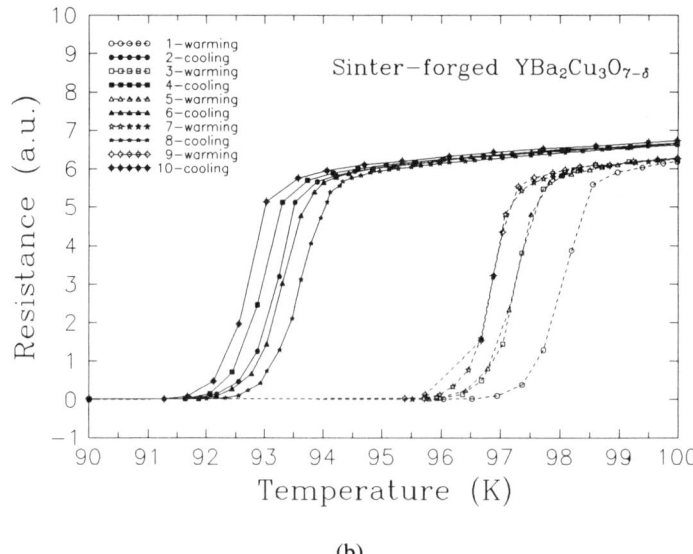

FIG. 16. (a) The velocity hysteresis of 15 MHz longitudinal sound waves propagating parallel to the forging axis. (b) The temperature dependence of resistance in the vicinity of T_c. The thermal loops correspond to those shown in (a). After Xu (1990).

5. Sintered High-T_c Superconductors

the existence of the hysteretic phenomenon. The experiment was performed over a period of days, with the sample staying at liquid-helium temperature during overnight shifts. The first warm-up was stopped at approximately 200 K and then each subsequent warm-up temperature was reduced by about 15 K. It is clearly shown in Fig. 16a that the hysteresis loops are shrunk as the up-limit warm-up temperature is reduced. In order to monitor T_c, resistance measurements were performed simultaneously with the velocity measurements, the purpose of which was to find if there existed any correlation between the hysteresis in velocity and the resistive transition of the superconductors. The results shown in Fig. 16b were that the T_c's did not change with the thermal cycle, and the transition temperatures were scattered within about 1 K of each other during either the cooling or the warming cycles, which are separated by 4 K because of thermal lag. These data show that the reported T_c enhancement is at least not related to the softer phase in the multiphase superconducting sample.

5. Sound Propagation in GdBa$_2$Cu$_3$O$_{7-\delta}$ and ErBa$_2$Cu$_3$O$_{7-\delta}$

Superconductivity was discovered in RBa$_2$Cu$_3$O$_{7-\delta}$, where R is a rare earth element other than yttrium, soon after YBa$_2$Cu$_3$O$_{7-\delta}$ was discovered. These superconductors are isomorphic compounds of YBa$_2$Cu$_3$O$_7$. Because of their similarity in almost all of their properties, studies of these compounds may assist in the understanding of the YBa$_2$Cu$_3$O$_7$ system.

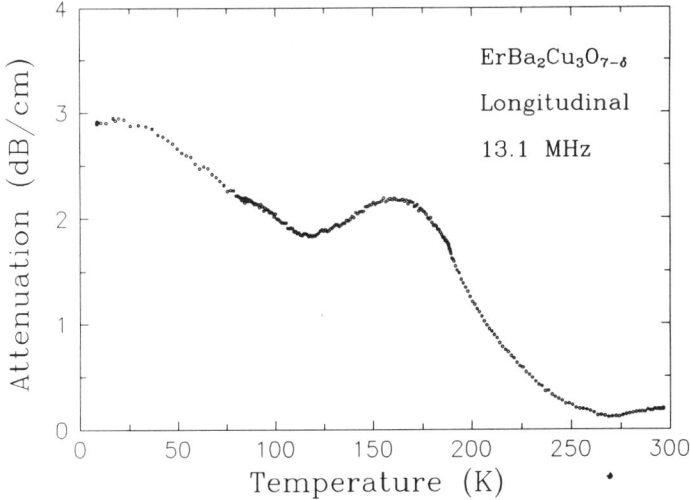

FIG. 17. Temperature dependence of longitudinal sound wave attenuation in ErBa$_2$Cu$_3$O$_{7-\delta}$ at 13.1 MHz. After Xu (1990).

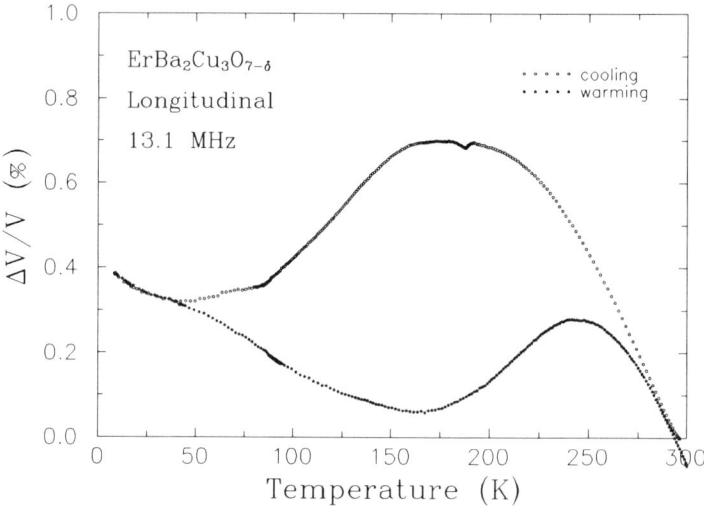

FIG. 18. Temperature dependence of longitudinal sound velocity change in $ErBa_2Cu_3O_{7-\delta}$ showing a large hysteresis. After Xu (1990).

Ultrasonic measurements were performed in $ErBa_2Cu_3O_{7-\delta}$ by Xu (1990) with longitudinal waves at 13.1 MHz. Figures 17 and 18 show the results for attenuation and velocity, respectively. T_c was determined to be 92 K, and the sample had a relative density of 90%. The attenuation curve looks similar to that for transverse waves propagating parallel to the forging axis with the polarization direction perpendicular to the forging axis in the singer-forged sample, as shown in Fig. 10a and 10b. There was one broad maximum in the whole temperature range, which was located at about 170 K. Frequency-dependent attenuation was not performed in this sample, but because of the similarities with the curves in Figs. 10a and 10b, it is believed that the maximum may be also a relaxation maximum resulting from similar mechanisms as those producing the maximum in the $YBa_2Cu_3O_{7-\delta}$ compound.

The velocity curve showed a large hysteresis whose behavior was similar to that observed in the $YBa_2Cu_3O_{7-\delta}$ compound discussed in Section 4.2. The magnitude of the hysteretic effect may depend on the sintering process and the size of the grains. As mentioned before, Ledbetter and Kim (1988) have also reported velocity hysteresis in both $HoBa_2Cu_3O_{7-\delta}$ and $EuBa_2Cu_3O_{7-\delta}$, with a smaller hysteresis effect in the former and a larger effect in the latter. The origin of the hysteresis in all these compounds may be similar to that in the $YBa_2Cu_3O_{7-\delta}$ compound.

Sound attenuation was also measured by Almond et al. (1989) in $GdBa_2Cu_3O_{7-\delta}$ for both longitudinal and transverse waves. Figure 19 shows the

5. Sintered High-T_c Superconductors

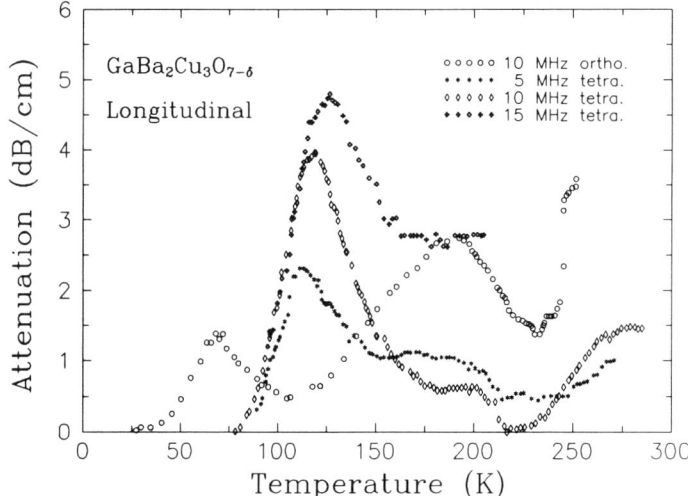

FIG. 19. The temperature dependence of the longitudinal wave attenuation in both the orthorhombic and tetragonal states of $GaBa_2Cu_3O_{7-\delta}$. After Almond et al. (1989).

longitudinal attenuation in both the orthorhombic and tetragonal phases. A peak below T_c at about 68 K and a broad maximum near 190 K were observed in the orthorhombic phase with 10 MHz waves. In the tetragonal phase, Almond et al. performed attenuation measurements at three different frequencies: 5, 10, and 15 MHz. The attenuation had a peak at about 113 K at 5 MHz, but it was shifted to about 118 K and 126 K at 10 MHz and 15 MHz, respectively. For transverse waves at 5 MHz in the orthorhombic phase, two peaks were again observed, one below and one above T_c, located approximately at 65 K and 170 K, respectively. They had smaller values relative to those measured at 10 MHz for longitudinal waves. For 10 MHz transverse waves in the tetragonal phase, one peak was observed at exactly the same position as that for 10 MHz longitudinal waves in the same phase.

A relaxation process appears to be responsible for the peaks in the temperature range between 110 K and 130 K in the tetragonal phase. The relaxation mechanism seemed unchanged for longitudinal waves and for transverse waves in this phase, since the peak at about 118 K stayed at the same position at the same frequency for both waves. If we assume this is true as well for the orthorhombic phase, then the peaks both below T_c and in the vicinity of 180 K may also be due to relaxation processes, since the peak below T_c was shifted to a slightly higher temperature, from 65 K to 68 K, for frequencies of 5 MHz and 10 MHz, respectively, and the peak at 170 K was shifted to 190 K. Both of these peaks were also observed in the $YBa_2Cu_3O_{7-\delta}$ compounds, and it has also been proposed that they are associated with relaxation processes, Sections 3 and 4. For a more thorough discussion of the attenuation measurements in $GdBa_2Cu_3O_{7-\delta}$, see Chapter IX.

Velocity measurements in $GdBa_2Cu_3O_{7-\delta}$ were performed by Brown et al. (1988) in two sintered polycrystalline samples, and by Saint-Paul et al. (1989) in a single-crystal sample. Brown et al. (1988) used a resonant frequency method near 320 KHz to measure the velocity change in samples with a relative density of 84% and a T_c of 94.5 K. The measurements showed a velocity discontinuity at T_c in a sample that had undergone several thermal cyclings between room temperature and 40 K. A discontinuity in the velocity slope was also observed and the slope had a smaller value below T_c. Saint-Paul et al. (1989) also discovered velocity and slope changes at T_c in their single crystal with $T_c = 92$ K with a pulsed echo method for 15 MHz transverse waves along the c-axis. The phenomena observed were again very similar to those observed in $YBa_2Cu_3O_{7-\delta}$ and discussed in Section 3.

In summary, the measurements in both $ErBa_2Cu_3O_{7-\delta}$ and $GdBa_2Cu_3O_{7-\delta}$ show that both attenuation and velocity behaved very similarly to the same properties in $YBa_2Cu_3O_{7-\delta}$, perhaps because the compounds have the same perovskite crystal structure. The results reaffirm those obtained on $YBa_2Cu_3O_{7-\delta}$ and are generally consistent with each other.

6. BiSrCaCuO and TℓBaCaCuO Superconducting Compounds

The BiSrCaCuO compound discovered by Maeda et al. (1988) and Chu et al. (1988) has a T_c that is higher than that of $YBa_2Cu_3O_{7-\delta}$. The TℓBaCaCuO compound discovered by Sheng and Hermann (1988) has an even higher T_c, 125 K. It was later determined that in these two compounds there are several phases. Each has a structure with a different number of Cu–O planes that lead to different critical temperatures. Their crystal structures and transition temperatures are shown in Appendix A. Studies on different phases with a different number of Cu–O planes would shed light on the origin of superconductivity in these materials.

A few research groups have investigated the ultrasonic properties in both BiSrCaCuO and TℓBaCaCuO compounds. But because of the technical limitations on separating different phases in the materials, and also because of the extremely poisonous nature of the Tℓ-based compounds, the number of research papers published on these two materials in the area of ultrasound is very limited as compared to that on $YBa_2Cu_3O_{7-\delta}$ materials. In addition, experiments were usually performed on Bi–2–2–1–2 or Tℓ–2–2–1–2 compounds that also have

5. Sintered High-T_c Superconductors

two Cu–O planes per unit cell, just as in $YBa_2Cu_3O_{7-\delta}$. The numbers after the elements in the previous sentence refer to the number of metallic atoms per formula unit in the same sequence as given in the title of this section.

As with $YBa_2Cu_3O_{7-\delta}$, discrepancies in the measurements are inevitable because of the nature of the materials. Wang et al. (1989a) observed three peaks in the attenuation of a Bi–2–2–1–2 phased single crystal at 250 K, 150 K, and 95 K for 7.5 MHz longitudinal waves in a direction 10° off the a or b-axis. Hu et al. (1989) observed two peaks at 250 K and 120 K in their attenuation data for 10 MHz longitudinal waves. These measurements were done on a mixed phase material with probably both 2–2–1–2 and 2–2–2–3 present. He et al. (1989) found peaks around 285 K, 200 K, and 125 K for 10 MHz longitudinal waves, in a 2–2–1–2 single-phase sample.

It seems that the peak around 250 K is again common to these measurements, similar to what was observed in $YBa_2Cu_3O_{7-\delta}$. Wang et al. (1989b) also propagated transverse waves at 5 MHz along the b-axis and found one broad peak around 250 K as well. In the data by Hu et al. (1989) for 5 MHz longitudinal waves, if one ignores the dramatic jump near 250 K, the maximum in attenuation is near 250 K. The mechanism that produced these peaks may be similar to the one discussed in Section 4 for the $YBa_2Cu_3O_{7-\delta}$ compounds. Actually, if they are due to interactions with excitations in the Cu–O planes, then since both $YBa_2Cu_3O_{7-\delta}$ and $Bi_2Sr_2Ca_nCu_{n+1}O_{6+2n}$, with $n = 1$, have two Cu–O planes per unit cell, one might expect similar effects in both systems.

In addition to those peaks observed below room temperature, Zeng et al. (1989) found, in their internal friction measurements, that two peaks appeared around 560 K and 400 K. For a thermally unstable sample, heat treatment improved the superconductivity, as evidenced by their resistivity measurements. At the same time, the relative amplitude of the internal friction peaks changed roles, with the higher-temperature peak being larger than the lower one after the improvement in T_c. The frequency dependence of the peaks showed an increase in temperature with increasing frequency, implying that a relaxation process was involved. But for a more stable sample sintered at 1,143 K, the peaks seemed frequency-independent. Elastic constants were determined in a Pb-doped Bi-SrCaCuO compound with normal phase 2–2–2–3 by Ledbetter et al. (1989) using ultrasound. Velocity measurements showed a general increase upon cooling (Xiang et al., 1989; Wang et al., 1989a) in single crystals. One exception is that Wang et al. (1989a) observed a softening around 250 K when propagating longitudinal waves in a direction at a small angle with respect to the a- or b-axis, but this was not observed in a direction perpendicular to the one above. Anisotropy in velocity in the basal plane was discovered by both Wang et al. (1989a) and Xiang et al. (1989), although it seems unlikely in this tetragonal structure material. The elastic constants and velocities for the Bi and $T\ell$ compounds are tabulated in Appendix B.

We will now discuss temperature-dependent attenuation and velocity measurements in the temperature range of the superconducting transition for high-T_c systems. As previously mentioned, an attenuation peak right below the superconducting transition was observed for two $YBa_2Cu_3O_7$ samples prepared by two groups with transducers of different materials (one was a 15 MHz quartz transducer [Xu et al., 1988], and the other was a 10 MHz $LiNbO_3$ transducer [Sun et al., 1988]). Although this peak in $YBa_2Cu_3O_7$ exhibited frequency dependence, a frequency-independent attenuation peak remained close to T_c with a smaller magnitude after a relaxation-type attenuation maximum was subtracted from the experimental data (Fig. 20). In measurements on samples of a 2-2-1-2 phase $T\ell$ compound (Sun et al., 1989b), an attenuation peak was also found close to T_c for longitudinal as well as shear waves. However, the peak in this $T\ell$ system stayed at the same temperature (within experimental errors) for 10, 30, and 67 MHz waves. Thus, it was frequency-independent, as shown in Fig.

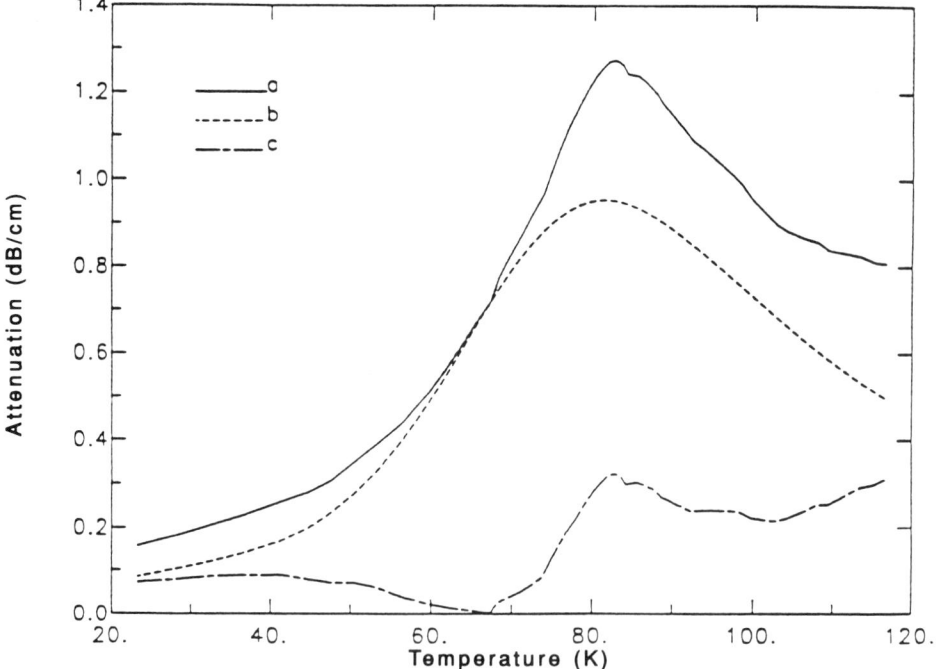

FIG. 20. An attenuation peak remains at a temperature close to T_c (curve c) when the calculated relaxation attenuation (curve b) is subtracted from the experimental results (curve a). Experimental error is within 5% in our measurements. After Sun et al. (1988).

5. Sintered High-T_c Superconductors

21. Similar features in attenuation data were reported by Pederson et al. (1989), who found a peak at 110 K for longitudinal waves at 3.2 MHz in their sample with a dominant 2-2-1-2 phase. Wang et al. (1989c) observed two distinct peaks at 90 K and 135 K in their internal friction measurements. They attributed the 135 K peak to a phaselike transition associated with the superconducting transition at 115 K and deduced that the 90 K peak might be associated with a low-temperature phase in the material. Figure 22 shows the shear wave velocity of $T\ell$–Ca–Ba–Cu–O as a function of temperature. A 4% change in velocity from room temperature to liquid helium temperature is observed. A slope change at T_c can be deduced if a straight line is passed through the data immediately above and below T_c. A similar slope change of the velocity vs. temperature curve has also been observed for $YBa_2Cu_3O_7$. However, the shape of the velocity curve in the $T\ell$-based samples is less steep above T_c than above T_c, while in $YBa_2Cu_3O_7$, the opposite effect is observed.

For further investigation of the possibility that an attenuation anomaly is directly associated with the transition to superconductivity, a specimen of Bi–Sr–Ca–Cu–O, which was identified to have 2-2-1-2 and 2-2-2-3 phases with the volume ratio 1:4, was measured. Figure 23 shows the attenuation as a function of temperature of longitudinal waves at 10 MHz for this sample. There are two maxima occurring near two T_c's, which are 20 K apart from each other. Temperature-dependent ac magnetic susceptibility data shown in Fig. 23 exhibit steps that may be interpreted as the superconducting transitions of the two phases

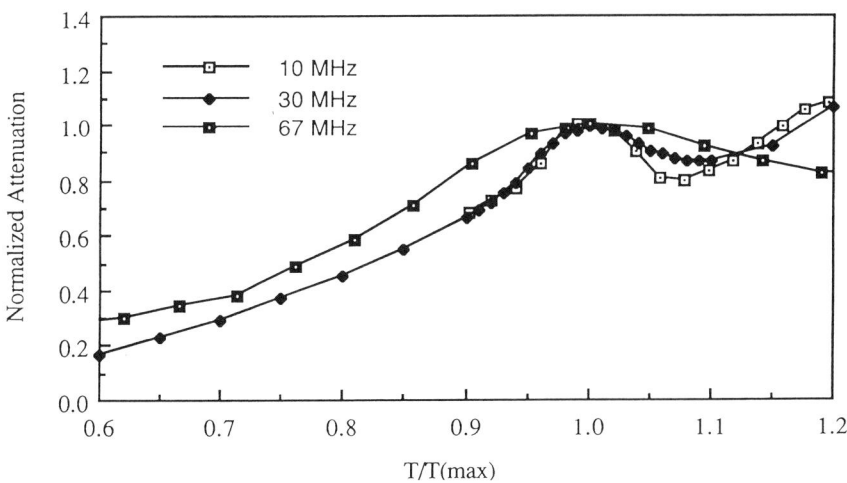

FIG. 21. Normalized attenuation vs. reduced temperature at 10, 30, and 67 MHz of $T\ell$–Ca–Ba–Cu–O. T(max) is the temperature where the attenuation maximum is centered. After Sun et al. (1990b).

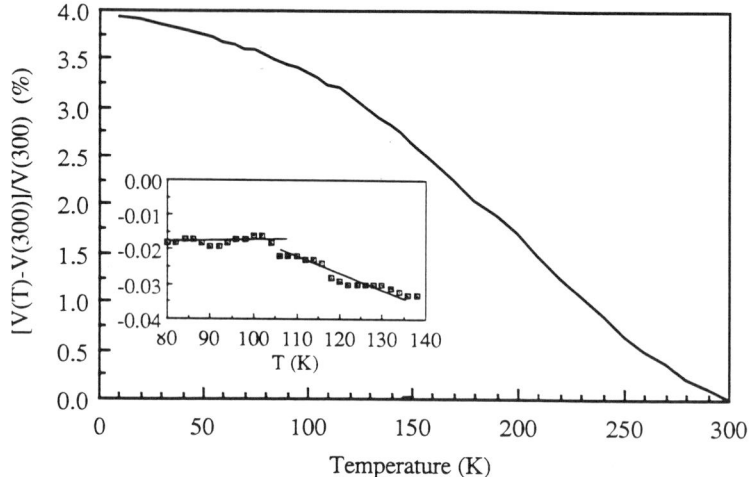

FIG. 22. Temperature dependent sound velocity change for Tℓ–Ca–Ba–Cu–O. The first derivative of this change with respect to T is shown in the inset. The solid straight lines are drawn for guidance to the eye. The experimental error in velocity change is 5%. After Sun et al. (1990b).

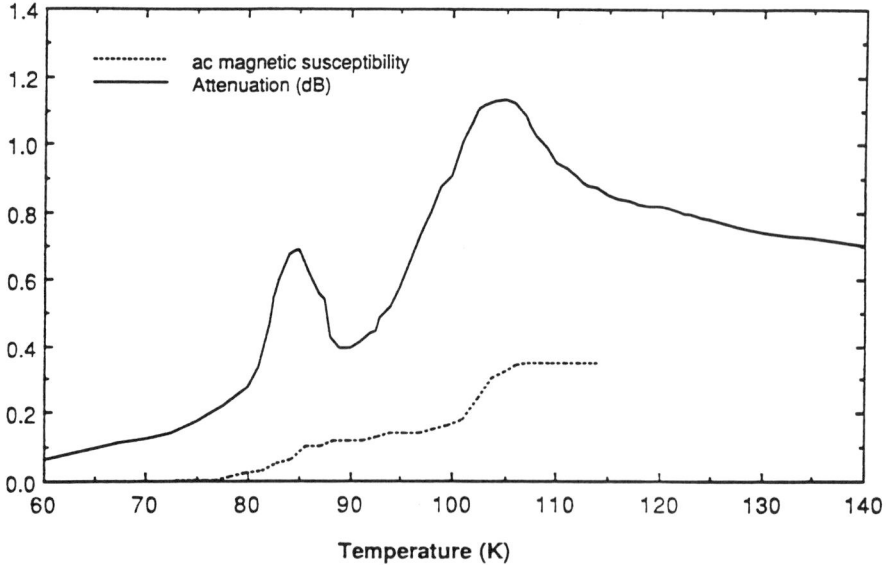

FIG. 23. Ultrasonic attenuation and ac magnetic susceptibility as a function of temperature for a two-phased Bi–Sr–Ca–Cu–O compound. After Sun et al. (1990a).

5. Sintered High-T_c Superconductors

occurring at 85 K and 105 K, respectively. Very possibly, the maxima are a consequence of these superconducting transitions. The velocity of longitudinal waves in this Bi–Sr–Ca–Cu–O sample was also determined between room temperature and 15 K by measuring the time interval between sound wave echoes. The longitudinal velocity increased with decreasing temperature for this polycrystalline sample. If the fractional change of the thermal expansion of the sample is negligible when compared with that of the sound propagation time, there is approximately a 3% increase in velocity from 295 K to 15 K. The fractional change of velocity is shown in Fig. 24. As was also found for the $YBa_2Cu_3O_7$ and the $T\ell$–Ca–Ba–Cu–O systems, there is a slope change of the velocity–temperature curve at T_c. The softening of the velocity at the superconducting phase transition was also observed for the Bi–Sr–Ca–Cu–O system. The time resolution for the velocity measurements in this sample may not have been good enough to allow quantitative determination of the decrease in velocity at T_c; nevertheless, a variation in the neighborhood of T_c is observable and becomes more apparent when the data are expanded in the vicinity of the two transition temperatures mentioned in the previous paragraph. (See inset of Fig. 24.) Therefore, by combining these experimental observations, we may conclude that the attenuation anomalies and the velocity variations occurring at T_c are common features of ultrasonic behavior near T_c among the multi-Cu–O–layered high-temperature superconductors, and that the onset of superconductivity does seem to directly affect the propagation of sound. A model of spin–phonon interaction

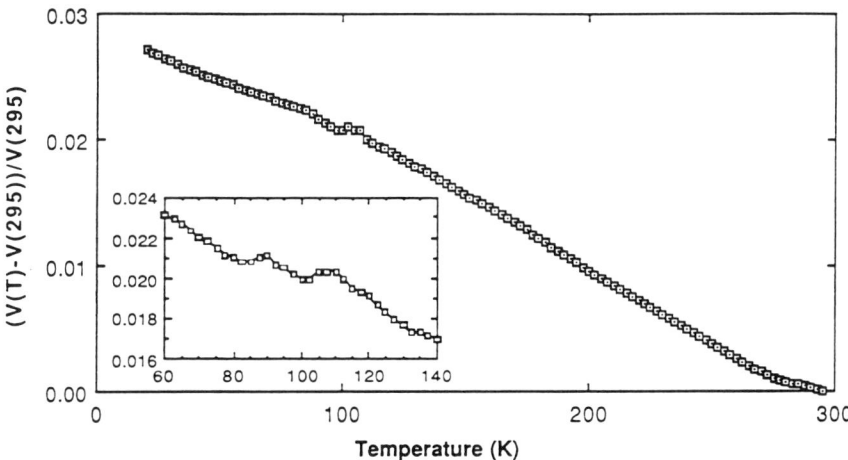

FIG. 24. Fractional change of sound velocity as a function of temperature for Bi–Sr–Ca–Cu–O. Inset shows the variation of velocity at temperatures near T_c. After Sun et al. (1990a).

(Sun and Levy, Chapter IV), which attributes the excess sound energy dissipation to the random force caused by thermal spin fluctuations on phonons, has been proposed to interpret these observations. It has also been proposed that the peak might be associated with a distortion of the oxygen ion displacement along the c-axis below T_c predicted by Kurihara (1989). In order to verify this explanation, temperature-dependent ultrasonic measurements in magnetic fields, as well as measurements on oriented crystals, will be necessary to provide correlations between the effects and the magnetic shifts in T_c and to isolate the effect of polycrystallinity on these ultrasound phenomena.

7. Sound Propagation in $B_{1-x}K_xBiO_3$

The critical temperatures of $YBa_2Cu_3O_{7-\delta}$ and $La_{2-x}Sr_xCuO_4$ superconductors are much higher than those of conventional superconductors, usually in the 90 K and in the 30 K range, respectively, but their crystal structure is perovskite, which is highly anisotropic and therefore introduces additional difficulties for both theoretical and experimental work when compared to conventional superconductors. On the other hand, the Ba–K–Bi–O superconductor system discovered by Matheiss et al. (1988) and identified by Cava et al. (1988) has a simple cubic structure in the superconducting phase and yet has T_c's above 20 K (Hinks et al., 1988a). The simplicity in structure may simplify both theoretical and experimental studies in the Ba–K–Bi–O system. The crystal structure is shown in Appendix A. Electron–phonon interaction is proposed to be the mechanism that produces superconductivity in the $Ba_{1-x}K_xBiO_3$ system, just as in conventional BCS superconductors. Hinks et al. (1988b) observed in the $Ba_{1-x}K_xBiO_3$ system a strong isotope effect that was nearly null in the $YBa_2Cu_3O_{7-\delta}$ system when they replaced ^{16}O by ^{18}O. The α index was found to be 0.45, as determined from the resistive transition. Attenuation and velocity measurements were performed as functions of both temperature and magnetic field in the $Ba_{1-x}K_xBiO_3$ systems with $x = 0.35$ and $x = 0.4$ for both longitudinal and transverse waves (Xu et al., 1990).

The high-density superconducting samples of $Ba_{1-x}K_xBiO_3$ were synthesized by a novel melt-processing method reported by Hinks et al. (1989). The preparation procedure starts with the stoichiometric oxide mixture of $BaBiO_3$, KO_2, and Bi_2O_3. The mixture is then melted in a Pt crucible under a N_2 gas flow, and the melt is poured into a cylindrical Cu mold 3.5 cm in diameter, and then cooled in a N_2 atmosphere to obtain a disklike pellet sample. Finally, the pellet is fired in N_2 up to 700°C and rapidly cooled to room temperature to completely

5. Sintered High-T_c Superconductors

remove any excess O_2, and then fired in an O_2 atmosphere up to 425°C and cooled to room temperature to obtain superconducting samples.

The superconducting transition temperature changes with the potassium concentration in the $Ba_{1-x}K_xBiO_3$ systems. The phase diagram showing the dependence of T_c on x is shown in Fig. 25. Starting from $x = 0.5$, T_c increases with decreasing potassium concentration x, until x equals 0.3, where T_c decreases precipitously with decreasing x. For x less than 0.3, the system is no longer superconducting, but becomes semiconducting.

The transverse wave attenuation at 8.5 MHz as a function of temperature is shown in Fig. 26. The attenuation initially drops rapidly upon cooling from room temperature. It levels off around 220 K. At low temperatures, it starts to increase and reaches a maximum, and then a plateau, with a maximum being at about 40 K and the plateau ending at about 29 K, which is the critical temperature of this Ba–K–Bi–O sample with $x = 0.4$. Below T_c, the attenuation does not decrease exponentially, but rather linearly. This is not what is expected from a conventional BCS-type superconductor, in which the attenuation decreases exponentially below T_c. This behavior looks similar to that observed in the heavy fermion superconductors such as UPt_3.

The velocity change in the Ba–K–Bi–O system is rather interesting. In the $YBa_2Cu_3O_{7-\delta}$ systems, the velocity increases monotonically upon cooling. In the Ba–K–Bi–O system, a broad maximum is found around 100 K, after cooling from room temperature. Below this temperature, the velocity first decreases,

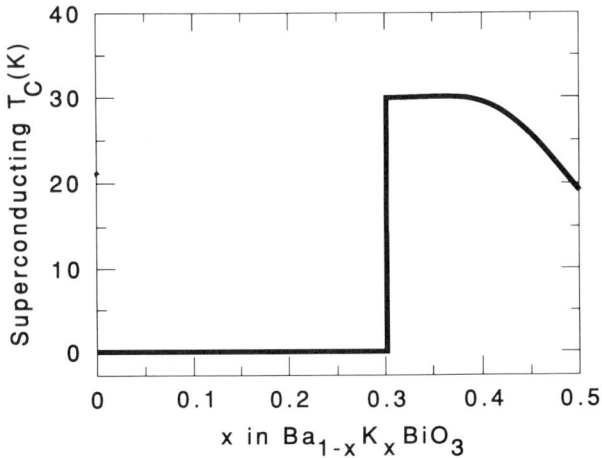

FIG. 25. Phase diagram for $Ba_{1-x}K_xBiO_3$ showing the dependence of the superconducting transition temperature on the concentration of potassium. Hinks et al. (1992).

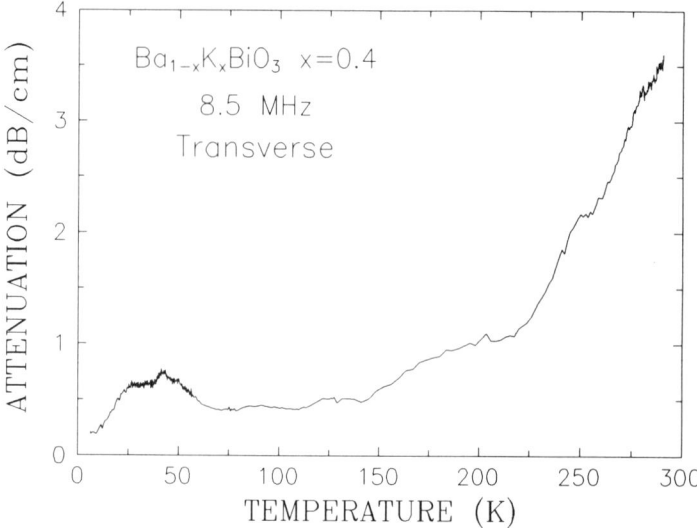

FIG. 26. Attenuation vs. temperature for 8.5 MHz transverse waves in $Ba_{1-x}K_xBiO_3$ with $x = 0.4$. A peak is located at low temperatures. After Xu *et al.* (1990).

reaching a minimum at about 40 K, and then increases down to our lowest temperature (1.9 K). Figure 27 displays the velocity change as a function of temperature. If we take a closer look at the velocity curve around T_c, which is shown in Fig. 28 and which was obtained from a higher-resolution measurement, we see clearly that a slope change in the velocity derivative with respect to temperature occurs right at T_c. This phenomenon is very similar to that observed in the YBaCuO systems, where slope changes are also observed.

One of the difficulties in ultrasonic measurements on the high-T_c superconductors has been to find consistent experimental results of attenuation that were directly related to the superconducting transition. Several investigators have observed peaks that have generally shifted to higher temperatures when higher-frequency measurements were performed. The one exception has been measurements on thallium high-T_c superconductors, wherein the peaks remained at T_c even when the frequency was increased by a factor of three. In our present samples it has not been possible to increase the frequency; therefore, we made measurements in a magnetic field at 17.2 K to see if a change in attenuation would be observed when the sample became normal. The results of these measurements are shown in Figs. 29, 30, and 31.

Although the data are noisy, the trends are clear. Figure 29 shows the attenuation of 8.5 MHz transverse waves as a function of magnetic fields at 17.2 K. Figure 30 displays the temperature variation during the measurements shown in

5. Sintered High-T_c Superconductors

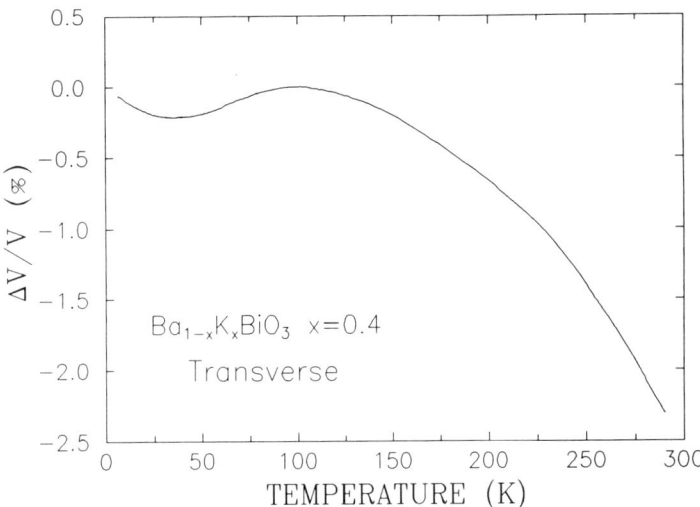

FIG. 27. Velocity vs. temperature for 8.5 MHz transverse waves in $Ba_{1-x}K_xBiO_3$ with $x = 0.4$. A maximum is observed at about 100 K, and a minimum is seen at about 40 K. After Xu et al. (1990).

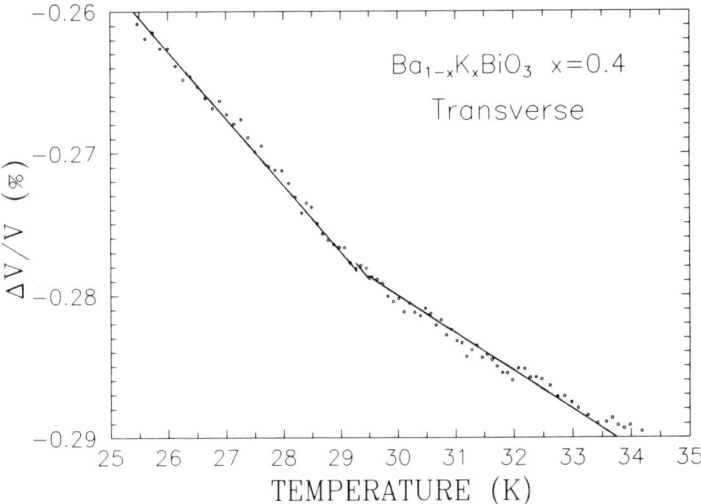

FIG. 28. Velocity vs. temperature on an expanded scale, for 8.5 MHz transverse waves in $Ba_{1-x}K_xBiO_3$ with $x = 0.4$. A slope change is seen at T_c. After Xu et al. (1990).

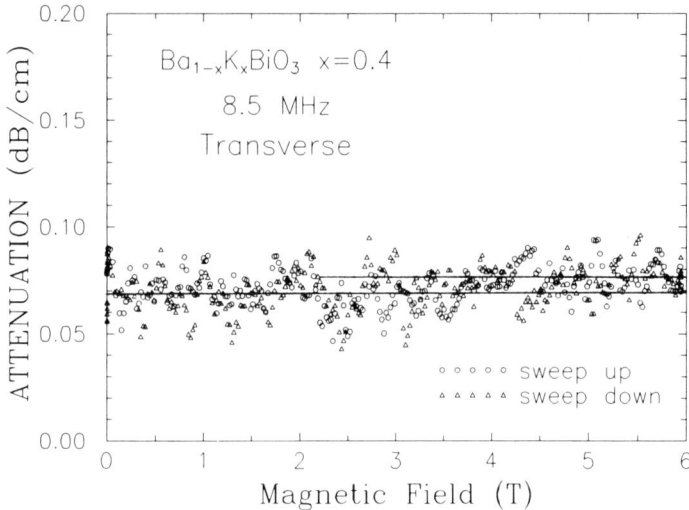

FIG. 29. Attenuation vs. magnetic field data. Straight lines have been drawn through the low-field and high-field data as an aid to the eye. The attenuation increases slightly in the high fields. After Xu (1990).

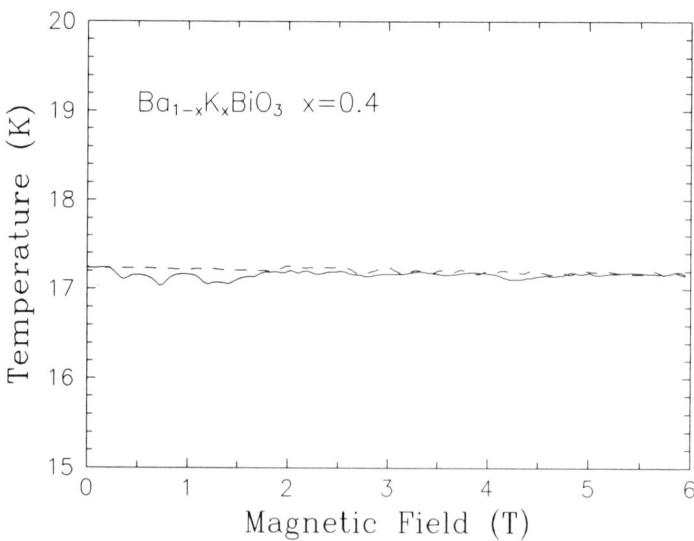

FIG. 30. Temperature variation as a function of magnetic fields for Figs. 29 and 31. The temperature was controlled at 17.2 K, with a maximum change of 0.2 K. After Xu (1990).

5. Sintered High-T_c Superconductors

Fig. 29 and Fig. 31. At low fields the attenuation stayed relatively unchanged. Although a small correction should be made because of an initial temperature drop, this correction is small, since the drop in temperature is only 0.2 K, which corresponds to a decrease in attenuation of 0.003 dB/cm at 17.2 K, according to Fig. 26. Above 2.5 Tesla, the attenuation started to increase with increasing fields. For $x = 0.4$, H_{c1} is several hundred Gauss, and H_{c2} is about 5 Tesla at 17.2 K (Batlogg et al., 1989); interestingly, the attenuation seemed to level off at about 5 Tesla.

Thus, it appears that attenuation increases almost imperceptibly in the mixed state of this BKBO sample. The total attenuation change from zero field to 6 Tesla that we may deduce from these noisy data is 0.005 dB/cm when the contribution from the temperature variation is taken into account. The straight lines in Fig. 29, are drawn as a guide to the eye. They are about 0.008 dB/cm apart. Thus, the total attenuation change from the superconducting state to the normal state (above H_{c2}) at 17.2 K in this BKBO system is at most 0.005 dB/cm. But from Fig. 26, we see that the total attenuation change is about 0.23 dB/cm from 17.2 K to 29 K. This change is almost two orders of magnitude larger than the attenuation change from the superconducting state to the normal state in the magnetic-field–dependent measurement. Therefore, we may say that the normal-state background attenuation must also decrease below 29 K. As small as it is, the observed change in attenuation in a magnetic field is still about one or two

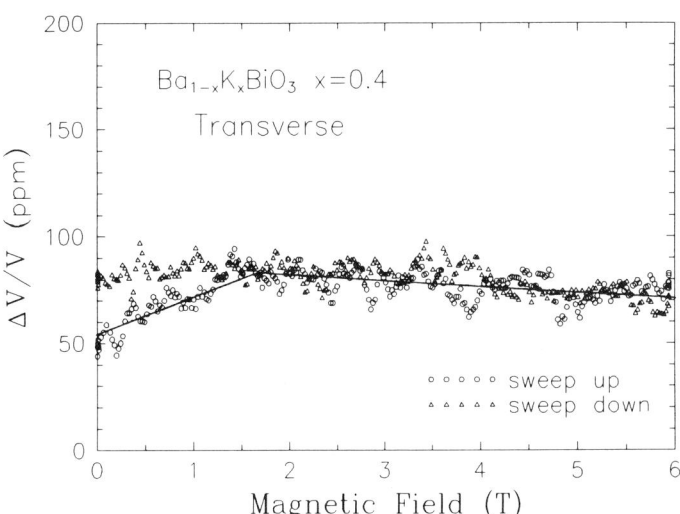

FIG. 31. Velocity vs. magnetic field data. Straight lines have been drawn through the low-field and high-field data as an aid to the eye. The velocity decreases in high fields above 1.5 Tesla. After Xu (1990).

orders of magnitude larger than what would be expected from electron–phonon interaction. Thus, it may be possible that the increase in attenuation is due to the motion of flux lines in the mixed state.

Measurements of the transverse velocity change as a function of magnetic fields at the same temperature (shown in Fig. 30) show that the initial velocity increases up to about 1.3 Tesla (Fig. 31). This initial increase is partially due to the initial drop in temperature in the sample, since a 0.2 K drop in temperature would correspond to an increase in 13 ppm in velocity, according to Fig. 27. Even after subtracting the contribution due to the drop in temperature, the velocity still increases with increasing fields up to 1.3 Tesla, above which the velocity curve shows a tendency to decrease with increasing fields up to 5 Tesla. There is basically no change in velocity above the H_{c2} of 5 T.

If we assume that there exist at least two vortex states in the mixed superconducting state of the BKBO system — a low-field crystalline state, and a high-field fluid state — we might be able to explain these observations. In low fields, the vortex lattice is stiffer, which contributes to the elastic constant, thus causing an increase in velocity. But in high fields, the fluid vortex state, the situation is different since the lattice becomes fluid and does not contribute as much to the elastic constant, leading to a decrease in velocity. When the field is swept down to zero, some flux lines would be trapped in the sample in the solid vortex state. This could explain why the final velocity is higher than the initial one.

The attenuation curve of 11.7 MHz longitudinal sound waves shows different behavior from that of transverse waves near room temperature (Fig. 32), where for longitudinal waves a large peak in attenuation is observed around 270 K, whose magnitude is about 14 dB/cm. At low temperatures, another peak—but rather a small one compared to the one at high temperature—is seen around T_c. This peak may be seen clearly in Fig. 33, which concentrates on the lower temperature range. The attenuation decreases below T_c, and the total change from T_c to 4 K is about 0.5 dB/cm.

The velocity change of longitudinal waves exhibited a much larger effect compared to that of transverse waves (Fig. 34). Although the features in both curves look similar, the total velocity change from room temperature to 4 K for longitudinal waves is about 8%, while that for transverse waves is only about 2.5%. This implies that the bulk modulus has a large change over this temperature range. The velocity has a broad maximum at around 80 K, and a minimum at around 50 K, below which it increases until 4 K. Room-temperature values for both longitudinal and transverse wave velocities are tabulated in Appendix B.

In a $Ba_{1-x}K_xBiO_3$ sample with $x = 0.35$, the attenuation of 55 MHz transverse waves was measured, and the results are shown in Fig. 35. The measurements were concentrated on the low-temperature part. When the sample was cooled,

5. Sintered High-T_c Superconductors

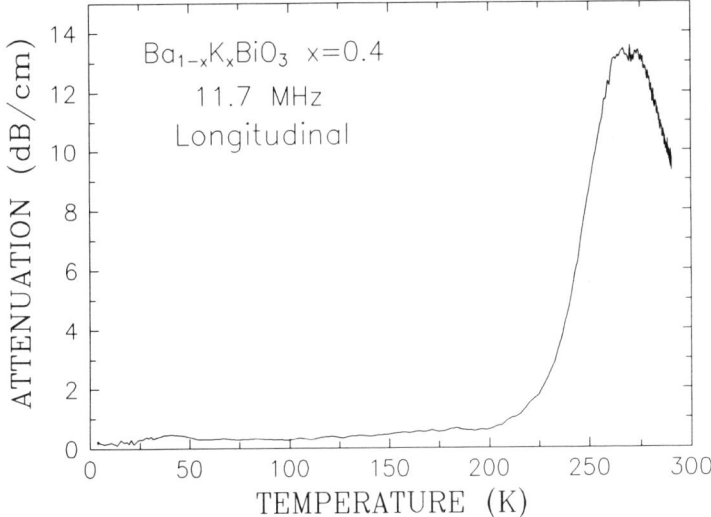

FIG. 32. Attenuation vs. temperature for 11.7 MHz longitudinal waves in $Ba_{1-x}K_xBiO_3$ with $x = 0.4$. A broad peak is centered about 270 K, and a small peak is seen at low temperatures. After Xu (1990).

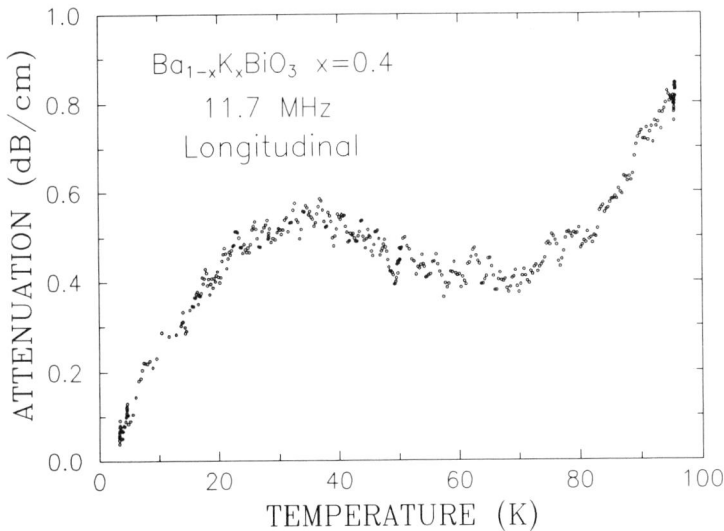

FIG. 33. Attenuation vs. temperature on an expanded scale at low temperatures for 11.7 MHz longitudinal waves. There is a broad maximum around 35 K, and the attenuation decreases below T_c. After Xu (1990).

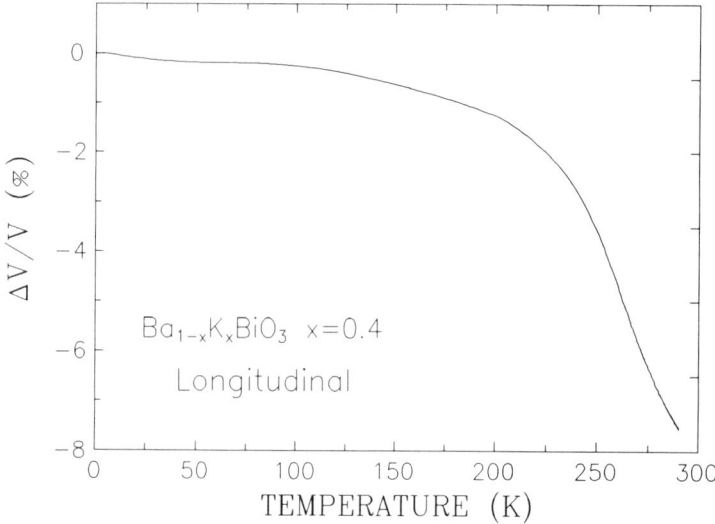

FIG. 34. Velocity vs. temperature for 11.7 MHz longitudinal waves in $Ba_{1-x}K_xBiO_3$ with $x = 0.4$. A maximum is observed at about 80 K, and a minimum is seen at about 40 K. After Xu (1990).

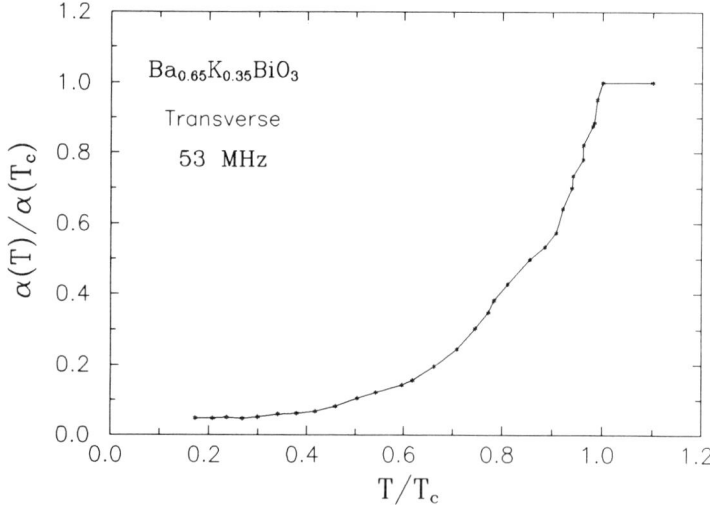

FIG. 35. Normalized attenuation vs. normalized temperature for 53 MHz transverse waves in $Ba_{1-x}K_xBiO_3$ with $x = 0.35$. After Xu (1990).

the attenuation started to drop below T_c and kept decreasing down to 4 K. Preliminary analysis showed that the decrease in attenuation below T_c is close to exponential. This is different from what was observed in the $Ba_{0.6}K_{0.4}BiO_3$ sample for both longitudinal and transverse waves, in which an almost linear dependence of the attenuation with respect to temperature was observed. In this sample with $x = 0.4$, it was shown that the linear decrease could not be directly associated with the transition to the superconducting state by using the magnetic field measurement results. Although magnetic field measurements have not been made on the sample with $x = 0.35$, it is expected that the exponential decrease in attenuation may also not be directly related to the superconducting state of this high-T_c superconductor.

8. Summary

Attenuation measurements in sinter-forged high-T_c superconductors have shown several features that appear to be common in all these systems. There is an attenuation maximum that occurs at around 250 K ± 50 K. This maximum is evident not only in those superconductors that contain CuO planes, but also in $Ba_{1-x}K_xBiO_3$, which does not, and which is the only high-T_c superconductor that has a simple cubic structure in the superconducting phase. In addition, measurements in sinter-forged $Y_1Ba_2Cu_3O_7$ appear to show that in this system, this maximum is associated with distortions of the CuO planes that carry superconducting currents. If this maximum is being produced by a relaxation process, then the activation energy of the temperature-dependent relaxation time for these peaks is about 650 ± 100 K (please see Appendix D; Xu, 1990). It should be pointed out that several investigators believe that these maxima in attenuation in the other systems may be associated with phase transitions that either are intrinsic to these systems or are produced by impurities. In some of the systems, there is an attenuation maximum at around 180 K. This maximum is absent in $Ba_{1-x}K_xBiO_3$. Measurements in sinter-forged $Y_1Ba_2Cu_3O_7$ show that this maximum moves up in temperature when the frequency of the sound waves is increased. Also, the maximum appears to be isotropic—namely, it is present when the propagation direction of the sound waves is either parallel or perpendicular to the CuO planes. The activation energy for the relaxation time associated with these peaks is about 1,100 ± 200 K (Xu, 1990). All of the systems, including $Ba_{1-x}K_xBiO_3$, usually exhibit a maximum around the superconducting transition temperature. Frequency measurements in a sintered sample of $Y_1Ba_2Cu_3O_7$ showed that the maximum was associated with a relaxation process with an activation energy of 450 K. Again, measurements in the sinter-forged samples of $Y_1Ba_2Cu_3O_7$ showed that this maximum was also associated with distortions of the CuO planes. The activation energies of these three sets of maxima are comparable to the energies postulated in several theoretical models for the virtual excitations that are coupling the superconducting Cooper pairs in these high-T_c systems. Furthermore, it is interesting to note that only one maximum was observed in the isomorph of the 123's $ErBa_2Cu_3O_7$, the one with an activation energy in the range of 1,100 K. This is the activation energy that V. Kresin (Chapter X) has postulated for the soft plasmons that may be responsible for the high T_c in these systems. It is somewhat surprising that it is the maximum that appears to be isotropically present in the sinter-forged samples. Our selection rules appear to indicate that these excitations are associated with the reservoir of electrons contained between the Cu–O planes. So we could deduce that it is these reservoir electrons that are responsible for the excitations that produce the attractive interaction which forms Cooper pairs in these high-T_c superconductors.

In conventional superconductors, the attenuation of sound waves drops precipitously below the superconducting transition temperature, and, as mentioned in Chapter I, the temperature dependence of the attenuation confirms the BCS (Bardeen *et al.*, 1957) energy gap model by yielding a fairly accurate determination of the temperature dependence of the energy gap. Therefore, most investigators have concentrated a considerable amount of effort on the attempt to find an attenuation signature at the superconducting phase transition. To date, although several investigators have found decreases in attenuation below the superconducting transition, none of these can be reliably ascribed to the appearance of the superconducting state. This should not be too surprising, since the expected effect at the frequencies of these measurements is in the range of 10^{-6} to 10^{-3} dB/cm. However, an attenuation signature at the superconducting transition temperature has been found. It is not a decrease in attenuation, but a peak in attenuation at T_c. The original evidence for such a peak was quite indirect. It came about after Sun *et al.* (1988) analyzed the relaxation peaks that appeared close to T_c. After subtracting theoretically generated relaxation attenuation curves from the data obtained at three different frequencies, they were still left with attenuation maxima centered around T_c (Fig. 20). The fact that these maxima did not shift with frequency made them very intriguing. But the fact that intricate data analysis was required to discover them was uncomfortable. Since then, however, measurements in Tℓ and Bi superconducting compounds, Figs. 21 and 22, have shown directly attenuation peaks at the superconducting transition that do not move with frequency. Thus, the attenuation signature appears to be a peak in attenuation at T_c. The mechanisms for the appearance of this peak are still under investigation. But we may speculate that this peak could be associated

5. Sintered High-T_c Superconductors

with superconducting fluctuations that are enhanced by the two-dimensional nature of the CuO planes that are responsible for superconductivity in these systems.

Sound velocity measurements exhibited varied behavior. The Young's modulus in $La_{0.85}Sr_{0.15}CuO_4$ exhibited a peak at around 200 K. The velocity of longitudinal waves in sintered $Y_1Ba_2Cu_3O_7$ increased monotonically as the temperature was lowered, with a distinct change in slope at T_c. Sensitive measurements in a single crystal show a distinctive drop in velocity at T_c that is associated with a velocity discontinuity. Because of the large crystallite size in the sinter-forged $Y_1Ba_2Cu_3O_7$ samples, hysteresis in the velocity curves was observed in the temperature range from 60 K to 250 K. This hysteresis was larger when the sound waves were propagating parallel to the c-axis than when they were propagating perpendicular to the c-axis. Longitudinal sound wave velocity curves in $ErBa_2Cu_3O_7$ exhibited similar hysteresis to that observed in sinter-forged $YBa_2Cu_3O_7$. Both longitudinal and transverse sound wave curves in $Ba_{1-x}K_xBiO_3$ increased monotonically as the temperature was lowered from room temperature. But the transverse wave velocity curve exhibited a broad minimum at around 35 K and a change in slope at T_c. The behavior of velocity in the high-T_c superconductors is covered in more detail in Chapters VI, VII, and VIII.

The ultrasonic data presented in this chapter have shown that there are distinct attenuation and velocity effects associated with the superconducting transition. Some of the velocity effects can be understood within the thermodynamic models that had previously been developed for conventional superconductors. The appearance of a peak in attenuation at T_c was somewhat unexpected, although in retrospect it should have been anticipated, since unsuccessful attempts have been made in the past to find such peaks in low-dimensional superconducting systems (Robinson et al., 1974). The relaxation peaks that have been observed, together with their selection rules determined from measurements in sinter-forged samples, may help to determine the nature of the excitations that are responsible for the coupling of the Cooper pairs that are responsible for the high transition temperature of the cuprate superconductors.

NOTE ADDED IN PROOF: Recent ultrasonic measurements (Li et al. 1992a,b) in low fields have shown an increase at 7 Oe in both attenuation, 0.2 db/cm, and velocity, 20 ppm, in melt textured $Y_1Ba_2Cu_3O_7$ at 77 K for shear waves travelling parallel to the c-axis with the magnetic field in the ab plane and perpendicular to the shear polarization direction. The attenuation change exhibits a maximum at around 55 K. These effects may be due to the interaction of the sound waves with the flux line lattice in the mixed state. The observed maximum may be caused by flux lines depinning or melting.

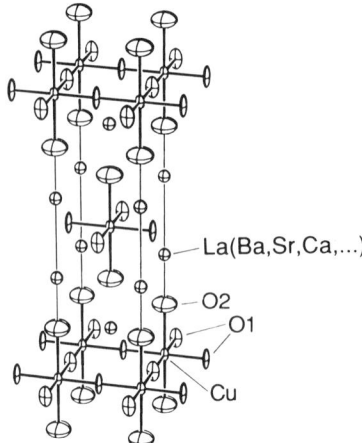

FIG. 36. Crystal structure for single layered $La_{2-x}R_xCuO_4$. After Schuller and Jorgensen (1989).

ACKNOWLEDGMENT

This work was supported by the Office of Naval Research.

Appendix A. Crystal Structure of High-T_c Superconductors

The crystal structures of the high-T_c superconductors are displayed in this appendix. Figure 36 shows the crystal structure for the single-layered $La_{2-x}R_xCuO_4$, where R may be Ba, Sr, or Cu. This compound has one Cu–O layer per cell identified by the Cu atom. Figure 37 shows the crystal structure for the double-layered $R_1Ba_2Cu_3O_{7-\delta}$, the now classical 1–2–3, where R can be Y or any of the rare-earth elements. In addition to the two Cu–O layers per unit cell identified by Cu2, and centered about the Y atom, there are Cu–O chains, along the b direction identified by Cu1. Figure 38 shows the crystal structures for $T\ell_m Ca_{n-1} Ba_2 Cu_n O_{2n+m+2}$ for $m = 1, 2$, and $n = 1, 2, 3$. These would be the same structures for $Bi_2Ca_{n-1}Sr_2Cu_nO_{2n+4}$ for $n = 1, 2$. Figure 40 plots the T_c's for the $T\ell$ and Bi compounds as a function of the number of Cu–O planes for $m = 2$ and $n = 1, 2, 3$. In Figs. 38 and 39 the Cu–O planes are identified by the octahedra for $n = 1$. The Cu atoms are located at the center of the octahedra, and the oxygen atoms at the apices. For $n = 2$, the Cu–O planes are identified by the pyramids and inverted pyramids. The Cu atoms are at the center of the bases of these pyramids. For $n = 3$, the Cu–O planes are identified by the pyramids, the inverted pyramids, and the plane rectangles between the bases of the pyramids. Oxygen atoms are at the apices and corners. Thus, $n = 1$ corresponds to one Cu–O plane per

5. Sintered High-T_c Superconductors

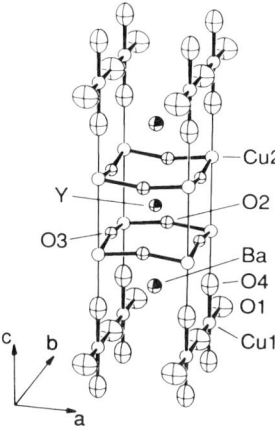

FIG. 37. Crystal structure for double-layered $R_1Ba_2Cu_3O_{7-\delta}$. After Schuller and Jorgensen (1989).

STRUCTURES OF SUPERCONDUCTING Tl-Ba-Ca-Cu-O

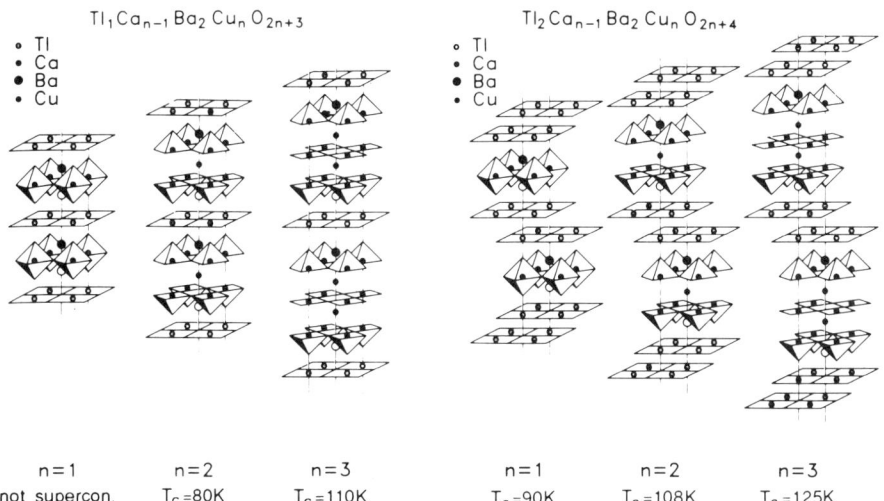

FIG. 38. Crystal structure for $Tl_mCa_{n-1}Ba_2Cu_nO_{2n+m+2}$ for $m = 1, 2$ and $n = 1, 2, 3$. The number of Cu–O layers per formula unit is given by n. These are similar to the crystal structures for $Bi_mCa_{n-1}Sr_2Cu_nO_{2n+m+2}$. After Parkin et al. (1988).

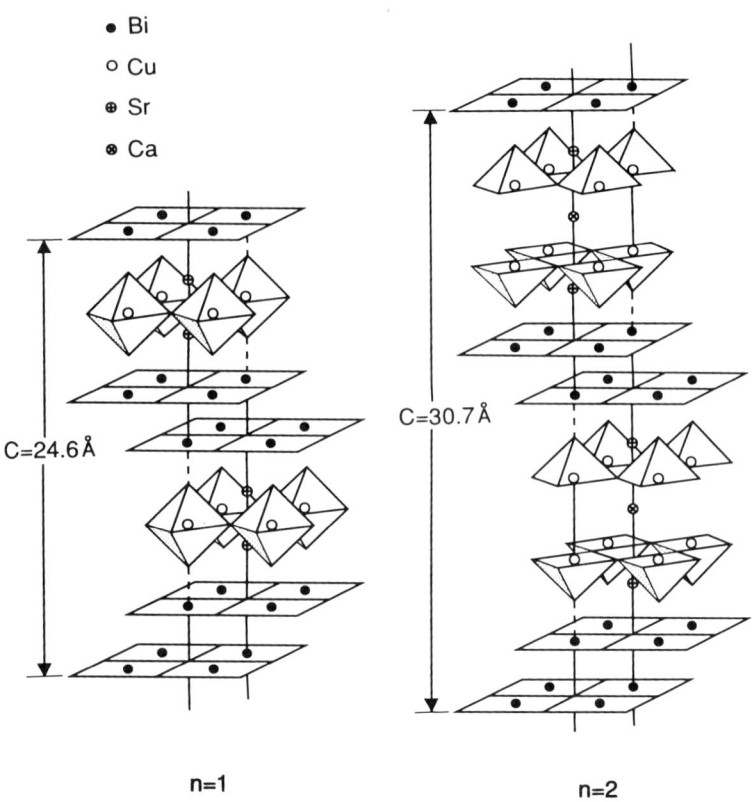

FIG. 39. Crystal structure for $Bi_2Ca_{n-1}Sr_2Cu_nO_{2n+4}$ for $n = 1, 2$. After Talvacchio (1989).

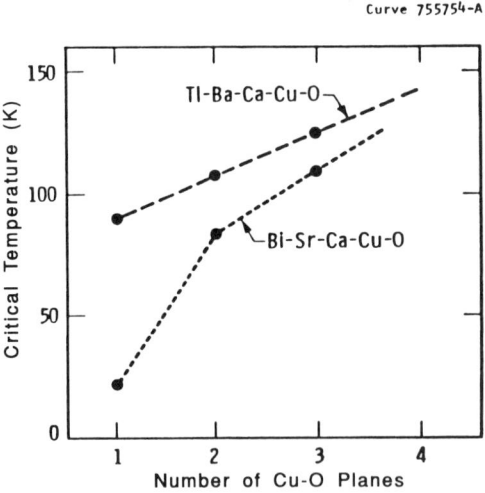

FIG. 40. Dependence of the superconducting transition temperatures on the number of Cu–O layers for both the $T\ell$ and the Bi compounds for $m = 2$.

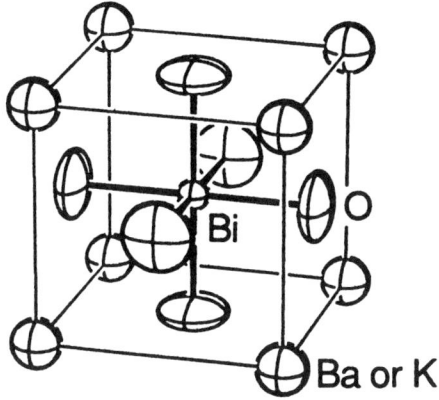

FIG. 41. Crystal structure for $Ba_{1-x}K_xBiO_3$. After Pei et al. (1990).

formula cell, $n = 2$ corresponds to two Cu–O planes, and $n = 3$ to three Cu–O planes. Figure 41 shows the crystal structure for $Ba_{1-x}K_xBiO_3$. This is the only high-T_c superconductor that does not contain any Cu atoms.

Appendix B. Selected Velocities and Elastic Constants for High-T_c Superconductors

In Table I, V_ℓ is the velocity of longitudinal waves, V_t is the velocity of transverse waves, d_{th} is the theoretical density of the ideal crystal, d is the density of the measured crystal, and f is the frequency of the ultrasonic waves used in the measurements.

The velocity in the ideal crystal V_I may be estimated from the measured velocity V_M by using the relation

$$V_I = \left(\frac{d_{th}}{d}\right)^{1/2} V_M.$$

$YBa_2Cu_3O_7$ is orthorhombic. Therefore, it has nine independent elastic constants. The sinter-forged sample discussed in Section 4 is axially symmetric about the c-axis. This additional symmetry reduces the number of elastic constants to five, which is the same number required for a tetragonal or a hexagonal crystal. We have measured three of these constants and give the results in Table II.

TABLE I

SELECTED VELOCITIES AND DENSITIES IN SINTERED HIGH-T_c SUPERCONDUCTORS

Sample		v_ℓ (10^3 m/s)	v_t (10^3 m/s)	d (g/cm^3)	d_{th} (g/cm^3)	f (MHz)	Ref.[i]
$La_{2-x}Si_xCuO_4$	$x = .15$	5			7.006	10	1
	$x = .15$	5.5				24	2
	$x = .14$	5.1				5	3
	$x = .12$	4.7				15	4
YBCO[a]		3.9				5	4
		4.0	2.5			5	5
		4.5		5.29		13	6
		4.1		5.23	6.37	12	7
		4.4	2.9	4.97		10	8
		3.8				11	9
		4.1	2.5			~15	10
YBCO[a] (sinter-forged)		4.4[e]	2.7[f]			12/8.5	
				6.25	6.37		11
		3.8[g]	2.7[h]			12/8.5	
$ErBa_2Cu_3O_{7-\delta}$		4.65		5.4		13.1	11
$Ba_{1-x}K_xBiO_3$	$x = 0.4$		2.6	7.1	7.502	8.5	11
	$x = 0.5$				7.363		
$T\ell$[b] 2212		5.1	3.7		7.46		12
Bi[c] 2212		4.5/4.1	2.5			7.5/5	13
2212		2.8				10	14
(BiPb)SrCaCuO 2223		3.7	2.3	3.76	6.51		15
2223		2.9				10	14
		2.7	2.0	5.4		13/9	16
BiSrCaCO	80% 2223	3.8		3.9	6.57	10	17
	80% 2212						

[a] $Y_1Ba_2Cu_3O_7$
[b] $T\ell_2Ba_2CaCu_2O_8$
[c] $Bi_2Sr_2Ca_1Cu_2O_8$
[d] $(BiPb)_2Sr_2Ca_2Cu_3O_{10}$
[e] $q\|a,b$
[f] $q\|a,b$; $\varepsilon\|a,b$
[g] $q\|c$
[h] $q\|c$; $\varepsilon\|a,b$

[i] References:
1. Bishop et al. (1987a).
2. Fossheim et al. (1987).
3. Horie et al. (1987a).
4. Bhattacharya et al. (1988a).
5. Almond et al. (1987a).
6. Suzuki et al. (1988).
7. Choi et al. (1989).
8. Ewart et al. (1987).
9. Laegreid et al. (1988).
10. Lang et al. (1988).
11. Xu, M.-F. (1990).
12. Sun et al. (1989b).
13. Wang et al. (1989a,b).
14. He et al. (1989).
15. Ledbetter et al. (1989).
16. Plechacek and Dominec (1990).
17. Sun et al. (1990).

5. Sintered High-T_c Superconductors

TABLE II

ELASTIC CONSTANTS FOR ORIENTED $YBa_2Cu_3O_7$ (AFTER XU, 1990)

C_{11} (dyn/cm^2)	C_{22} (dyn/cm^2)	C_{33} (dyn/cm^2)	d (g/cm^3)	d_{th} (g/cm^3)
1.17×10^{12}	0.86×10^{12}	0.43×10^{12}	6.05	6.37

Appendix C. Temperature Position of Attenuation Peaks for Sinter-Forged $YBa_2Cu_3O_7$ Samples

TABLE III

ATTENUATION PEAK POSITION VS. PROPAGATION CONFIGURATION AND FREQUENCY
(AFTER XU, 1990)

Propagation Configuration	f (MHz)		Attenuation Peak Position (K)	
Long. $q \perp c$, $\varepsilon \perp c$	12.0	70	180	250
Long. $q \| c$, $\varepsilon \| c$	12.0	—	180	—
Trans. $q \| c$, $\varepsilon \perp c$	14.5	—	180	—
	41.0	—	195	—
Trans. $q \perp c$, $\varepsilon \perp c$	6.5	—	160	—
	10.1	—	180	—

Appendix D. Activation Energies for Relaxation Times Associated with Relaxation Attenuation Peaks

TABLE IV

ACTIVATION ENERGIES FOR RELAXATION TIMES ASSOCIATED WITH RELAXATION ATTENUATION PEAKS[a]

Systems	80 K Peak	180 K Peak	250 K Peak
Single-phase YBCO (15 MHz), Long.	450 K		750 K
Sinter-forged YBCO (12 MHz), $q \| a, b$; $\varepsilon \| a, b$	380 K	840 K	480 K
Sintered-forged YBCO (14.5 MHz), $q \| c$; $\varepsilon \| c$		1,210 K	
Sinter-forged YBCO (41 MHz), $q \| c$; $\varepsilon \| c$		1,360 K	
$ErBa_2Cu_3O_{7-\delta}$ (13.1 MHz), long.		890 K	

[a] The first column describes the system and the sound wave polarization. The next three columns give the activation energies in units of temperature that were obtained using the procedure described in Figs. 13a and 13b. The averages of these values for each of the peaks at the three different temperatures are the values that are used in the summary, Section 8. (After Xu, 1990.)

References

Almond, D. P., Lambson, E. F., Saunders, G. A., and Hong, W. (1987a). *J. Phys.* **F17**, L261–L266.
Almond, D. P., Lambson, E., Saunders, G. A., and Wang, H. (1987b). *J. Phys.* **F17**, L221–L224.
Almond D. P., Wang, Q., Freestone, J., Lambson, E. F., Chapman, B., and Saunders, G. A. (1989). *J. Phys. Cond. Matter.* **1**, 5993–5996.
Axe, J. D., Moudden, A. H., Hohlwein, D., Cox, D. E., Moudenbaugh, A. R., and Xu, Youwen (1989). *Phys. Rev. Lett.* **62**, 2751–2754.
Bardeen, J., Cooper, L. N., and Schrieffer, J. R. (1957). *Phys. Rev.* **108**, 1175–1204.
Batlogg, B., Cava, R. J., Schneemeyer, L. F., and Espinosa, G. P. (1989). *IBM J. Res. Develop.* **33**, 208–214.
Baum, H. P., Ph.D. thesis, University of Wisconsin at Milwaukee, Milwaukee, 1990 (unpublished).
Baum *et al.* (1990) p. 249.
Bednorz, J. G., and Müller, K. A. (1986). *Z. Phys.* **B64**, 189–193.
Bhargava, R. N., Herko, S. P., and Osborne, W. N. (1987). *Phys. Rev. Lett.* **59**, 1468–1471.
Bhattacharya, S., Higgins, M. J., Johnston, D. C., Jacobson, A. J., Stokes, J. P., Goshorn, D. P., and Lewandowski, J. T. (1988b). *Phys. Rev. Lett.* **60**, 1181–1184.
Bhattacharya, S., Higgins, M. J., Johnston, D. C., Jacobson, A. J., Stokes, J. P., Goshorn, D. P., and Lewandowski, J. T. (1988a). *Phys. Rev.* **B37**, 5901–5904.
Bishop, D. J., Gammel, P. L., Ramirez, A. P., Cava, R. J., Battlogg, B., and Reitman, E. A. (1987a). *Phys. Rev.* **B35**, 8788–8790.
Bishop, D. J., Ramirez, A. P., Gemmel, P. L., Batlogg, B., Reitman, E. A., Cava, R. J., and Millis, A. J. (1987b). *Phys. Rev.* **B36**, 2408–2410.
Bourne, L. C., Zettl, A., Chang, K. J., Cohen, Marvin, L., Stacy, Angelica M., and Ham, W. K. (1987). *Phys. Rev.* **B35**, 8785–8787.
Breazeale, M. A., and Jiang, W. H. (1990). In "IEEE 1990 Ultrasonics Symposium Proceedings" (B. R. McAvoy, ed.). IEEE, New Jersey, Catalog No: 90CH2938-9, pp. 1297–1300.
Brown, S. E., Migliori, A., and Fisk, Z. (1988). *Solid State Commun.* **65**, 483–486.
Calemczuk, R., Bonjour, E., Henry, J. Y., Forro, L., Ayache, C., Jurgens, M. J. M., Rossat-Mignod, J., Barbara B., Burlet, P., Couach, M. A. Khoder, F., and Salce, B. (1988). *Physica* 960–961, C153–155.
Cannelli, G., Cantelli, R., Cordero, F., Costa, G. A., Ferretti, M., and Olcese, G. L. (1987). *Phys. Rev.* **B36**, 8907–8909.
Cannelli, G., Cantelli, R., Cordero, F., Costa, G. A., Ferretti, M., and Olcese, G. L. (1988). *Europhys. Lett.* **B36**, 271–276.
Cava, R. J., Batlogg, B., Krajewski, J. J., Farrow, R. C., Rupp, L. W., Jr., White, A. E., Short, K. T., Peck, W. F., Jr., and Kometani, T. Y. (1988). *Nature* **332**, 814–816.
Chen, J. T., Wenger, L. E., McEwan, C. J., and Logothetis, E. M. (1987). *Phys. Rev. Lett.* **58**, 1972–1975.
Choi, P. K., Koizumi, H., Takagi, K., and Suzuki, T. (1989). *Solid State Commun.* **70**, 1175–1178.
Chu, C. W., Bechtold, J., Gao, L., Hor, P. H., Huang, Z. J., Meng, R. L., Sun, Y. Y., Wang, Y. Q., and Xue, Y. Y. (1988). *Phys. Rev. Lett.* **60**, 941–943.
Crawford, M. K., Fourneth, W. E., McCarron, E. M. II, Harlow, R. L., and Moudden, A. H. (1990). *Science* **250**, 1390–1394.
Esquinazi, P., Luzuriaga, J., Durán, C., Esparza, D. A., and D'Ovidio, C. (1987). *Phys. Rev.* **B36**, 2316–2318.
Ewert, S., Guo, S., Lemmens, P., Stellmach, F., Wynants, J., Arlt, G., Bonnenberg, D., Kliem, H., Comberg, A., and Passing, H. (1987). *Solid State Commun.* **64**, 1153–1156.

Fischer, H. E., Watson, S. K., and Cahill, D. G. (1988). *Comments on Condensed Matter Phys.* **14**, 65–127.
Flemming, R. M., Batlogg, B., Cava, R. J., and Rietman, R. A. (1987). *Phys. Rev.* **B35**, 7191–7194.
Fossheim, K., Lagreid, T., Sandvold, E., Vassenden, F., Müller, K. A., and Bednorz, J. G. (1987). *Solid State Commun.* **63**, 531–533.
Francois, M., Junod, A., Yvon, K., Fischer, P., Capponni, J. J., Strobel, P., Marezio, M., and Hewat, A. W. (1988). *Physica* **C152–155**, 962–963.
Golding, B., N. Birge, O., Haemmerle, W. H., Cava, R. J., and Rietman, E. (1987). *Phys. Rev.* **B36**, 5606–5608.
Golding, B., Haemmerle, W. H., Schneemeyer, L. F., and Waszczak, J. V. (1988). *Proceedings of the IEEE 1988 Ultrasonics Symposium* (B. R. McAvoy, ed.) IEEE New Jersey Cat. No 88CH2578-3, 1079–1083.
He, Yusheng, Baiwen Zhang, Sihan Lin, Jiong Xiang, Yongming Lou, and Haoming Chen (1987). *J. Phys.* **F17**, L243–L248.
He, Yusheng, Jiong Xiang, Xin Wang, Aisheng He, Jincang Zhang, and Fanggao Chang (1989). *Phys. Rev.* **B40**, 7384–7386.
Hikami, S., Hirai, T. and Kagashima, S. (1987). *Jpn. J. Appl. Phys.* **26**, L314.
Hikata, A. McKenna, M. J., Elbaum, C., Kershaw, R., and Wold, A. (1989). *Phys. Rev.* **B40**, 5247–5250.
Hinks, D. G. (1992). Private communication.
Hinks, D. G., Dabrowski, B., Jorgensen, J. D., Mitchell, A. W., Richards, D. R., Shiyou Pei, and Shi Donglu (1988a). *Nature* **333**, 836–838.
Hinks, D. G., Richards, D. R., Dabrowski, B., Marx, D. T., and Mitchell, W. (1988b). *Nature* **335**, 419–421.
Hinks, D. G., Mitchell, A. W., Zheng, Y., Richards, D. R., and Dabrowski, B. (1989). *Appl. Phys. Lett.* **54**, 1585–1587.
Horie, Y., and Mase, S. (1989). *Solid State Commun.* **69**, 535–538.
Horie, Y., Fukami, T., and Mase, S. (1987a). *Solid State Commun.* **63**, 653–656.
Horie, Y., Terashi, Y., Fukuda, H., Fukami, T., and Mase, S. (1987b). *Solid State Commun.* **64**, 501–504.
Horn, P. M., Keane, D. T., Held, G. A., Jordan-Sweet, J. L., Kaiser, D. L., Holtzberg, F., and Rice, T. M. (1987). *Phys. Rev. Lett.* **59**, 2772–2775.
Hu, Jiankai, Senkui Zhang, Qianlin Zhang, Weili Gai, Tingzhang Deng, Liangkun Zhang, Yusheng He, and Jiong Xiang (1989). *Physica* **C162–164**, 444–445.
Jerome, D., and Kang, W. (1988). *J. Appl. Phys.* **63**, 4005–4008.
Klochko, V. S., Makarov, V. I., Tkachenko, V. F., Voronov, A. P., and Zavaritsky, N. V. (1989). In "High Temperature Superconductivity from Russia" (A. I. Larkin and N. V. Zavaritsky, eds.). World Scientific Publishing Co., p. 263–269.
Kresin, V. Z. (1987a). *In* "Novel Superconductivity" (S. A. Wolf and V. Z. Kresin, eds.). Plenum, New York, pp. 309–322.
Kresin, V. Z. (1987b). *Solid State Commun.* **63**, 725–727.
Kurihara, S. (1989). *Phys. Rev.* **B39**, 6600–6606.
Lagreid, T., and Fossheim, K. (1988). *Europhys. Lett.* **6**, 81–88.
Lagreid, T., Fossheim, K., Sandvold, E., and Julsrud, S. (1987). *Nature* **330**, 637–638.
Laegreid, T., Fossheim, K., and Vassenden, F. (1988). *Physica* **C153–155**, 1096–1099.
Lang, M., Lechner, T., Riegel, S., Steglich, F., Weber, G., Kim, T. J., Lüthi, B., Wolf, B., Rietschel, H., and Wilhelm, M. (1988). *Z. Phys.* **B69**, 459–463.
Ledbetter, H. M., and Kim, S. A. (1988). *Phys. Rev.* **B38**, 1857–1860.
Ledbetter, H. M., Kim, S. A., Goldfarb, R. B., and Togano, K. (1989). *Phys. Rev.* **B39**, 9689–9692.

Lemmens, P., Stellmach, F., Ewert, S., Guo, S., Wynants, J., Arlt, G., Comberg, A., Passing, H., and Marbach, G. (1988). Physica **C153–155**, 294–295.
Li, Z. X., Levy, M., Sarma, B. K., Salem-Sugui, S., and Shi, D. (1992). *Bull. Amer. Phys. Soc.* **37**, 229.
Li, Z. X., Levy, M., Sarma, B. K., Salem-Sugui, S., Shi, D., Crabtree, G. W. (1992) In "Proceedings of the 1992 Applied Superconductivity Conference," (to be published).
Luthi, B., Wolf, B., Kim, T., Grill, W., and Renker, B. (1987). *Jpn. J. Appl. Phys.* **26**, 127.
Maeda, H., Tanaka, Y., Fukutomi, M., and Asano, T. (1988). *Jpn. J. Appl. Phys.* **4**, L209–L210.
Makarov, V. I., Klochko, V. S., Zavaritski, N. V., and Petrov, S. V., (1987). *Sov. Phys. JETP Lett.* **46**, S129–S131.
Mason, W. P. (1958). "Physical Acoustics and the Properties of Solids." Van Nostrand, Princeton, New Jersey.
Mattheiss, L. F., Gyorgy, E. M., and Johnson, Jr., D. W. (1988). *Phys. Rev.* **B37**, 3745–3746.
Müller, V., de Groot, K., Maurer, D., Roth, Ch., Rieder, K. H., Eickenbusch, E., and Schöllhorn, R. (1987). *Proc. 18th Int. Conf. on Low Temp. Phys., Jpn. J. Appl. Phys.* **26**, 2139–2140.
Parkin, S. S. P., Lee, V. Y., Nazzal, I., Savoy, R., Huang, T. C., Gorman, G., and Beyers, R. (1988). *Phys. Rev.* **B38**, 6531–6537.
Pederson, D. O., El Ali, A., Sheng, Z. Z., and Hermann, A. M. (1989). *Phys. Rev.* **B 40**, 7313–7315.
Pei, Shiyou, Jorgensen, J. D., Dabrowski, B., Hinks, D. G., Richards, D. R., and Mitchell, A. W. (1990). *Phys. Rev.* **B41**, 4126–4141.
Plechacek, Vladimir, and Dominec, Jiff (1990). *Solid State Commun.* **74**, 633–635.
Qian, Y. J., Xu, M.-F., Shenstrom, A., Baum, H-P., Ketterson, J. B., Hinks, D., Levy, M., and Sarma, B. K. (1987). *Solid State Commun.* **63**, 599–602.
Robinson, D. A., Maki, K., and Levy, M. (1974). *Phys. Rev. Lett.* 709–712.
Robinson, Q., Georgopoulos, P., Johnson, D. L., Marcy, H. O., Kannewurf, C. R., Hwu, S.-J., Marks, T. J., Poeppelmeier, K. R., Song, S. N., and Ketterson, J. B. (1987). *Adv. Ceram. Mater.* **2**, 380–383.
Saint-Paul, M., Tholence, J. L., Noël, H., Levet, J. C., Potel, M., and Gougeon, P. (1989). *Solid State Commun.* **69**, 1161–1163.
Schirber, J. E., Ginley, D. S., Venturini, E. L., and Morosin, B. (1987). *Phys. Rev.* **B35**, 8709–8710.
Schuller, I. K., and Jorgensen, J. D. (1989). *Materials Research Society Bulletin* **14**, 27–30.
Sheng, Z. Z., and Hermann, A. M. (1988). *Nature* **332**, 55–58.
Shi, X. D., Yu, R. C., Wang, Z. Z., Ong, N. P., and Chaikin, P. M. (1989). *Phys. Rev.* **B39**, 827–830.
Song, S. N., Robinson, Q., Hwu, S.-J., Johnson, D. L., Poeppelmeier, K. R., and Ketterson, J. B. (1987). *Appl. Phys. Lett.* **51**, 1376–1378.
Srinivasan, R., Girirajan, K. S., Ganesan, V., Radhakrishnan, V., and Rao, G. V. S. (1988). *Phys. Rev.* **B38**, 889–892.
Sun, K. J., Winfree, W. P., Xu, M.-F., Sarma, B. K., Levy, M., Caton, R., and Selim, R. (1988). *Phys. Rev.* **B38**, 11988–11991.
Sun, K. J., Sorbello, R. S., and Levy, M. (1989a). *Phys. Rev.* **B40**, 2133–2137.
Sun, K. J., Winfree, W. P., Xu, M.-F., Levy, M., Sarma, B. K., Singh, A. K., Osofsky, M. S., and Le Toruneau, V. M. (1989b). *Physica* **C162–164**, 446–447.
Sun, K. J., Parker, F. R., Winfree, W. P., Syed, H. I., Meng, R. L., Sun Y. Y., Hor, P. H. Chu, C. W. (1990a). In "IEEE 1990 Ultrasonic Symposium Proceedings," (B. R. McAvoy, ed.) IEEE, New Jersey, Catalog No 90CH2938-9, pp. 1293–1296.
Sun, K. J., Winfree, W. P., Xu, M.-F., Sarma, B. K., Singh, A. K., Osofsky, M. S., and Le Toruneau, V. M. (1990b), *Phys. Rev.* **B42**, 2569–2572.
Suzuki, M., Okuda, Y., Iwasa, I., Ikushima, A. J., Takabatake, T., Nakazawa, Y., and Ishikawa, M. (1988). *Jpn. J. Appl. Phys. Ltrs.* **27**, L308–L310.

5. Sintered High-T_c Superconductors

Takagi, H., Uchida, S., Kichino, K., Kitazawa, K., Fueki, K., and Tanaka, S. (1987). *Jpn. J. Appl. Phys.* **26**, L320.
Talvacchio, J. (1989). Private communication.
Testardi, L. R. (1973). In *"Physical Acoustics,"* Vol. X (W. P. Mason and R. N. Thurston, eds.). Academic Press, New York, pp. 193–296.
Testardi, L. R. (1975a). *Rev. Mod. Phys.* **47**, 637–648.
Testardi, L. R. (1975b). *Phys. Rev.* **B12**, 3849–3854.
von Molnar, S., Torressen, A., Kaiser, D., Holtzberg, F., and Penney, T. (1988). *Phys. Rev.* **B37**, 3762–3765.
Wang, Yening, Huimin Shen, Jinsong Zhu, Ziran Xu, Min Gu, Zhongmin Niu, and Shifang Zhang (1987). *J. Phys.* **C20**, L665–L668.
Wang, Yening, Jin Wu, Jinsong Zhu, Huimin Shen, Yifeng Yan, and Zhongxian Zhao (1989a). *Physica* **C162–164**, 454–455.
Wang, Yening, Xiaohua Chen, Huimin Shen, and Linhai Sun (1989c). *Physica* **C162–164**, 456–457.
Waynert, J. A., Salvo, Jr., H., and Levy, M. (1974). *Phys. Rev.* **B10**, 1859–1864.
Wu, M. K., Ashburn, J. R., Torng, C. J., Hor, P. H., Meng, R. L., Gao, L., Huang, Z. J., Wang, Y. Q., and Chu, C. W. (1987). *Phys. Rev. Lett.* **58**, 908–910.
Wu, Ting, Lagreid, T., Fossheim, K., Axe, J. D., and Hiaaka, Y. (1989). *Physica* **C162–164**, 448–449.
Xiang, X.-D., Chung, M., Brill, J. W., Hoen, S., Pinsukanjana, P., and Zetl, A. (1989). *Solid State Commun.* **69**, 833–836.
Xu, M.-F., Ph.D. thesis, University of Wisconsin at Milwaukee, Milwaukee, 1990 (unpublished).
Xu, M.-F., Baum, H-P., Schenström, A., Sarma, B. K., Levy, M., Sun, K. J., Toth, L. E., Wolf, S. A., and Gubser, D. U. (1988). *Phys. Rev.* **B37**, 3675–3677.
Xu, M.-F., Bein, D., Wiegert, R. F., Sarma, B. K., Levy, M., Zhao, Z., Adenwalla, S., Moreau, A., Robinson, Q., Johnson, D. L., Hwu, S.-J., Poeppelmeier, K. R., and Ketterson, J. B. (1989). *Phys. Rev.* **B39**, 843–846.
Xu, M.-F., Qian, Y. J., Sun, K. J., Zheng, Y., Ran, Q., Hinks, D., Sarma, B. K., and Levy, M. (1990). *Physica* **B165–166**, 1281–1282.
Zeng, W. G., Zhang, J. X., Lin, G. M., Du, C. L., Fung, P. C. W., and Siu, G. G. (1989). *Solid State Commun.* **70**, 333–335.
Zhao, A., Adenwalla, S., Moreau, A., Ketterson, J. B., Robinson, Q., Johnson, D. L., Hwu, S.-J., Poeppelmeier, K. R., Xu, M.-F., Hong, Y., Wiegert, R. F., Levy, M., and Sarma, B. K. (1989). *Phys. Rev.* **B39**, 721–724.
Zhao, Z. X., Chen, L. Q., Yang, Q. S., Huang, Y. Z., Chen, G. H., Tang, R. M., Liu, G. R., Cui, G. C., Chen, L., Wang, L. Z., Geo, S. Q., Li, S. L., and Bi, J. Z. (1987). *Kexue Tongbao* **32**, 661.

—6—
Sound Velocity Studies of Ceramic High-Temperature Superconductors

S. BHATTACHARYA

NEC Research Institute, Princeton, New Jersey

1. Introduction 303
2. Theory 306
   2.1. Mean-Field Analysis 306
   2.2. Fluctuation Effects 309
3. Sound Velocity in Ceramic Samples 311
   3.1. La–Sr–Cu–O 313
   3.2. La–Ba–Cu–O 317
   3.3. Y–Ba–Cu–O (1–2–3) 320
4. Results near T_c 328
5. Magnetic Field Dependence 337
6. Conclusions 342
   6.1. General Sample-Related Problems 342
   6.2. Outlook for Future Work 343
   References 345
   Bibliography 346

1. Introduction

Since the discovery of high-temperature superconductivity in the lanthanum–barium cuprates by Bednořz and Muller (1986), nearly every conceivable measurement has been performed on this class of materials. Sound propagation studies are not an exception. In fact, sound velocity anomalies were among the first anomalous results obtained in these materials. In conventional superconductors, sound propagation studies have primarily focused on the attenuation, since it is the more useful quantity. It had provided one of the first measurements of the energy gap and a direct confirmation of the BCS theory. In unconventional superconductors such as the heavy-fermion systems, they have been useful in determining the symmetry of the order parameter through the observation of nodes (points or lines) in the highly anisotropic energy gap. Sound attenuation studies have also been performed extensively on the cuprates as well; they are reviewed in the previous chapter.

Elastic (mechanical) measurements, either through sound propagation or through vibrating-reed-type measurements, are not routinely made in conventional superconductors for the simple reason that the condensation energy per particle is about $(kT_c)^2/E_F$ (where E_F is the fermi energy), approximately 10^{-6} times the typical elastic energy (which is of the order of E_F per particle). The net effect in sound velocity is of the order of a few parts per million (Alers and Waldorf, 1962) and can be studied only with the most precise of techniques in the most carefully prepared single crystals. These measurements were used in conventional superconductors to test the thermodynamic relationships that exist among the various second derivatives of the free energy, i.e., the compressibility, the specific heat and the thermal expansion coefficient, as well as the thermodynamic derivatives of the critical field and the reversibility of the Meissner effect. For a review of these studies, see Chandrasekhar (1969). These measurements have yielded information on the microscopic parameters involved in the BCS theory, such as the strain dependence of the electronic density of states, electron–electron interactions etc. For a review, see Seraphim and Marcus (1962).

In the A15 compounds, superconductivity occurs close to a structural transition. These systems have been studied more extensively with respect to their elastic properties in order to provide a detailed understanding of how the proximity to the structural transition affects superconductivity in general, and the role of soft phonons in particular (Testardi, 1973).

Extensive sound velocity measurements have been made in the cuprate superconductors for a variety of reasons. First, as is obvious from their structures, they are extremely anisotropic, with strong two-dimensional character. Thus, the coupling of superconductivity to structural distortions could also be highly anisotropic. Measurement of the sound velocity anomaly at the superconducting transition for sound waves with different polarizations and propagating in different directions would, in principle, provide information on this particular aspect of superconductivity. Second, there are structural transitions in these systems too, such as a tetragonal-to-orthorhombic transition, that occur above T_c. This suggested possible similarities with the A15 compounds, and soft mode effects being responsible for a possible phonon-mediated superconductivity (Fossheim et al., 1987) also for these systems.

In spite of substantial efforts, the advances in our understanding of the central issues in the study of sound propagation have been slow, hampered by technical problems. Preparation of large and homogeneous single crystals has proven to be extremely difficult. Moreover, these systems, and especially their superconducting properties, have turned out to be extremely sensitive to the composition, such as the oxygen stoichiometry. Coupled with the fact that the superconducting coherence length is extremely short in these materials, there are serious concerns

6. Ceramic High-Temperature Superconductors

about sample homogeneity even in small single crystals. Because these inhomogeneities do not produce much change in the lattice constants, some of these issues, such as possibility of phase separation in scales small compared to grain sizes (1 mm) but large compared to the coherence length (10 Å), cannot be easily resolved by the usual procedures of neutron or x-ray diffraction. Thus, detailed quantitative comparison of data obtained in different laboratories has proven to be quite difficult.

Much of the early measurements of elastic properties, using ultrasound techniques or vibrating-reed techniques, were performed on ceramic samples. This has obviously prevented us from obtaining information about the anisotropy. A partial solution can in principle be obtained from the so-called sinter-forged samples (Zhao *et al.*, 1989) that yields a nearly uniaxial symmetry, whereby the behavior in the Cu–O layers, as opposed to perpendicular to them, can indeed be sorted out.

On the theoretical side, the primary difficulty in finding an explanation of the high-temperature superconductivity has been in obtaining a consistent description of the normal state from which the superconducting condensation occurs. There is no consensus yet on issues as central as whether the normal state can even be described as a Fermi liquid. It has also been found recently in the LaBaCuO system that small, and what would otherwise be considered apparently insignificant, structural rearrangements (Axe *et al.*, 1989) in the Cu–O planes in the normal state at temperatures well above T_c result in nearly a total loss of superconductivity. These results underscore the need to explore the nature of the normal state in great detail. It has been suggested (Bhattacharya *et al.*, 1988a) that there is evidence, in both sound propagation characteristics as well as in other properties, of transitions and/or anomalies of yet unknown origins in these systems that occur above T_c. Measurements of sound velocity anomalies have been valuable in identifying their existence. In some instances, single-crystal samples were found to confirm earlier results obtained in some ceramic samples. In these cases, the information obtained in ceramic measurements can be assumed to be intrinsic, albeit with caveats, as we discuss later in this article.

In this chapter the sound velocity measurements in ceramic samples are reviewed. The literature is already vast; a detailed review of most of the work is already difficult. The review is not intended to be a summary of the enormous range of measurements reported. Instead, the review attempts to describe some of the main results and their significance. At the end of the chapter a list of references has been provided for the reader. In many instances several papers have reported similar results; in those cases a typical behavior has been discussed. In many instances there is a lack of consensus in what effects are considered intrinsic and, even when they are, how to interpret them. Controversies and uncertainties have also been discussed. There has also been a great deal of confusion regarding the thermodynamic anomalies at T_c. Therefore, we have included a summary of the thermodynamics of the phase transition worked out recently by Millis and Rabe (1988). Earlier studies are contained in Testardi (1975).

2. Theory

2.1. Mean-Field Analysis

The superconducting transition in conventional superconductors is an excellent example of a mean-field second-order transition. This is so because the superconducting coherence length, i.e., the distance between the electrons forming the Cooper pair, is very large, of the order of a few thousand angstroms. There are many pairs in between the members of any pair, and thus the conditions of a mean-field theory are well satisfied. Indeed, it can be shown, through what is called the Ginzburg criterion, that in an ordinary superconductor, the typical reduced temperature range over which critical fluctuation effects are observable is restricted to an immeasurably small range of about 10^{-15} (Ma, 1976).

The situation with the oxide superconductors is quite different. The typical coherence length is, in addition to being anisotropic, quite small, about 10 Å. This results in a considerably larger fluctuation regime. Although the true critical regime is still rather small, gaussian fluctuations are expected to be readily observable. In any event, a mean-field analysis is a good starting point for a discussion of the transition.

We start with the mean-field free energy, which is the sum of the normal state and the superconducting state contributions. This free energy allows us to consider three types of singularities: (1) a discontinuity in the temperature derivatives of certain lattice parameters, (2) a discontinuity in the magnitudes of certain sound velocities, and (3) a discontinuity in the temperature derivatives of certain sound velocities. The superconducting part of the free energy contains an explicit coupling to the lattice that makes these evaluations possible. In what follows, we closely follow the analysis performed by Millis and Rabe. We summarize the main results only; for details we refer the reader to the original paper.

The normal state contribution to the free energy can be written as an expansion in the symmetry-invariant combinations of the lattice strains ε_{ij} of the equilibrium structure at T_c. In both the two well-known classes of materials, the overall crystal structure at T_c is orthorhombic.

Ignoring the smooth background temperature dependences of the expansion coefficients, the normal state free energy is given by

6. Ceramic High-Temperature Superconductors

$$F_N(\varepsilon_{ij}) = \tfrac{1}{2}\varepsilon_i C_{ij}\varepsilon_j + \tfrac{1}{2} c'_i \varepsilon_i'^2 + \Gamma_{ijk}\varepsilon_i\varepsilon_j\varepsilon_k + \Lambda_{ij}\varepsilon_i\varepsilon_j'^2 + \sigma_i\varepsilon_i + \ldots \quad (1)$$

Here ε_i and ε'_i represent the diagonal and off-diagonal elements of the strain tensor, C_{ij} and c'_i are the elastic moduli, Γ and Λ are the anharmonic elastic coefficients, and σ_i are the externally applied stresses. The various sound velocities are related to the various combinations of the elastic moduli; the sound velocity along i is given by v_i:

$$\rho v_i^2 = C_{ii} \quad (2)$$

The superconducting contribution to the free energy is given by

$$F_S = -\tfrac{1}{2} N(T - T_c)^2 [1 - A(1 - T/T_c) + \ldots]. \quad (3)$$

The coupling of superconductivity to the lattice is caused by the strain dependence of T_c and N, given by

$$T_c = T_{c0}[1 + \alpha_i e_i + \tfrac{1}{2}\varepsilon_i \Delta_{ij}\varepsilon_j + \tfrac{1}{2}\Delta'_i \varepsilon_i^2 + \ldots], \quad (4a)$$

$$N = N_0[1 + \beta_i \varepsilon_i + \ldots], \quad (4b)$$

with the following definitions of the various parameters:

$$\alpha_i = d\ln T_c/d\varepsilon_i; \qquad \Delta_{ij} = (1/T_{c0})\, d^2 T_c/d\varepsilon_i d\varepsilon_j; \quad (5)$$
$$\Delta'_i = (1/T_{c0})\, d^2 T_c/d\varepsilon_i'^2; \qquad \beta_i = d\ln N/d\varepsilon_i.$$

The superconducting state contribution to the specific heat is the temperature derivative of the free energy at constant strain given by

$$C_V = NT[1 + 3A(T - T_c)/T_c]. \quad (6)$$

The specific heat jump at T_c is then given by

$$\Delta C = C_V(T_c+) - C_V(T_c-) = N_0 T_{c0}. \quad (7)$$

For the discontinuity in the temperature derivative at T_c, we obtain

$$(dC/dT)(T = T_c+) - (dC/dT)(T = T_c-) = -3N_0 A. \quad (8)$$

In order to obtain the anomalies in the elastic constants and the lattice parameters, we expand F_S in powers of $(T - T_c)$ and ε. The total free energy is $F = F_N + F_S$, where

$$F_S = -\tfrac{1}{2} N_0 (T - T_{c0})^2 \{1 - A(1 - T/T_{c0})\} - \tfrac{1}{2} N_0 T_{c0}^2 [-2(T/T_{c0} - 1)$$
$$\times \alpha_i \varepsilon_i + \{\alpha_i \alpha_j + (T/T_{c0} - 1)(3A\alpha_i\alpha_j - 2\alpha_i\beta_j - \Delta_{ij})\varepsilon_i\varepsilon_j\} \quad (9)$$
$$- (T/T_{c0} - 1)\Delta'_j \varepsilon'_j].$$

The discontinuity in C_{ij} is given by

$$\Delta C_{ij}/C_{ij} = C_{ij}^{-1}[C_{ij}(T_c+) - C_{ij}(T_c-)] = (\Delta C\, T_{c0}/C_{ij})\,\alpha_i\alpha_j. \qquad (10)$$

Similarly, the discontinuity in the sound velocity is given by

$$\Delta v_i/v_i = -(\Delta C\, T_{c0}/2C_{ii})\alpha_i^2. \qquad (11)$$

Note that the dimensionless parameter $(\Delta C\, T_{c0}/C_{ij})$ is the ratio of the superconducting condensation energy to the elastic energy and is a small quantity $\sim 10^{-4}$. Thus, it is sufficient to obtain terms up to the leading nontrivial order of this parameter.

The equilibrium strains for $T < T_c$ are obtained by minimizing the free energy with respect to ε_i and ε_i'. The discontinuity in the logarithmic temperature derivative of the strain at T_c is given by

$$\Delta_1\varepsilon_i = T_{c0}\,[(d\varepsilon_i/dT)_{T_c+} - (d\varepsilon_i/dT_{T_c-}] = -\Delta C\, T_{c0}\, C_{ij}^{-1}\, \Lambda_{ij}. \qquad (12)$$

This relation, in conjunction with (11), yields the well-known Pippard–Buckingham–Fairbank equation for an isotropic system that has been used extensively in evaluating the thermodynamic properties near the superfluid transition in liquid He^4. This equation expresses the relationships among the discontinuities in the three second derivatives of the free energy, namely, the specific heat, the thermal expansion coefficient, and the compressibility. Note also that the change in the velocity is strictly opposite of that in the specific heat, independent of the sign of α_i since it appears in a quadratic form, as can be seen in (11).

Finally, one can also obtain the anomalies in the temperature derivatives of the elastic constants as well. We replace ε_i by $\varepsilon_i^0 + \delta\varepsilon_i(T)$ in the free energy and obtain the logarithmic temperature derivative of C_{ij} from the term of order $(T - T_{c0})(\varepsilon_i^0)^2$:

$$\Delta_1 C_{ij} = \Delta C\, T_{c0}\,[3A\alpha_i\alpha_j - 2\alpha_i\beta_j - \Lambda_{ij} + 6\Gamma_{ijk}C_{kl}^{-1}\alpha] \qquad (13)$$

and

$$\Delta_1 C_i' = -\Delta C\, T_{c0}\,[\Lambda_i' + 2C_{jk}^{-1}\,\alpha_k\,\Lambda_{ji}]. \qquad (14)$$

Based on this analysis, we obtain the qualitative features expected at the superconducting transition in both the longitudinal and transverse sound velocities, as well as in the specific heat. They are shown in Fig. 1. Note that the discontinuity in the longitudinal sound velocity is given by α_i, which is related to a first derivative of T_c with respect to the strain, i.e., to the linear strain dependence of T_c. The discontinuities in the slopes of the sound velocities are related also the sound derivatives of T_c, such as Λ_i', which relates to the quadratic strain dependence of T_c. Thus, it is impossible, in principle, to separate these depen-

6. Ceramic High-Temperature Superconductors

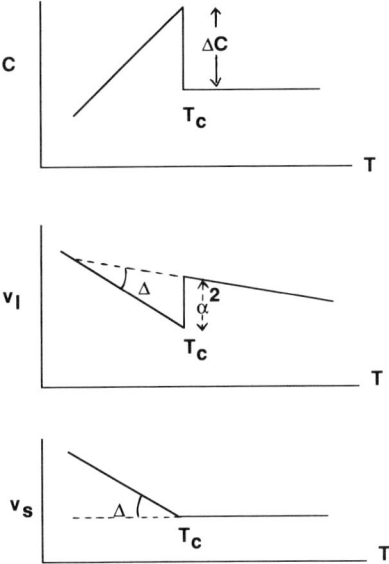

FIG. 1. Typical behavior of the specific heat and the sound velocities, longitudinal and shear, at the superconducting transition. The solid lines represent the mean field behavior; the discontinuity in longitudinal sound is determined by the first strain derivative of T_c, such as α, while the slope changes are controlled by the second derivatives, such as the Δ values.

dences by measuring the discontinuities in both the magnitude and the temperature derivatives. Note also that no information on the sign of α_is can be obtained from sound velocities, since it occurs in (11) in the quadratic form, as noted above; the velocity must soften at T_c. (Thus, an increase in the sound velocity at T is forbidden by this analysis, and reports to that effect [Migliori et al., 1988] should be considered in a context different from what is discussed here.) But the discontinuity in the temperature derivatives of the strain can yield such information, as can be seen in (12).

2.2. Fluctuation Effects

In the previous section, results were obtained from a mean-field free energy. It is nevertheless clear that the relationships between the thermal expansion coefficient, the sound velocity, and the specific heat are mandated by thermodynamics, independent of the precise form of the free energy. The fact that the sound velocity or the thermal expansion coefficient acquires a discontinuity is caused by the fact that the specific heat acquires this behavior. In other words, the nonanalyticity of all the other parameters is related to that of the specific heat.

As we discussed earlier, the superconducting transition in conventional systems is a good example of a mean-field transition due to the large value of the bare correlation length. But in the cuprate superconductors, the correlation length is considerably smaller, and therefore fluctuation effects may be more easily discernible. Note that from the symmetry of the order parameter, the superconducting transition falls into the universality class of the x–y model. In the three-dimensional x–y model, studied extensively and precisely in the case of the superfluid transition in helium, the specific heat shows the famous lambda anomaly. An analysis of the data yields a weak, actually logarithmic, divergence of the specific heat, given by

$$C_p \sim [A_\pm/\alpha] \, [t]^{-\alpha}, \tag{15}$$

where α approaches zero and t is the reduced temperature $[t = (T - T_c)/T_c]$. This would then imply that both the sound velocity and any expansion coefficient would have the following behavior:

$$\rho v_i^2(T) \sim c_{ii} - (A_\pm/\alpha)(2 - \alpha)(1 - \alpha) \, [t]^{-\alpha} \, \alpha_i^2, \tag{16a}$$

$$\varepsilon_i(T) \sim -(2 - \alpha)[A_\pm/\alpha] \, C_{ij}^{-1} \, [t]^{1-\alpha} \, \alpha_i, \tag{16b}$$

as can be obtained simply by the appropriate differentiations of the most singular part of the free energy functional of the form

$$f(T, \varepsilon) \sim [A_\pm/\alpha] \, [t]^{2-\alpha} + \ldots, \tag{17}$$

where the factor A is of dimension $N_0 T_c^2$. The temperature range over which this critical fluctuation will be observable can be roughly estimated by the Ginzburg criterion, which sets a reduced temperature scale t_G. True critical behavior is expected to be observable for $t \ll t_G$. In this case we get

$$t_G \sim [\gamma_d]^{2/(4-d)} \tag{18}$$

where d is the spatial dimension, with the coefficients in $d = 2$ and 3 are given by

$$\gamma_3 \sim 8\pi \, \Delta C \, \xi_0^3, \tag{19a}$$

$$\gamma_2 \sim 4\pi \, \Delta C \, \xi_0^2 \, a, \tag{19b}$$

where ΔC is the mean field jump in the specific heat and ξ_0 is the bare correlation length in the mean field theory. The latter equation is valid for a set of uncoupled two-dimensional layers with spacing a.

6. Ceramic High-Temperature Superconductors

It should be remembered, though, that all deviations from a true mean field behavior are not necessarily caused by critical fluctuations. In fact, the presence of Gaussian corrections (obtained by retaining terms up to quadratic in the order parameter in the Landau–Ginzburg free energy; for details, see Ma, 1976) to the mean field behavior is most likely to be observed in this class of materials as well. In this case one finds

$$C = \Delta C \, \gamma d \, [t]^{(d-4)/2} \quad (t > t_G), \tag{20a}$$

$$C = \Delta C \, \{1 + (\gamma d \, [t]^{(d-4)/2}/2^{(d-2)/2}\} \quad (-t > t_G). \tag{20b}$$

Again, as we have seen in the previous section, thermodynamic relations would imply similar anomalies in the sound velocity and thermal expansion coefficients as well. The sound velocity in the Gaussian approximation is given by

$$(\delta v_i/v_i) = [-\Delta C \, T_c \, \alpha_i^2/c_{ii}] \, \gamma_d \, [t]^{(d-4)/2} \quad (t < t_G), \tag{21a}$$

$$(\delta v_i/v_i) = [-\Delta C \, T_c \, \alpha_i^2/c_{ii}] \, \{1 + [\gamma_d \, [t]^{(d-4)/2}/2^{(d-2)/2}\} \quad (-t > t_G), \tag{21b}$$

in the two respective cases. Following the arguments presented by Millis and Rabe, we arrive at an estimate of t_G for these materials to be 10^{-4}, or about 0.01 K. But Gaussian fluctuations are observable to a reduced temperature of 10^{-1}, or about 10 K above T_c. At the time of this writing, available results suggest that the observed fluctuations are Gaussian in character. It should be noted, however, that because of questions of disorder, these effects are difficult to ascertain unambiguously, particularly in ceramic samples and often in so-called single crystals as well.

It should be noted that the thermodynamic relationships between the various second derivatives of the free energy are valid regardless of whether a mean-field transition is appropriate or fluctuation effects are present. Figure 1 shows schematically the behavior of the specific heat and the sound velocities, longitudinal and transverse, near T_c. In the mean-field case, one can separate the effects in the sound velocity from those in its temperature derivative. This cannot be done easily in the presence of fluctuations, since both the (longitudinal) sound velocity and its temperature derivative are proportional to the fluctuation contribution of the specific heat. Thus, both would show precursor effects and the temperature dependence of v_1, for example, will be a combination of both.

3. Sound Velocity in Ceramic Samples

Ceramic samples suitable for sound propagation studies or vibrating-reed measurements are usually prepared by sintering compacted powder. The samples

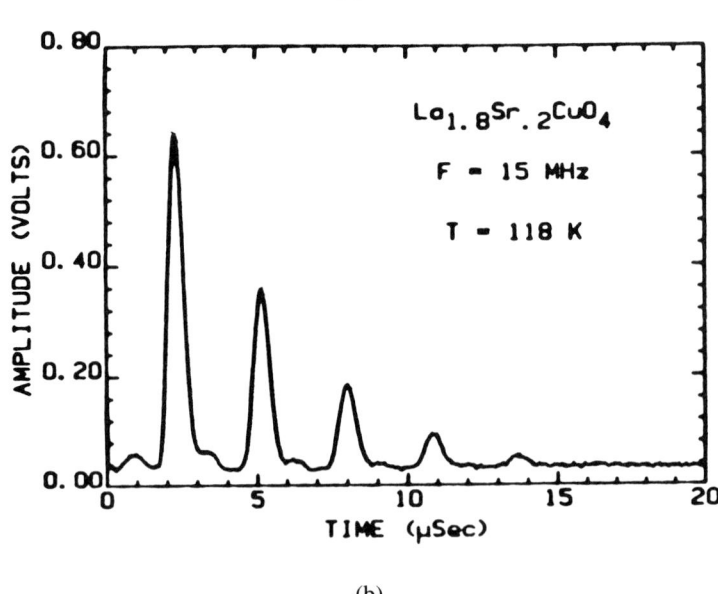

FIG. 2. (a) A photograph of a well-sintered ceramic pellet used by Bhattacharya et al. for sound propagation measurements. The surface has been polished to a mirror finish. (b) A typical echo train in the sample above.

usually need further oxygenation after sintering. In this class of samples, the data are extremely sensitive to the levels of porosity. For detailed discussions of some of these issues, see the article by Almond in this volume. It is not at all clear whether any useful data can be obtained on samples that are highly porous. Thus, a high-density sample that is extremely well sintered is essential. In Fig. 2a we show a picture of the sample on which measurements were made by Bhattacharya *et al.* (1988a, b). The high density (>95%) and excellent sintering allowed for a mirror finish to be obtained after grinding and polishing. This also allows good bonding of the transducer to the sample. A reasonably clean and exponentially decaying echo train can be observed, as is shown in Fig. 2b.

3.1. La–Sr–Cu–O

We start with the more widely studied member of the 2–1–4 class of materials. It was discovered by Vaknin et al. (1987) that the end member of this family, i.e., La_2CuO_4, is an antiferromagnetic insulator. The Néel temperature depends sensitively upon the oxygen deficiency. In addition, two-dimensional antiferromagnetic fluctuations are also observed in this system (Shirane *et al.*, 1987). Upon doping with divalent Sr in the place of the trivalent La, which is equivalent to doping the systems with holes, the magnetism depletes, and superconducting behavior with a transition around 30 K emerges. In addition to the antiferromagnetic transition, the system also exhibits a tetragonal-to-orthorhombic structural transition. Figure 3 shows the phase diagram of this system. The exact nature of the phase diagram remains somewhat uncertain, both with respect to how the T–O (tetragonal–orthorhombic) line merges with the T_c, line and also with respect to the locus of the T_c line. For the latter, it is not yet entirely clear whether T_c actually decreases upon increasing x or whether the superconducting fraction decreases (Fleming *et al.*, 1987) while T_c remains the same. We will see in what follows that depending on which scenario is correct, one arrives at drastically different ways of analyzing the data. In any event, T_c nearly peaks around $x = 0.15$ and depletes above $x = 0.25$, where the system is a metal but not a superconductor. The emergence of the superconducting state out of an antiferromagnetic insulator and disappearance into a metal that is not a superconductor remains the central mystery of the problem and awaits the development of the correct theoretical picture.

The temperature dependence of the Young's modulus and the Q^{-1} of a ceramic sample is shown in Fig. 4. The most remarkable feature of the data is the extremely large softening of the Young's modulus around 200 K, accompanied by a large enhancement of damping. The velocity and damping of longitudinal

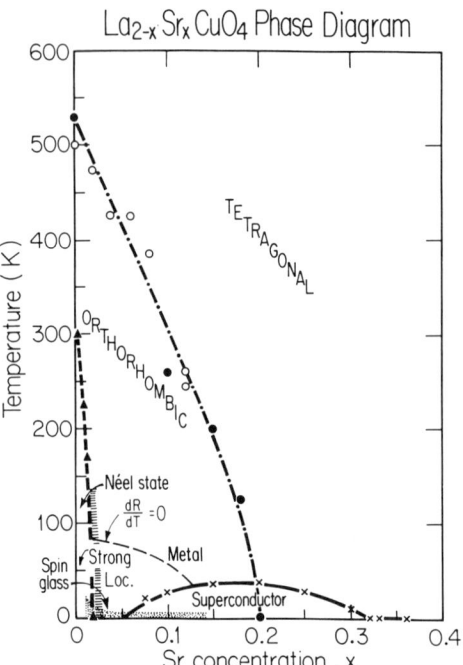

FIG. 3. Phase diagram of the 2-1-4 sample $La_{2-x}Sr_xCuO_{4-y}$. (From Birgeneau and Shirane (1988). (Reprinted by permission from "Physical Properties of High Temperature Superconductors," World Scientific Publishing)

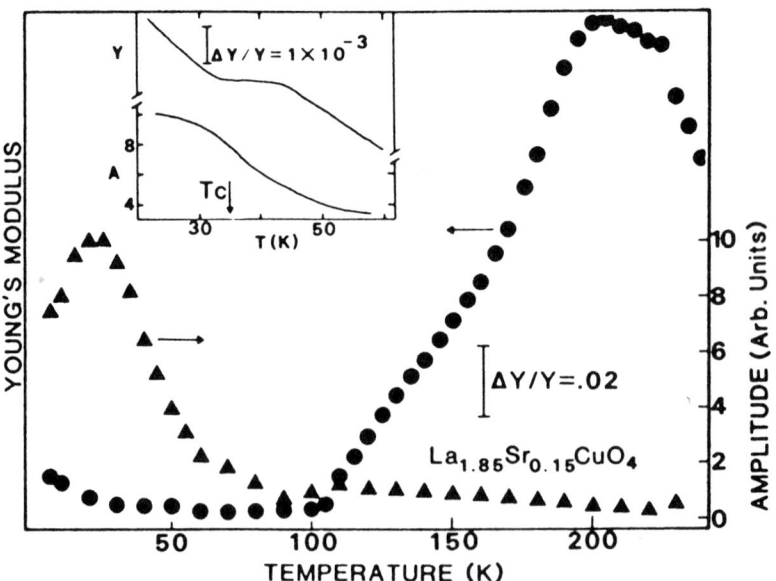

FIG. 4. Temperature dependence of the Young's modulus and the damping in the LaSrCuO system (Bourne et al.).

6. Ceramic High-Temperature Superconductors

ultrasound is shown in Fig. 5, which shows nearly the same effect. Figure 6 shows a comparison between the longitudinal and transverse sound velocity in a sample of the same family. Clearly, the effect is even more pronounced for the transverse case. Although this large softening was reported by several groups, the origin of this effect was not entirely clear in the early works. However, it is now established (Lee *et al.*, 1990) that this softening is caused by the aforementioned tetragonal-to-orthorhombic transition. They conjecture that the elastic modulus C_{66} is responsible for this softening. Recent single-crystal data (Migliori *et al.*, 1990) have confirmed this prediction.

Note that the results are somewhat unusual in the sense that the sound velocity does not recover below the transition, but remains low. Indeed, the measured sound velocity at liquid helium temperatures is smaller than that at room temperature. This has been discussed by Lee *et al.* (1990), who attribute this lower modulus to the motion of domain walls. The same mechanism would also yield much larger acoustic losses, compatible with the significantly greater damping observed in Fig. 4.

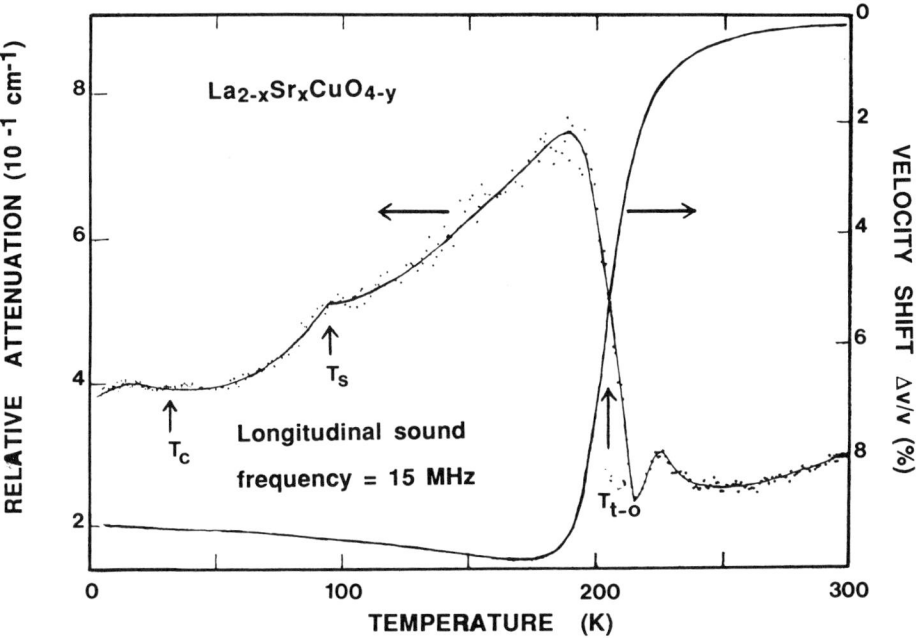

FIG. 5. Temperature dependence of the velocity and attenuation of longitudinal ultrasound (Bhattacharya *et al.*).

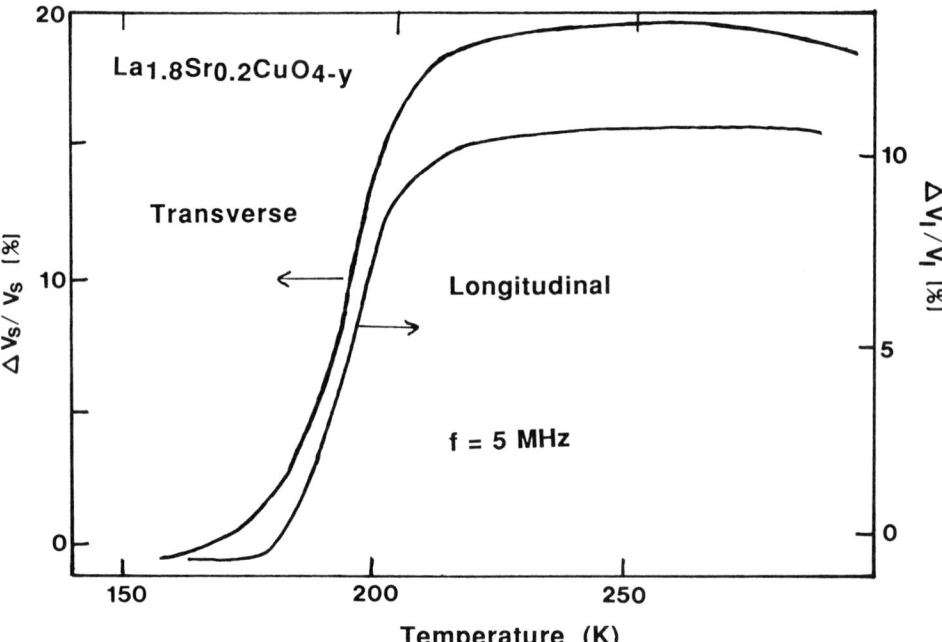

Fig. 6. Temperature dependence of longitudinal and transverse ultrasound velocities near the tetragonal-to-orthorhombic transition in LaSrCuO (Bhattacharya et al.).

A great deal of early work focused on this anomaly and conjectured that this soft phonon mode is responsible for the elevated superconducting transition temperature. That the explanation is not quite so was soon borne out by the absence of such a soft-mode behavior in the 1–2–3 material, which has a T_c about three times as large. A tetragonal-to-orthorhombic transition occurs in this material also, albeit at much higher temperatures (Tallon et al., 1988), which shows the conventional sound velocity anomaly to be described later. But an anomalous softening is not observed in this system.

Below T_0, the tetragonal-to-orthorhombic phase transition, the sound velocity increases with decreasing T, as is usual. However, two anomalies are observed at low temperatures, one at T_c and the other at 95 K. Following the nomenclature of Bhattacharya et al. (1988a, b), we refer to the latter as T_s. This anomaly was first observed in the damping by Horie et al. (1987). Both of these anomalies are characterized by sound velocity anomalies in the form of an apparent change of slope. These slope changes at T_c and at T_s are even more magnified for the case of the transverse sound, as shown in Fig. 7. Clearly, the anomaly at T_c is likely to be a result of the superconducting phase transition itself. However, much of the early studies of these materials assumed that a sound velocity

6. Ceramic High-Temperature Superconductors

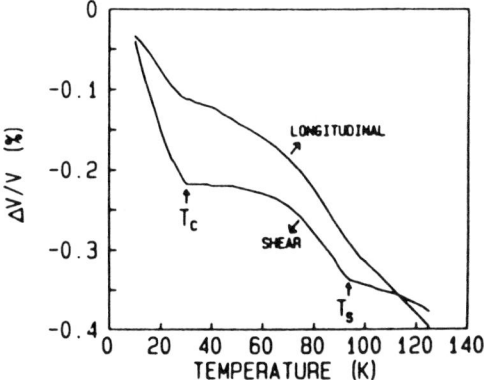

FIG. 7. Temperature dependence of the longitudinal and transverse sound velocity at low temperatures in LaSrCuO. The sound velocity anomalies at T_c and T_s are much more pronounced in the transverse case.

anomaly or a lattice strain anomaly is a direct result of a separate "structural anomaly," distinct from superconductivity. It is this misunderstanding of the consequences of superconductivity itself and the effects mandated by thermodynamics (described in the previous section) that led to many confusing analyses of the data. In any event, it is quite clear from the data shown here that this system has measurably large sound velocity anomalies at T_c and T_s, although the identity of the latter remains unknown. We will return to these issues later.

3.2. La–Ba–Cu–O

This other member of the 2–1–4 family is the original system reported by Bednorz and Muller in their first paper reporting the discovery of high-temperature superconductivity. This material has not received nearly the same amount of attention from the sound propagation community, although it appears that in some respects this original material may be the most interesting of all the cuprate superconductors.

Many of the early data on this material were obtained by Fossheim and Laegrid (for a review of their results, see Fossheim and Laegrid, 1989). Their data, analogous to Fig. 5, are shown in Fig. 8. It is obvious that the gross features of the data are essentially identical. That two different materials of the same family show similar data in two different laboratories is an unusual occurrence in this field and deserves special mention. As in the previous material, here, too, one sees a gigantic velocity softening accompanied by a large increase in damping; indeed, even the small features of the data associated with this T–O

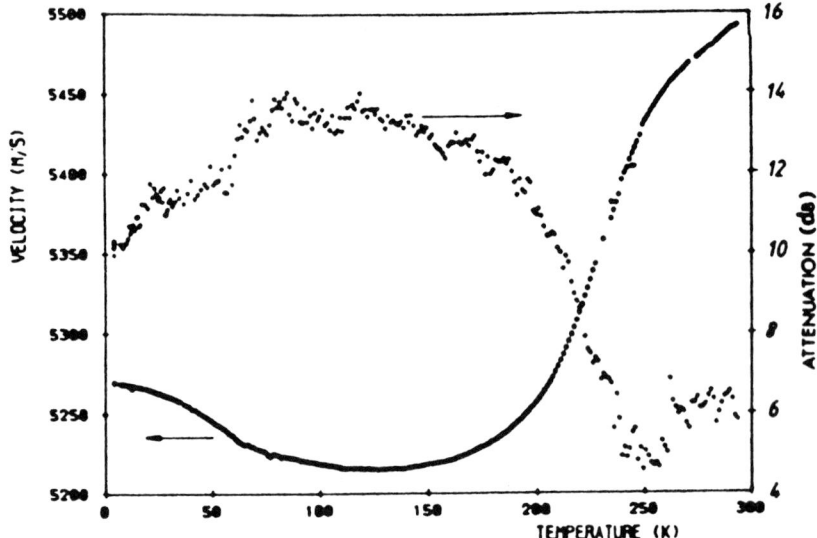

FIG. 8. Temperature dependence of the velocity and attenuation of longitudinal ultrasound in the LaBaCuO system (Fossheim et al.). (Copyright 1989 by International Business Machines Corporation; reprinted with permission)

transition are similar in these two figures. The tetragonal-to-orthorhombic transition is also well studied in this material. At this time, there is no reason to conclude that the anomaly is any different in this material than in the previous material.

Fossheim and his colleagues also noted that there is yet another anomaly below T_0 but above T_c. For simplicity we will call it T_s (which is T^* in the notation of these authors), though the reader is forewarned that there is yet no concrete evidence that they are the same in the two materials. Figure 9a shows the variation of velocity and damping in the vicinity of T_s. Unfortunately, in contrast to the previous sample, no data are available for the transverse sound. In any event, we find again a weak velocity anomaly at T_s accompanied by a rapid drop in damping, similar to what has been observed in LaSrCuO.

The behavior near T_c in this material is significantly different from that in the previous material. Early studies did not reveal any anomaly at T_c. Subsequently, the existence of a jump discontinuity of magnitude similar to the previous system was judged from the data. However, the sound velocity decreased with decreasing temperature below T_c, as shown in Fig. 9b, a behavior that has not been seen in any other system. The origin of this difference remains uncertain.

To summarize, both members of the 2–1–4 class show two distinct anomalies in the sound velocity and damping above T_c: a gigantic velocity softening

6. Ceramic High-Temperature Superconductors

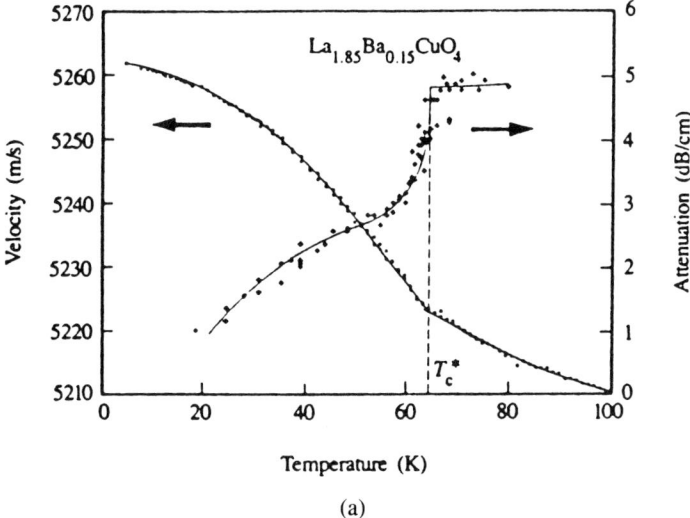

(a)

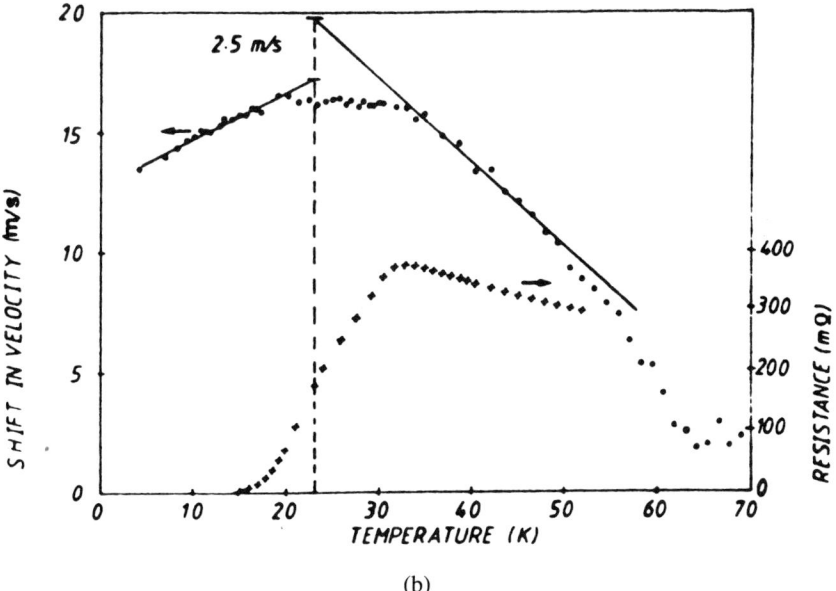

(b)

FIG. 9. (a) Temperature dependence at low temperatures of the velocity and damping of longitudinal ultrasound (Fossheim et al.) in the vicinity of a second anomaly above T_c. (b) Sound velocity anomaly at T_c in the LaBaCuO system (Fossheim et al., 1989). Note that the slope change in the velocity is of the sign opposite to that in the previous case, which is allowed by thermodynamics. (Copyright 1989 by International Business Machines Corporation; reprinted with permission)

dominated by a shear modulus that occurs at the tetragonal-to-orthorhombic transition; and a second, and much weaker, anomaly at T_s, which suggests the existence of yet another transition. But it is not at all clear whether they are the same in the two materials, or if there is any relationship between them and superconductivity.

A striking result has been reported recently by Axe et al. (1989) on this material. They observe that there is indeed a subtle structural transition at T_s that results in a reversal to a low-temperature tetragonal (LTT) structure that is very similar to the orthorhombic structure, except for a small buckling of the Cu–O planes with wave vector Q with equal components along (110) and $(\bar{1}\text{-}10)$. As the fraction of this phase grows, superconductivity depletes. The phase diagram showing the variation of T_c and the LTT fraction is shown in Fig. 10. Thus, it is clear that a small and subtle structural difference produces a very large effect on superconductivity.

A further, and somewhat disturbing, conclusion is that there is a possibility in the doped systems that the samples are multiphased. It is conceivable that the phase transition at T_s in these systems is related to a subsystem that has an intrinsically low T_c, or no superconductivity at all. In that case, T_s has no bearing on superconductivity except as a signature of the phase separation. This is particularly significant in view of much experimental work on aspects of magnetism in this class of materials. If indeed the system is multiphased, there remains a possibility that magnetism is operative in the nonsuperconducting phases present in the system and may not represent intrinsic magnetism associated with superconductivity. It is far from clear that the aforementioned scenario is correct, but it underscores the importance of these issues. It is nevertheless also clear that whatever the correct theoretical model may be, it has to explain this extreme dependence of superconductivity on some subtleties of the structure and immense robustness with respect to some other types of structural differences. The La–Ba (2–1–4) system appears to be ideally suited to a detailed inquiry aimed at answering these questions.

3.3. Y–Ba–Cu–O (1–2–3)

The canonical member of this class is the celebrated Y-compound, $YBa_2Cu_3O_{7-\delta}$, discovered by Wu et al. (1987). This material has been studied most extensively by almost any technique conceivable. Elastic and ultrasonic properties are not an exception. There is no indication yet that there is much difference between this system and any other member of this family with a different rare earth replacing Y.

This material is a stoichiometric compound, unlike the doped 2–1–4 system.

6. Ceramic High-Temperature Superconductors

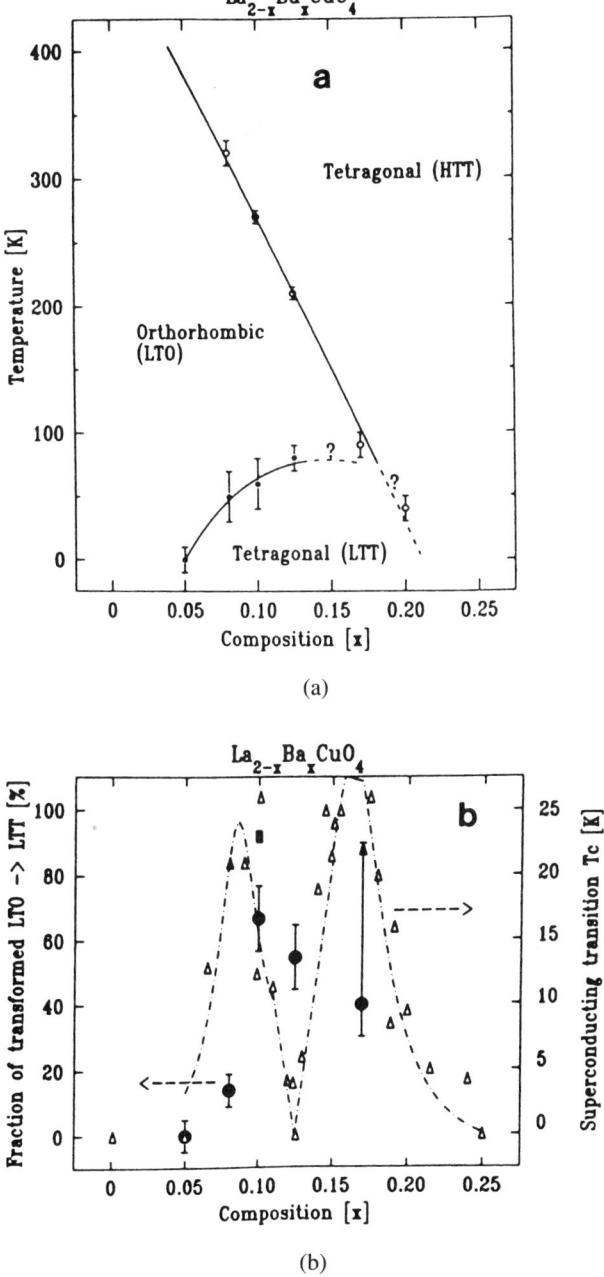

FIG. 10. (a) Phase diagram of LaBaCuO. (b) Variation of T_c and fraction of transformed LTT phase (Axe *et al.*).

The oxygen content can be varied here, again amounting to changing the level of doping, which controls the phase diagram. Here, too, the carriers are holes, as judged from the Hall conductivity. The phase diagram is shown in Fig. 11. As δ, the oxygen deficiency, increases, T_c changes slightly until it drops to a plateau of $T_c = 60$ K at a higher O-deficiency. Upon a further decrease in the O content, superconductivity completely disappears, and the system becomes an antiferromagnetic insulator at $O_{6.5}$, with all the features seen before in the lanthanum cuprate case. This shows the close resemblance between this system and the previous one in their proximity to an antiferromagnetic insulator. In contrast to the 2–1–4 class, however, one cannot obtain the highly doped metal (which is not a superconductor) in this family of compounds.

This material also undergoes a tetragonal-to-orthorhombic structural transition, but at much higher temperatures, ~900 K. Measurements by Tallon *et al.* (1988) are shown in Fig. 12. In contrast with the previous systems, however, the soft mode behavior is conventional. The Young's modulus softens at the transition, recovering immediately below. Similarly, the damping shows a pronounced peak at the transition, but at lower temperatures, it recovers to its high-temperature values. Why these two systems behave so differently remains unknown at this stage.

Most of the sound velocity measurements have been performed at lower temperatures, typically between room temperatures and liquid He temperatures. But in this particular temperature range, there is an enormous range of sound velocity and damping data reported so far. The variation is quite large, leading to sig-

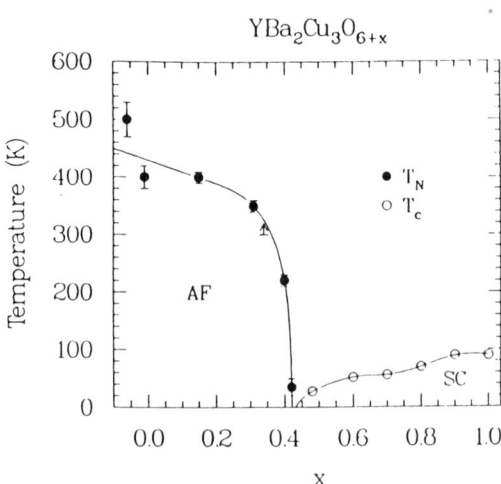

FIG. 11. Phase diagram of YBCO (Birgeneau and Shirane, 1988). (Reprinted by permission from "Physical Properties of High Temperature Superconductors," World Scientific Publishing)

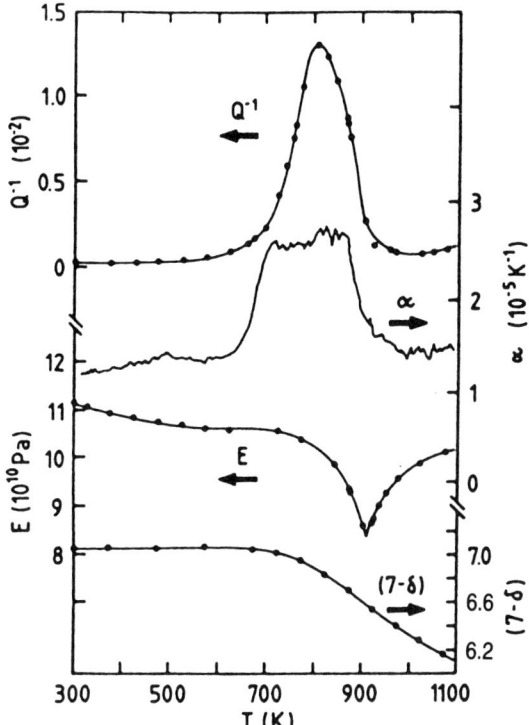

FIG. 12. Soft mode behavior near the tetragonal-to-orthorhombic transition in YBCO (Tallon et al., 1988).

nificant confusion about what is intrinsic to the system and what is spurious. Early measurements led to considerable divergence of opinion among various groups. In most cases, little attention was paid initially to the oxygen stoichiometry of the samples, which is now known to be extremely significant. In the absence of such a characterization of the samples, a comparison of data obtained by various groups is nearly impossible, even if the effects of porosity are carefully accounted for.

A second difficulty has been a consistent reporting of a strongly hysteretic behavior of sound velocity in some samples of these materials. Until recently, it has been reported only in ceramic samples and not in single crystals, leading to suspicions that the effect is related to intergranular changes in the vicinity of structural transitions. Subsequently, it has been reported in single crystals as well (for details on the single crystals, see the chapter by Golding in this volume). This behavior has been reported both in ordinary ceramic samples and in the so-called singer-forged samples. On the other hand, nonhysteretic data have also been reported by various other groups. In the latter cases, the grain sizes tend to be a few microns, considerably smaller than the previous set, although why this difference leads to such a large change in the data remains obscure.

In spite of the fact that many measurements have been made in this material, there are relatively few studies with the required precision and with sufficient characterization of the sample for the data to be useful or for a comparison of results from different laboratories to be meaningful.

The temperature dependences of the sound velocity measured by two different groups are shown in Fig. 13a and 13b, respectively, once again because they are the same, which is an exceptional situation. Clearly, a pronounced change of slope is observed near T_c in the sound velocity even in such an expanded scale. In many measurements of this class of materials, too, there are reports of other sound velocity anomalies and other phase transitions. There are persistent reports of a phase transition around 240 K. Many measurements of sound velocity and damping show anomalies at this temperature, although one does not observe the softening as in the 2–1–4 family. Structural measurements have reported first-order–like transitions with changes in lattice constants. Although there have been suggestions of oxygen vacancy ordering, there is no consensus on this issue yet. Some measurements of calorimetry, e.g., by Laegreid et al. (1988) have revealed the presence of a large specific-heat anomaly at this temperature. The authors also report that the fraction undergoing this transition is the superconducting fraction, as evidenced by the concomitant increase of the anomaly at both transitions. The same study also shows data, which are not discussed, that show the presence of a second specific-heat anomaly at a lower temperature, around 160 K. Indeed, many sound velocity measurements in both single crystals and in ceramic samples have reported such an anomaly. The first such report, by Bhattacharya et al. (1988a, b), showed an anomaly in both longitudinal and transverse velocities, shown in Fig. 14, once again in the form of slope changes in velocity. But the exact temperature varies substantially, ranging from 108 K in some single crystals (Saint-Paul et al., 1989) to about 160 K, depending possibly on the oxygen stoichiometry. Bhattacharya et al. (1988a, b) attributed this anomaly to T_s, in analogy with the 2–1–4 materials. In this case, too, there is no compelling evidence that the transition is of the same origin as in the 2–1–4 family of compounds. Once again, this transition was also found to be significantly affected by the oxygen stoichiometry, similar to the trends observed in the LaSrCuO system.

There is a special class of ceramic samples available in this family that deserves special mention. These are the so-called sinter-forged materials. They are prepared in such a way that the sintered material has the c-axis (i.e., the axis perpendicular to the Cu–O planes) of the individual crystallites nearly aligned.

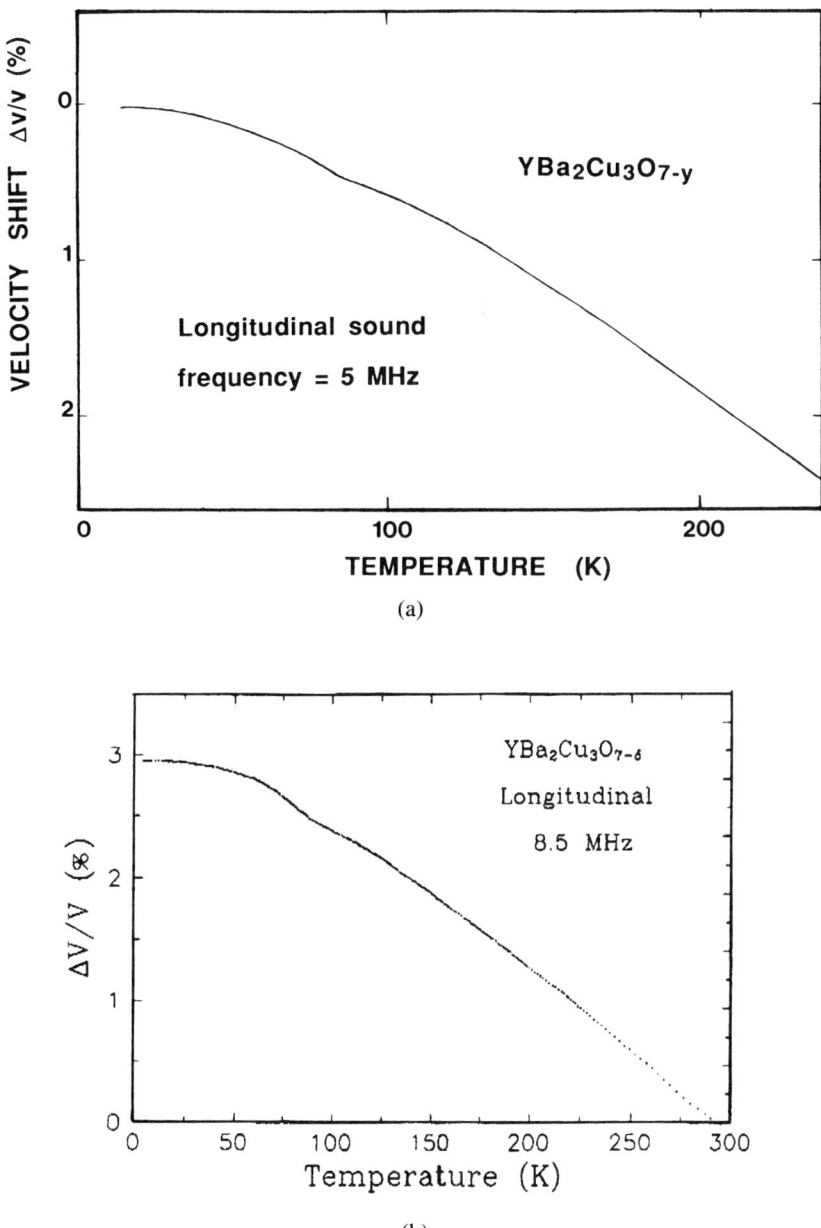

FIG. 13. (a) Temperature dependence of longitudinal ultrasound in YBCO (Bhattacharya et.al.). (b) Same as above (Xu et al.).

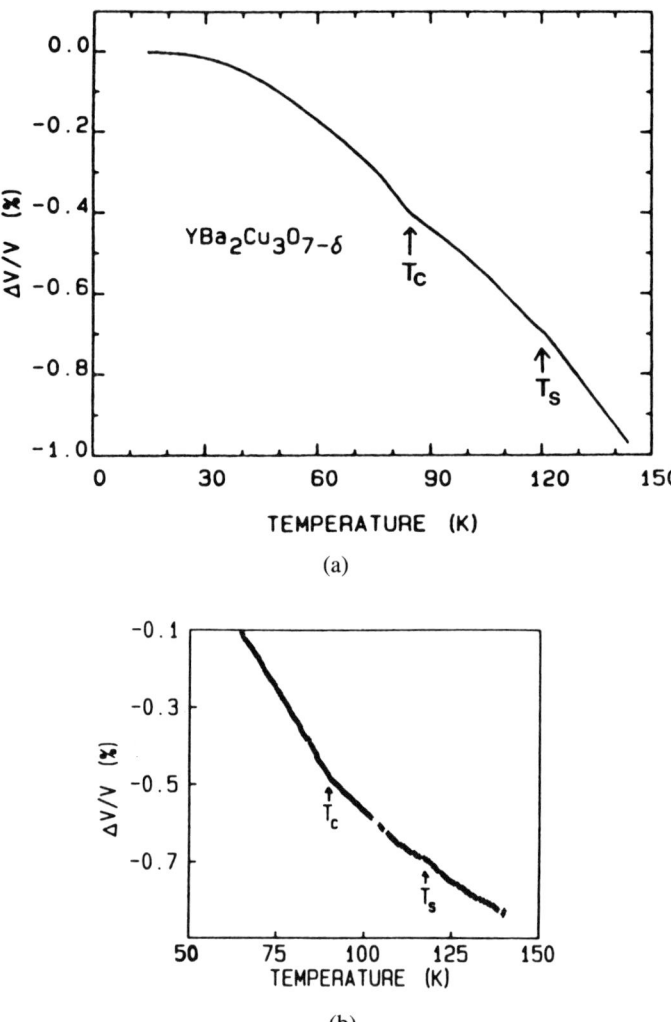

FIG. 14. Variation of (a) longitudinal and (b) transverse sound velocity in YBCO near T_c (Bhattacharya et al., 1988a,b).

As a result, the samples display uniaxial symmetry. These are the only known samples where the anisotropic sound velocity has been measured. Table I shows the room-temperature values of the sound velocities with different polarizations and propagation vectors. The longitudinal modulus is considerably softer along the c-axis than in the a–b plane. The shear modulus for layers shearing against each other is also measured. The sound velocity in these systems is found to be

6. Ceramic High-Temperature Superconductors

TABLE I

ANISOTROPIC SOUND VELOCITIES IN SINTER-FORGED CERAMIC SAMPLES OF YBaCuO
(FROM ZHAO ET AL. (1989)[a]

Polarization	Propagation		
	a	b	c
a	$(C_{11}/\rho)^{1/2}$ 4.4	$[(C_{11} - C_{12})/\rho]^{1/2}$	$(C_{44}/\rho)^{1/2}$ 2.66
b	$[(C_{11} - C_{12})/\rho]^{1/2}$	$(C_{11}/\rho)^{1/2}$ 4.4	$(C_{44}/\rho)^{1/2}$ 2.66
c	$(C_{44}/\rho)^{1/2}$ 2.66	$(C_{44}/\rho)^{1/2}$ 2.66	$(C_{33}/\rho)^{1/2}$ 3.77

[a] All velocities are in 10^5 cm/s.

strongly hysteretic, similar to what is found in other ceramics. The results are shown in Fig. 15.

To summarize, the detailed temperature dependences of the sound velocities show that there are indeed other anomalies above the superconducting transition temperatures. This is a nontrivial result in the following sense. Much of the

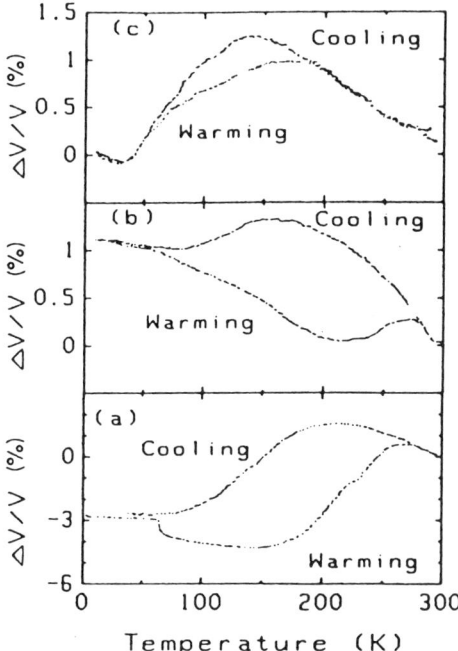

FIG. 15. Temperature dependence and hysteris of found velocity in sinter-forged YBCO (Zhao et al.).

theoretical activity centers around finding an adequate description of the normal state from which the superconducting transition occurs. It is generally viewed that the normal state describes the room-temperature state of the material, except when the tetradonal-to-orthorhombic phase transition occurs below room temperature. If these sound velocity anomalies are not spurious, then it is clear that there are other subtle rearrangements occurring between room temperature and T_c. Thus, it is imperative that these transitions be studied carefully. If real, these states represent the true normal state from which the condensation actually occurs.

At the time of this writing there are few data on the other two important classes of samples, i.e., the bismuth-based samples and the thallium-based samples. The latter compound has the highest reported value of T_c in this class of materials. Also, there are no data available on the so-called n-type superconductors, where the charge carriers are electrons rather than holes, as in the more common variety. Very recent measurements of the Bi-based samples show the presence of a similar anomaly around 220 K, again similar to what has been observed in the 1-2-3 family.

4. Results near T_c

Measurements on ceramic samples are basically of two types: (1) propagation of either longitudinal or transverse sound, and (2) measurements of the Young's modulus by the vibrating-reed method. A ceramic sample can be thought of as an isotropic object for which the velocities of longitudinal and transverse sound are given by

$$\rho v_L^2 = B + 4G/3, \tag{22a}$$

$$\rho v_S^2 = G, \tag{22b}$$

where B and G are some effective bulk and shear modulus of the effective isotropic medium. In the vibrating-reed method, the effective Young's modulus of an isotropic sample is given by

$$Y = 3\,GB/(B + G/3). \tag{23}$$

The relation between these moduli and the elastic constants of the individual crystallites constituting the ceramic material is not straightforward. Effective medium approximations assuming continuity of stress or strain across grain boundaries yield the bounds on their values. The Voigt notation, assuming constant strain in the crystallites, yields

$$B \leq B_V = \tfrac{1}{9} \{\Sigma\, c_{ii} + \Sigma\, c_{ij}\}, \tag{24a}$$

$$G \leq G_V = \tfrac{1}{15} \{\Sigma\, c_{ii} - \tfrac{1}{2} \Sigma\, c_{ij} + 3 \Sigma\, c_i'\}. \tag{24b}$$

6. Ceramic High-Temperature Superconductors

Using the Reuss approximation, i.e., constant stress in the crystallites, we obtain a lower bound for the elastic moduli:

$$B \geq B_R = [\Sigma c_{ii}^{-1} + \tfrac{1}{2} \Sigma c_{ij}^{-1}]^{-1}, \qquad (25a)$$

$$G \geq G_R = 15[4 \Sigma c_{ii}^{-1} - 2 \Sigma c_{ij}^{-1} + 3 \Sigma c_i'^{-1}]. \qquad (25b)$$

These bounds span a range of values substantially larger than the magnitudes of the anomalies in question. Furthermore, this uncertainty depends on the degree of anisotropy of the underlying crystallites specific to the relevant crystal symmetry. If, however, the anisotropies are not too large, it might be possible to estimate the changes of the intrinsic elastic constants from the measured changes in the effective moduli of the isotropic medium. For the convenience of the estimates, we will use the Reuss approximation, which yields the following relations at T_c:

$$\Delta B_R = -\Delta C\, T_c (B_R\, d \ln T_c/dP)^2, \qquad (26a)$$

$$\Delta G_R = \tfrac{2}{15} \Delta C\, T_c (G_R/B_R)^2\, [3\, \Sigma\{B_R\, d \ln T_c/d\sigma_i\}^2 \qquad (26b)$$

$$- \{B_R\, d \ln T_c/dP\}^2].$$

The important point to note is that although the intrinsic shear moduli do not have any discontinuity, the effective shear modulus does. By symmetry, the discontinuity in G_R vanishes if all the strain derivatives of T_c are equal. In the Reuss approximation, the discontinuities in the longitudinal and transverse sound velocities are

$$[\Delta v_L/v_L] = \tfrac{1}{2}[\Delta C\, T_c/B][1 + 4G/3B]^{-1} \{[B\, d \ln T_c/dP]^2 + 8G^2/45B^2 \qquad (27a)$$

$$\times [3\, \Sigma\, (B \ln T_c/d\sigma_i)^2 - (B \ln T_c/dP)^2]\},$$

$$[\Delta v_S/v_S] = \tfrac{1}{15} (\Delta C\, T_c/B)(G/B)\{3\, \Sigma\, (B \ln T_c/d\sigma_i)^2 - (B \ln T_c/dP)^2\}. \qquad (27b)$$

In addition to these discontinuities, there are also discontinuities in the temperature derivatives of the sound velocities. The logarithmic temperature derivatives of the elastic moduli B and G at T_c can be obtained by using the relations

$$\Delta_1 \ln B = 2[(1 + 4G/3B)\, \Delta_1 \ln v_L - \tfrac{4}{3}(G/B)\, \Delta_1 \ln v_S], \qquad (28a)$$

$$\Delta_1 \ln G = 2\Delta_1 \ln v_S. \qquad (28b)$$

It is possible in principle to obtain measures of these quantities from the data, as was shown by Bhattacharya et al. (1988a).

Before we embark upon an analysis of the experimental results, it is important to note the special situation that arises in these systems. The sound velocity in these materials is strongly temperature-dependent at temperatures near T_c, i.e., they have a strongly varying background that is presumably unrelated to superconductivity. In conventional superconductors the situation is much simpler for two reasons. First, as we have discussed in Section 2, the contributions of the electrons and phonons are rather small at the low temperatures. Therefore, the background does not pose a serious problem. Second, magnetic fields in excess of H_{c2} are accessible in the laboratory; thus, the system can be driven to its normal state and the background behavior easily ascertained (Alers and Waldorf, 1962).

In the cuprate systems, neither is possible. Studying the closely related antiferromagnetic insulators such as $LaCuO_4$ for the 2–1–4 systems or $YBa_2Cu_3O_6$ for the 1–2–3 family is of uncertain value in determining the background behavior, since the insulator has entirely different electronic properties. Because of the extremely large values of the upper critical fields (>MG), one cannot drive the system normal at temperatures more than a fraction below T_c. The uncertainties of the background behavior has led to entirely different views of the temperature dependence of the sound velocity below T_c.

We first discuss the 2–1–4 family of compounds. The only measurement to date to have observed a definite drop at T_c is by Bishop *et al.* (1987a). Their data are shown in Fig. 16. Although this result has not been reproduced by any other group, it is significant in its observation of the thermodynamic anomaly. A more typical result is shown in the inset of Fig. 4, due to Bourne *et al.* (1987), who report the behavior of the Young's modulus near T_c obtained by a vibrating-reed method. As the temperature is lowered towards T_c, the modulus flattens out, and below T_c it grows very rapidly. In this case the thermodynamic discontinuity is imagined to be rounded (presumably due to inhomogeneities), superimposed on a T-dependent background. Thus, the rounded thermodynamic discontinuity is given by the difference at T_c between the two extrapolated lines from above and below T_c. Note that this rounded discontinuity is of the order of a few hundred parts per million. This is nearly two orders of magnitude larger than what is typically observed in conventional superconductors.

A quantitative analysis of the data is complicated by the fact that crucial parameters such as the Meissner fraction are rarely reported in the sound propagation studies. In the absence of such information, in addition to other parameters such as the oxygen stoichiometry, comparison of bulk anomalies such as the specific heat and sound velocities is nearly impossible except at the order-of-magnitude level. The most comprehensive sound velocity measurement in this system has been performed by Bhattacharya *et al.* (1988a,b) and Higgins *et al.* (1989). We discuss their results in some detail. They measured the longitudinal and the transverse sound velocity near the transition shown in Figs. 5 and 6. From these results they also obtained the temperature dependence of the

6. Ceramic High-Temperature Superconductors

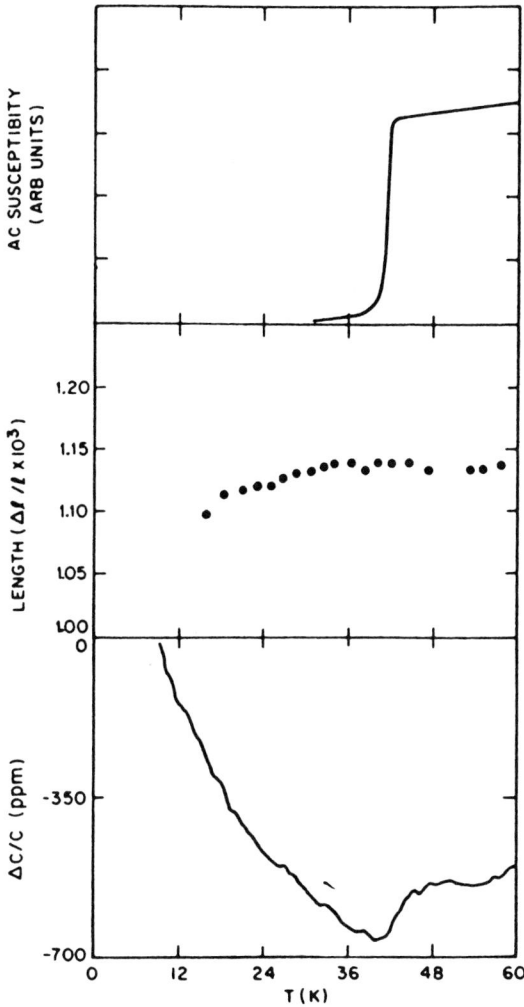

FIG. 16. Softening of sound velocity in LaSrCuO (Bishop et al., 1987).

effective B shown in Fig. 17a. The inset shows the region close to T_c, which clearly shows a smeared jump discontinuity. For a comparison, the specific heat data are shown in Fig. 17b. The striking similarities between the two sets of data would seem to eliminate any origin of the sound velocity anomaly as anything other than the superconducting transition, so long as the specific heat anomaly is thought to have the same origin. For a more detailed comparison, we consider the quantity $1/T(\Delta B/B)$ shown in Fig. 17c obtained from the data of Higgins et al. (1989).

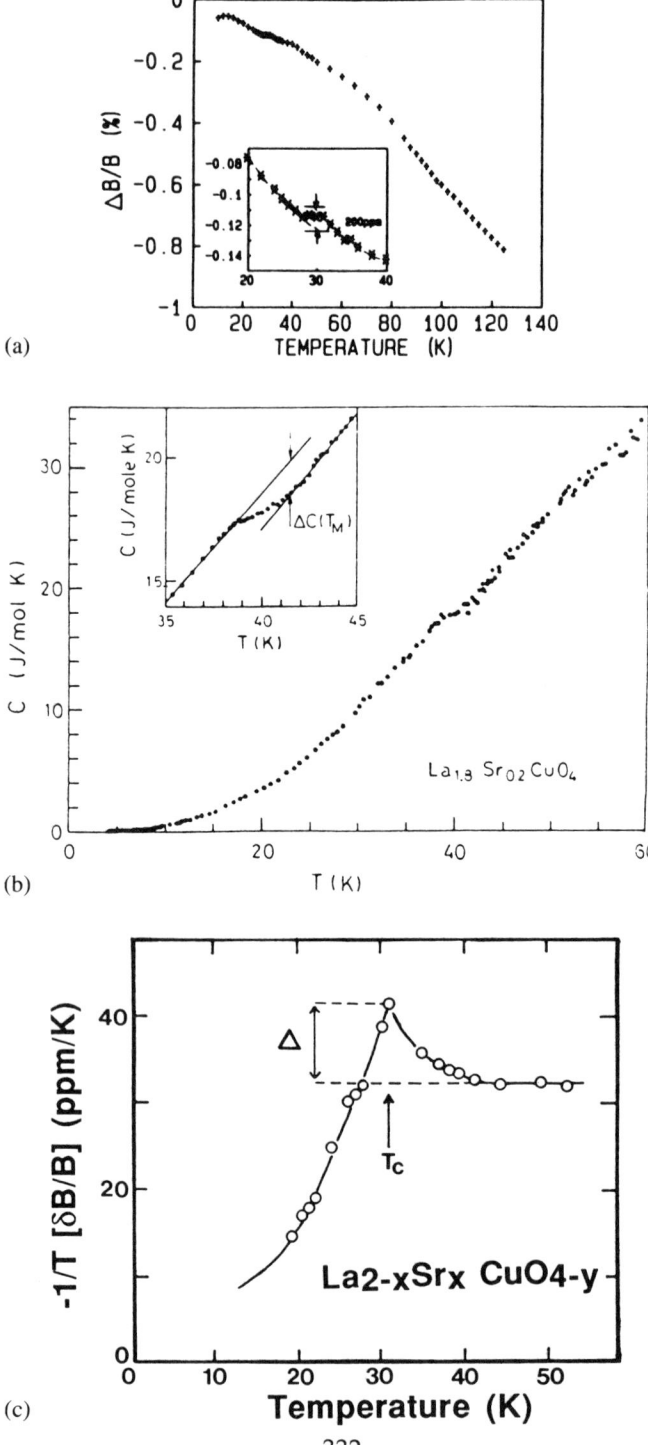

6. Ceramic High-Temperature Superconductors

While it is conceivable that there are significant fluctuation effects, it is premature to analyze the data in these terms, since disorder can mimic fluctuation effects. Instead we perform an effective mean-field analysis. This is justified by the fact that the thermodynamic relationships do not depend on the specific form. Therefore, one can extract an effective mean-field jump discontinuity for all the relevant parameters and test their self-consistency. Table II summarizes the various parameters measured independently. The thermodynamic analysis, Eq. (26a), then yields a value of $(\Delta B/B) \sim 300$ ppm, which compares well with the measured value of ~ 350 ppm, in view of the uncertainties involved.

Returning to the data on the temperature derivatives, it is obvious that the change in the derivative in the longitudinal sound is due primarily to the change in the derivative in G, as seen in the shear wave velocity. There is a small change in the derivatives of the effective bulk modulus. The data yield the following values: $\Delta_1 \ln v_L = 9.4 \times 10^{-4}$ and $\Delta_1 \ln v_S = 2.1 \times 10^{-3}$. Using these values one can obtain estimates of the following quantities:

$$\Delta_1 \ln B = 0.6 \times 10^{-3} \quad \text{and} \quad \Delta_1 \ln G = 4.2 \times 10^{-3}.$$

This analysis (Millis and Rabe, 1988) shows that the large change of slope in the sound velocities is indeed caused by the slope change of G, as first proposed by Bhattacharya et al. (1988a).

We now return to the $YBa_2Cu_3O_{7-\delta}$ system. In this case, the behavior of the sound velocity below T_c has been variously described as (1) a stiffening of the velocity presumably above the normal-state background; (2) a recovery of the velocity toward the normal-state behavior, it having softened considerably above

TABLE II

PARAMETERS DETERMINING THE MAGNITUDE OF THE EFFECT OF SUPERCONDUCTIVITY ON THE ELASTIC PROPERTIES OF THE HIGH-T_c SUPERCONDUCTORS[a]

	T_c (K)	v_1 (10^5 cm/s)	v_s	ΔC (mJ/cm^3 K)	$B(d \ln T_c/dP)$
$La_{1.8}Sr_{.2}CuO_{4-y}$	35	5.2	3.2	11	14
$YBa_2CU_3O_7$	90	4.2	2.6	50	1.3

[a]The sound velocities (at 300K) are obtained from Bhattacharya et al. (1988), and the rest from Millis and Rabe (1988).

FIG. 17. (a) Temperature dependence of the effective bulk modulus near T_c (Bhattacharya et al.). (b) Temperature dependence of specific heat (Nieva et al., 1987). (c) T-dependence of $\{-1/T[dB/B]\}$ in LaSrCuO (from Higgins et al., 1989).

T_c; and (3) the onset of a relaxation process, unrelated to superconductivity, but fortuitously coincident with it. At this time there is no resolution between the first two possibilities because we cannot obtain the normal state significantly below T_c. It should, however, be pointed out that in all conventional superconductors studied to date, the normal-state moduli are larger than the superconducting-state values. Experimental results of Higgins et al. (1989) also point in that direction. But the absence of any field dependence far above T_c suggests that there may be little precursor softening as well.

The third possibility is much more serious and deserves careful attention. Note that the elastic modulus of a metal is given by

$$M(T) = M_0 - A T^2 - B T^4, \qquad (29)$$

where the quadratic contribution comes from the electrons and the quartic term from the phonons. Obviously, the background variation is much more strongly T-dependent for the 1–2–3 material than for the 2–1–4 material. This is readily observed in the specific heat, for instance. In order to suppress the background contribution, the quantity $\{-1/T(\Delta v_L/v_L)\}$ versus temperature is shown in Fig. 18a; the data are taken from Bhattacharya et al. (1988b). It shows a striking similarity with the behavior of the specific heat, shown in Fig. 18b from the data of Laegreid et al., (1988). Moreover, the effective jump discontinuity ($dv_L/v_L \sim -50$ ppm) is in excellent agreement with the estimates obtained from Table II, once again making it highly improbable that anything other than superconductivity is responsible for the sound velocity anomaly.

To conclude the discussion of the thermodynamic anomaly, we consider the effects on the transverse sound velocity. Figure 19a shows the temperature dependence of the shear wave velocity near T_c in the same system, taken from the data in Bhattacharya et al. (1988a), shown in Fig. 13 over a larger temperature range. The slope discontinuity at T_c is obvious. This is symptomatic of transitions and not of relaxation effects. Figure 19b shows the same quantity plotted near T_c in the superconductor BaKBiO system (Xu, 1990). This material is a copper-free superconductor with no antiferromagnetism in close proximity to superconductivity. Recent theoretical work suggests that it is a conventional superconductor with phonon-mediated pairing. That one sees the same behavior at T_c in this system as well shows that a cause unrelated to superconductivity is unlikely.

From the experimental data one can also evaluate the jump discontinuities in the temperature derivatives of the sound velocities. The results yield $\Delta_1 \ln v_L = 5.8 \times 10^{-3}$ (Bhattacharya et al., 1988a) and 4.2×10^{-3} (Bishop et al., 1987b) and $\Delta_1 \ln v_S = 5.9 \times 10^{-3}$ (Bhattacharya et al., 1988a). Thus, in contrast to the 2–1–4 system, the discontinuities in B and G are comparable:

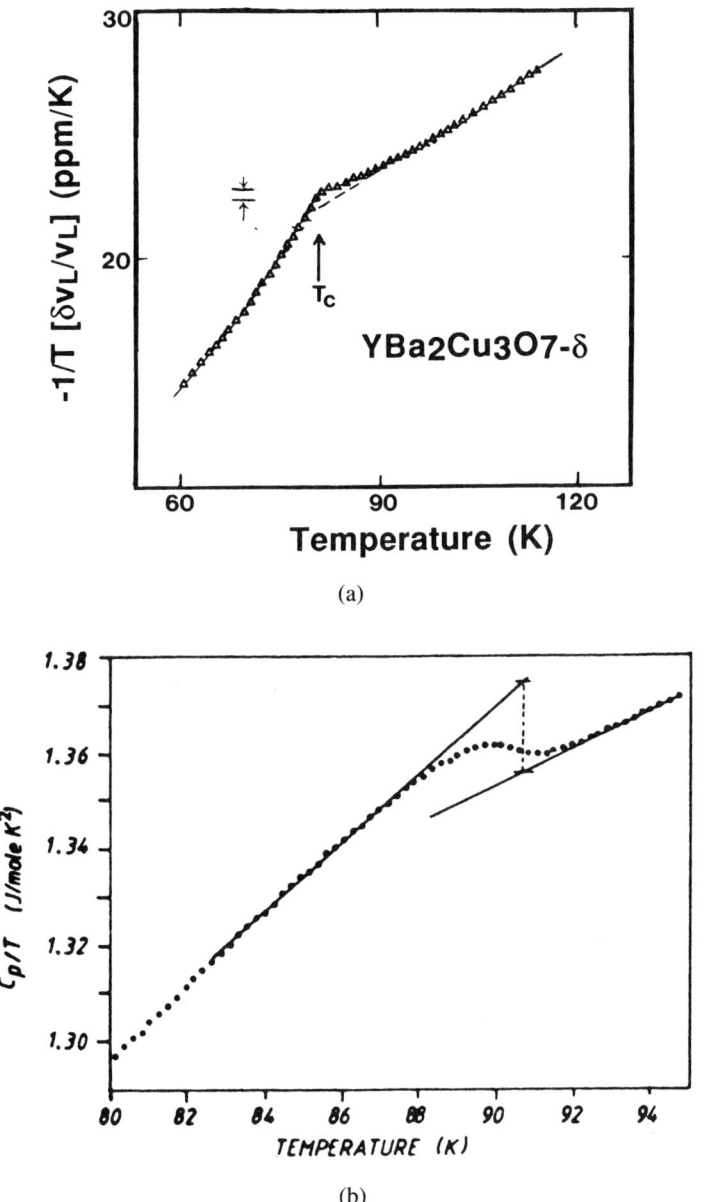

FIG. 18. (a) T-dependence of $\{-1/T[\Delta v_L/v_L]\}$ in YBCO (from Bhattacharya et al., 1988b). (b) T-dependence of specific heat in ceramic YBCO (Laegreid et al., 1988).

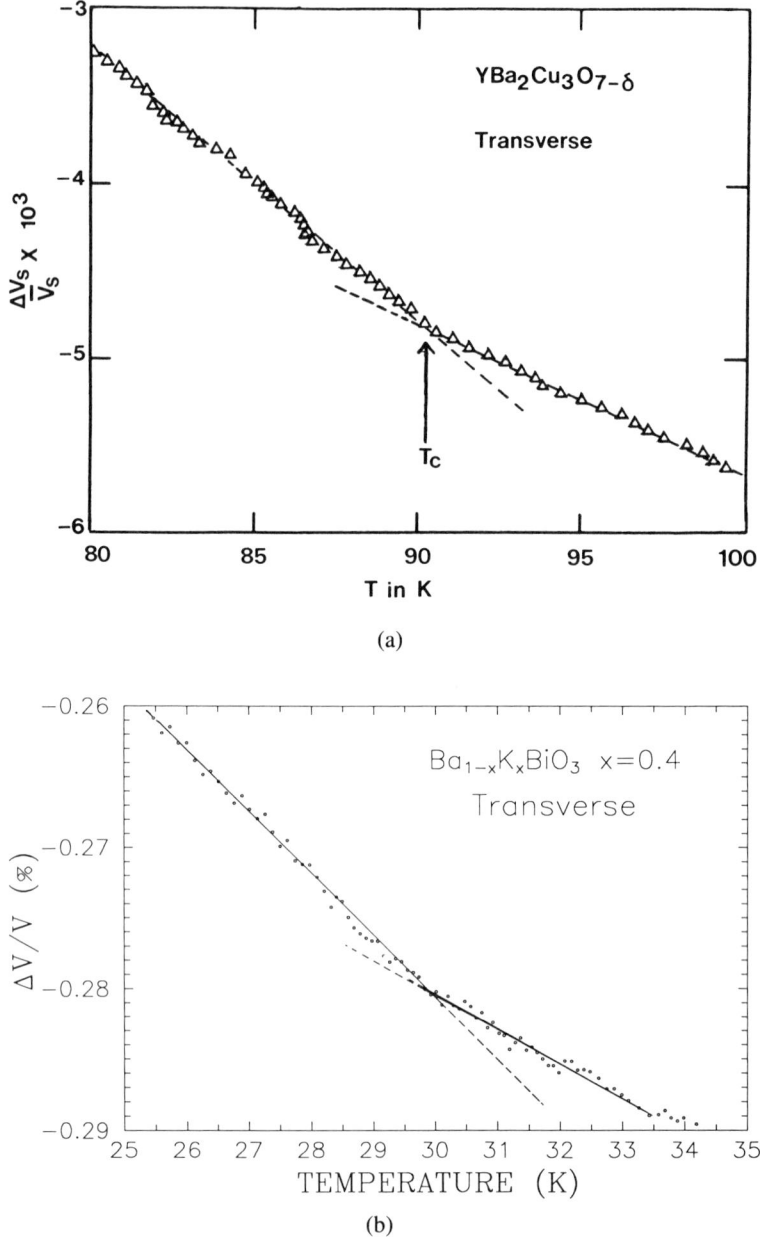

FIG. 19. (a) T-dependence of transverse sound velocity in YBCO near T_c (Bhattacharya et al., 1988a). (b) T-dependence of transverse sound velocity in BaKBiO near T_c (Xu et al., 1990).

$$\Delta_1 \ln B = 1.1 \times 10^{-2}$$

$$\Delta_1 \ln G = 1.2 \times 10^{-2}$$

Based on the mean field theory and effective medium estimates, Millis and Rabe have analyzed these results, for which we refer the reader to the original paper. Here we review their final conclusions. They observe that for reasonable estimates of the various parameters such as $d \ln T_c/dP$, d^2T_c/dP^2, $d \ln N_0/dP$, etc., the change in slope of the effective shear modulus is extremely large. Although the discrepancy can be a result of a small error in the estimate of the first of the aforementioned parameters (which enters into the estimates quadratically), there is another, more interesting possibility. They suggest that the large value derives from the second derivatives $B^2/T_c(d^2T_c/d\sigma_i^2)$ or $B^2/T_c(d^2T_c/d\sigma_i d\sigma_j)$, which have to be $\sim 5 \times 10^3$ and 5×10^2 for the 2–1–4 and 1–2–3 systems, respectively. These numbers are orders of magnitude larger than what is observed in conventional superconductors. For example, the dimensionless slope discontinuity, i.e., $\Delta_1 \ln v_s$, is $\sim 6 \times 10^{-4}$ in the cuprate superconductor in Fig. 19a, while in the copper-free superconductor it is an order of magnitude smaller (Fig. 19b).

5. Magnetic Field Dependence

One of the advantages of a conventional superconductor is, paradoxically, the low value of T_c and the reasonably accessible value of H_{c2}, which helps in evaluating the normal-state behavior. At the typical temperature of about 10 K, the effects of phonons are substantially frozen out and the carrier contributions dominate. In the cuprate superconductors, the high value of T_c implies that the phonon contribution to many parameters is extremely large. In principle, these effects can be subtracted if one can drive the system normal by applying a magnetic field, since the phonon bath is identical in these two cases. In this respect the cuprates present the second hurdle. These are extreme type-II superconductors. The small value of the coherence length implies a large value of the upper critical field, H_{c2}, the extrapolated value exceeding megagauss range, greater than any other type type-II superconductor known and thus beyond the range of any laboratory superconducting magnet. This means that any available magnetic field cannot keep the system normal down to temperatures lower than about a tenth below T_c.

Nevertheless, the measurements in the presence of a magnetic field are expected to produce at least one important conclusion. As mentioned earlier, whether the observed sound velocity anomalies are related to the superconducting transition or merely to some other spurious effect or to an unrelated relaxation phenomenon can indeed be settled by these measurements. Unfortunately, there are only a few of these measurements. The first measurement was made by Bourne et al (1987), who measured the effect of a magnetic field on the elastic constant by the vibrating-reed method. Although the experimental results are not easily interpretable because of eddy current effects, one can nevertheless observe the qualitative results on the 2–1–4 system. The results are shown in Fig. 20. In the absence of a normalization, the relationship between them remains unclear. But that the behavior near the transition is affected by the field is obvious.

By far the most detailed study of the effects of a magnetic field has been performed by Higgins *et al.* (1989) in the 2–1–4 sample. They followed their earlier work by measuring both the longitudinal and the transverse sound velocity. These measurements were made on a sintered ceramic pellet of high density of $La_{1.8}Sr_{0.2}CuO_{4-y}$. The sample had a T_c of 32 K from zero-resistance and zero thermopower and an onset temperature of 35 K from Meissner data, with a Meissner fraction of 45%.

Figures 21a, 21b, and 21c show the results of the measurements of the sound velocities for zero field and at a high field (6.5 T). Two effects near T_c are immediately obvious. First, the magnetic field suppresses the anomaly (i.e., the velocity at the plateau is higher than in the zero-field case) and smears the transition as well. Second, the transition is shifted to lower temperature by an amount $\delta T = 4$ K, yielding a value for $dH_{c2}/dT = 1.6$ T/K, which is in excellent agreement with independently measured values. Furthermore, the correlation

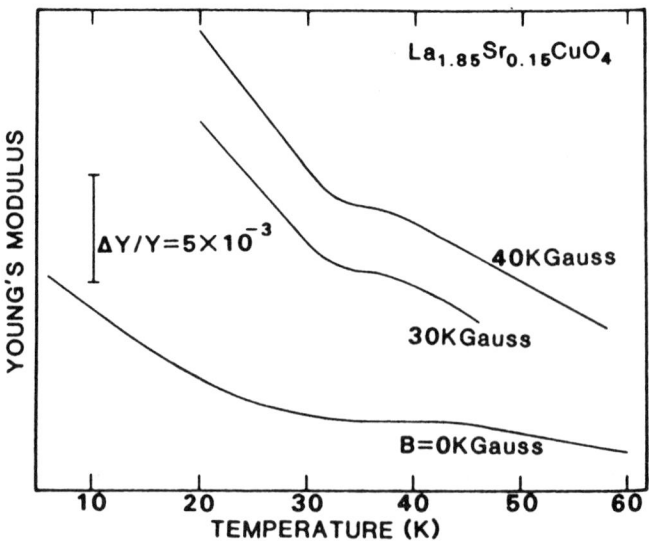

FIG. 20. Magnetic field dependence of the Young's modulus in LaSrCuO (Bourne *et al.*, 1987).

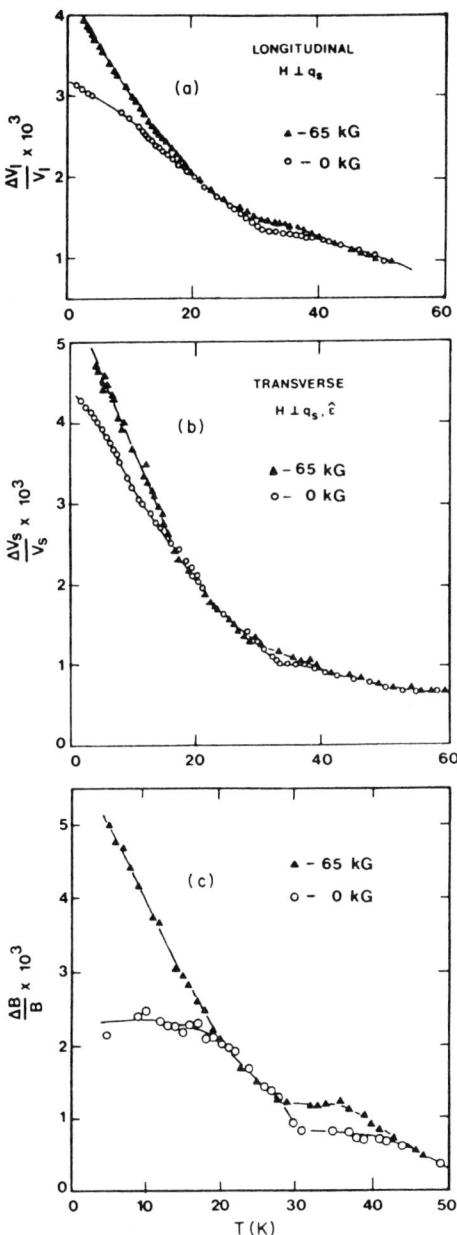

FIG. 21. T-dependence of (a) longitudinal, (b) transverse, and (c) bulk modulus in zero field and at $H = 6.5$ Tesla in LaSrCuO (Higgins et al., 1989).

with specific heat suggests that the specific-heat data would be quite different too. At the time of this writing there are no magnetic-field–dependent specific heat measurements in this system. But the trend is in excellent agreement with the data on the 1–2–3 system (Salamon, *et al.*, 1988).

From these measurements one can also obtain information about the mixed state of the superconductor as well. In the cuprate superconductors in particular, a great deal of work has gone into the study of the Abrikosov vortex lattice. The work of Higgins et al (1989) also addressed these issues. They note the surprising result that the velocity with and without a field is the same over a large temperature range below T_c, below which they become different. They attribute this behavior to pinning of the vortex lattice. Figure 22 shows the field dependence of the transverse wave velocity at two temperatures. At the lower temperature, the velocity increases linearly with field as it does in conventional type II superconductors (Alers and Waldorf, 1962). But above a certain field, H_{cr}, the increase changes to a nearly square-root dependence. At a higher temperature, the square-root behavior is observable up to a smaller field. Above another field, called H_{irr}, the irreversibility field, the velocity starts to decrease. They attribute this to a depinning of the lattice. They also find that the sound velocity itself becomes hysteretic below H_{irr}, which would appear to confirm that the effects are indeed due to pinning of the vortex lattice. H_{cr} is called a crossover field. These authors attribute it to a crossover between two different types of pinning, strong and weak, either as a competition between the rigidity of the lattice and the pinning potential, or between two types of pinning centers.

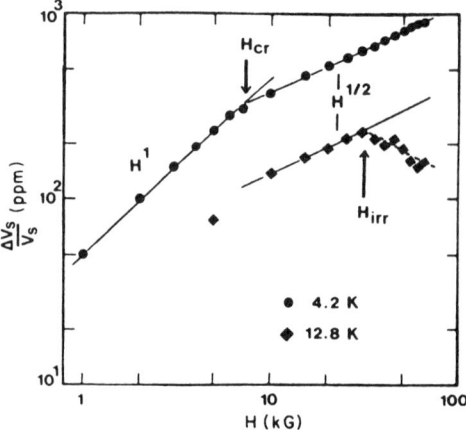

FIG. 22. Field dependence of transverse sound velocity at different temperatures (Higgins, *et al.*, 1989).

6. Ceramic High-Temperature Superconductors

Using these measurements as well as measurements of the reversible and irreversible magnetization, these authors obtain a pinning phase diagram both classes of superconductors, shown in Fig. 23. For the 1–2–3 system, only the magnetization measurements were performed. A few features of this result deserve comment. First, the so-called irreversibility field has been measured in great detail in both single crystals and ceramics in the 1–2–3 family. At these fields the average spacing between the flux lines is smaller than the typical grain size, and it is not surprising that the ceramics yield the same results as the crystals for the field perpendicular to the Cu–O planes. It is also interesting to note that while the irreversibility field is strongly temperature-dependent, the crossover field is not. This also points to a geometric matching effect for the latter case. That the crystals do not show any signature of this field also suggests that the effect is unique to the ceramics. Thus, the crossover between intergranular and intragranular pinning appears to be a likely explanation. Since the grain boundaries are very different from the grains themselves, it is not surprising that the

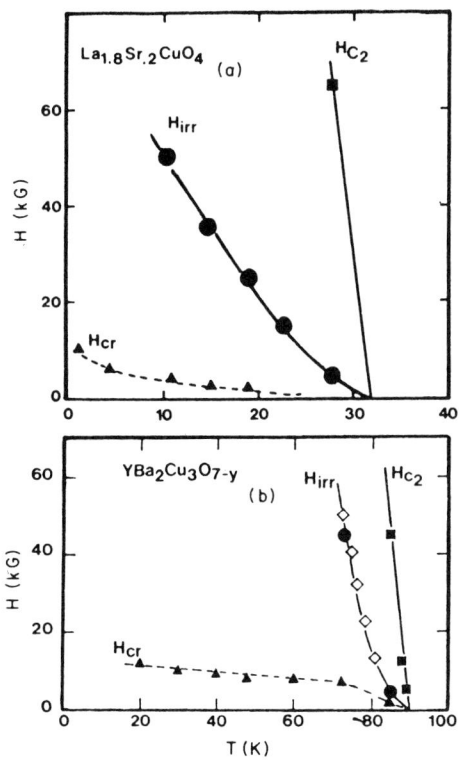

FIG. 23. Pinning phase diagram in LaSrCuO and in YBCO (Higgins, *et al.*, 1989).

pinning potential is considerably stronger. Thus, the crossover may also be between a strong-pinning and a weak-pinning regime.

Magnetic field dependence of sound velocity has been studied in some of the 1–2–3 samples as well, but the results are quite contradictory. Some measurements obtain an effect due to the field (Golding et al., 1988), while others (Horie et al., 1989) do not. So at this stage the situation remains quite controversial and confusing for this family of materials.

6. Conclusions

6.1. GENERAL SAMPLE-RELATED PROBLEMS

In spite of the vast literature that has accumulated in every aspect of the cuprate superconductors, several sample-related problems have proved difficult to overcome. These problems have led to ambiguities that relate not to details, but to the heart of the problem, i.e., identifying the mechanism responsible for superconductivity. All these problems have their origin in the form of disorder. In the 2–1–4 class of materials, the superconductors are obtained by doping the stoichiometric end member La_2CuO_4. The phase diagram shown in Fig. 1 makes it clear that the superconductivity does not set in until the doping exceeds a minimum value. In other words, the superconductors are well separated from the antiferromagnetic insulator. It has nevertheless been known for a long time that the properties of the antiferromagnet itself are extremely sensitive to the O-defect level. The Néel temperature can be changed drastically by a small change in O. Furthermore, it is also known that small changes in the doping (either by changing O or the bivalent atom—strontium, for example) drastically alter the superconducting properties. It is common experience that the samples of the doped material rarely display full Meissner effect, implying that nominally the same composition may contain enough variation to destroy superconductivity totally. To make matters even more complex, the lattice parameters are not different enough between the superconducting sample and the nonsuperconducting analogue for the difference to be resolved easily by neutron or x-ray diffraction measurements. It is possible, for example, to suggest that in the 2–1–4 material there is really a unique composition for high T_c; any material made at intermediate composition is actually phase-separated between the insulator and the superconductor, yielding the varied levels of Meissner fraction. This hypothesis calls into question any interpretation of the presence of magnetism in any sample of intermediate composition as an intrinsic effect. This problem has also plagued any detailed quantitative analysis of the data.

The sound velocity measurement is a bulk measurement, just as the specific

6. Ceramic High-Temperature Superconductors

heat is. Therefore, any quantitative comparison, such as that required for a thermodynamic analysis, must be performed with full knowledge of the superconducting fraction and, at the same time, behavior of the nonsuperconducting fraction as well. Alternatively, measurements need to be performed on the same sample so that self-consistency checks can be made.

A second and somewhat more detailed problem is related to the important question of fluctuations. It is now generally believed that superconductivity is associated with the Cu–O layers and is dominantly two-dimensional. In addition, the short coherence length implies that fluctuation effects may be much larger in this system than in conventional superconductors. Although subtle issues, such as the difference between Gaussian fluctuations and critical fluctuations (if any), remain to be tested out explicitly, a great of work has gone into the study of fluctuation effects themselves. It is in this specific area that the role of disorder in these materials needs to be understood in the greatest detail. It is suggested in many studies that the oxygen stoichiometry may be quite different in the grains themselves, as opposed to the grain boundary. Therefore, the ceramic samples can be viewed as consisting of grains of superconductor separated by weak links. The true macroscopic transition will occur in two steps: first an intragranular transition, and at a lower temperature, an intergranular one through Josephson coupling among the grains. The onset of the long-range order develops throughout the temperature interval between the two. This scenario is conceptually different from the case of a single-phase sample with no islands or weak links with a large precursor fluctuation regime. Therefore, it is premature to analyze the data obtained from ceramic samples in terms of the fluctuations without having the role of disorder clearly sorted out. The reader is warned that this does not imply that fluctuation analysis is necessarily incorrect. Indeed, it has often been observed that the paraconductivity data above T_c in single-crystal 1–2–3 material for conduction along the planes are the same as that on good quality ceramic samples.

6.2. OUTLOOK FOR FUTURE WORK

The sound velocity measurements in ceramic samples of high-T_c superconductors have yielded some valuable information. First, they have demonstrated the effects of soft-mode behavior associated with the structural phase transitions. However, at this time the role of the soft-mode behavior in superconductivity remains unclear. In this respect, it would be particularly valuable to study the sound velocity in the 2–1–4 system, where the structural transition occurs in close proximity to T_c, where one can hope to find similarities with and differences from the A15 superconductors. Second, there are recurrent reports of other transitions above T_c in this class of materials. Sound velocity anomalies have been observed near these transitions, which have been referred to as T_s in this review, following the nomenclature of Bhattacharya et al. (1988,a,b), although there is insufficient evidence to suggest that they are the same in all systems. There are many other physical properties that show distinct anomalies at these temperatures (see Bhattacharya et al., 1988a). The nature of these transitions and what role they play for superconductivity remain uncertain. This is further complicated by the observation that at least in one system, namely, LaBaCuO, a subtle change in the structure, which is accompanied by a sound velocity anomaly, produces a profound effect on superconductivity. The sound velocity studies have been critical in establishing their existence.

Third, the presence of sound velocity anomalies at T_c itself has been a subject of considerable importance. Although this review takes the position that these anomalies are intrinsic to this class of systems, this view is by no means unanimous. A competing view, supported by early single-crystal data that reported no anomaly near T_c, is that most measurements in ceramic samples are spurious (See, for example, Allen et al. 1988). Since then, however, better-quality (presumably more homogeneous, particularly in the oxygen content) single crystals have become available. These crystals show data consistent with some of the ceramic data that are discussed more extensively in this review. Indeed, it is perhaps clear now that carefully prepared ceramic samples can be made homogeneous more easily than their single-crystal analogues. Considerable confusion had also surrounded early analyses of the thermodynamic anomalies, in that the distinctions between longitudinal and tranverse sound, between the bulk and shear moduli and also the Young's modulus (measured in typical vibrating reed measurements), and the implications of the thermodynamic anomalies for the various strain derivatives of T_c were not sufficiently clearly understood, inspite of the relatively early arrival of an excellent theoretical analysis (Millis and Rabe, 1988).

The analysis by Millis and Rabe of the existing data suggests that while the first strain derivatives of T_c are quite conventional in their magnitude, some of the second derivatives may be unusually large. If this conjecture is correct, then detailed experimental work should focus on the identification of the specific derivative(s). Due to the large inherent anisotropy of these systems, it is reasonable to assume that although they are observable in ceramic samples, these effects are also inherently anisotropic. Clearly, the identity of the specific polarization(s) and the strain component(s) that are responsible for this anomalous behavior can only come from single-crystal data. When available, these would hopefully provide a stringent constraint on the possible mechanisms of superconductivity in this class of materials. Therefore, it is likely that the usefulness

6. Ceramic High-Temperature Superconductors

of ceramic materials is quite limited at this rather mature stage of the field; they would be useful in locating anomalies and overall qualitative behavior in these materials. But the future must belong to good-quality, homogeneous, and carefully characterized single crystals.

References

Alers, G. A., and Waldorf, D. L. (1962). *IBM J. Res. Dev.* **6**, 94.
Allen, P. B., Fisk, Z., and Migliori, A. (1988) *in* "Physical Properties of High Temperature Superconductors," Vol 1. (Edited by D. M. Ginsberg) World Scientific.
Axe, J. D., *et al.* (1989). *Phys. Rev. Lett* **62**, 2751.
Bednorz, J. G., and Muller, K. A. (1986). *Z. Phys.* **B64**, 189.
Bhattacharya, S., Higgins, M., Johnston, D., Jacobson, A., Stokes, J., Lewandowski, J., and Goshorn, D. (1988a). *Phys. Rev. B.: Condens. Matter* **37**, 5901.
Bhattacharya, S., Higgins, M., Johnston, D., Jacobson, A., Stokes, J., Goshorn, D., and Lewandowski, J. (1988b). *J. Phys. Rev. Lett.* **60**, 1181.
Birgeneau, R. J., and Shirane, G. (1989). *In* Physical Properties of High-Temperature Superconductors, Vol. 1 (D. M. Ginsberg, ed.). World Scientific, Singapore.
Bishop, D., Gammel, P., Ramirez, A., Cava, R., Batlogg, B., and Rietman, E. (1978a). *Phys. Rev. B.: Condens. Matter* **35**, 8788.
Bishop, D. J., Ramirez, A. P., Gammel, P. L., Batlogg, B., Rietman, E. A., Cava, R. J., and Millis, A. J. (1987b). *Phys. Rev.* **B36**, 2408.
Bourne, L. C., Zettl, A., Chang, K. J., Cohen, M. L., Stacy, A. M., and Ham, W. K. (1987). *Phys. Rev. B35*, 8785.
Chandrasekhar, B. S. (1969). *In* "Superconductivity," Vol. 1 (R. D. Parks, ed.). Marcel Dekker, New York.
Fleming, R. M., Batlogg, B., Cava, R. J., and Rietman, E. A. (1987). *Phys. Rev.* **B35**, 7191.
Fossheim, K., and Laegreid, T. (1989). *IBM J. Res. Dev.* **33**, 365.
Fossheim, K., Laegreid, T., Sandvold, E., Vassenden, F., Mueller, K., and Bednorz, J. (1987). *Solid State Commun.* **63**, 531.
Golding, B., Haemmerle, W. H., Schneemeyer, L. F., and Waszczak, J. V. (1988). *Proceedings of the IEEE Ultrasonics Symposium, Vol. 2, 1079*.
Higgins, M., Goshorn, D., Bhattacharya, S., and Johnston, D. (1989). *Phys. Rev. B* **40**, 9393.
Horie, Y., Fukami, T., and Mase, S. (1987). *Solid State Commun.* **63**, 653.
Horie, Y., Terashi, Y., and Mase, S. (1989). *Solid State Commun.* **63**, 653.
Laegreid, T., Fossheim, K., and Vassenden, F. (1988). *Physica C* **153–155**, 1096.
Lee, W. K., Lew, M., and Nowick, A. S. (1990). *Phys. Rev.* **B41**, 149.
Ma, S. K. (1976). "Modern Theory of Critical Phenomena." W. A. Benjamin, Inc., Reading, Massachusetts.
Migliori, A., Chen, T., Alavi, B., and Gruner, G. (1988). *Solid State Commun.* **63**, 827.
Migliori, A., Visscher, W. M., Wong, S., Brown, S. E., Tanaka, I., Kojima, H., and Allen, P. B. (1990). *Phys. Rev. Lett.* **64**, 2458.
Millis, A., and Rabe, K. (1988). *Phys. Rev. B* **38**, 8908.
Nieva, G., Martinez, E. N., de la Cruz, F., Esparza, D. A., and D'Ovidio, C. A. (1987). *Phys. Rev. B* **36**, 8780.
Saint-Paul, M., Tholence, J., Noel, H., Levet, J., Potel, M., and Gougeon, P. (1989). *Solid State Commun.* **69**, 1161.
Salamon, M. B., Inderhees, S. E., Rice, J. P., Pazol, B. G., Ginsburg, D. M., and Goldenfeld, N. (1988). *Phys. Rev.* **B38**, 885.
Seraphim, D. P., and Marcus, P. M. (1962). *IBM J. Res. Dev.* **6**, 94.

Shirane, G., Endoh, Y., Birgeneau, R. J., Kastner, M. A., Hidaka, Y., Oda, M., Suzuki, M., and Murakami, T. (1987). *Phys. Rev. Lett.* **59**, 1613.
Tallon, J., Schuitema, A., and Tapp, N. (1988). *Appl. Phys. Lett.* **52**, 507.
Testardi, L. R. (1973). In "Physical Acoustics," Vol. 10 (W. P. Mason, ed.), Academic Press, New York.
Testardi, L. R. (1975). *Phys. Rev.* **B12**, 3849.
Vaknin, D., Sinha, S. K., Moncton, D. E., Johnston, D. C., Newsam, J. M., Satinya, C. R., and King, H. E. (1987). *Phys. Rev. Lett.* **58**, 2802.
Wu, M. K., Ashbourn, J. R., Torug, C. J., Hor, P. H., Meng, R. L., Gao, L., Huang, Z. J., Dang, Y. Q., and Chu, C. W. (1987). *Phys. Rev. Lett.* **58**, 908.
Xu, M., (1990). Ph.D. Thesis, University of Wisconsin.
Zhao, Z., Adenwalla, S., Moreau, A., Ketterson, J. B., Robinson, Q., Johnson, D. L., H·wa, S. J., Poeppelmeier, R. K., Xu, M. F., Hong, Y., Wiegert, R. F., Levy, M., and Sarma, Rimat K. (1989). *Phys. Rev. B* **39**, 721.

Bibliography

I. Results on the (1-2-3) System

Al-Kheffaji, A., Cankurtaran, M., Saunders, G., Almond, D., Lambson, E., and Draper, R. (1989). *Philos. Mag. B* **59**, 487.
Almond, D., Lambson, E., Saunders, G., and Hong, W. (1987). *J. Phys. F: Met. Phys.* **17**, L261.
Almond, D., Saunders, G., and Lambson, E. (1988). *Supercond. Sci. Technol.* **1**, 163.
Anshukova, N., Vorob'ev, G., Golovashkin, A., Ivanenko, O., Kazei, Z., Krynetskii, I., Levitin, R., Mil, B., Mitsen, K., and Snegirev, V. (1987). *Pis'ma Zh. Eksp. Teor. Fiz.* **46**, 373.
Bar'yakhtar, V., Varyukhin, V., and Strongin, S. (1989). *Dopov Akad. Nauk Ukr. RSR, Ser. A: Fiz.-Mat. Tekh. Nauki,* **3**, 53.
Bridge, B. (1989). *J. Mater. Sci. Lett.* **8**, 695.
Bridge, B., and Round, R. (1989). *J. Mater. Sci. Lett.* **8**, 691.
Calemczuk, R., Bonjour, E., Henry, J., Forro, L., Ayache, C., Jurgens, M., Rossat-Mignod, J., Barbara, B., Burlet, P., et al. (1988). *Physica C* **153-155**, 960.
Cannelli, G., Cantelli, R., Cordero, F., Costa, G., Ferretti, M., and Olcese, G. (1988). *Europhys. Lett.* **6**, 271.
Deng, T., Zhang, L., Gu, H., Xiao, Z., and Chen, L. (1989). *Jpn. J. Appl. Phys.* **28**, 39.
Deng, T., Zhang, L., Huicheng, G., Xiao, Z., and Chen, L. (1988). *Chin. Phys. Lett.* **5**, 461.
Du, J., Jiang, J., Wang, X., and Yin, H. (1988). *Wuli Xuebao* **37**, 1556.
Gaiduk, A., Zherlitsyn, S., Prikhod'ko, O., Fil, V., Seminozhenko, V., Nesterenko, V., and Pershin, S. (1988). *Fiz. Nizk. Temp.* **14**, 718.
Guillon, F., and Kemberg-Sapieha, J. (1988). *Phys. Lett. A.* **131**, 315.
Golovashkin, A., Danilov, V., Ivanenko, O., Leitus, G., Mitsen, K., Perepechko, I., Karpinksii, O., and Shamrai, V. (1987). *Novel Supercond., Proc. Int. Workshop Novel Mech. Supercond.* **883**,.
Golovashkin, A., Danilov, V., and Ivanenko, O., Mitsen, K., Perepechko, I. (1989). *Dokl. Adad. Nauk SSSR* **305**, 589.
Golovashkin, A., Danilov, V. k., Ivanenko, P., Mitsen, K., and Perepechko, I. (1987). *Pis'ma Zh. Eksp. Teor. Fix.* **46**, 273.
He, Y., Zhang, B., Lin, S., Xiang, J., Lou, Y., Chen, H. (1987). *J. Phys. F: Met. Phys.* **17**, L243.
Hoen, S., Bourne, L., Kim, C., and Zettl, A. (1988). *Phys. Rev.f B.* **38**, 11949.
Horie, Y., Terashi, Y., Fukuda, H., Fukami, T., and Mase, S. (1987). *Solid State Commun.* **64**, 501.

Horie, Y., Terashi, Y., and Mase, S. (1989). *J. Phys. Soc. Jpn.* **58,** 279.
Ivanov, A., Volkova, L., Grigut, O., Popovich, A., Revenko, Y., Svistunov, V., Tarkenkov, V., Tsymbal, L., and Cherkasov, A. (1988). *Fiz. Nizk. Temp* **14,** 202.
Kobelev, N., Kondakov, S., and Soifer, Ya. M. (1989). *Fiz. Tverd. Tela* **31,** 57.
Laegreid, T., and Fossheim, K. (1988). *Europhys. Lett.* **6,** 81.
Lagreid, T., Fossheim, K., and Vassenden, F. (1988). *Physica C,* **153–155,** 1096.
Lang, M., Lechner, T., Riegel, S., Steglich, F., Weber, G., Kim, T., Luethi, B., Wolf, B., Rietschel, H., and Wilhelm, M. (1988). *Z. Phys. B: Condens. Matter.* **69,** 459.
Ledbetter, H., Austin, M., Kim, S., Datta, T., and Violet, C. (1987). *J. Mater. Res.* **2,** 790.
Levy, M., Xu, M., Baum, H., Schenstron, A., Qian, Y., Sun, K., and Sarma, B. (1987). *Ultrason. Symp. Proc.* **2,** 1151.
Levy, M., Xu, M., Sarma, B., Zhao, Z., Adenwalla, S., Robinson, Q., and Ketterson, J. (1988). *Ultrason. Symp. Proc.* **2,** 1097.
Mase, S., Horie, Y., Yasuda, T., Kusaba, M., and Fukami, T. (1988). *J. Phys. Soc. Jpn.* **57,** 607.
Mueller, V., De Groot, K., Maurer, D., Roth, C., and Rieder, K. (1987). *Physica* **B+C 148,** 296.
Mueller, V., De Groot, K., Maurer, D., Roth, C., Rieder, K., Eickenbusch, E., and Schoellhorn, R. (1987). *Jpn. J. Appl. Phys., Part 1* **26,** 2139.
Muller, V., Maurer, D., Roth, C., Hucho, C., Winau, D., De Groot, K., Eickenbusch, H., and Schoellhorn, R. (1988). *Physica C* **153–155, 280.**
Nishihara, H., Hayashi, K., Okuda, Y., and Kajimura, K. (1989). *Phys. Rev. B* **39,** 7351.
Patel, N., Sarkar, P., Troczynski, T., Tan, A., and Nicholson, P. (1987). *Adv. Ceram. Mater.* **2,** 615.
Ramachandran, V., Ramadass, G., and Srinivasan, R. (1988). *Physica C.* **153–155,** 278.
Round, R., and Bridge, B. (1987). *J. Mater. Sci. Lett.* **6,** 1471.
Saint-Paul, M., Tholence, J., Monceau, P., Noel, H., Levet, J., Potel, M., Gougeon, P., and Capponi, J. (1988). *Solid State Commun.* **66,** 641.
Srinivasan, R., Ramachandran, V., Seshadri, A., and Ramadass, G. (1987). *Pramana* **29,** L603.
Suzuki, M., Okuda, Y., Iwasa, I., Ikushima, A., Takabatake, T., Nakazawa, Y., and Ishikawa, M. (1988). *Jpn. J. Appl. Phys.* **27,** 50.
Suzuki, M., Okuda, Y., Iwasa, I., Ikushima, A., Takabatake, T., Nakazawa, Y., and Ishikawa, M. (1988). *Physica C,* **153–155,** 266.
Xu, M., Bein, D., Wiegert, R., Sarma, B., Levy, M., Zhao, Z., Adenwalla, S., Moreau, A., Robinson, Q., et al. (1989). *Phys. Rev. B* **39,** 843.
Xu, M., Bein, D., Hong, Y., Sarma, B., Levy, M., Zhao, Z., Adenwalla, S., Moreau, A., Robinson, Q., et al. (1989). *J. Less-Common Met.* **149,** 447.
Xu, M., Schenstrom, A., Hong, Y., Bein, D., Sarma, B., Levy, M., Zhao, Z., Adenwalla, S., Moreau, A., et al. (1989). *IEEE Trans. Magn.* **25,** 2414.
Yoshimoto, M., Tanabe, S., Soga, N. (1987). *Chem. Lett.* **11,** 2193.
Zettl, A., Bourne, L., Creager, W., Crommie, M., and Hoen, S. (1989). *Synth. Met.* **29,** F723.

II. RESULTS ON THE (2-1-4) SYSTEM

Bishop, D., Gammel, P., Ramirez, A., Batlogg, B., Cava, R., and Millis, A. (1987). *Novel Supercond., Proc. Int. Workshop Novel Mech. Supercond.* **659,**
Burkhanov, A., Gudkov, V., Zhevtovskikh, I., Kozhevnikov, V., Nasish, V., Podgornykh, S., Startsev, V., Tkach, A., Ustinov, V., et al. (1987). *Fiz. Met. Metalloved.* **64,** 397.
Esquinazi, P., and Duran, C. (1988). *Physica C,* **153–155,** 1499.
Esquinazi, P., Luzuriaga, J., Duran, C., Esparza, D., and D'Ovidio, C. (1987). *Phy. Rev. B.: Condens. Matter* **36,** 2316.
Fukami, T., Horie, Y., Yasuda, T., Ono, F., Cheng, M., Takano, S., and Mase, S. (1987). *Physica B+C* **148,** 521.

Gaiduk, A., Zherlitsyn, S., Panfilov, A., Puzilov, V., Stepanenko, A., Fil, V., and Chernyi, A. (1987). *Fiz. Nizk. Temp.* **13,** 653.
Horie, Y., Fukami, T., and Mase, S. (1987). *Solid State Commun.* **63,** 653.
Horie, Y., Fukami, T., and Mase, S. (1987). *Jpn. J. Appl. Phys. Part 1* **26,** 1125.
Luethi, B., Wolf, B., Kim, T., Grill, W., and Renker, B. (1987). *Jpn. J. Appl. Phys. Part 1* **26,** 1127.
Makarov, V., Zavaritskii, N., Klochko, V., Voronov, A., and Tkachenko, V. (1988). *Pis'ma Zh. Eksp. Teor. Fix.* **48,** 326.
Mase, S., Yasuda, T., Horie, Y., and Fukami, T. (1988). *Solid State Commun.* **65,** 477.
Mase, S., Yasuda, T., Horie, Y., Kusaba, M., and Fukami, T. (1988). *J. Phys. Soc. Jpn.* **57,** 1024.
Zhang, M., Zheng, D., Tao, R., Pu, F., and Zhou, S. (1988). *Physica C,* **153–155,** 255.

III. Results on Low Temperature Properties

Esquinazi, P., Duran, C., Fainstein, C., Nunez Regueiro, M. (1988). *Phys. Rev. B: Condens. Matter* **37,** 545.
Golding, B., Birge, N., Haemmerle, W., Cava, R., and Rietman, E. (1987). *Phys. Rev. B: Condens. Matter* **36,** 5606.
Hikata, A., McKenna, M., Elbaum, C., Kershaw, R., and Wold, A. (1989). *Phys. Rev. B* **40,** 5247.
McKenna, M., Hikata, A., Takeuchi, J., Elbaum, C., Kershaw, R., Wold, A. (1989). *Phys. Rev. Lett.* **62,** 1556.
Nunez Regueiro, M., Esquinazi, P., Izbaizky, M., Duran, C., Castello, D., and Lauzuriaga, J. (1988). *Ann. Phys.* **13,** 401.
Nunez Regueiro, M., Esquinazi, P., Izbizky, M., Duran, C., Castello, D., Luruziaga, J., and Neiva, G. (1988). *Physica C* **153–155,** 1016.

— 7 —
Acoustic Studies of Single-Crystal High-T_c Superconductors

BRAGE GOLDING

Department of Physics and Astronomy,*
Michigan State University,
East Lansing, Michigan
and
AT&T Bell Laboratories, Murray Hill, New Jersey

1. Introduction 349
2. The High-T_c Superconductors 351
   2.1. Structural Aspects 351
   2.2. Crystals and Crystal Growth 352
3. Acoustic Methods for Small Single Crystals 352
   3.1. Audio Frequencies: Vibrating-Reed Technique 353
   3.2. Radio Frequencies: Bulk Mechanical Resonance 354
   3.3. Radio and Microwave Frequencies: Plane Waves 355
   3.4. Thermally Excited Phonons: Brillouin Scattering 357
4. Single-Crystal Experiments and Results 357
   4.1. Normal State above T_c 357
   4.2. The Superconducting Transition Region 364
   4.3. The Superconducting State 370
5. Summary and Outlook 376
   Acknowledgments 378
   References 378

1. Introduction

The discovery of superconductivity in copper oxides by Bednorz and Mueller (1986) and the subsequent synthesis of related materials with superconducting critical temperatures T_c above 90 K (Wu *et al.*, 1987; Sheng and Hermann, 1988) have revolutionized condensed matter science. Nevertheless, six years after this seminal event, there appears to be no general agreement on the mechanism responsible for cuprate superconductivity. It is accepted that Cooper pairs

*Present address

exist, but that their pairing is most likely mediated by elementary excitations other than phonons that form the "glue" in conventional superconductivity. This belief, however, does not preclude the potential usefulness of experimental studies of acoustic phonons in novel superconductors. Indeed, ultrasonic studies of heavy fermion metals (Sarma *et al.*, this volume, Chapter 3) in which superconductive pairing is nonphononic have been influential in developing a picture of their unconventional superconductivity.

Unlike conventional metals, sound propagation in high-T_c material is complex and reflects the remarkably wide range of phenomena exhibited by multiconstituent oxides. Several factors conspire to exacerbate the difficulties in extracting clear results from acoustic investigations: large anisotropies arising from the 2-D planar structural elements, low crystalline symmetry, sensitivity to oxygen stoichiometry, impurities, crystallographic twins, magnetic interactions, flux lattice phenomena, structural instabilities, and numerous phonon branches, to mention a few.

In practice, the bulk of acoustic investigations of high-T_c superconductors have been performed on ceramic samples comprising aggregated micron-sized crystallites. Such experiments generally probe spherically averaged properties (unless special techniques to align the crystallites are implemented). Acoustic wavelengths are generally much greater than the crystallite dimensions, since elastic scattering restricts the uppermost frequencies at which appreciable transmission is possible.

The increasing availability of single crystals of high-T_c superconductors has been crucial in developing an understanding of their subtle properties, particularly in the normal state. Single crystals have presented opportunities, as well as new challenges, for investigations of the acoustic properties of these novel systems. Whereas ceramics have been relatively easy to synthesize, the successful growth of crystals requires not only highly developed scientific intuition, but also the touch of the artisan. It should be noted that control of stoichiometry is generally more difficult in crystals as compared to microcrystalline ceramics. Generally speaking, the larger the crystal, the more likely it is to be inhomogeneous. As a result, at this time there have been only a handful of acoustic and ultrasonic studies on high-T_c single crystals.

The overall plan of this chapter is simply to consolidate the information from the relatively sparse single crystal experiments that have been published since 1987. As usual, the difficulty in assessing the quality of the information (and of the materials) is profound. No attempt has been made to correlate the single-crystal results with the overwhelming amount of data on ceramics. The interested reader is referred to other chapters in this volume as preparation for that task.

7. Single-Crystal High-T_c Superconductors

2. The High-T_c Superconductors

2.1. STRUCTURAL ASPECTS

2.1.1. $YBa_2Cu_3O_{7-\delta}$

The compound $YBa_2Cu_3O_{7-\delta}$, otherwise referred to as "123" or "YBCO," is undoubtedly the most thoroughly investigated high-T_c superconductor at this time. Although not the first high-T_c material to be discovered, its transition temperature above 90 K and its relatively straightforward synthesis account for the enormous amount of attention devoted to it. The structure of $YBa_2Cu_3O_7$ at room temperature is an orthorhombic, tripled perovskite (Capponi *et al.*, 1987) with a unit cell containing two Cu–O_2 layers ("planes") and one Cu–O layer ("chain"). The crystals generally possess extensive [110] twins, as well as other more complex microstructures.

As oxygen is removed from the stoichiometric $\delta = 0$ compound, it preferentially vacates the chain sites (Cava *et al.*, 1987), where the Cu is fourfold coordinated. For $\delta \approx 0.3$–0.5 T_c drops to approximately 60 K (Beyers *et al.*, 1989). Ordering of the chains occurs, with alternate rows depleted of oxygen. For $\delta \gtrsim 0.6$, the structure is tetragonal, and the material is no longer superconducting, but forms an antiferromagnetic insulator.

2.1.2. $La_{2-x}Sr_xCuO_{4-y}$

The stoichiometric compound La_2CuO_4 is body-centered tetragonal (I4/mmm, K_2NiF_4 structure) above 530 K, where it transforms on cooling to an orthorhombic (Cmca) structure (Longo and Raccah, 1973). These crystals are also extensively twinned at room temperature (Chen, 1989). The sixfold oxygen-coordinated copper ions, describing a distorted octahedron, form square planar networks separated by two staggered La–O layers. The stoichiometric compound is an antiferromagnetic Mott insulator with $T_N \approx 300$ K.

As La^{3+} is replaced with Sr^{2+} (or Ba^{2+}), the doping on the cation lattice introduces holes into the conduction band. Upon doping, the tetragonal–orthorhombic transition decreases, approaching room temperature for $x \approx 0.1$ (Fleming *et al.*, 1987). Concurrently, the introduction of carriers induces an insulator-to-metal transition and the appearance of superconductivity. The highest T_cs (~ 30 K) occur near $x = 0.15$.

2.1.3. Other High-T_c Superconductors

For completeness, we mention other important classes of high-T_c materials, although many fewer acoustic investigations have been made with them. These include the Tl-based superconductors, such as $Tl_2Ba_2Ca_2Cu_3O_{10}$ (Tl-2223), which has the highest known T_c, near 124 K (Sheng and Hermann, 1988). The Bi-based superconductor $Bi_2Sr_2CaCu_2O_6$ (Bi-2212) possesses a T_c near 85 K (Maeda *et al.*, 1988). These materials form highly two-dimensional structures and can be synthesized with variable numbers of Cu–O layers.

Superconductivity occurs in the T'-structure 214 compounds, of which $Nd_{2-x}Ce_xO_{4-y}$ is the prototype. This phase is tetragonal with space group I4/mmm, and a T_c near 20 K has been reported for $x = 0.15$ (Tokura *et al.*, 1989). This material possesses the distinction of being an electron-doped superconductor as shown, for example, by Hall effect measurements.

2.2. Crystals and Crystal Growth

A description of crystal growth methods is beyond the scope of this review. Nevertheless, there are at least two aspects of the preparation of high-T_c single crystals that are critical to acoustic measurements: crystal dimensions and crystalline homogeneity. The bulk of all studies on high-T_c materials have been carried out on ceramic samples, as a result of the difficulty in growing high-quality single crystals.

When crystals are available, the choice of acoustic measurement techniques is dictated by the shapes and sizes of the crystals. Standard pulse echo ultrasonic methods for the 10–100 MHz region, wherein a thin piezoelectric plate is glued to one of a pair of flat parallel surfaces, have been used on several occasions. It is difficult to assure plane wave propagation in small crystals, particularly when reflections from lateral faces create interference. As described in Section 3, low-frequency acoustic resonance methods are effective for crystals with regular geometric shapes, whereas plane wave methods only require a pair of parallel opposite faces. The growth of single crystals using flux methods can result in crystals with natural crystallographic faces suitable for either approach.

Determination of oxygen stoichiometry and homogeneity is particularly important for all high-T_c materials because of the sensitivity of T_c to oxygen content. Particular attention has been paid to achieving homogeneity in flux-growth materials by extended oxygen annealing (Schneemeyer, 1990).

3. Acoustic Methods for Small Single Crystals

The availability of single crystals of the high-T_c superconductors has been restricted generally to (1) thin reeds, since the crystals' morphology is platelike, and (2) small, naturally faceted, 3-D parallelepipeds. The former are suitable for acoustic measurements that excite flexural modes of thin bars, whereas the latter may allow the propagation of plane waves, provided that the sample dimensions are appreciably larger than an acoustic wavelength.

3.1. Audio Frequencies: Vibrating-Reed Technique

The vibrating-reed method utilizes the flexural (or torsional) motion of a thin bar of rectangular cross-section that is clamped rigidly at one end (Barmatz and Golding, 1974). The opposite end is free to oscillate at the natural mechanical resonance frequency of the bar, given by

$$f_n = (d/2\pi\sqrt{12})(k_n/\ell)^2 v_E. \tag{1}$$

Here, ℓ is the length of the reed, d is the thickness, and the k_n are constants with values 1.875 and 4.694 for $n = 1$ and 2, respectively. The quantity v_E is the Young's modulus sound velocity, which is related to the Young's modulus E and the mass density ρ by $v_E = (E/\rho)^{1/2}$.

It is possible also to excite a torsional mode by an extension of this method. A small wire is attached to the free end of the reed so that it is orthogonal to the reed's axis. Electrodes are positioned to exert a torque on the wire, twisting the reed about its axis.

A common arrangement for exciting and detecting the flexural resonance is shown in Fig. 1. The reed is attached to a massive block with a solder or conductive epoxy adhesive. The free end is excited electrostatically with a small electrode driven at frequency $f/2$. The oscillation of the reed is detected at the opposite electrode, which is used as a dc-biased capacitive microphone. A tuned receiver at f, usually a preamplifier followed by a phase-sensitive lock-in amplifier, functions as the detector. The mechanical Q of the reed can be determined from the half-power points of the resonance as the oscillator is swept through the resonant frequency f_R. It is convenient to phase-lock the oscillator to f_R by using the dispersive component of the amplifier to control the oscillator frequency. As an external variable is changed—for example, the temperature or magnetic field—f_R and the amplitude of the resonance can be tracked continuously.

Two studies of $YBa_2Cu_3O_7$ have been made utilizing the vibrating-reed technique. Shi et al. (1989) have reported measurements of a single crystal. The reed surfaces corresponded to the crystallographic axes, with the c-axis perpendicular to the thin dimension. Since the crystal is microtwinned, the other directions are the a- and b-axes, which cannot be distinguished. The flexural mode resonated at a fundamental frequency of 20 kHz, which leads to an estimated Young's modulus of 2,200 kbar at room temperature (Shi, private communication).

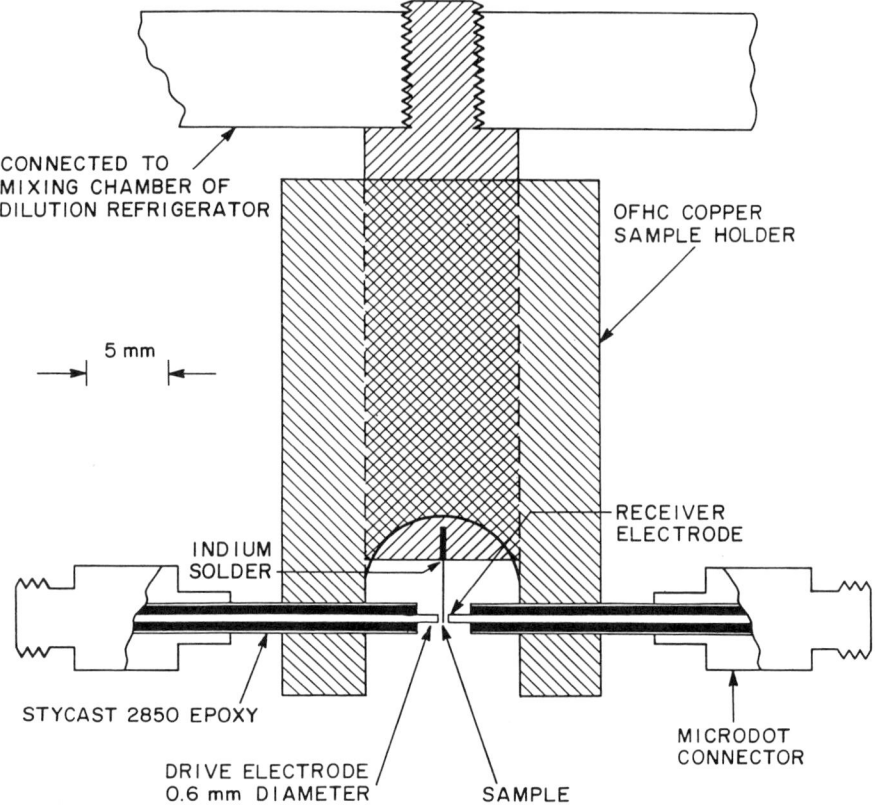

FIG. 1. Schematic drawing of a vibrating-reed sample holder (Golding et al., 1988). The sample, shown in cross-section, is coated with a thin Au film and soldered with In into the holder. A sliding outer body allows the electrodes to be positioned at the reed's tip. The sliding coaxial electrodes can be positioned arbitrarily close to the sample.

Golding et al. (1988a, 1991) have measured the flexural mode in a somewhat longer crystal affording lower frequencies. Both the fundamental resonance at 3.4 kHz and the first overtone at 22.0 kHz were studied. The sample orientation was the same as that used by Shi et al. (1989) (see Fig. 11).

3.2. RADIO FREQUENCIES: BULK MECHANICAL RESONANCE

This technique has been applied to single crystals of La_2CuO_4 (Migliori et al., 1990a) and $La_{2-x}Sr_xCuO_{4-y}$ (Migliori et al., 1990b). The sample is a rectangular parallelepiped, mechanically excited at a point of low symmetry, i.e., a corner.

7. Single-Crystal High-T_c Superconductors

For samples with dimensions of a few millimeters, typical resonant frequencies range from 0.5 to 1.5 MHz. The resonant oscillations of the crystal are detected at an opposite corner with a broadband transducer. A prescription for obtaining the complete set of elastic constants from the resonant frequencies, the sample dimensions, and the mass density has been given. A more complete explanation of the technique is given in this volume.

3.3. RADIO AND MICROWAVE FREQUENCIES: PLANE WAVES

For samples with lateral dimensions ≥2–3 mm, it is practicable to generate plane waves using bonded piezoelectric transducers. The sample is prepared with a set of flat and parallel faces, and one or two piezoelectric plates are attached with a thin layer of adhesive. A short pulse excites the plate at its resonant frequency and launches an ultrasonic pulse into the crystal. With a single transducer, the wave is reflected from the opposite face and is detected by the transducer. The pulse-echo technique is thus a time-domain method for measuring sound velocity and attenuation. Either longitudinal or transverse sound can be generated by an appropriately chosen orientation of the transducer material.

Measurements of sound propagation in $YBa_2Cu_3O_7$ have been reported by Saint-Paul et al. (1988, 1989) at frequencies from 15 to 600 MHz, and by Kim et al. (1990) at frequencies from 8 to 53 MHz. Zavaritsky et al. (1989) have reported measurements at 50 MHz on a La_2CuO_4 single crystal.

Studies of microwave sound propagation at 1 GHz in $YBa_2Cu_3O_7$ have been reported by Golding et al. (1988a, 1991). In this work, thin film transducers of ZnO were sputtered directly onto the surfaces of a small single crystal with typical dimensions 0.2–0.3 mm. The small crystals necessitated extremely short pulses, of order 20 ns. The experimental arrangement is shown in Fig. 2. The sample is mounted on the groundplane of a microstripline substrate. The 1 GHz excitation pulses are brought to the sample by 50-ohm lines and a small wire that contacts the ZnO surface. The contact area, 0.05 mm in diameter, is sufficiently small that reflections from lateral surfaces are completely absent. An example of the decay of the acoustic reflections is shown in Fig. 3.

The sample and contacts are surrounded by a pair of pickup coils that are used to monitor the superconducting transition via flux expulsion at T_c. The assembly can be rotated by ±90° in an external magnetic field of a 9 T superconducting solenoid. A dilution refrigerator provides temperatures as low as 5 mK. In spite of the extensive twinning, observed by an optical microscope, there was no evidence for scattering of the 1 GHz ultrasonic waves, nor was there any evidence that relaxational dissipation due to acoustic-twin boundary coupling plays any significant roles at these frequencies.

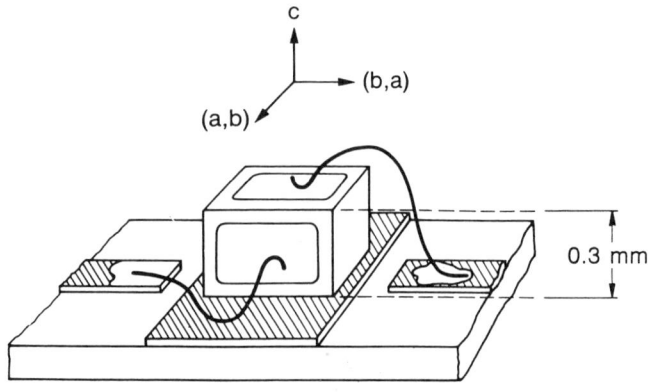

YBa$_2$Cu$_3$O$_7$ SINGLE CRYSTAL

FIG. 2. Arrangement of electrodes on a YBa$_2$Cu$_3$O$_7$ single crystal with dimensions approximately 0.2–0.3 mm on a side. Piezoelectric ZnO films ~1 μm thick have been sputtered onto two orthogonal faces. The superconducting sample is mounted onto the microstripline and grounded with silver paint. Contacts to the ZnO are made with In-tipped Cu wires.

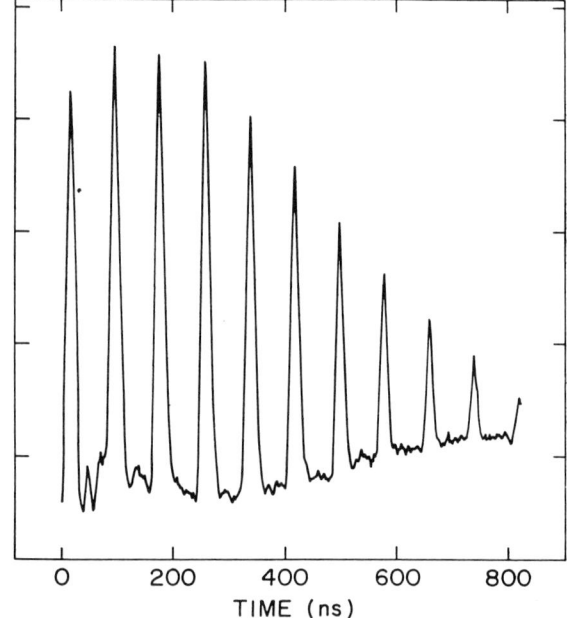

FIG. 3. Ultrasonic reflections for 0.96 GHz longitudinal waves propagating along the c-axis in YBa$_2$Cu$_3$O$_7$ at 80 K. The sample is shown in Fig. 2; the c-axis sample length is 0.22 mm (Golding et al., 1988a). (© 1988 IEEE)

7. Single-Crystal High-T_c Superconductors

3.4. THERMALLY EXCITED PHONONS: BRILLOUIN SCATTERING

Inelastic scattering of nearly monochromatic laser light by thermal phonons (Fleury, 1970) can provide information on surface acoustic modes. The frequency shift of the laser can be measured by high-resolution Fabry–Perot interferometers, and the sound velocity obtained from the Brillouin expression (Baumgart et al., 1989).

Since the resolution is limited to approximately a few percent, this method has been most useful in obtaining absolute velocities for surface phonons. However, Brillouin scattering can probe phonons in the 10–20 GHz regime, making it the highest-frequency technique utilized to date.

4. Single-Crystal Experiments and Results

4.1. NORMAL STATE ABOVE T_c

4.1.1. Elastic Moduli

The elastic stiffness c_{ij} can be calculated from sound velocities v_{ij} via the relationships $c_{ij} = \rho v_{ij}^2$, where ρ is the mass density. In $YBa_2Cu_3O_7$, sound propagation has been studied in microtwinned crystals so that the effective symmetry is tetragonal. As a result, propagation in the a or b directions cannot be distinguished. Table I gives a compilation of the complete elastic moduli for $YBa_2Cu_3O_7$ (Ledbetter and Lei, 1991). Partial sets of elastic moduli have been obtained from geometrical resonances of rectangular parallelpipeds of La_2CuO_4 (Migliori et al., 1990a) and $La_{2-x}Sr_xCuO_{4-y}$ ($x = 0.14$) (Migliori et al., 1990b) and are given in Table II.

There have been some systematic studies on the effect of oxygen stoichiometry on sound velocities in single crystals and ceramics. The velocity of Rayleigh waves propagating in the (001) plane in tetragonal (oxygen-deficient) and in orthorhombic $YBa_2Cu_3O_7$ was studied (Aleksandrov et al., 1990). It was claimed that oxygenated crystals (as well as materials with Dy, Ho, Zr, and Tm substituted

TABLE I

MEASURED AND ESTIMATED ORTHORHOMBIC ELASTIC STIFFNESSES (IN GPa) OF $YBa_2Cu_3O_7$ AT 300 K[a]

	c_{11}	c_{22}	c_{33}	c_{44}	c_{55}	c_{66}	c_{12}	c_{13}	c_{23}	B
$YBa_2Cu_3O_7$	223	244	138	61	47	97	37	89	93	115

[a]The self-consistent values have been calculated by Ledbetter and Lai (1991) based on linear compressibilities (Aleksandrov et al. 1988), phonon dispersion curves from neutron scattering (Reichardt et al. 1988) and from GHz longtudinal sound propagation (Golding et al. (1988). B is the bulk modulus.

TABLE II

MEASURED ELASTIC STIFFNESSES (IN GPA) FOR La_2CuO_4 (MIGLIORI ET AL. 1990a) AND $La_{1.86}Sr_{0.14}CuO_4$ (MIGLIORI ET AL. 1990b) AT 297 K

	c_{11}	c_{22}	c_{33}	c_{44}	c_{55}	c_{66}	c_{12}	c_{13}	c_{23}	B
$LaCuO_4$	172	171	200	65.6	65.8	96.8	90	73	73	113
$La_{1.86}Sr_{0.14}CuO_4$	248	—	204	67.4	—	58.3	48	—	65	—

for Y) had a 10–15% larger Rayleigh velocity. In a study of ceramic samples, the bulk modulus and Poisson ratio decreased by about 20% as δ was varied from 0.5 to 0 (Lemmens et al., 1990). A stiffening of the transverse mode as δ increased from 0.4 to 0 of about 20% was noted elsewhere. It thus appears to be difficult to draw any unambiguous conclusions from the studies. The variation in the ceramic density with different preparation conditions and treatment may be an important factor.

In summary, the static elastic stiffnesses of the high-T_c superconductors appear to be unremarkable. The elastic properties are anisotropic, but characteristic of three-dimensionally bonded solids—in contrast, for example, to the large anisotropy evidenced in the dc conductivity. It is possible that BSCCO may be more anisotropic; complete sets of elastic constants have not yet been obtained for this material.

4.1.2. Temperature Dependence of Sound Velocities and Damping

4.1.2.1. Low-Frequency Young's Modulus There have been two investigations of the temperature-dependent Young's modulus in single-crystal $YBa_2Cu_3O_7$. Using a vibrating reed technique, Shi et al. (1989) reported measurements on a crystal with dimensions $1 \times 0.2 \times 0.05$ mm^3 at a frequency near 20 kHz. Their results for the Young's modulus sound velocity are shown in Fig. 4b for temperatures between 270 and 4.2 K. As in most crystals, the sound velocity has negative slope with respect to T, as expected from normal anharmonicity. There appears to be a distinct slope change in the vicinity of 240 K, but a very slight slope change at 90 K near T_c (this will be discussed in more detail in Section 4.2). The overall velocity stiffening on cooling over this range is greater than 4.5%.

Similar measurements have been reported by Golding et al. (1988b). The Young's modulus velocity of a vibrating-reed sample of dimensions $2.5 \times 0.2 \times 0.02$ mm^3, oriented similarly to the sample of Shi et al. (see Section 3.1). The fundamental resonance occurred near 3.4 kHz. The velocity shown in

7. Single-Crystal High-T_c Superconductors

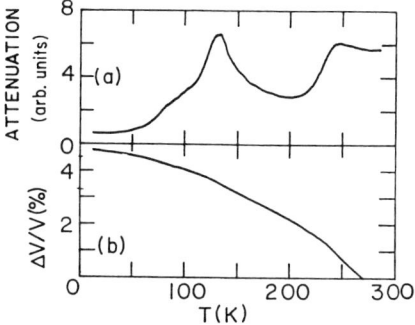

FIG. 4. (a) Damping of 20 kHz flexural mode for a single crystal reed of $YBa_2Cu_3O_7$. (b) Relative change of the Young's modulus sound velocity (Shi *et al.*, 1989).

Fig. 5 has many features in common with Shi *et al.* (1989). There is a distinct slope change near 240 K and a weak cusp near 90 K. A significant difference exists, however, in the overall stiffening: only 1.7% from 275 to 1.5 K. The reason for this large discrepancy between the two investigations is not understood. The frequency difference is an unlikely explanation. Measurements on the second overtone of this crystal at 22 kHz for temperatures below 90 K show very little difference with the 3.4 kHz data (see Section 4.1.4). It is possible that differing

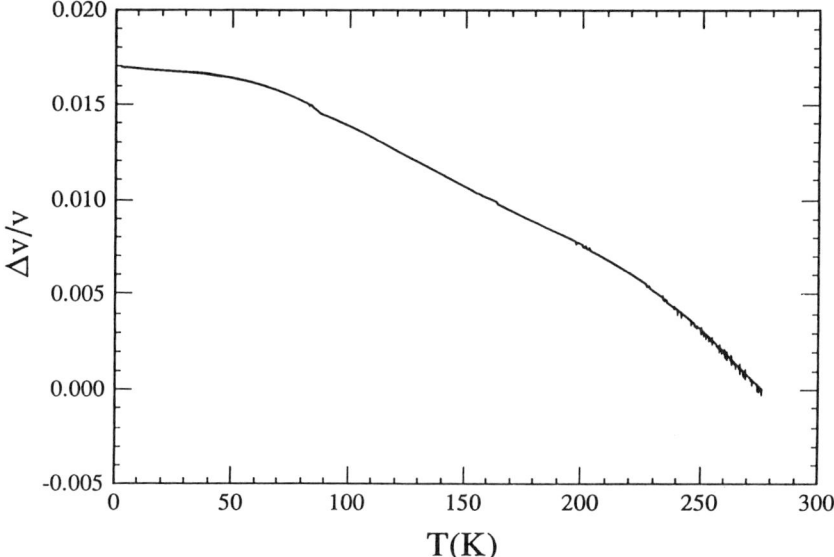

FIG. 5. Relative change of the 3.4 kHz Young's modulus velocity for a single crystal of $YBa_2Cu_3O_7$. Note the softening that takes place above the superconducting transition at 88 K.

stoichiometry, principally oxygen content, may play some role, although such a large effect on anharmonicity would be surprising. This issue needs to be clarified in future investigations.

4.1.2.2. Low-Frequency Damping Above T_c, there are two prominent features in the damping. As noted by Shi *et al.* (1989) and shown in Fig. 4a, there is a feature near 240 K associated with the velocity anomaly. A second prominent feature occurs near 130 K. The behavior observed by Golding *et al.* (1988) in Fig. 6 is appreciably different. The 240 K feature is less prominent, and the damping possesses a quasi-exponential temperature-dependent background. A peak occurs at 115 K at 3.4 kHz; measurements at 22 kHz on the same crystal show that this peak temperature has shifted to 125 K. Such behavior is consistent with a picture of strain coupling to a defect located in a low symmetry site. For a relaxational process, the peak occurs when $\omega = \tau^{-1}$, where ω is the angular excitation frequency and τ^{-1} is the defect hopping rate. For a thermally activated process, $\tau^{-1} \sim \exp(-\Delta E/kT)$, where ΔE is the over-the-barrier hopping energy. Using only the two frequencies, ΔE is estimated to be ~ 0.2 eV.

4.1.2.3. Longitudinal Plane Wave Velocities Longitudinal modes excited at 1 GHz have been reported by Golding *et al.* (1988a). Propagation along the *c*-axis and in the *a–b* plane is shown in Fig. 7. Very little structure is visible in these data. The feature at 240 K is not visible, and evidence of changes at T_c

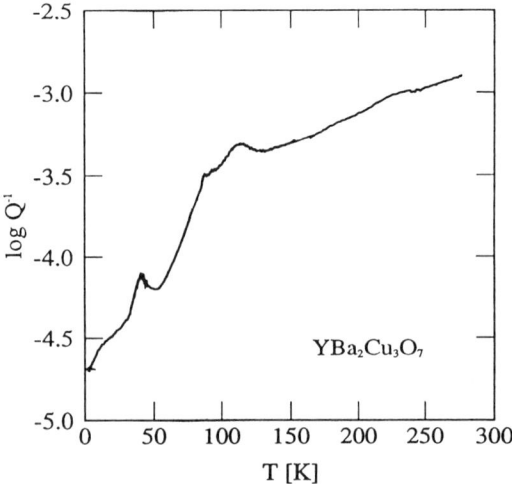

FIG. 6. Q^{-1} (damping) of 3.4 kHz flexural modes for a single crystal of YBa$_2$Cu$_3$O$_7$ (Golding *et al.*, 1988b). Note the logarithmic scale for Q^{-1}.

7. Single-Crystal High-T_c Superconductors

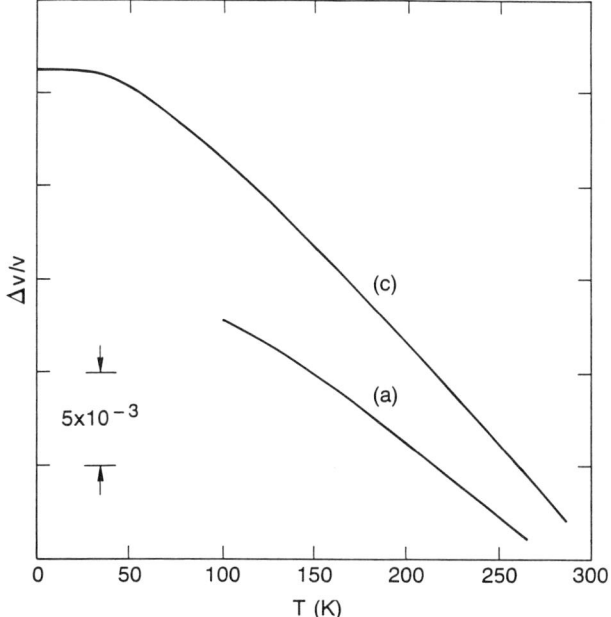

FIG. 7. Relative change in velocity of longitudinal modes propagating along the c-axis and in the ab-plane. The frequency is 0.96 GHz (Golding et al., 1988a).

is not easily discerned when data are displayed over this relatively large region of temperature. The c-axis mode is more anharmonic than the a-axis. The velocity increase on cooling from room temperature to 4.2 K is about 2% for the c-axis mode, about a factor of two larger than an estimate for the comparable ab-plane mode.

Measurements at much lower frequencies (5 MHz) (Kim et al., 1990) have shown unusual hysteresis when samples are cooled from room temperature and then heated. Differences of several percent are observed with a dependence on thermal history. Rather than stiffening on cooling from room temperature, the longitudinal velocity associated with the c_{33} mode softens by several percent from 200 K to approximately 50 K, Fig. 8a. On warming, hysteretic behavior is manifested by a velocity several percent lower than the cooling curve over much of the range. The results are sample-specific and point to ferroelasticlike relaxational processes associated with defects, possibly twin boundaries. Such phenomena were first noted in fine-grained, ceramic samples and appear to be closely related (Ewert et al., 1987; Lemmens et al., 1990). An alternate physical mechanism involving linear coupling to optical phonons has also been proposed (Horie et al., 1989). At high frequencies (above 10^8 Hz, perhaps), the defects

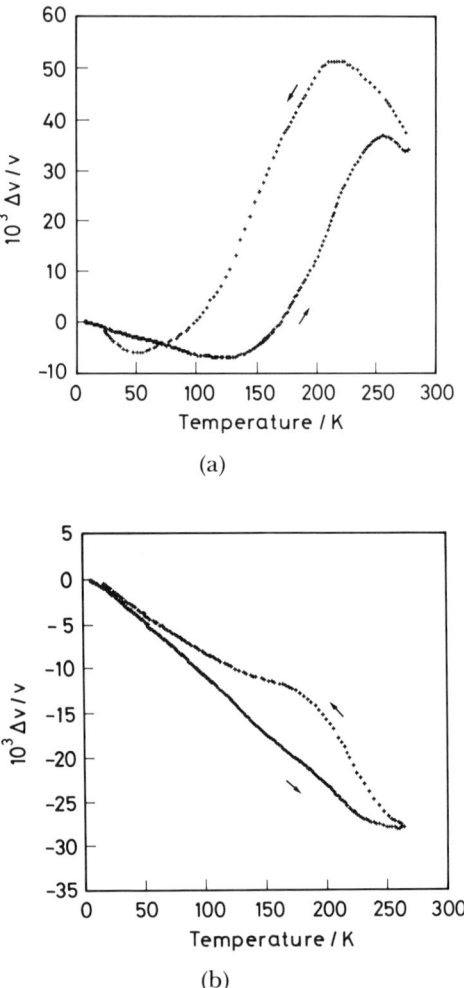

FIG. 8. (a) Temperature-dependent sound velocity at 5 MHz for propagation along the c-axis in $YBa_2Cu_3O_7$. Note the striking hysteretic effect that appears in this single crystal (Kim et al., 1990), similar to phenomena observed in ceramic $YBa_2Cu_3O_7$ (Ewert et al., 1987). (b) Temperature dependence of relative velocity and attenuation of the transverse c_{44} mode at 8 MHz (Kim et al., 1990). Note the lack of features at $T_c \approx 90$ K for this $YBa_2Cu_3O_7$ crystal.

cannot follow the rapidly oscillating strain field and cannot contribute to dispersion or dissipation.

It has been noted that the temperature coefficients of the velocity associated with c_{33} near 100 K show large variations among different investigations (Kim et al., 1990). For example, $(1/v)(dv/dT)$ at 100 K (Saint-Paul et al., 1989) is

7. Single-Crystal High-T_c Superconductors

reported to be greater than 2×10^{-4}/K. A value of 8×10^{-5}/K is reported at 53 MHz (Kim et al., 1990), whereas the data in Fig. 7 (Golding et al., 1988a) yield a slope close to 9×10^{-5}/K. There is a clear discrepancy in the slope reported by Saint-Paul et al. (1989) with the two other similar results.

4.1.2.4. Transverse Plane Waves: Velocities The temperature dependence of the c_{44} mode, in which transverse waves are propagated along the c-axis, is shown in Fig. 8b (Kim et al., 1990). Hysteretic behavior is clearly evidenced in the data at 8 MHz, with nearly 1% difference on heating and cooling. These measurements suggest a total stiffening for this shear mode of about 2.5% between 260 K and 4.2 K. An equal change is noted for the c_{11} mode at 8 MHz.

The temperature dependence of the transverse mode propagating in the ab-plane along [110] with polarization in the plane has been reported by Saint-Paul and Henry (1989). Since this mode shears the Cu–O plane, it may shed some light on the intraplanar coupling. Figure 9 shows data at 48 MHz from room temperature to 4.2 K. The overall change for his mode is quite small, $\approx 0.8\%$, similar to the longitudinal mode in the ab-plane. Interestingly, there is no evidence for anomalous behavior in the vicinity of T_c where anomalies in the lattice parameters of $YBa_2Cu_3O_7$ have been reported (Horn et al., 1987).

4.1.2.5. Transverse Attenuation The attenuation of the c_{44} mode with propagation along the c-axis and in the ab-plane have been reported (Kim et al., 1990). An interesting feature of the c-axis transverse data, Fig. 10, is the appearance of two absorption peaks at 125 and 220 K at a frequency of 8 MHz. Note that the background is hysteretic. These peaks show a correspondence to

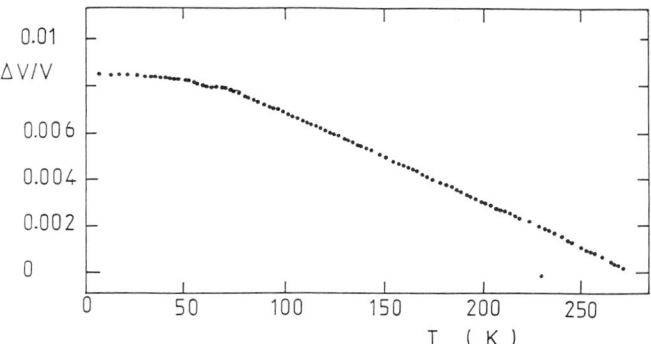

FIG. 9. Temperature dependence of the sound velocity change for the transverse mode associated with $(c_{11}-c_{12})/2$ at 48 MHz in $YBa_2Cu_3O_7$ (Saint-Paul and Henry, 1989). (© 1989 Pergamon Press plc., Reprinted with permission)

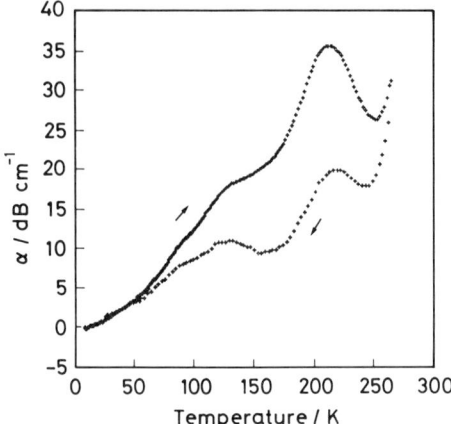

FIG. 10. Attenuation of the c_{44} mode measured by propagating 8 MHz transverse phonons along the c-axis of $YBa_2Cu_3O_7$. Note the extreme hysteretic effects on thermal cycling. The characteristic absorption maxima seen at lower frequencies in Figs. 4 and 6 are present (Kim *et al.*, 1990).

lower-frequency data (see Figs. 4 and 6), as well as to ceramic results at MHz frequencies. For the c_{66} mode, the attenuation is monotonically decreasing from room temperature without any clear sign of the attenuation peaks.

4.2. The Superconducting Transition Region

4.2.1. Elastic Mean-Field Theory Near T_c

The superconducting transition can be regarded as an example of a second-order mean-field transition. The elastic anomalies and their connection to thermodynamic singularities that occur at the transition have been extensively discussed (Testardi, 1973; Millis and Rabe, 1988). The predictions of the model relate the mean field discontinuity in the specific heat C_p at T_c to the following:

(a) discontinuities in compressional elastic moduli and sound velocities;
(b) discontinuities in the temperature derivatives of the elastic moduli and sound velocities; and
(c) discontinuities in the temperature derivatives of the lattice parameters, i.e., thermal expansion coefficients.

Adopting the notation of Millis and Rabe (1988), the elastic free energy in the normal state is

7. Single-Crystal High-T_c Superconductors

$$F_n(\vec{\varepsilon}) = \frac{1}{2}\varepsilon_i c_{ij}\varepsilon_j + \frac{1}{2}\tilde{c}_i\tilde{\varepsilon}_i^2 + \Gamma_{ijk}\varepsilon_i\varepsilon_j\varepsilon_k + \Lambda_{ij}\varepsilon_i\tilde{\varepsilon}_j^2 + \sigma_i\varepsilon_i, \quad (2)$$

where $i, j, k = 1,2,3$; ε_i represents diagonal strains ε_{xx}, ε_{yy}, ε_{zz}, $\tilde{\varepsilon}_i$ represents off-diagonal strains ε_{xy}, ε_{yz}, ε_{xz}; c_{ij} and $\tilde{c}_i$ are the orthorhombic elastic stiffnesses; γ_{ijk} and λ_{ij} are third-order elastic moduli; and the σ_i are the applied stresses. The superconducting transition temperature strain dependence can be expressed as

$$T_c = T_{c0}\left(1 + \alpha_i\varepsilon_i + \frac{1}{2}\varepsilon_i\Delta_{ij}\varepsilon_j + \frac{1}{2}\tilde{\Delta}_i\tilde{\varepsilon}_i^2 + \ldots\right), \quad (3)$$

where $\alpha_i \equiv d\ln T_c/d\varepsilon_i$, $\Delta_{ij} \equiv (1/T_{c0})d^2T_c/d\varepsilon_i d\varepsilon_j$, and $\tilde{\Delta}_i \equiv (1/T_{c0})d^2T_c/d\tilde{\varepsilon}_i^2$. By expanding the free energy in powers of $T-T_{c0}$ and ε, one finds to second order that the jump in the elastic stiffness at T_c is

$$\frac{\Delta c_{ij}}{c_{ij}} = \frac{1}{c_{ij}}[c_{ij}(T_c^+) - c_{ij}(T_c^-)] = \frac{\Delta C_p T_c}{c_{ij}}\alpha_i\alpha_j, \quad (4)$$

where ΔC_p in the specific-heat jump at T_c. Since the elastic discontinuity is small in practice ($<10^{-4}$), the corresponding jump in sound velocity for mode i is

$$\frac{\Delta v_i}{v_i} = \frac{\Delta C_p T_{c0}\alpha_i}{2 c_{ij}}. \quad (5)$$

For orthorhombic symmetry or higher, the $\tilde{c}_{jk}$ do not exhibit a discontinuity at T_c.

The temperature derivatives of the elastic moduli show discontinuities at T_c given by

$$T_{c0}\left\{\frac{d}{dT}c_{ij}\bigg|_{T_{c0}^+} - \frac{dc_{ij}}{dT}\bigg|_{T_{c0}^-}\right\} = \Delta C_p T_{c0}[3A\alpha_i\alpha_j - 2\alpha_i\beta_i - \Delta_{ij} + 6\Gamma_{ijk}c_{kl}^{-1}\alpha_l] \quad (6)$$

and

$$T_{c0}\left\{\frac{d\tilde{c}_i}{dT}\bigg|_{T_{c0}^+} - \frac{d\tilde{c}_i}{dT}\bigg|_{T_{c0}^-}\right\} = -\Delta C_p T_{c0}[\tilde{\Delta}_i + 2c_{jk}^{-1}\alpha_k\Lambda_{ji}], \quad (7)$$

where A and β_i are coefficients in the free energy expansion near T_c. The thermal expansivity is discontinuous at T_c and is given by

$$T_{c0}\left\{\frac{d\varepsilon_i}{dT}\bigg|_{T_c^+} - \frac{d\varepsilon_i}{dT}\bigg|_{T_c^-}\right\} = -\Delta C_p T_{c0} c_{ij}^{-1}\alpha_j. \quad (8)$$

The preceding expressions are derived from mean-field theory and neglect critical fluctuations. The evidence for fluctuation phenomena is beyond the scope of this chapter. The interested reader is referred to Millis and Rabe (1988) and Inderhees et al. (1988, 1991).

4.2.2. Sound Velocity Experiments near T_c

4.2.1.1. Low-Frequency Measurements Vibrating-reed experiments at 20 kHz of the Young's modulus have been reported for $YBa_2Cu_3O_7$ crystal with a superconducting onset temperature of 92 K (Shi et al., 1989), Fig. 11. Examination of the data shows that there is clearly a slope difference above and below the "plateau" region near 90 K. In addition, there is a smeared discontinuity, most clearly seen by extrapolating the background velocities to T_c. This offset is associated with the mean-field discontinuity discussed earlier. Since the Young's modulus distortion contains shear and compressional strain, the jump at T_c must be associated with the compressional strain. The discontinuity at T_c may not appear obvious in the raw data, since the transition is probably broadened by several K. Extrapolation of the T-dependent velocity to T_c ($\approx$89 K) from above and below leads to a discontinuity $\delta E/E = 190 \times 10^{-6}$. If T_c were chosen to be higher than 89 K, an even greater apparent discontinuity would result. The discontinuity in the temperature derivative of E is determined to be 35×10^{-6}/K. The same authors described the study of a torsional mode near T_c. No discontinuity of the shear velocity was seen, but a slope discontinuity of 74×10^{-6}/K was reported.

A very similar experiment on the Young's modulus velocity was conducted by Golding et al. (1988b), as illustrated by the data in Fig. 12. Sound velocities

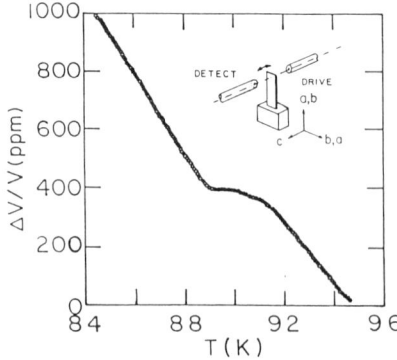

FIG. 11. Detail of transition region of Fig. 4. Inset shows sample geometry for flexural excitation (Shi et al., 1989).

7. Single-Crystal High-T_c Superconductors

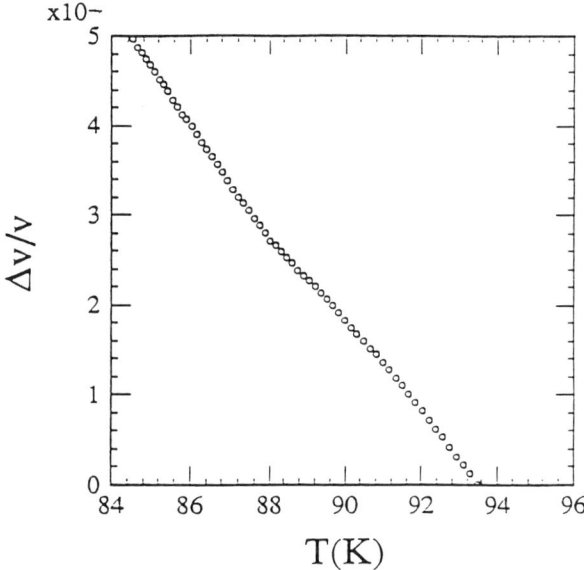

FIG. 12. Detail of transition region for single crystal YBa$_2$Cu$_3$O$_7$ at 22 kHz (Golding *et al.*, 1988). Note the different vertical scales between these results and those in Fig. 11.

near T_c are shown for measurements at 22 kHz, but essentially identical results were found at 3.4 kHz. As noted earlier, this investigation found a background slope more than a factor of two smaller than Shi *et al.* (1989). In Fig. 12, the apparent discontinuity at T_c, taken as 88.7 K, is approximately $40 \pm 5 \times 10^{-6}$, almost a factor of five smaller than the results by Shi *et al.*, but more consistent with high-frequency measurements discussed in the next section. The discontinuity of the temperature derivatives of the velocity is 15×10^{-6}/K.

4.2.2.2. High-Frequency Measurements Measurements of the sound velocity discontinuity at 1 GHz for longitudinal modes propagating along the *c*-axis and the twinned *ab*-axis have been made (Golding *et al.*, 1988a). As shown in Figures 13 and 14 (open circles, lower data) for the *c*-axis mode, an extremely sharp anomaly is observed at 87.3 K. Extrapolation of the background to this temperature results in a discontinuity $\Delta v/v = 30 \pm 5 \times 10^{-6}$. Note that the background above T_c is not particularly well represented by a linear function of T. There is appreciable curvature, indicating enhanced softening at least 5 K above T_c. Such behavior is most likely to be a manifestation of superconducting fluctuations seen in the specific heat (Inderhees *et al.*, 1988; Millis and Rabe, 1988). Also shown in Fig. 14 are *ab*-plane data, also at 1 GHz. Although the

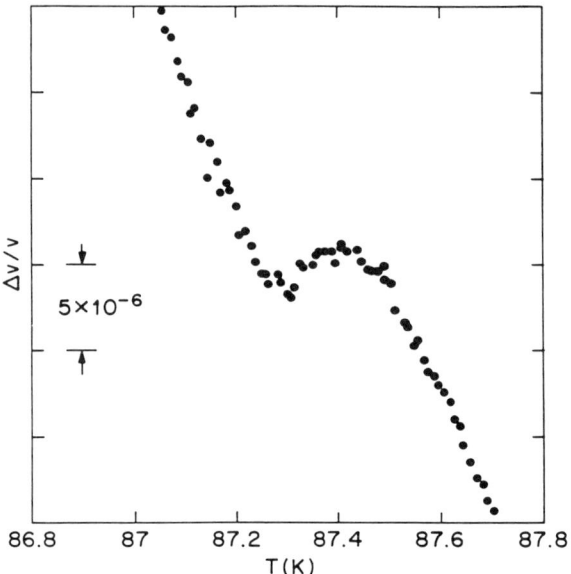

FIG 13. Highly expanded view of transition region shown in Fig. 7, for 0.96 GHz c-axis propagation in $YBa_2Cu_3O_7$. On this scale the discontinuity at $T_c \cong 87.3$ is readily apparent.

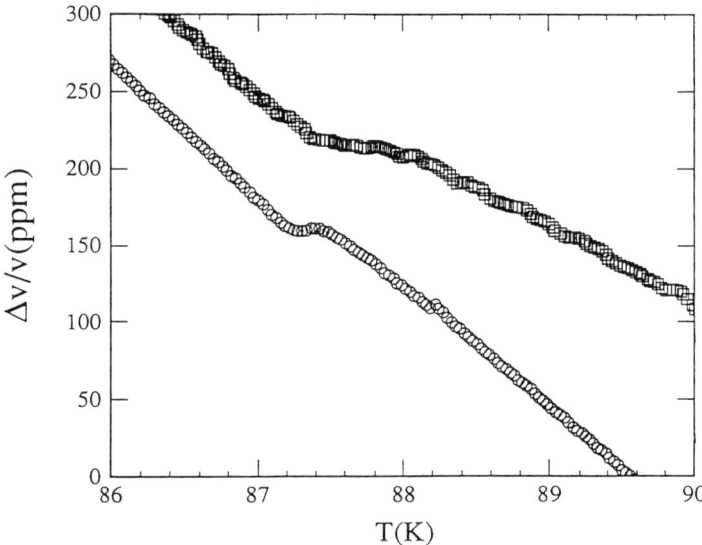

FIG. 14. Expanded view of the transition region in $YBa_2Cu_3O_7$ for c-axis (circles) and ab-plane (squares) propagation at 0.96 GHz. Note the additional broadening of the transition for the ab-plane mode.

T_c is the same, the transition is appreciably broadened, even though the measurements were made on the same crystal.

In the GHz experiments, an extremely small volume of $YBa_2Cu_3O_7$ is sampled by the sound beam, approximately 0.05 mm in diameter by 0.2 mm. The acoustically excited region is relatively homogeneous in comparison to the larger crystals utilized in vibrating-reed and lower-frequency plane wave experiments. Nevertheless, some evidence for the existence of large-scale inhomogeneity can be inferred by comparing c-axis with ab-axis data for the same crystal. As noted above, the transition broadening is enhanced by at least a factor of two to three for ab-propagation. Since incomplete oxygenation is the most likely origin of a distribution in T_c, and since oxygen diffuses most slowly along the c-direction, one would conclude that c-axis propagation samples a very small composition gradient. Although the crystal was annealed in oxygen for four weeks after growth, this was apparently not enough to fully equilibrate the oxygen concentration on a scale of 0.1 mm. When the velocity backgrounds are extrapolated to 87.3 K, a discontinuity $(25 \pm 5) \times 10^{-6}$ is obtained for ab-propagation, not appreciably different from the c-axis value.

Other measurements of the discontinuity for c-axis propagation have been subsequently reported. Measurements at 15 MHz (Saint-Paul et al., 1989) have been reported for a crystal with T_c of 89 K. A discontinuity for this mode of 50×10^{-6} is obtained. Measurements at 53 MHz (Kim et al., 1990) for c-axis propagation and 9 MHz for ab-plane propagation have also been described. In both cases, discontinuities of less than 50×10^{-6} can be estimated, although with appreciable uncertainty. It is interesting to note that it was possible to obtain a reasonable value for the jump at T_c in spite of an enormous hysteresis appearing during thermal cycling—for example, see Fig. 9.

It is of some significance to calculate the strain dependence of T_c from Eq. (5) and the experimental results. Golding et al. (1988a), using $\Delta C_p = 50$ mJ/cm^3/K from Junod et al. (1988) and their 1 GHz data, find $|\alpha_c| = 1.41$ and $|\alpha_{ab}| = 1.79$. The result for strain in the ab-plane represents an average over the a- and b-axes because of the microtwinned sample. Note that it is not possible to determine the sign of the strain derivatives. Nevertheless, these results are remarkable. They show no evidence for the large anisotropy exhibited by almost all other properties of $YBa_2Cu_3O_7$. Particularly interesting is that T_c is not particularly sensitive to strain parallel to the c-axis. In view of the proposal (Anderson, 1990) that T_c should depend strongly (possibly exponentially) on interplanar coupling in the RVB model, these results take on additional significance.

The left-hand side of Eq. (6) for the logarithmic temperature derivatives of the compressional elastic moduli can also be evaluated. One finds 1.9×10^{-3} and 4.9×10^{-3}, obtained respectively from c-axis and ab-axis sound propagation.

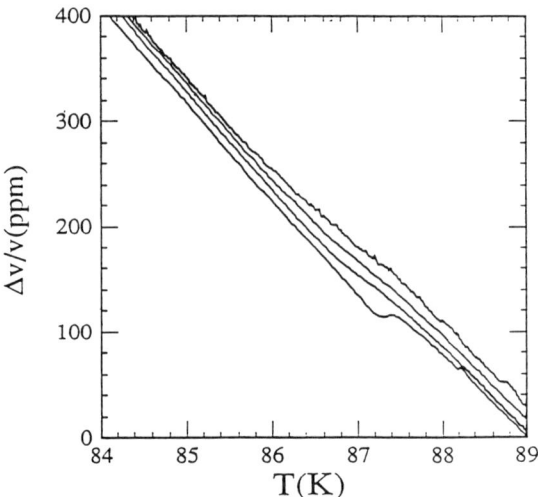

FIG. 15. Effect of a perpendicular magnetic field on the velocity discontinuity at T_c in YBa$_2$Cu$_3$O$_7$. The lowest curve is $H = 0$. The field is increased by 1 T for each consecutive curve. Note the rounding of the discontinuity and the shift of the midpoint to lower temperature as the field is increased (Golding *et al.*, 1988a). (© 1988 IEEE)

These numbers compare favorably to estimates derived from ceramic data and quoted by Millis and Rabe (1988).

The discontinuity in the thermal expansivity can be obtained from Eq. (8). The expansivity jump along the c-axis is 0.49×10^{-6}/K, whereas the average ab-axis jump is 0.38×10^{-6}/K. These values are quite small and are consistent with, but below, the resolution of x-ray or neutron lattice parameter measurements (David *et al.*, 1987).

The influence of external magnetic fields has been studied by Golding *et al.* (1988). Results shown in Fig. 15 for c-axis propagation indicate that T_c shifts to lower temperature as H increases. The shift is consistent with the phase boundary (Worthington *et al.*, 1987). $dH_{c2}/dT = -0.8$ T/K. In contrast to typical superconductors, the discontinuity is broadened by unusually small magnetic fields. A similar phenomenon has been noted in calorimetric measurements near T_c by Inderhees *et al.* (1991) and attributed to finite-size effects due to a restricted coherence length $\xi^\perp$.

4.3. THE SUPERCONDUCTING STATE

In this section we discuss several features of sound propagation in superconductors below T_c. In conventional superconductors, great attention has been

7. Single-Crystal High-T_c Superconductors

devoted to understanding the influence of the superconducting gap on attenuation of sound by electrons. For a BCS superconductor, the ratio of superconducting to normal electronic absorption is given by (Schrieffer, 1981)

$$\frac{\alpha_s}{\alpha_n} = \frac{2}{1 + \exp[\Delta(T)/k_B T]}, \quad (7)$$

provided that $h\nu \ll k_B T_c$, where ν is the phonon frequency. A short electronic scattering length l_e leads to a very small α_n. In the $ql_e \ll 1$ limit, α_n varies as ν^2, and it has been estimated that for YBa$_2$Cu$_3$O$_7$, $\alpha_n/\nu^2 \sim 10^{-2}$ cm^{-1} GHz^{-2} (Golding et al., 1988a). Thus, a determination of $\Delta(T)$ via Eq. (7) is difficult because the absolute electronic absorption change below T_c is small and other temperature-dependent absorption mechanisms may mask the electronic one. Note that even though it would appear advantageous to attempt high-frequency experiments, it would be beneficial only if other nonelectronic background contributions vary more weakly than ν^2. To date there have been no acoustic experiments with a realistic claim to observing electronic absorption in high-T_c materials.

In the following sections, we focus on absorption occurring below T_c that appears to be largely relaxational. At temperatures below 10 K, dispersive phenomena similar to that observed in glassy materials have been observed. Study of these tunneling excitations, as well as other defects, may be able to clarify electron–phonon interactions in these substances.

4.3.1. Absorption Anomalies Below T_c

There have been extensive reports of attenuation peaks below T_c in YBa$_2$Cu$_3$O$_7$. Figure 4a shows a relatively weak shoulder that straddles the region at 90 K near T_c. A similar but somewhat sharper peak at the same temperature is seen in Fig. 6. In spite of the large differences that exist between the two studies, these, as well as other absorption maxima, are qualitatively similar. At lower temperatures, additional structure in the absorption has been discovered (Golding et al., 1988b), Fig. 16. The plots show $Q^{-1}(T)$ for the fundamental and overtone of a vibrating-reed crystal. Two additional peaks are seen: a higher-temperature peak between 40 and 50 K that shifts to higher temperature as the frequency increases, and a low-temperature peak at 15 K that appears only weakly frequency-dependent. Note also that the background damping continues to decrease to the lowest temperature reached in this experiment, with Q approaching 50,000.

Attenuation maxima below T_c have been noted in higher-frequency studies of a longitudinal mode propagating in the basal plane (Saint-Paul and Henry, 1989), Fig. 17. A very strong absorption peak moves upward in temperature as the

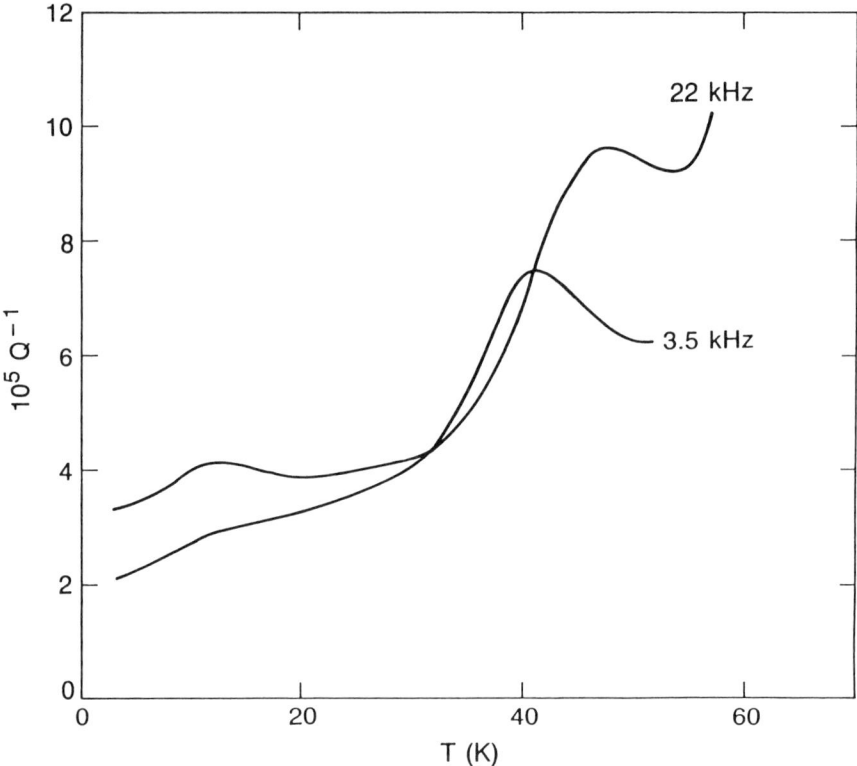

FIG. 16. Q^{-1} for single-crystal reed of $YBa_2Cu_3O_7$ at the fundamental and overtone frequencies of 3.4 and 2.2 kHz. Note the location of the damping peaks, as well as the low background as $T \to 0$.

frequency increases from 15 to 600 MHz. The attenuation has been analyzed as a relaxation process with a single relaxation rate. An activation energy E/k_B = 480 K and a pre-exponential attempt frequency of $\sim 10^{12}$ s^{-1} have been extracted by this analysis. It is suggested that the relaxation involves motion of oxygen modes, possibly involving the O4 apical oxygen. However, there is as yet no strong evidence linking any of the relaxational processes to specific optical modes or hopping entities in $YBa_2Cu_3O_7$.

4.3.2. Low-Temperature Sound Velocities

As $T \to 0$, it is usually expected that anharmonic processes will eventually freeze out and that the temperature dependence of the elastic moduli will disappear, i.e., $dc_{ij}/dT \to 0$ in this limit. Although this situation will obtain for highly perfect crystals, excitations arising from disorder can alter the scenario. The

7. Single-Crystal High-T_c Superconductors

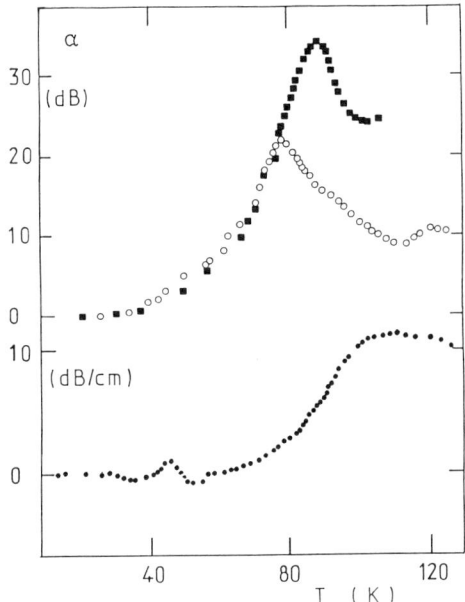

FIG. 17. Attenuation maxima for longitudinal waves propagating along (110) in $YBa_2Cu_3O_7$. Frequencies: 15 MHz, solid circles; 300 MHz, open circles; 600 MHz, squares (Saint-Paul and Henry, 1989). Note how the peak shifts to higher temperature as the frequency increases. (© 1989 Pergamon Press plc., Reprinted with permission)

most familiar examples occur in insulating and metallic glasses. On cooling, dv/dT, which has negative slope above a few kelvins, can change its sign and vary as log T at lower temperatures. At much lower temperatures, another sign reversal takes place. These phenomena can be understood in terms of resonant and relaxational interactions between phonons and two-level tunneling systems, which can be thought of as highly anharmonic local modes. The tunneling entities are atoms or groups of atoms that quantum-mechanically tunnel between two local potential minima. Disorder induces a distribution of tunneling energies, and a typically flat spectrum gives rise to a linear temperature-dependent contribution to the specific heat C_p (Phillips, 1981). The tunneling system C_p is usually smaller than the normal electronic C_p, but can be observed in superconductors at temperatures well below T_c (Graebner et al., 1977). The interested reader is referred to the literature for more details (Black, 1981; Hunklinger and Raychaudhuri, 1986). Competing relaxational and resonant contributions create a maximum in the sound velocity in glasses. Such a maximum, and thus a similarity to glasses, was noted in $YBa_2Cu_3O_7$ below 1 K in a ceramic vibrating-reed sample (Golding et al., 1987) for measurements at a single frequency. It was reasoned that the tunneling levels interacted primarily with phonons and not with conduction electrons, as would be expected for a superconductor for $T \ll T_c$. It was noted that oxygen atoms are the most likely candidates for a tunneling system in $YBa_2Cu_3O_7$.

More recently, data have been reported for ceramic $YBa_2Cu_3O_7$ (McKenna et al., 1989) that show a velocity maximum that is independent of frequency from 5 to 25 MHz. Since phonon relaxation predicts that the temperature of the velocity maximum T_m should scale as $\omega^{1/3}$, whereas electron relaxation should be essentially frequency-independent, McKenna et al. concluded that an appreciable density of unpaired conduction electrons exists below T_c. If true, this would suggest gapless behavior and a non-BCS ground state. However, it is generally accepted (Batlogg, 1990) that pairing in $YBa_2Cu_3O_7$ is into an s-wave state with a uniform gap.

In order to understand this issue, it is clearly necessary to study the frequency dependence over a sufficiently wide range. Figure 18 shows vibrating-reed velocities for ceramic $YBa_2Cu_3O_{7-\delta}$ with $\delta = 0$ and 0.3. The samples were of particularly high quality obtained from the same materials used previously to establish the $\delta = 0.3$, 60 K plateau for oxygen-deficient $YBa_2Cu_3O_{7-\delta}$ (Cava et al., 1987). The $\delta = 0$ sample shows a peak at $T_{max} \approx 50$ mK, whereas the

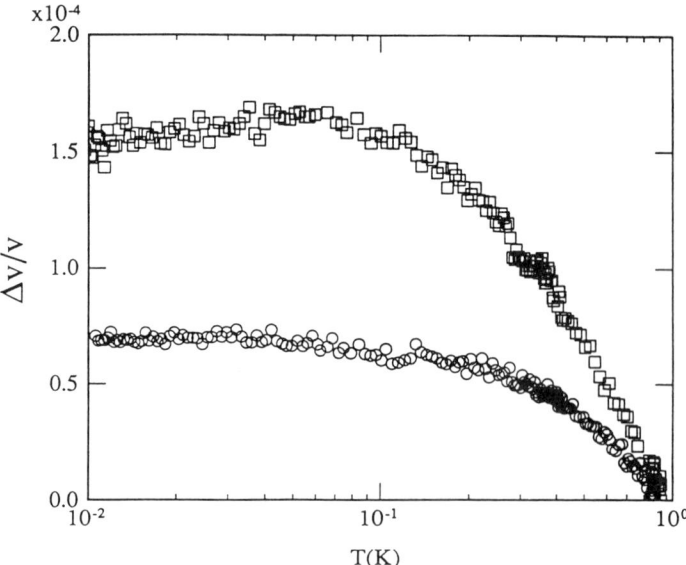

FIG. 18. Velocity shift below 1 K for vibrating reeds of $YBa_2Cu_3O_{7-\delta}$ ceramics. Squares: $\delta = 0$; circles, $\delta = 0.3$. Note the clear maximum that occurs near 50 mK for the $\delta = 0$ sample, which signals a crossover from resonant to relaxational regimes as T increases (Golding et al., 1991).

7. Single-Crystal High-T_c Superconductors

$\delta = 0.3$ sample ($T_c \approx 60$ K) exhibits an extremely shallow maximum near 30 mK. This suggests that tunneling centers possess a higher density of states when in highly oxygenated $YBa_2Cu_3O_{7-\delta}$. The differences in T_{max} between these results and the earlier study (Golding *et al.*, 1987) reflect sample quality, as the early sample had been exposed to atmospheric H_2O and CO_2.

Evidence for a velocity maximum for *single-crystal* propagation has been observed (Golding *et al.*, 1988a). Figure 19 shows the peak near 3 K for *c*-axis propagation at 1 GHz. Below T_m, a quasi-logarithmic temperature dependence exists.

It is now possible to combine ceramic and single crystal data into a single graph of T_{max} vs. frequency. Figure 20 shows such a plot spanning nearly six decades in frequency. The dotted line indicates the dependence expected for phonon-mediated relaxation. The data of McKenna *et al.* fall close to this line, particularly at their lower frequencies. These results argue convincingly that phonons are relaxing the tunneling levels, and that conduction-electron–mediated relaxation need not be invoked.

The conclusion that emerges from these low-temperature experiments is that even high-quality single crystals, as well as ceramic samples of $YBa_2Cu_3O_7$,

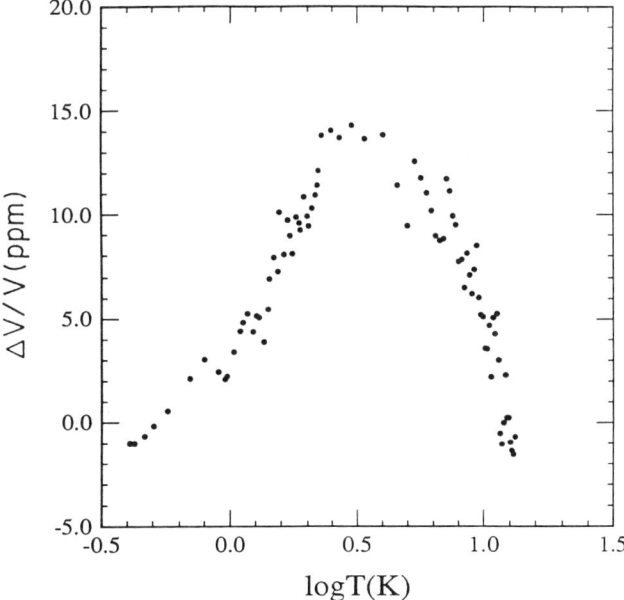

FIG. 19. Sound velocity shift at 0.96 GHz for single-crystal $YBa_2Cu_3O_7$ with *c*-axis propagation. A similar feature, *viz.* a maximum at 3 K, is observed also for *ab*-propagation.

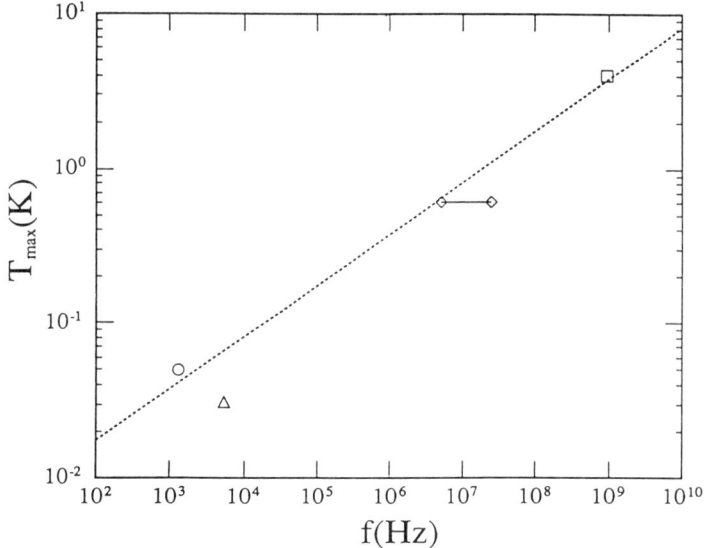

FIG. 20. Plot of T_{max}, the temperature of the velocity maximum in YBa$_2$Cu$_3$O$_7$, for several experiments carried out over a 10^6 range of frequencies. Circle and triangle, $\delta = 0$ and $\delta = 0.3$ described in Fig. 18; diamond (McKenna et al., 1989); square, single crystal depicted in Fig. 19.

contain broad distributions of low-lying tunneling excitations. The modes associated with the disorder couple strongly to phonons, and their density of states is enhanced when the chain sites are highly occupied.

5. Summary and Outlook

The experimental investigation of high-T_c superconductors via acoustic and ultrasonic methods is clearly a promising, but presently immature, field of study. As noted in the introduction to this chapter, there are several valid reasons for the relatively sparse information derived from single crystals. The most obvious difficulty is the absence of well-characterized, reasonably large, and homogeneous crystals amenable to "conventional" ultrasonic techniques. Nevertheless, progress has been made by developing or modifying the techniques to accommodate the small samples presently available.

Some areas in which useful information has emerged are (1) reliable data on static elastic stiffnesses and sound velocities, (2) elastic properties in the vicinity of T_c, and (3) relaxational processes responsible for multiple absorption peaks. Rather than summarizing results already described within the chapter, it may be

7. Single-Crystal High-T_c Superconductors

more appropriate to mention some areas that require greater attention—as well as to describe topics that were neglected in this review.

Critical Phenomena. The short coherence lengths in 2-D superconductors are responsible for enhanced order-parameter fluctuations near T_c. A careful analysis of sound velocities near T_c that incorporates corrections to mean-field behavior would be valuable and complementary to previous specific heat studies (Inderhees et al., 1991).

Oxygen Stoichiometry. The role of oxygen, or oxygen vacancies, in defect relaxational processes has not been assessed. Acoustic methods should be extremely sensitive to local defect motions and configurations. Investigations of the δ-dependence of relaxation strengths in single crystals can help in distinguishing among several potential contributors—for instance, bistable apical oxygens, plane vs. chain sites, the charge state of copper, optical modes, coupling to carriers, and the role of disorder.

Magnetism. The planar structures of the layered copper oxides provide ideal systems for studies of 2-D, $s = \frac{1}{2}$ physics. In addition, substitution with transition metal or lanthanide elements has led to several examples of ordered magnetic states. If spin–phonon couplings have typical magnitudes, acoustic methods could be used to study magnetic phase transitions and magnetic critical phenomena in reduced dimensionality.

Flux Lattices. In an external magnetic field, effects involving cooperative behavior of vortices and their pinning have received a great deal of attention. Low-frequency acoustic methods have been instrumental in detecting transitions between states of the flux lattice (Gammel et al., 1988). An article by Chien et al. (1991) contains references to earlier work.

Structural Transitions. Phase transitions driven by the condensation of soft phonon modes occur in several of the perovskite-related phases. The best studied example is the tetragonal–orthorhombic transition in $La_{2-x}Sr_xCuO_{4-y}$ (Fossheim and Laegreid, 1989; Migliori et al., 1990b).

Non-cuprate Oxides. The discovery of superconductivity near 30 K in K-doped $BiBaO_3$ has raised questions concerning the necessity for copper-based magnetic correlations for high-T_c superconductivity. This is currently an intensely debated question, and ultrasonic measurements of absorption arising from electron–phonon interactions should be valuable in understanding the electronic excitations.

There is, as yet, no compelling explanation for the phenomenon of high-T_c superconductivity. Since extraordinarily rich physics is still being uncovered in the search for a mechanism, one may wish to view this fact as a benefit—and not yet a liability. There is still time for a definitive experiments, possibly using physical acoustics, to provide the long-sought answers.

ACKNOWLEDGMENTS

I thank J.E. Graebner and A.J. Millis for helpful conversations, B. Batlogg, R.J. Cava, and L.F. Schneemeyer, for samples used in my experiments, and W.H. Haemmerle for his invaluable technical assistance over many years.

References

Aleksandrov, I., Goncharov, A., and Stishov, S. (1988). *JETP Lett.* **47**, 428.
Aleksandrov, V. V., Velichkina, T. S., and Voronkova, V. I., Yakovlev, I. A., and Yanovskii, V. K. (1990). *Solid State Commun.* **73**, 559.
Anderson, P. W. (1990). "High-Temperature Superconductivity—The Los Alamos Symposium." Edited by D. Bedell, D. Coffey, D. Meltzer, D. Pines, and J. R. Schrieffer. Addison-Wesley, Redwood City, California.
Barmatz, M., and Golding, B. (1974). *Phys. Rev.* **B9**, 3064.
Batlogg, B. (1990). "High-Temperature Superconductivity—The Los Alamos Symposium." Edited by D. Bedell, D. Coffey, D. Meltzer, D. Pines, and J. R. Schrieffer. Addison-Wesley, Redwood City, California.
Baumgart, P., Blumenroder, S., Erle, A., Hillebrands, B., Günterodt, G., and Schmidt, H. (1989). *Solid State Commun.* **69**,1135.
Bednorz, J. G. and Mueller, K. A. (1986). *Z. Phys.* **B64**, 189.
Beyers, R., Ahn, B. T., Gorman, G., Lee, V. Y., Parkin, S. S. P., Ramirez, M. L., Roche, K. P., Vazquez, J. E., Gur, T. M., and Huggins, R. A. (1989). *Nature* **340**, 619.
Black, J. L. (1981). "Glassy Metals I." Edited by H.-J. Güntherodt and H. Beck, Springer-Verlag, Berlin.
Capponi, J. J., Chaillout, A. W., Hewat, P., Lejay, M., Marezio, M., Nguyen, N., Raveau, B., Soubeyroux, J. L., Tholence, J. L., and Tournier, R. (1987). *Europhysics Lett.* **3**, 1301.
Cava, R. J., Batlogg, B., Chen, C. H., Rietman, E. A., Zahurek, S. M., and Werder, D. (1987). *Phys. Rev.* **B36**, 5719.
Chen, C. H. (1989). In "Physical Properties of High-Temperature Superconductors," (D. M. Ginsberg, ed.) World Scientific Press.
Chien, T. R., Jing, T. W., Ong, N. P., and Wang, Z. Z. (1991). *Phys. Rev. Lett.* **66**, 3075.
David, W. I. F., Edwards, P. P., Harrison, M. R., Jones, R., and Wilson, C. C. (1987). *Nature* **331**, 245.
Ewert, S., Guo, S., Lemmens, P., Stellmach, F., Wynants, J., Arlt, G., Bonnenberg, D., Klim, H., Comberg, A., and Passing, H. (1987). *Solid State Commun.* **64**, 1153.
Fleming, R. M., Batlogg, B., Cava, R. J. and Rietman, E. A. (1987). *Phys. Rev.* **B35**, 7191.
Fleury, P. A. (1970). "Physical Acoustics." Vol. VI. Academic Press, New York.
Fossheim, K., and Laegreid, T. (1989). *IBM J. Res. Dev.* **33**, 365.
Gammel, P. L., Schneemeyer, L. F., Waszczak, J. V., and Bishop, D. J. (1988). *Phys. Rev. Lett.* **61**, 1666.
Golding, B., Birge, N. O., Haemmerle, W. H., Cava, R. J., and Rietman, E. (1987). *Phys. Rev.* **B36**, 5606.
Golding, B., Haemmerle, W. H., Schneemeyer, L. F., and Waszczak, J. V. (1988a). *IEEE Ultrsonics Symposium Proc. 1988*, 1079.
Golding, B., Haemmerle, W. H., and Schneemeyer, L. F. (1988b). *J. Acoust. Soc. Am.* **84**, S109.
Golding, B., Haemmerle, W. H., and Schneemeyer, L. F. (1991). (unpublished).
Graebner, J. R., Golding, B., Schutz, R. J., Hsu, F. S. L., and Chen, H. S. (1977). *Phys. Rev. Lett.* **39**, 1480.
Horie, Y., Terashi, Y., and Mase, S. (1989). *J. Phys. Soc. Jpn.* **58**, 279.

Horn, P. M., Deane, D. T., Held, G. A., Jordan-Sweet, J. L., Kaiser, D. L., Holtzberg, F., and Rice, T. M. (1987). *Phys. Rev. Lett.* **59**, 2772.

Hunklinger, S., and Raychaudhuri, A. K. (1986). *Prog. in Low Temp. Phys.* **9**, 265.

Inderhees, S. E., Salamon, M. B., Goldenfeld, N., Rice, J. P., Pazol, B. G., Ginsberg, D. M., Liu, J. Z., and Crabtree, G. W. (1988). *Phys. Rev. Lett.* **60**, 1178.

Inderhees, S. E., Salamon, M. B., Rice, J. P., and Ginsberg, D. M. (1991). *Phys. Rev. Lett.* **66**, 232.

Junod, A., *et al.* (1988). *Physica* **C152**, 50.

Kim, T. J., Kowalewski, J., Assmus, W., and Grill, W. (1990). *Z. Phys.* **B78**, 207.

Ledbetter, H. and Lei, M. (1991). *J. Mater. Res.* **6**, 2253.

Lemmens, P. *et al.* (1990). *Proc. of the Euro. Mat. Res. Soc.*, Strasbourg.

Longo, J. M., and Raccah, P. M. (1973). *J. Sol. St. Chem.* **6**, 526.

Maeda, H., Tanaka, Y., Fukutomi, M., and Asano, T. (1988). *Jpn. J. Appl.Phys.* **27**, L209.

McKenna, M. J., Hikata, A., Takeuchi, J., Elbaum, C., Kershaw, R., and Wold, A. (1989). *Phys. Rev. Lett.* **62**, 1156.

Migliori, A., Visscher, W. M., Brown, S. E., Fisk, Z., Cheong, S.-W., Alten, B., Ahrens, E. T., Kubat-Martin, K. A., Maynard, J. D., Huang, Y., Kirk, D. R., Gillis, K. A., Kim, H. K., and Chan, M. H. W. (1990a). *Phys. Rev.* **B41**, 2098.

Migliori, A., Visscher, W. M., Wong, S., Brown, S. E., Tanaka, I., Kojima, H., and Allen, P. B. (1990b). *Phys. Rev. Lett.* **64**, 2458.

Millis, A. J., and Rabe, K. M. (1988). *Phys. Rev.* **B38**, 8908.

Phillips, W. A. (1981). "Amorphous Solids: Low-Temperature Properties." Springer-Verlag, New York.

Reichardt, W., Pintschovius, L., and Hennion, B. (1988). *Supercond. Sci. Technol.* **1**, 173.

Saint-Paul, M., Tholence, J. L., Monceau, P., Noel, H., Levet, J. C., Potel, M., Gougeon, P., and Capponi, J. J. (1988). *Solid State Commun.* **66**, 641.

Saint-Paul, M., Tholence, J. L., Noel, H., Levet, J. C., Potel, M., and Gougeon, P. (1989). *Solid State Commun.* **69**, 1161.

Saint-Paul, M., and Henry, J. Y. (1989). *Solid State Commun.* **72**, 685.

Schneemeyer, L. F. (1990). "Chemistry of Superconducting Materials." Edited by T. A. Vanderah. Noyes Press.

Schrieffer, J. R. (1964). "Theory of Superconductivity," Benjamin, New York.

Sheng, Z. Z., and Hermann, A. M. (1988). *Nature* **332**, 55.

Shi, X. D., Yu, R. C., Wang, Z. Z., Ong, N. P., and Chaikin, P. M. (1989). *Phys. Rev.* **B39**, 827.

Testardi, L. R. (1973). "Physical Acoustics," Vol. X. Academic Press, New York.

Tokura, Y., Takagi, H. and Uchida, S. (1989). *Nature* **337**, 345.

Worthington, T. K., Gallagher, W. J., and Dinger, T. R. (1987). *Phys. Rev. Lett.* **59**, 1160.

Wu, M. K., Ashburn, J. R., Tong, C. J., Hor, P. H., Wong, R. L., Gao. L., Huang, Z. J., Wang, Y. Q., and Chu, C. W. (1987). *Phys. Rev. Lett.* **58**, 908.

Zavaritsky, N. V., Samoilov, A. V., Yurgens, A. A., Klochko, V. S., and Makarov, V. I. (1989). *Physica* **C162**, 562.

—8—
Ultrasonic Measurements of Elastic Constants in Single Crystals of La_2CuO_4

J. D. MAYNARD AND M. J. McKENNA
Department of Physics, The Pennsylvania State University, University Park, Philadelphia

A. MIGLIORI AND WILLIAM M. VISSCHER
Los Alamos National Laboratory, Los Alamos, New Mexico

1. Introduction 381
2. Development of the Small-Sample Resonant Ultrasound Technique 382
3. Measurements on Single-Crystal Samples of La_2CuO_4 397
4. Measurements in Superconducting $La_{1.86}Sr_{0.14}CuO_4$ 401
   Acknowledgments 407
   References 407

1. Introduction

One of the most important measurements for the study of superconducting phase transitions is the velocity and attenuation of ultrasound. Because the elastic moduli, and therefore the sound velocities, are directly related to the second derivative of the Gibbs free energy with respect to stress, ultrasound provides a direct thermodynamic probe of phase transitions. Additional information, such as the coupling of phonons with electrons, relaxation mechanisms, etc., may also be determined from the attenuation of ultrasound.

If the elastic constants of a superconductor are measured as a function of temperature, certain features in the vicinity of the transition temperature must be present. Many non-oxide superconductors with high transition temperatures have transitions that are accompanied by structural instabilities or structural transitions. In some cases the structural transition may be arrested by the onset of superconductivity, but the structural instability remains (Testardi, 1964). Such instabilities are undoubtedly important to the theory of superconductivity. For conventional Bardeen–Cooper–Schreiffer (BCS) superconductors, the increase of the superconducting transition temperature near a structural instability is understood because the electron pairing mechanism involves strong electron–phonon coupling. For the new oxide superconductors, the role of the lattice is uncertain, as evidenced by the inconclusive isotope shifts (Batlogg et al., 1987). Nevertheless, the measurement of the elastic constants is important because, as a derivative of the free energy, the elastic constants are a sensitive probe of the environment in which the electrons pair.

In order to develop a basic understanding of the intrinsic properties of the new oxide superconductors, it is preferable to make measurements on single-crystal bulk samples. A problem has been that single-crystal samples are typically very small, only a few hundred microns in size. Such a small sample size makes conventional ultrasound measurements difficult, if not impossible (except at GHz frequencies where some structural effects are masked). In this chapter we describe the development of a new small-sample resonant ultrasound apparatus that can determine all of the independent elastic constants of a single-crystal sample a few hundred microns in size in a single measurement. Detailed measurements of all nine of the elastic constants of a single untwinned crystal of undoped La_2CuO_4, showing anomalies in the vicinity of the superconducting transition temperature, and measurements of $La_{1.86}Sr_{0.14}CuO_4$ near the orthorhombic–tetragonal transition will be presented.

2. Development of the Small-Sample Resonant Ultrasound Technique

Before describing the small-sample resonant ultrasound technique, it is appropriate to review the connection between the elastic constants of a solid and the propagation of ultrasound (Truell et al., 1969), and to establish some notation.

For a three-dimensional elastic solid, one may use indices (i, j, etc.) that can take on the values 1, 2, and 3, referring to x, y, and z coordinate directions, and write the strain as

$$\varepsilon_{ij} \equiv \frac{1}{2}\left(\frac{\partial \psi_i}{\partial x_j} + \frac{\partial \psi_j}{\partial x_i}\right), \tag{1}$$

where ψ_i is the displacement in the ith direction. The stress, σ_{ij}, is a force per unit area acting on a surface element, where the first index refers to the coordinate direction of a component of the force, and the second index refers to the coordinate direction of the unit normal to the surface element. Hooke's law is written as

$$\sigma_{ij} = c_{ijkl}\varepsilon_{kl}, \tag{2}$$

where summation over repeated indices is implied. For a small volume element, $dV = dx_1 dx_2 dx_3$, the net force in the i-direction is $(\partial \sigma_{ij}/\partial x_j)dV$, and Newton's law may be written as

8. Single Crystals of La_2CuO_4

$$\frac{\partial \sigma_{ij}}{\partial x_j} = \rho \frac{\partial^2 \psi_i}{\partial t^2}, \qquad (3)$$

where ρ is the mass density.

The symmetric nature of the definitions, and the assumption that the elastic energy must be quadratic in the strains, reduces the number of independent elements of c_{ijkl} from 81 to 21. Additional symmetries of a particular crystalline group will further reduce the number of independent elastic constants; in particular, the orthorhombic symmetry of the oxide superconductors result in only nine independent elastic constants.

To determine the modes of sound propagation one must solve Eqs. (1) through (3) with some specified boundary conditions. Because of the tensor nature of the equations, the relation between particle displacement and the direction of wave propagation is quite complicated. To tackle the complexity and make a connection between ultrasound measurements and the elastic constants, two approaches may be taken. The first, conventional (Truell et al., 1969) approach is to note that if one had a sample with a large (infinite) plane surface that was perpendicular to one of the principal axes of the elastic tensor, and if a plane wave could be launched from that surface, then the tensor equations would uncouple, and a longitudinal wave or one of two transverse waves could propagate independently. In this case, for each wave, the relationship between the sound velocity and the independent elastic constants is fairly straightforward. The second, or resonance, approach involves the use of a computer to numerically solve for the resonance frequencies of a rectangular solid with stress-free boundary conditions. This second, quite new, approach (Demarest, 1969; Ohno, 1976; Ohno et al., 1986) is the one employed in the small-sample resonant ultrasound technique. Although the conventional pulse technique is used almost universally, the resonance technique has significant advantages, as we discuss below.

The conventional technique for measuring ultrasound and the elastic constants in solids has not changed significantly in many years. An illustration of the conventional technique is shown in Fig. 1. A solid sample is cut and polished so that two of the faces are parallel to each other, and perpendicular to a principal axis of the elastic tensor, based on the crystal symmetry as determined with x-ray diffraction or other technique. A piezoelectric (usually quartz or lithium niobate) transducer with plated electrodes is then bonded to the sample face with coupling grease, epoxy, etc. The sample and transducer have sizes on the order of centimeters, with the transducer thinner than the sample. An ultrasonic pulse of 10–100 cycles at a resonant frequency of the transducer is launched, and after a time delay an echo is received. The time delay, roughly equal to the time-of-flight, may be precisely measured using phase-sensitive detection. From the

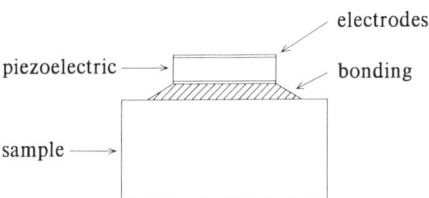

FIG. 1. Principal elements of the conventional pulse-echo method of measuring ultrasound velocity and determining elastic constants.

delay and the measured distance between the parallel faces, the speed of the ultrasound and some combination of the elastic constants may be determined.

As compared to an ideal (infinite plane wave) situation, there are some problems with the conventional technique. For centimeter-sized samples, they may be either negligible or manageable. However, for very small samples, only a few hundred microns in size, the problems become severe. The problems are the following: (Truell *et al.*, 1969):

1. Transducer ringing. The transducer is an acoustically active element in the system, and by driving it at resonance, a significant amount of acoustic energy is concentrated in it rather than in the sample, as indicated by ringing in the transducer. The ringing of the transducer complicates the determination of the time-of-flight.

2. Effects of bonding and electrodes. For very small samples, the bonding and electrode materials may be significant acoustic elements in the system, and these, like the transducer, may mask the properties that one wishes to measure.

3. Alignment and parallelism of the faces. For a very small sample, aligning and polishing parallel faces becomes more difficult. Without parallel faces, part of the pulse may interact with the side of the sample, resulting in sidewall effects, as discussed below. Lack of alignment severely complicates the determination of the elastic constants.

4. Beam diffraction and sidewall effects. In the ideal situation, one must launch an infinite plane wave in order to uncouple the different sound modes. In reality the transducer must have a finite size, and hence the beam that it launches is not an ideal plane wave, but must have side lobes. These side lobes will not travel with the theoretical speed of the uncoupled model. Furthermore, the sample itself has a finite lateral size, and the waves in the side lobes can reflect off the sidewalls of the sample, be converted into other modes, and reenter

8. Single Crystals of La_2CuO_4

the main beam. All of these effects conspire to produce errors in the time-of-flight measurement.

5. The pulse-echo technique requires a measurement of a time interval, which is difficult to do as accurately as, for example, a frequency measurement. Nearly all of the effects listed above, with the addition of external noise, make it difficult to determine an accurate time-of-flight. Furthermore, if a sample is only 250 μm in length with a sound speed of 5,000 m/s, then the time-of-flight would be only 100 ns, and the pulse would have to be only ~10–20 ns wide; such short pulses, requiring gigahertz frequencies, are difficult to generate and measure, and special equipment and techniques are required (Truell, et al. 1969). In addition, sample attenuation, which increases with frequency, makes it difficult to detect the echos of these high-frequency pulses.

6. With one aligned and polished face, one can measure at most three independent sound modes, or equivalently three relations involving the elastic constants. To do this, the transducer must be driven in a longitudinal and two shear modes, and possibly a different or realigned transducer must be used. In order to measure all of the independent elastic constants (e.g., 21 for triclinic symmetry), the entire measurement must be repeated for different aligned and polished faces (at least seven times for the triclinic case). In practice this is rarely done.

Some of the problems that are more severe for small samples may be partially alleviated by using very small thin-film transducers (Golding et al., 1988). By contrast, these problems do not exist or are dramatically reduced when the resonance technique is used.

In the resonance technique, one uses a sample that is cut and polished into a rectangular parallelepiped. The sample does not have to be aligned with the principal axes, but in practice knowledge of the actual orientation is essential. Two transducers are weakly coupled to the sample; one is driven at some frequency f, and the other monitors the response of the sample. The drive frequency is swept, and a plot of the response amplitude as a function of f shows peaks at resonance frequencies f_n with widths Δf. From a sufficiently large set of resonance frequencies, or spectrum, the elastic constants may be determined. The advantages of the techniques are the following:

1. By resonating the sample, the sample itself acts as a natural amplifier with a gain equal to the quality factor, $Q = f_n/\Delta f$, of the resonance. This amplification occurs before the transduction, minimizing effects of noise from the transducers.

2. The transducers may, and in fact should, be small and nonresonant, so that the problem of transducer ringing is eliminated. A small transducer will not significantly load the sample and shift the resonant frequencies. The transducer should have a high resonant frequency, below which it will have the broadband response necessary for finding the resonances in the sample.

3. Because a large signal amplitude is made possible by resonating the sample, the transducers need not be strongly coupled, and in fact the transducers should be weakly coupled. By having the transducers weakly coupled, the resonances of the sample are not significantly altered by loading or other transducer effects. Since nearly all of the acoustic energy is stored in the sample, changes that occur with temperature, etc., reflect properties of the sample rather than those of the transducer, bonding, or electrodes. Thus, artifacts of the transducer, bonding, electrodes, etc., are minimized.

4. There are no undesirable sidewall effects. The sidewalls and all possible types of sound propagation are all accounted for in the normal modes and resonant frequencies of the sample.

5. The sample need not be precisely dimensioned. Shaping and orienting are very helpful, but not absolutely essential. One always measures more resonant frequencies than independent elastic constants, so that the elastic constants are overdetermined; the sample dimensions and orientation may be used as adjustable parameters in fitting the resonance spectrum.

6. All of the elastic constants may be determined with a single measurement. There is no need to drive transducers in different modes, remount transducers, or reorient the sample.

7. The method employs continuous-wave excitation and phase-sensitive detection, and the measurement of frequency is virtually the easiest and most precise of all types of measurements.

With all of the advantages listed above, and essentially no intrinsic disadvantages, one might wonder why the resonant technique is not used instead of the prevalent pulse technique. The reason is that the resonant technique requires considerable numerical computation in order to convert the measured resonant frequencies to elastic constants. (The computational algorithm is described in a later section.) In the pulse technique, the different sound modes are decoupled by launching plane waves in an oriented sample; in the resonant technique, the

8. Single Crystals of La_2CuO_4

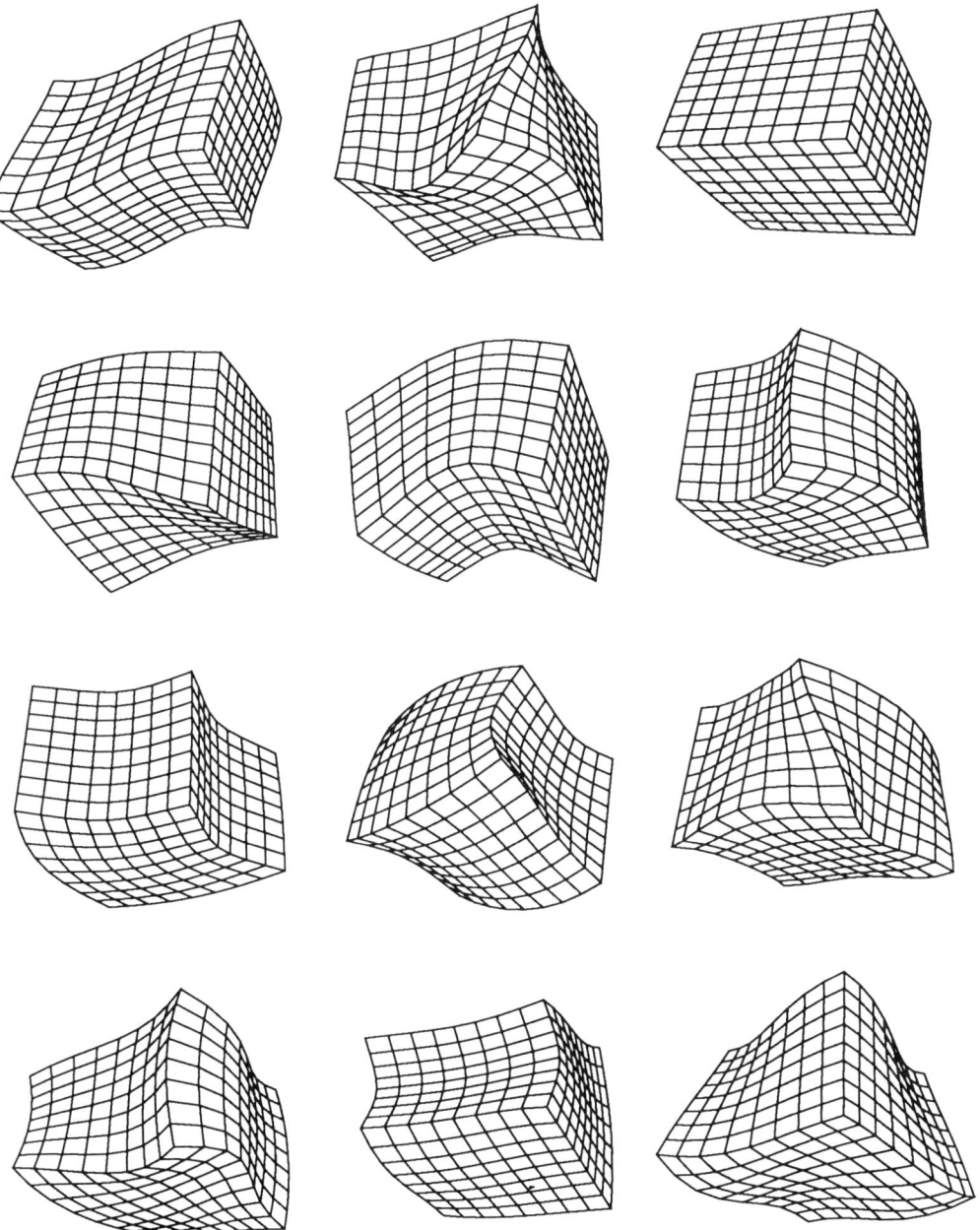

FIG. 2. Schematic representation of some of the normal modes of vibration of an elastic cube with stress-free boundary conditions.

decoupling must be accomplished by analysis with a powerful computer. As an illustration of the complexity, Fig. 2 shows some of the normal modes of an elastic cube; the modes are nontrivial and not as easily analyzed as are plane waves. In the past, mainframe computers have been relatively inaccessible and expensive, and the pulse technique has been the more convenient option. Now, however, desktop computers and workstations are sufficiently powerful to accomplish the calculations required for the resonant technique in a reasonable amount of time, and there is every reason to expect the resonant technique to replace the pulse method.

The resonance technique requires that the transducers be small and broadband. If one has a sample that is only a few hundred microns in size, then one would like to have transducers that are only a few microns in size (at least in thickness). While this appears to be a difficult criterion, such transducers are readily fabricated, thanks to the availability of piezoelectric plastic film transducer material. A particular material, polyvinylidene fluoride (PVDF), is inexpensive and commercially available (Kynar Piezo Film, Penwalt Corp., 900 First Avenue, King of Prussia, Pennsylvania, 19406) in sheets as thin as 9 microns. Some physical properties of this material, along with those of more traditional piezoelectric materials, barium titanate and quartz, are presented in Table I. It can be seen that PVDF has excellent properties in all categories; it should be noted that the low quality factor of PVDF is a strong advantage for the resonant ultrasound technique, because broadband response is desired.

Very small ultrasound transducers, appropriate for making measurements on samples only a few hundred microns in size, are easily fabricated with the PVDF film. The procedure is as follows:

1. PVDF film is available already plated with metal for electrodes. We have found that the factory metallization is not sufficiently robust for the small transducer fabrication. The problem (for the small transducers) is that a slight abrasion, electrochemical reaction, etc., may result in an open circuit to the active area

TABLE I

COMPARISON BETWEEN CONVENTIONAL PIEZOELECTRIC MATERIALS AND FLEXIBLE PIEZOELECTRIC POLYVINYLIDENE FLUORIDE (PVDF) FILM

	$BaTiO_3$	Quartz	PVDF
Electric field/pressure (VmN^{-1})	5	50	200
Quality factor Q	10^3	10^5	10
Acoustic impedance ρc (kg $m^{-2}s^{-1}$)	30	15	3
Frequency response	MHz	MHz	GHz

8. Single Crystals of La_2CuO_4

of the transducer (as described below). If the PVDF film has factory metallization, it may be removed with a 1:5 diluted (by weight) NaOH solution followed by a 1:5 diluted (by volume) nitric acid solution.

2. The PVDF film may be cleaned in an ultrasonic cleaner with a liquid soap solution, followed by a distilled water rinse.

3. A ~2 cm square sheet of 9 μm thick PVDF film is metallized with aluminum (1,000 Å thick) on parts of both sides, as illustrated in Fig. 3a, using a vacuum evaporator. With a knife-edge mask, the PVDF is plated on one side up to a line slightly beyond the middle of the PVDF sheet. On the opposite side, the sheet is plated up to a line that overlaps the line on the opposite side by ~500 μm. For this evaporation, it is necessary to use a small filament, separated some distance from the PVDF sheet, so that the PVDF is not excessively heated by the evaporation.

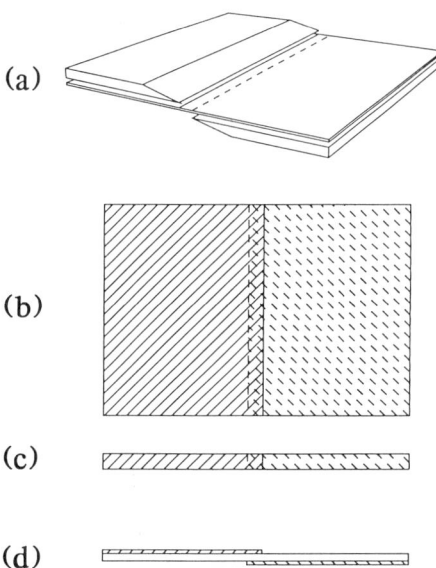

FIG. 3. Illustration of the steps in fabricating a small (500 × 500 × 9 μm) broadband (100 MHz) ultrasound transducer using a piezoelectric PVDF plastic film. (a) Knife-edge masks for evaporation of aluminum metallization. The masks and the PVDF film are flipped in order to metallize both sides of the film. (b) View of the metallized film illustrating overlap of the aluminum metallization on the two surfaces of the film. (c) A transducer strip cut from the metallized PVDF film. The 500 × 500 μm square overlap region forms the active area of the transducer, and the rest of the metallization forms the leads. (d) A side view of the transducer strip showing the overlap region with the thickness (9 μm actual) greatly exaggerated.

4. After the metallization, the PVDF sheet is cut into ~500 μm wide strips, as shown in Fig. 3b and 3c, using a razor blade. The aluminum plating, extending from either end of the strip to overlap in the middle, forms the two leads for the transducer; the region of overlap in the center, ~500 μm square, forms a capacitor with the piezoelectric film sandwiched in between; this is the active area of the transducer. It should be noted that the transducer and the leads are only 9 μm thick.

The transducer strips, as described above, are not mounted directly to the sample, but instead are mounted in an apparatus as shown in Fig. 4. This apparatus contains two insulating blocks, each of which carries a pair of beryllium–copper spring metal prongs. Two PVDF transducer strips are mounted under tension between the prongs as shown in Fig. 4; the ends of the transducer strips are attached to the prongs, mechanically and electrically, with conducting epoxy. Coaxial leads for the transducers are connected using the screws that secure the prongs. The active areas of the transducer strips are centered in the region between each pair of prongs, and the sample is mounted between the two active areas. The resonance measurement employs one of the transducers as a drive, and the other as a monitor.

The sample is oriented so that contact with the sample is made at two corners. This is done for two reasons. First, the corners are the mechanically weakest points of the sample, so that mounting at the corners provides the weakest coupling to the transducers and the least shift in the resonance frequencies due

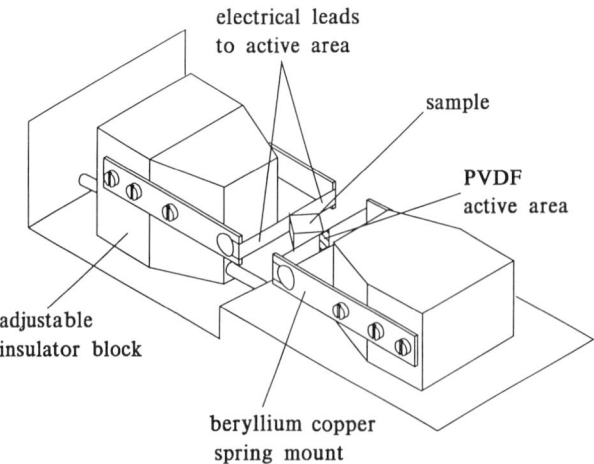

FIG. 4. Schematic drawing of the small-sample resonant ultrasound apparatus. The PVDF active areas make contact with the sample at its corners.

8. Single Crystals of La_2CuO_4

to transducer loading. Second, the corners are points of low symmetry, so that nearly all of the normal modes of the sample can be driven and detected from the corners. In some cases, the acoustic coupling between the sample and the transducers may be increased by placing a very small dot of silicone oil between the sample corner and the point of contact with the transducer strips.

To make a measurement, one simply drives one transducer with a sinusoidal voltage (amplitude ~5 V), sweeps the frequency, and monitors the response with the other transducer. For increased sensitivity, the response may be monitored with phase-sensitive detection. The resonant frequencies of the sample, f_n, are identified as the frequencies where there are rapid changes in the response amplitude and phase.

When the drive frequency is swept through a resonance, the acoustic signal goes through a maximum and also undergoes a 180 degree phase shift. Thus, as the frequency is swept, the acoustic signal constructively and destructively interferes with the electrical crosstalk between driven and monitoring transducers, producing a small wiggle on the crosstalk amplitude. If the crosstalk amplitude is slowly varying at frequencies away from the acoustic resonance, it can be eliminated by taking a derivative of the receiver amplitude with respect to frequency; the derivative also enhances the resonance wiggle. Taking the derivative is accomplished by using a frequency modulation (FM) mode for the drive and a lock-in amplifier for the receiver.

The method for measuring the resonances with the FM technique is illustrated in Fig. 5. The modulation $\cos(g\pi\Delta ft)$ is provided by the reference output of the lock-in amplifier, with $\Delta f \cong 100$ Hz. This signal is applied to the phase modulation input of the frequency synthesizer, whose center frequency is set at $f \cong$ 1–10 MHz. The output of the synthesizer, $\cos[2\pi ft + \cos(2\pi\Delta ft)]$ (essentially frequency modulated between $f - \Delta f$ and $f + \Delta f$ at a frequency Δf) is applied to the drive transducer. The signal from the receive transducer is amplified, detected, and monitored with the lock-in amplifier. As the center frequency of the synthesizer is slowly swept, the presence of a resonance within Δf of the

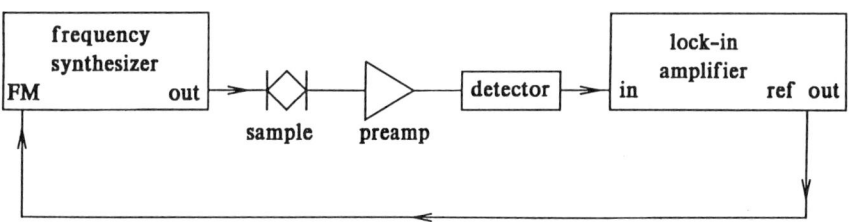

FIG. 5. Diagram of the electrons used with the frequency-modulation (FM) technique for dealing with electromagnetic crosstalk in the small-sample ultrasound apparatus.

center frequency causes a rapid variation of the receive transducer signal, which is sensitively detected by the lock-in amplifier.

In our most recent research, we have found that the resonance signals are sufficiently strong so that the phase modulation technique is not necessary. We simply use phase-sensitive detection (using a high-frequency lock-in at the drive frequency rather than at a modulation frequency) and adjust the phase of the reference so that the crosstalk is nulled over a reasonable frequency range.

Typical resonance frequency measurements are presented in Figs. 6 and 7, which show lock-in output versus frequency for a small quartz crystal sample used for testing the technique (as will be described later). It can be seen that the resonance frequencies may be readily determined. From the expanded view in Fig. 7, the Q of the resonance can be determined. Typical values for the Q of these resonances are ~4,000 at room temperature. These values, which are quite good considering the size of the samples and the presence of silicone coupling fluid, increase at lower temperatures (77 K).

As already mentioned, a difficulty with the resonance method is the computational problem of determining the elastic constants from the spectrum of resonance frequencies. One needs to find solutions to the differential equation (3),

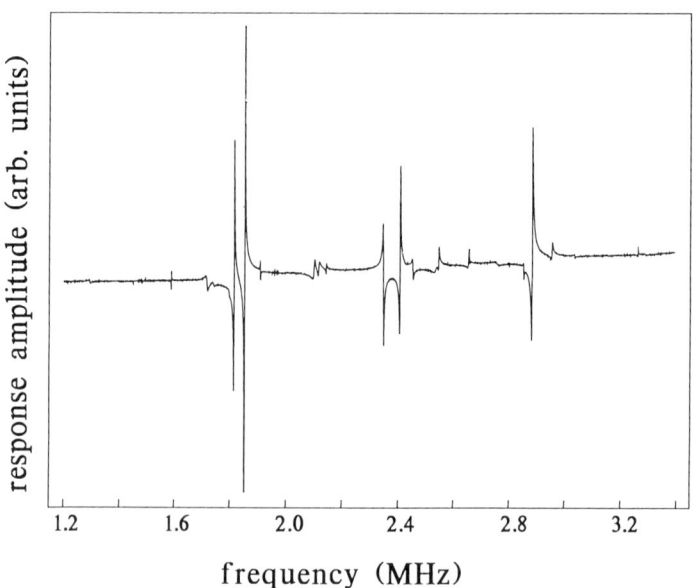

FIG. 6. Response amplitude measured at the monitor transducer as a function of the frequency of the drive transducer signal. Resonance frequencies are readily measured with high accuracy.

8. Single Crystals of La_2CuO_4

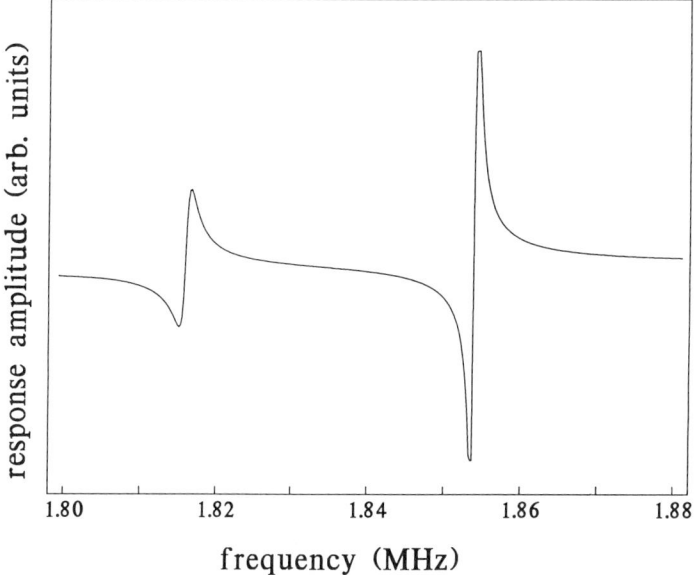

FIG. 7. Expanded view of the two resonances from Fig. 6. The quality factory (Q) is ~4,000. With the high signal-to-noise, the resonance frequency may be determined to a few parts per million.

subject to the assumption of vanishing surface traction at the surface of the sample,

$$\sigma_{ij} n_j = 0, \quad i = 1, 2, 3, \tag{4}$$

where the n_j are the components of the unit vector normal to the surface. If a time dependence proportional to $\cos(\omega t + \phi)$ is assumed, then solutions to Eqs. (3) and (4) exist only for discrete values of ω, which are the resonant frequencies.

The boundary value problem expressed with Eqs. (3) and (4) can be replaced by a single variational problem (Morse and Fesbach, 1953a; Ekstein and Schiffman, 1956). That is, the displacement field ψ_i that satisfies the elastic wave equation (3) and boundary condition (4) can be shown to be a function for which the integral

$$L = \int_V (\rho \omega^2 \psi_i \psi_i - c_{ijkl} \varepsilon_{ij} \varepsilon_{kl}) dV \tag{5}$$

is stationary. The integration is performed over the volume of the sample. This may be demonstrated as follows:

Since c_{ijkl} is symmetric in k and l, Eq. (2) may be rewritten as

$$\sigma_{ij} = c_{ijkl} \frac{1}{2} \left(\frac{\partial \psi_k}{\partial x_l} + \frac{\partial \psi_l}{\partial x_k} \right) = c_{ijkl} \frac{\partial \psi_k}{\partial x_l}. \tag{6}$$

Using this result and the sinusoidal time dependence of ψ_i, Eq. (3) becomes

$$\frac{\partial \sigma_{ij}}{\partial x_j} = c_{ijkl} \frac{\partial^2 \psi_k}{\partial x_j \partial x_l} = \rho \frac{\partial^2 \psi_i}{\partial t^2} = -\rho \omega^2 \psi_i, \tag{7}$$

or

$$\rho \omega^2 \psi_i + c_{ijkl} \frac{\partial^2 \psi_k}{\partial x_j \partial x_l} = 0. \tag{8}$$

Using Eq. (6), the stress-free boundary condition may be written as

$$\sigma_{ij} n_j = c_{ijkl} \frac{\partial \psi_k}{\partial x_l} n_j = 0. \tag{9}$$

We can again use the symmetry of c_{ijkl} and rewrite L as

$$L = \int_V \left(\rho \omega^2 \psi_i \psi_i = c_{ijkl} \frac{\partial \psi_i}{\partial x_j} \frac{\partial \psi_k}{\partial x_l} \right) dV. \tag{10}$$

We now examine the variation of L, δL, due to a small change in the displacement field $\delta \psi_i$ away from the solution:

$$\delta L = 2 \int_V \left(\rho \omega^2 \psi_i \delta \psi_i - c_{jkl} \frac{\partial \psi_k}{\partial x_l} \frac{\partial \delta \psi_i}{\partial x_j} \right) dV. \tag{11}$$

The second term in Eq. (11) can be integrated by parts to yield

$$\delta L = 2 \int_V \left(\rho \omega^2 \psi_i + c_{ijkl} \frac{\partial^2 \psi_k}{\partial x_l \partial x_j} \right) \delta \psi_i dV - 2 \int_S \left(c_{ijkl} \frac{\partial \psi_k}{\partial x_l} n_j \right) \delta \psi_i dS, \tag{12}$$

where the last term is now an integral over the surface of the sample.

At the stationary point we require that $\delta L = 0$. Then, since $\delta \psi_i$ is arbitrary and behaves independently in the volume integral and in the surface integral, the integrands in the brackets in the volume and surface integrals must vanish independently. The vanishing of the expressions in the brackets yields the elastic wave equation (8), as well as the boundary condition, Eq. (9).

In order to find approximate solutions for which L is stationary, the Rayleigh–Ritz method (Morse and Fesbach, 1953a) may be used. A solution is expanded as a linear combination of N orthonormal basis functions:

$$\vec{\psi} = \sum_{p=1}^{N} a_p \vec{\Phi}_p. \tag{13}$$

Then L becomes

$$L = (\rho \omega^2 \delta_{pq} - \Gamma_{pq}) a_p a_q, \tag{14}$$

8. Single Crystals of La_2CuO_4

where δ_{pq} is the Kronecker delta, and where

$$\Gamma_{pq} = \int_V c_{ijkl} \frac{\partial \Phi_{pi}}{\partial x_j} \frac{\partial \Phi_{qk}}{\partial x_l} dV. \tag{15}$$

Requiring that the derivative of L with respect to the coefficients a_p vanish yields the $N \times N$ matrix eigenvalue problem

$$(\Gamma_{pq} - \rho\omega^2 \delta_{pq}) a_q = 0. \tag{16}$$

The existence of solutions a_p requires that the determinant of the matrix in brackets vanish; the solution of the resulting Nth-order polynomial in $\rho\omega^2$ yields N resonant frequencies $f_n = \omega_n/2\pi$, $n = 1$ to N. The N eigenvectors $(a_p)_n$ for the matrix problem yield the normal modes, as expressed with Eq. (13).

In practice, good convergence of the approximate solutions may be obtained using basis functions formed from normalized Legendre polynomials P_λ (Demarest, 1969):

$$\Phi_{pi} = (L_1 L_2 L_3)^{1/2} P_\lambda(x_1/L_1) P_\mu(x_2/L_2) P_\nu(x_3/L_3) e_i, \tag{17}$$

where the subscript p now represents three subscripts λ, μ, and η; $2L_1$, $2L_2$, and $2L_3$ are the dimensions of the sample; and e_i is a unit vector in the x_i direction.

In order to obtain a good approximation to the correct solutions and resonant frequencies, it may be necessary to use Legendre polynomials up to order 10 and as many as 1,500 basis functions, depending on how many eigenmodes one wishes to obtain. However, it is not necessary to solve the eigenvalue problem for the full 1,500 × 1,500 matrix; because of the symmetry of the elastic tensor and the odd or even parity of the Legendre polynomials, the matrix can be put into block diagonal form, with as many as eight blocks. The eigenvalue problem for each block can be solved independently, with significant savings in computation time. The details of the manipulation of the basis functions may be found in the literature (Ohno, 1976; Ohno et al., 1986).

The eigenvalue problem outlined above yields the resonance frequencies, f_n, given the elastic constants, c_{ijkl}. For the resonance technique, what is needed is the inverse. In order to solve the inverse problem, an iteration scheme is used. Note that only the independent elastic constants need to be considered—as few as two for an isotropic material, up to 21 if it is triclinic. The independent elastic constants will be indicated simply as c_α. The steps in the iteration scheme are as follows:

1. Make some estimate of the independent elastic constants of the material; previous data on the material or a similar material may be used.

2. Use estimated c_α in the eigenvalue problem to calculate the expected frequency f_n^0 to the correct normal mode.

3. Use the expected fequencies f_n^0 to try to assign the experimental frequencies f_n to the correct normal mode.

4. Choose a "figure of merit"

$$F = \sum_n (f_n - f_n^0)^2. \qquad (18)$$

Note that modes that do not have observable experimental frequencies should not be included in the sum.

5. Use some multidimensional minimization scheme to find the point in C_α space at which the sum of squares F is minimized. The Levenberg–Marquardt method (see Press et al., 1986) works better than others we have tried.

6. Compare the frequencies f_n' computed in step 5 with the f_n. Often some frequencies are missed in the measurement; sometimes spurious ones are observed. It is here that operator intervention, guided by hunches and intuition, is often necessary. One then returns to step 3, with f_n' replacing f_n^0, and repeats the process. Steps 3, 4, and 5 are reiterated until one obtains a satisfactory fit (an acceptably small value of F).

The main difficulty with the whole procedure lies in step 3. It is not always easy to assign each experimental resonance frequency to the correct mode.

In order to test the small-sample ultrasound apparatus, the data acquisition procedure, and the computer analysis, we have used a small, oriented, single-crystal sample of quartz, a material whose elastic constants are available in the literature. It is interesting to note that there is a significantly wide range of values that have been reported. Table II lists some of the reported values (Vigoureux and Booth, 1950; Cady, 1946; Voigt, 1928; Bechmann, 1958; Mason, 1950).

Our quartz test sample (Valpey-Fisher Corp., 75 South Street, Hopkington, Massachusetts 01748) was a rectangular parallelepiped, with dimensions $L_1 \cong$ 1.0 mm, $L_2 \cong 1.4$ mm, and $L_3 \cong 1.3$ mm, with the 1, 2, and 3 axes oriented with the standard quartz "x," "y," and "z" axes, respectively. The sample measurement and analysis were as described above, except that the sample dimensions were varied slightly ($\sim 1\%$) from the nominal values in order to improve the fit with the experimental resonance frequencies. The values of the elastic constants determined with the resonant ultrasound technique are presented in the last row in Table II. It can be seen that the values are in good agreement with the values from the literature. It should be noted that our sample is significantly smaller than any other sample used to obtain such data.

8. Single Crystals of La_2CuO_4

3. Measurements on Single-Crystal Samples of La_2CuO_4

Single crystals of La_2CuO_4 were grown from a CuO flux, quenched, and later annealed at 650 K in N_2 for a short time and cooled to 300 K over 10 hours. The ultrasound sample was oriented to within 1° of its principal axes via x-rays. Neither x-ray nor polarized-light scattering revealed any signs of twinning. In principle, the presence of twins cannot be ruled out; however, we note that twins in fractured samples are very apparent optically. Sample dimensions were 1.73 × 1.76 × 0.70, all ±0.01 mm, with the b axis parallel to the small dimension and normal to the Cu–O plane. The a and c axes were each oriented parallel to an edge, although they were not distinguished. After the N_2 anneal process, the measured density was 7.03 g/cm^3, compared to a theoretical density of 7.08 g/cm^3, suggesting reasonably good oxygen stoichiometry. The Neel temperature $T_N = 305$ K for the ultrasound sample, and $T_N = 316$ K for the specific-heat sample, confirming the stoichiometry.

Measurements of the elastic constants of the La_2CuO_4 sample were made as described for the crystalline quartz test sample. A typical resonance spectrum is shown in Fig. 8. In our results, we have not corrected for the unknown but almost certainly anisotropic thermal expansion, but we expect it to be less than 3×10^{-6}/K below 45 K (Lang et al., 1987).

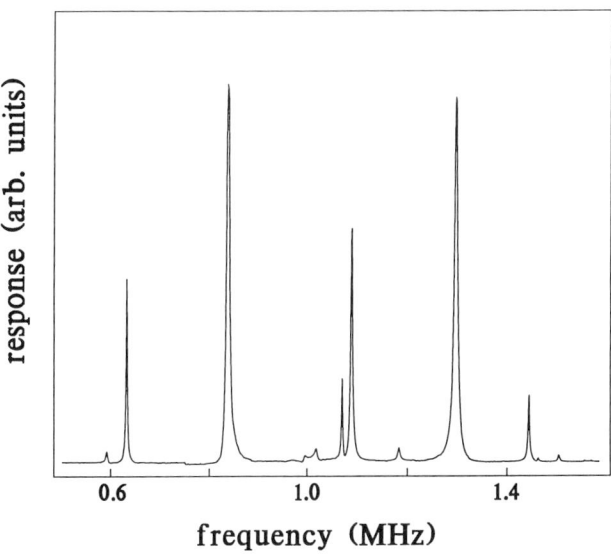

FIG. 8. Typical room-temperature spectrum for the undoped La_2CuO_4 sample described in the text.

Table III contains a summary of the determination of the nine elastic constants (orthorhombic symmetry) at 310 K, 297 K, and 44 K. Because we did not distinguish between the a and c axes, we have arbitrarily assigned the a axis to the "1" subscript and the c axis to the "2" subscript of the c_{ij}. The resonant frequencies themselves were determined using a fitting algorithm from raw data exhibiting quality factors Q on the order of 1,000. Because we have observed Qs on the order of 3×10^4 for fused silica, our sample (not the apparatus) must be responsible for the high dissipation. Unfortunately, we cannot rule out defect structures in general; thus, the high dissipation may not be of intrinsic interest.

The bulk modulus B for our sample can be extracted from the c_{ij}. We obtain $B = 1.126 \times 10^{12}$ dyn/cm^2 ± 0.5% at 310 K, 1.125×10^{12} dyn/cm^2 at 297 K, and 1.138×10^{12} dyn/cm^2 at 44 K. Ledbetter et al. (1989) measured high density powder samples and obtained very good agreement with our results. They found $B = 1.145 \times 10^{12}$ dyn/cm^2 at room temperature, with a monotonic increase to 1.159×10^{12} dyn/cm^2 as the temperature was lowered. The average compressional wave speed (5.05 km/s) we derive from our data is in very good agreement with results on dense ceramics (5.04 km/s, by Kim et al., 1988). Because c_{66} is larger than c_{44} or c_{55}, we find our average shear wave speed (3.25 km/s) to be substantially different than that for ceramics (2.84 km/s, by Ledbetter et al., 1989). It is not unreasonable to expect the shear wave speed to be sensitive to details of the sintering process. Also included in Table III are the results of a molecular dynamics simulation carried out by Allan and Mackrodt (1988). Most significant is that the measured in-plane shear modulus c_{66} is much larger than their calculated value. The discrepancy demonstrates a fundamental failing of the expected force laws obtained from simple chemical considerations. This is not surprising in light of the predicted metallic nature of this material (Kasowski

TABLE II

VALUES OF THE ELASTIC CONSTANTS OF QUARTZ

Source	C_{ij} (Mbar)					
	C_{11}	C_{33}	C_{44}	C_{12}	C_{13}	C_{14}
Cady[a]	8.75	10.77	5.73	0.76	1.51	−1.72
Voigt[b]	8.55	10.57	5.71	0.73	1.44	−1.69
Bechmann[c]	8.67	10.72	5.79	0.70	1.19	−1.79
Mason et al.[d]	8.61	10.71	5.87	0.51	1.05	−1.83
This work	8.59	10.65	5.75	0.52	0.99	−1.68

[a]Cady (1946).
[b]Voigt (1928).
[c]Bechmann (1958).
[d]Mason (1950).

8. Single Crystals of La_2CuO_4

TABLE III

ELASTIC CONSTANTS OF La_2CuO_4 IN MBAR (10^{12} DYN/CM^2)[a]

T	C_{11}	C_{22}	C_{33}	C_{23}	C_{13}	C_{12}	C_{44}	C_{55}	C_{66}	Run #
310 K	1.722	1.716	2.000	0.732	0.728	0.892	0.652	0.658	0.971	2
297 K	1.719	1.712	2.000	0.731	0.727	0.904	0.656	0.658	0.968	1
44 K	1.688	1.668	2.000	0.728	0.714	1.000	0.705	0.660	1.036	1
	1.99	1.84	1.90	0.70	0.65	0.65	0.65	0.64	0.66	Ref.

[a] The last row is from Allan and Mackradt (1988), but we have taken the liberty of correcting an obvious error by a factor of two in several of their numbers. Note that data at 310 K were taken after the sample was removed and reinserted. We estimate our errors as follows. The rms frequency deviation in the fitting procedure was 6 kHz. We used a varying number of resonances (typically about 15) approximately uniformly distributed from 0.6 MHz to 2.2 MHz. Thus the average fitting error is about 0.4%. Errors in sample dimensions and density were typically of order 1% as stated in the text. Thus, our precision is 0.4% or better, and our absolute accuracy is about 1%.

et al., 1991) from simple band structure calculations. We do not know the reason for the discrepancies, although they may be a result of correlation effects that are not included in the simulations. The related antiferromagnetic properties (Cheong et al., 1988) are also not included in the simulations, but magnetoelastic effects away from phase transitions are generally very weak compared to the differences shown in Table III.

In Figs. 9 and 10 we show sets of data representing the variation of some resonant frequencies with temperature. The frequencies plotted in Fig. 10 are

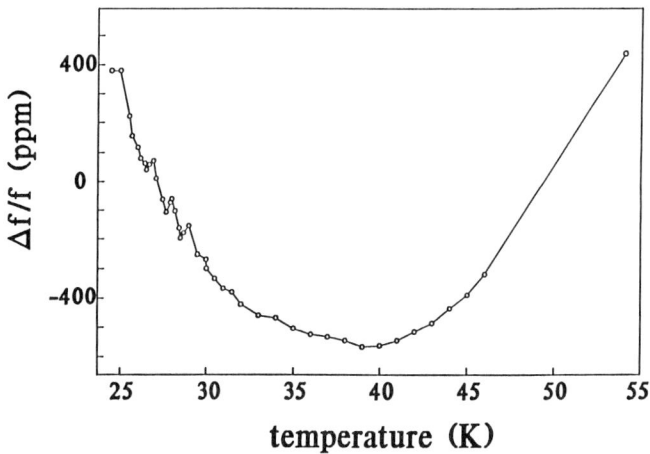

FIG. 9. The frequency of a resonance dependent primarily on c_{11} and c_{22} as a function of temperature. Temperature stability was 0.1 K, and flexible 9 μm PVDF transducers were used. The frequency error bar is about 25 ppm.

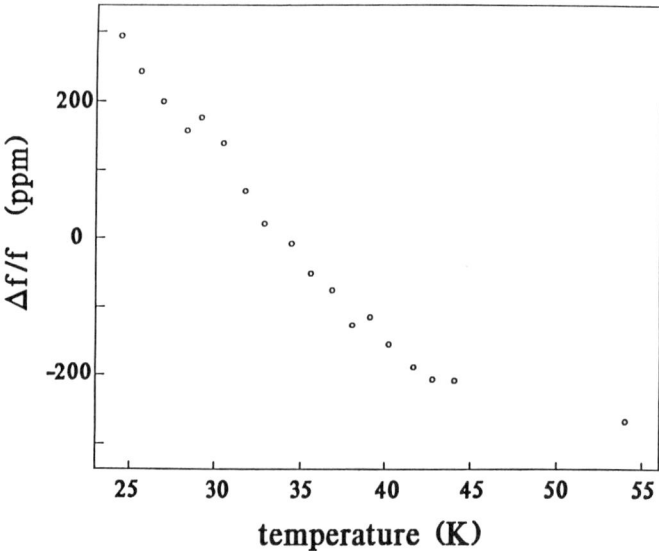

FIG. 10. The frequency of a resonance dependent only on c_{66} as a function of temperature. The temperature stability was 0.8 K, and rigid transducers were used. The frequency error bar is less than 25 ppm.

dependent almost exclusively on c_{66}, whereas those plotted in Fig. 9 are primarily affected by changes in c_{11} and c_{22}. These data were not deconvoluted to obtain the variation in elastic constants because the deconvolution process produces additional scatter, making the trends harder to see. However, from our fits at 310 K, 297 K, and 44 K, we know the dependence of each frequency on each elastic constant at these temperatures. Between 25 K and 54 K the dependence of frequencies on elastic constants is very close to the 44 K values. Note the features near 29 K in the compressional modes, and the change in slope at 38 K for both compression and shear.

In discussing the broader ultrasound minimum, we begin by emphasizing several key points. First, our sample is a single untwinned crystal at room temperature that shows no surface or bulk superconductivity when cooled. Second the sound velocities for compressional waves normal to the Cu–O plane, $\sqrt{c_{33}/\rho}$, and for shear waves both displacement vector and propagation vector in the Cu–O planes, $\sqrt{c_{66}/\rho}$, are significantly higher than data for dense ceramics (Kim *et al.*, 1988), indicating strong anisotropy. Third, the system appears nearly tetragonal ($c_{44} = c_{45}$ hr, $c_{11} = c_{22}$, etc.) at 297 K, but becomes increasingly orthorhombic as it is cooled. Taken together, these qualities establish that the interesting Cu–O plane produces anisotropic elastic effects, that the in-plane anisotropies increase on cooling, and that elastic and specific-heat features appear in the nonsuperconductor at a temperature near T_c in the superconductor.

8. Single Crystals of La_2CuO_4

Others have also seen unusual features near T_c in both superconducting and nonsuperconducting samples. For example, Lang et al. (1987) observe breaks in slope of the thermal expansion coefficient near 37 K in both La_2CuO_4 and fully superconducting $La_{1.85}Sr_{0.15}CuO_4$. Bhattacharya et al. (1988) observe a strong stiffening of the shear modulus of superconducting ceramic $LA_{1.8}Sr_{0.2}CuO_{4-y}$ below T that is similar to ours in the nonsuperconductor, but they do not see the same broad local minimum in the compressional modulus. It thus appears that some of the observed thermodynamic and conductivity features present near T_c do not require the presence of superconductivity.

A minimum in the sound velocities is not typical of crystalline solids. It is common, however, in systems with low-frequency relaxation processes. Such systems include structural glasses (Hunklinger et al., 1981) such as fused silica, which has a minimum in its sound velocities near 40 K caused by an attenuation peak there. Other glasses (Hunklinger and Schickfus, 1981) have similar minima between 10 K and 100 K. Such effects also occur in ferromagnets (Heeger et al., 1961), where the activation of domain wall motion strongly influences the temperature dependence of the elastic properties.

For fused silica, a disorder-induced attenuation peak at 40 K makes the sound velocities decrease at temperatures above 4 K, and then increase above the attenuation peak. Thermal activation of the Si–O–Si bond disorder seems to be responsible for the decrease. The increase at higher temperatures is not understood (Hunklinger and Schickfus, 1981), though it seems to be a universal feature of glasses. It is, therefore, compelling to look for an attenuation mechanism sensitive to temperature in the La_2CuO_4 system. One possibility is the oxygen sliding mode for which Cohen et al. (1989a,b) have calculated the potential. They find the potential to be highly anharmonic and nearly flat-bottomed, with a first excited state at about 60 K. These qualities could produce large oxygen motions strongly coupled to acoustic phonons, and therefore an attenuation mechanism peaking with thermal occupation of the optical mode. There is some direct evidence for large oxygen motions in neutron scattering measurements (Egami et al., 1987). Should such a strongly temperature-dependent attenuation mechanism be responsible for the ultrasound results, then the attenuation will depend on frequency as well. This suggests that studies of the ultrasonic attenuation may clarify the processes we observe in the elastic moduli.

4. Measurements in Superconducting $La_{1.86}Sr_{0.14}CuO_4$

Two single crystals of $La_{1-x}Sr_xCuO_4$—the first prepared by Tanaka and Kojima (1989), was $2.890 \times 2.816 \times 2.405$ mm with $x = 0.14$; the second was prepared by Fisk (see Allen et al., 1989) with $x = 0.10$—were studied using resonant ultrasound. The sample with $x = 0.14$ had a superconducting transition temperature of 36 K, and a structural phase transition temperature (SPT) T_s of 223 K, while the sample with $x = 0.10$ had a T_s of 330 K, consistent with the phase diagram for this compound (see Birgeneau and Shirane, 1989). The structural phase transition in this system is associated with tilting of the oxygen octahedra in the CuO plane. This mode is a zone-edge (x-point) phonon studied by neutron scattering (Boni et al., 1988). The order parameter is then the static tilt angle and has, therefore, two components, one in the 110 direction the other in the 1$\bar{1}$0 direction because the tetragonal structure above T_s allows by symmetry a tilt along either diagonal of the square face of the unit cell. Upon going from the symmetric (tetragonal) to the unsymmetrical (orthorhombic) phase, the orthorhombic a-axis can have two possible orientations, and therefore twins develop. Because of the formation of twins, the background ultrasonic attenuation in the 1 MHz range, the range of interest here, is enormous, and no reliable resonance data could be obtained for either sample below its T_s (and therefore near T_c either). However, above T_s, very high-quality data for both samples were obtained, enabling a complete description of the SPT, the determination of numerically accurate parameters for the double-well potential responsible for the soft mode, and a surprising connection to the mechanism of high-temperature superconductivity—all because of the accuracy of the zone-center (acoustic) measurements made possible by resonant ultrasound spectroscopy and its intrinsically low sensitivity to Landau–Khalatnikov damping at a second-order phase transition.

To understand the implications of the measurements, note that the SPT is accompanied by a rotation by 45 degrees of the unit-cell principal axes. This is easily seen by imagining the square tetragonal basal plane in the symmetric phase being stretched at two diagonally opposite corners into a diamond shape at T_s. The orthorhombic principal axes consist of a rectangle inscribed (in reciprocal space) inside the diamond and rotated 45 degrees. Now consider the original basal plane tetragonal axes both above and below T_s. Above T_s, they form a square; below, a diamond. The diamond shape is obtained from a square by applying a shear strain (c_{66}). For a phase transition, the development of orthorhombic symmetry is accompanied in this case by a spontaneous shear strain as well as a tilt of the octahedra. Thus, c_{66} must either couple to the order parameter or be the zone-center part of the dispersion curve ending at the X-point.

Therefore, measurements made at the Brilloiun zone center using ultrasound can be used to study the effects of what is principally a zone-edge phenomenon. To do this, we used resonant ultrasound spectroscopy, but with a different arrangement than described in the first part of this work. The arrangement used here, and shown schematically in Fig. 11, uses diamond lithium niobate composite transducers (US Patent Application S/N 406007, 1989). The construction

8. Single Crystals of La_2CuO_4

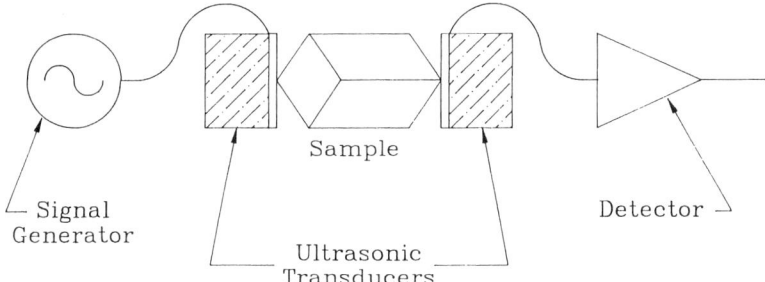

FIG. 11. Shown is a schematic of the sample mounting scheme for the $La_{2-x}Sr_xCuO_4$ measurements. The diamond/lithium niobate composite transducers make dry point contact with the sample.

is such that the low-frequency (or order 100 kHz) bending modes of a 30 MHz compressional-mode lithium niobate wafer are pushed to higher frequencies by bonding to a 1.5 mm diameter × 1.0 mm long diamond cylinder. This structure then has a lowest resonance of 4.3 MHz, higher than any of the frequencies we measured. These transducers then act to both transmit and receive, with dry contact made to the diagonally opposite corners of the sample, as shown in Fig. 11. The dry contact ensures weak coupling, minimal disturbance of the resonances by mass loading of the sample by either grease or the transducers themselves, and very reproducible (0.05%) resonances after removal and reinstallation of the sample. However, to ensure weak transducer–sample coupling, contact forces less than 0.5 g weight must be used. Thus, the signals are lower than for PVDF, and we therefore chose to use a custom-built, thermal-noise–limited superheterodyne receiver for detection.

Using this system we can fit elastic constants to our measured 20 or 30 frequencies with standard deviations of 0.05% or so, which appears to be the current limit for absolute accuracy for determination of shear moduli. However, because the remaining moduli determine resonant frequencies in a weaker way, the accuracy for c_{11}, for example, can be more than an order of magnitude lower. In Fig. 12 we show the two shear moduli determined with this system for the $x = 0.14$ sample vs. temperature, and the on-resonance motion for the resonances that are almost purely dependent on these moduli. All the other moduli were also determined (Migliori et al., 1990). Note that c_{44} (as well as everything but c_{66}) is expected to be uncoupled to the SPT, while c_{66} is expected to be strongly coupled, and this is indeed what we find. At the SPT, the material goes orthorhombic and twins. Resonances below the SPT are very poor. The contrast above and below T_s is shown in Fig. 13; similar behavior was found in the $x = 0.10$ sample as well.

The softening of c_{66} down to the SPT was originally fitted to 2-D Gaussian

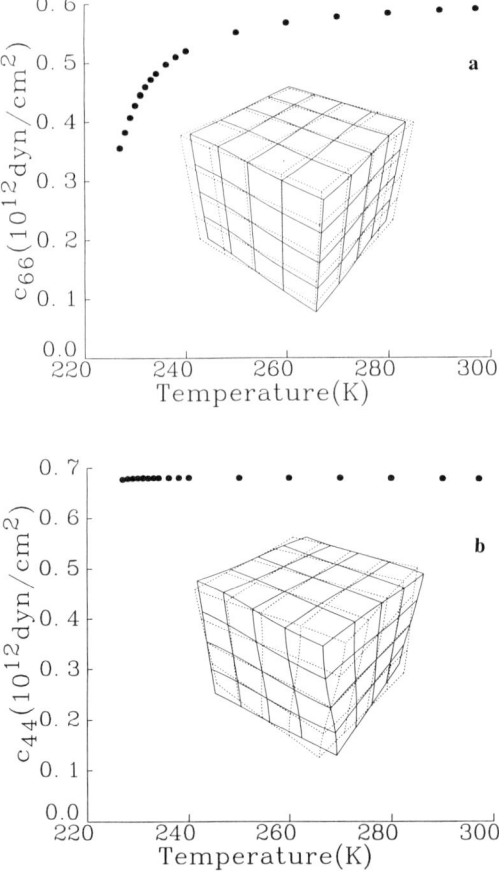

FIG. 12. The two shear moduli for $La_{2-x}Sr_xCuO_4$ with $x = 0.14$, showing clearly that c_{66} (a) couples to the SPT while c_{44} (b) does not. The insets show the strains for each of the appropriate resonances used to obtain the moduli.

fluctuations (Migliori et al., 1990). However, it appears that a self-consistent phonon approximation (Pytte and Feder, 1969) is closer to the physics, and describes well the behavior of other similar materials. The SPA uses a simple quartic–quadratic double-well potential for the appropriate zone-edge soft phonon to describe the temperature at which the free energy forces an SPT by approximating the anharmonic part of the potential with a thermal average over $\langle u_i^2 \rangle$, where u_i is a phonon displacement coordinate. Then a simple temperature-dependent linear theory is used to determine the entire dispersion curve for that phonon self-consistently. Knowing from neutron scattering data (Boni et al., 1988) that the x-point soft mode and c_{66} are part of the same dispersion curve,

8. Single Crystals of La$_2$CuO$_4$

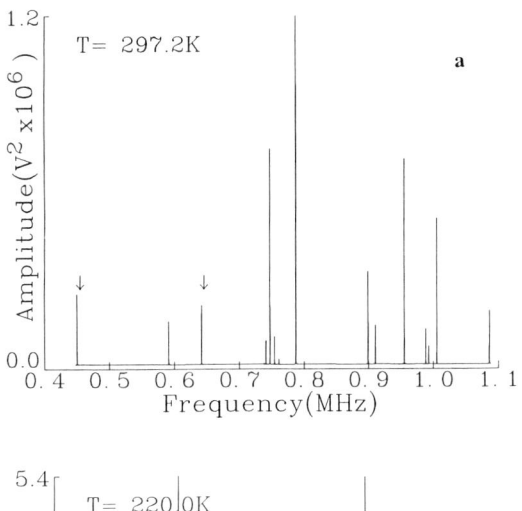

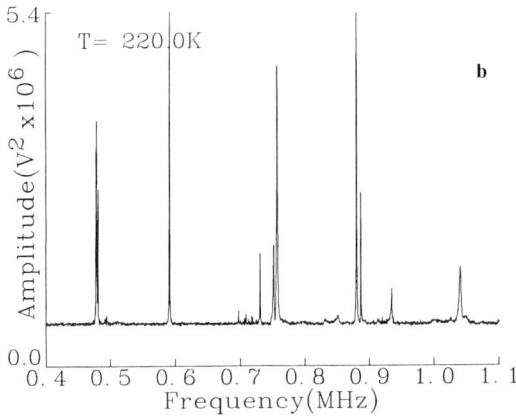

FIG. 13. A portion of the resonance spectrum of La$_{2-x}$Sr$_x$CuO$_4$ with $x = 0.14$; (a) 3 K above the SPT; (b) 20 K below the SPT. The data below the SPT clearly reflect the presence of twins.

and establishing the correct temperature dependence of c_{66}, enables application of this theory. The temperature dependence of c_{66} must be Curie–Weiss for things to work. In Fig. 14, we show, for example, a Curie–Weiss plot of c_{66} for $x = 0.14$. A similar quantity fit is obtained for $x = 0.10$, displaying the accuracy obtainable from the resonant ultrasound and unavailable from neutrons. For the $x = 0.10$ sample, however, an unoriented and very small (1.0 mm) randomly shaped flake was used. This is possible because we knew that only c_{66} softened near T_s from our study of the $x = 0.14$ sample, and so we could assume that any frequency shift reflected this. Therefore, we merely chose a strongly temperature-dependent resonance and tracked it. The size and shape of

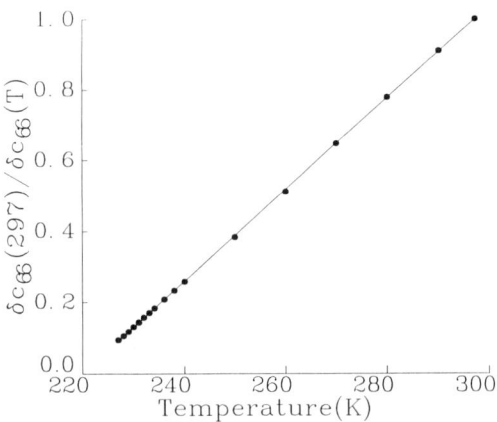

FIG. 14. A Curie–Weiss plot of the data for the $La_{2-x}Sr_xCuO_4$ samples with $x = 0.14$, showing that a critical exponent of unity clearly describes the phase transition.

the sample prevent any other ultrasound technique from working. Another advantage of RUS is seen in the magnitude of the drop in c_{66}. Other high-quality pulse echo measurements at 15 MHz show only 8% softening near T_s, while we observe nearly a factor of two at 0.6 MHz. This is primarily associated with the limited relaxation rate of the order parameter as the SPT is approached. For long-wavelength excitations (ultrasound), this relaxation rate is frequency-independent and described by the Landau–Khalatnikov damping mechanism.

Having established that the SPT is Curie–Weiss–like for two compositions, and applying the self-consistent phonon approximation (Bussmann-Holder et al., 1991a), we obtain the numerical constants for the double-well potential. We can then examine the behavior of the system well below T_s by expanding about one of the minima into which it must fall upon becoming orthorhombic, and plot the effective anharmonicity vs. x by extrapolating to other compositions. Note that this anharmonicity is primarily a result of the unique space-filling qualities of the oxygen $2p$ orbital (Bussmann-Holder et al., 1991b; Benedek et al., 1987) that is present in all the high-T_c materials, and therefore leads to a violation of Migdal's theorem.

In Fig. 15 we plot the oxygen octahedral tilt potentials vs. x, where the anharmonicity at the bottom of the double wells increases drastically just where the superconducting transition peaks. Because of the electronic origin of the anharmonicity, the usual limitations on phonon-derived T_cs are bypassed, suggesting that perhaps anharmonic electron–phonon coupling is a viable mechanism for high-temperature superconductivity. Should this be the case, an isotope effect of less than 0.5, or even negative (Bussmann-Holder et al., 1991a), is possible and might explain the variation observed in it as well.

8. Single Crystals of La₂CuO₄

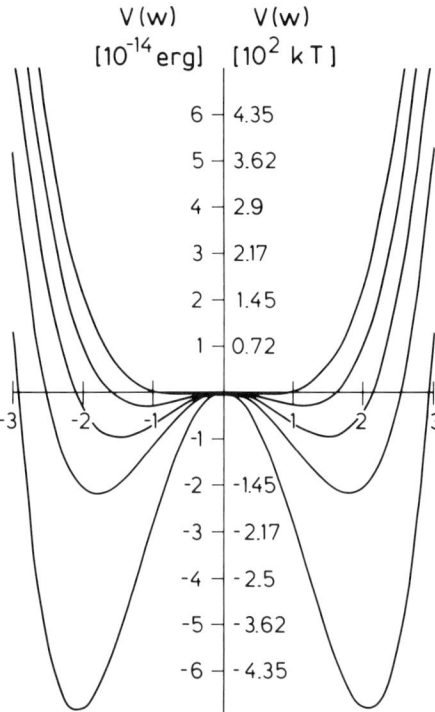

FIG. 15. A plot of the double-well potential of the relevant soft made, determined by the self-consistent phonon theory and the resonant ultrasound data.

ACKNOWLEDGMENTS

Work performed at Los Alamos under the auspices of the United States Department of Energy. Work performed at the Pennsylvania State University under NSF Grants DMR 9000549 and ONR.

References

Allan, N. L., and Mackrodt, W. C. (1988). *Mat. Res. Soc. Symp. Proc.* **99**, 797–798.
Allen, P., Fisk, Z., Migliori, A. (1989). *In* "Physical Properties of High Temperature Superconductors I" (D. M. Ginsberg, ed.). World Scientific, Singapore, p. 213.
Axe, J. D., Moudden, A. H., and Hohlwein, D., Cox, D. E., Mohanty, K. M., Moodenbaugh, A. R., and Youwen, Xu (1989). *Phys. Rev. Lett.*, **62**, 2751–2754.
Batlogg, B., Kourouklis, G., Weber, W., Cava, R. J., Jayaraman, A., White, A. E., Short, K. T., Rupp, L. W., and Rietman, E. A. (1987). *Phys. Rev. Lett.* **59**, 912–914.
Bechmann, R. (1958). *Phys. Rev.* **110**, 1060–1061.
Benedek, G., Bussmann-Holder, A., and Bilz, H. (1987). *Phys. Rev. B* **36**, 630–638.
Bhattacharya, S., Higgins, M. J., Johnston, D. C., Jacobson, A. J., Stokes, J. P., Lewandowski, J. T., and Goshorn, D. P. (1988). *Phys. Rev. B* **37**, 5901–5904.

Birgenau, R. J., and Shirane, G. (1989). In "Physical Properties of High Temperature Superconductors I," (D. M. Ginsberg, ed.). World Scientific, Singapore, p. 151.
Boni, P., Axe, J. D., Shirane, G., Birgeneau, R. J., Gabbe, D. R., Jensen, H. P., Kastner, M. A., Peters, C. J., Picone, P. J., and Thurston, T. R. (1988). Phys. Rev. B **38**, 185–194.
Bussmann-Holder, A., Migliori, A., Fisk, Z., Sarrao, J. L., and Cheong, S.-W., (1991a). Phys. Rev. Lett., **67**, 512–515.
Bussmann-Holder, A., Bishop, A. R., and Batistic, I. (1991b). Phys. Rev. B, **43**, 13728–13731.
Cady, W. G. (1946). "Piezoelectricity," Vol. 1, pp. 134–137. Dover, New York.
Cheong, S.-W., Fisk, Z., Willis, J. O., Brown, S. E., Thompson, J. P., Remeika, J. P., Cooper, A. S., Aikin, R. M., and Schifert, D. (1988). Solid State Commun. **65**, 111–114.
Cohen, R. E., Pickett, W. E., Boyer, L. L., and Krakauer, H. (1989a). Phys. Rev. Lett. **60**, 817–820.
Cohen, R. E., Pickett, W. E., and Krakauer, H. (1989b). Phys. Rev. Lett. 62, 831–834.
Demarest, H. H. (1969). J. Acoust. Soc. Am. **49**, 768–775.
Egami, T., Dmowski, W., and Jorgensen, J. D. (1987). Rev. of Solid State Science **1**, 101–106.
Ekstein, H., and Schiffman, T. (1956). J. Appl. Phys. **27**, 405–412.
Golding, B., Haemmerle, W. H., Schneemeyer, L. F., and Waszcak, J. V. (1988). In "Proceedings of the 1988 Ultrasonics Symposium." Institute of Electrical and Electronics Engineers, New York, pp. 1079–1083.
Heeger, A. J., Beckman, O., and Portis, A. M. (1961). Phys. Rev. **123**, 1652–1660.
Hunklinger, S., and Schickfus, M. V. (1981). In "Amorphous Solids: Low Temperature Properties" (W. A. Phillips, ed.). Springer-Verlag, New York.
Kasowski, R. V., Hsu, W. Y., and Herman, F. (1991). Solid State Commun., to be published.
Kim, T. J., Luthi, B., Schwarz M., Kuhnberger, H., Wolf, B., Hampel, G., Nikl, D., and Grill, W. (1988). J. Magn. Mater. **76–77**, 604–606.
Lang, M., Steglich, F., Schefzyk, R., Lechner, T., Spille, H., Rietchel, W., Goldacker, W., and Renker, B. (1987). Euro. Phys. Lett. **4**, 1145–1149.
Ledbetter, H. M., Kim, S. A., and Violet, C. E. (1989). Masters' theses, HTSC, Stanford, California.
Mason, W. P. (1950). "Piezoelectric Crystals and Their Applications to Ultrasonics." Van Nostrand, New York.
Migliori, A., Visscher, W. M., Wong, S., Brown, S. E., Tanaka, S., Kojima, H., and Allen, P. B. (1990). Phys. Rev. Lett. **64**, 2458–2461.
Morse, P., and Fesbach, H. (1953a). "Methods of Theoretical Physics." McGraw-Hill, New York, pp. 1130–1137.
Morse, P., and Fesbach, H. (1953b). Ibid., pp. 1114–1118.
Ohno, I. (1976). J. Phys. Earth **24**, 355–379.
Ohno, I., Yamamoto, S., Anderson, O. L., and Noda, J. (1986). J. Phys. Chem. Solids **47**, 1103–1108.
Press, W. H., Flannery, B. P., Teukolsky, S. A., and Vetterling, W. T. (1986). "Numerical Recipes, the Art of Scientific Computing." Cambridge University Press, Cambridge, pp. 523 ff.
Pytte, E., and Feder, J. (1969). Phys. Rev. **187**, 1077–1088.
Tanaka, I., and Kojima, H. (1989). Nature (London) **337**, 21–22.
Testardi, R. L. (1964). In "Physical Acoustics: Principles and Methods," Vol. XIII (W. P. Mason and R. N. Thurston, eds.). Academic Press, New York, pp. 637–648.
Truell, R., Elbaum, C., and Chick, B. B. (1969). "Ultrasonic Methods in Solid State Physics." Academic Press, New York.
Suzuki, T. (1990). Hiroshima University, private communication.
Vigoureux, P., and Booth, C. F. (1950). "Quartz Vibrators and Their Applications," 3rd Ed. His Majesty's Stationery Office, London, pp. 49–51.
Voigt, W. (1928). "Lehrbuch der Kristallphysik." Teubner, Leipzig.

—9—
A Rationalisation of the Diversity in the Elastic Response of Polycrystalline Superconducting Oxides

D. P. ALMOND
School of Materials Science, University of Bath, Claverton Down, Bath, United Kingdom

1. Introduction 409
2. Review of Initial Work 410
   2.1. Contrasting Characteristics 410
   2.2. Effect of Sample Oxygen Content 412
   2.3. Internal Stress 413
   2.4. Porosity 417
3. A Comparison of the Ultrasonic Characteristics of Superconducting and Non-superconducting Material 418
   3.1. Rationale 418
   3.2. Comparison of Sound Velocities 418
   3.3. Anelastic Relaxation in the Tetragonal State 420
   3.4. Anelastic Relaxation in the Orthorhombic, Superconducting State 423
   3.5. Relaxation Strengths 426
   3.6. Anelasticity and Oxygen Diffusion 427
   3.7. Magnetic Phenomena 428
4. High-Temperature Anomalies 428
5. Conclusions 429
   Acknowledgments 431
   References 431

1. Introduction

The very extensive literature of high-temperature superconductivity contains numerous examples of what have proved to be preliminary results and premature claims. The excitement and pressure "to be first" at the time of the early discoveries was not conducive to methodical scientific research. The ultrasonics community was not immune from these influences. A particular problem was that the new superconductivity was so unexpected that no one knew what to expect. Consequently, any effect observed at or near the superconducting transition temperature, T_c, initially was reported as being, or possibly being, a characteristic of the new superconducting state.

When it became possible to compare the ultrasonic results obtained by various research groups throughout the world on differently prepared samples, a large diversity in the data was revealed. To this day there remain differences about the magnitudes of the sound velocities at room temperature, and their temperature dependences near T_c. The aims of this article are to review the ultrasonic data and to offer explanations for some of the diversity. The review will concentrate on the most widely studied of the new superconductors, $YBa_2Cu_3O_{7-x}$ (YBCO) and its related compounds. The review will be devoted mainly to results obtained from polycrystalline samples, since single-crystal studies are being dealt with elsewhere in this volume. The review will, however, incorporate low-frequency internal friction measurements, as these considerably extend the narrow frequency range accessible to workers using the higher-frequency ultrasonic pulse-echo techniques. The clarification of the frequency dependence that is then afforded proves crucial to the interpretation of the data.

2. Review of Initial Work

2.1. CONTRASTING CHARACTERISTICS

The early published ultrasonic studies of YBCO were predominantly of the temperature dependence of the velocity of sound. Workers were looking for evidence of the type of mode-softening that characterised the A15 superconductors (Testardi, 1973). The published data can be divided into two groups: that which showed large variations in sound velocity and significant hysteresis with temperature (Almond et al., 1987; Muller et al., 1987; Lang et al., 1987; Ledbetter and Kim, 1988), and that in which no hysteresis was found and in which much smaller thermal anomalies were recorded at temperatures near T_c (Bishop et al., 1987; Horie et al., 1987; Bhattacharya et al., 1988).

The data obtained by Almond et al. (1987) shown in Fig. 1 is typical of the first group. Extensive hysteresis was found in the data obtained during cooling and warming. This hysteresis disappeared at low temperatures where a significant increase in velocity occured. The temperature of the onset of this hysteresis-free velocity enhancement did not, however, correspond to the sample T_c of 90 K. Similar hysteresis was reported by Muller et al. (1987a) and by Lang et al. (1988), but without so distinctive an increase in velocity at low temperatures. More recently, Ledbetter and Kim (1988) found hysteresis in $RBa_2Cu_3O_{7-x}$ with $R = $ Y, Ho, and Eu, and reported the effects to be larger in the Eu superconductor and lower in the Ho with respect to the original Y superconductor. They suggested that the temperatures 160 K and 70 K were significant, since at the first there was a change in sign of the velocity temperature gradient, and at the second

9. Elastic Response of Polycrystalline Superconducting Oxides 411

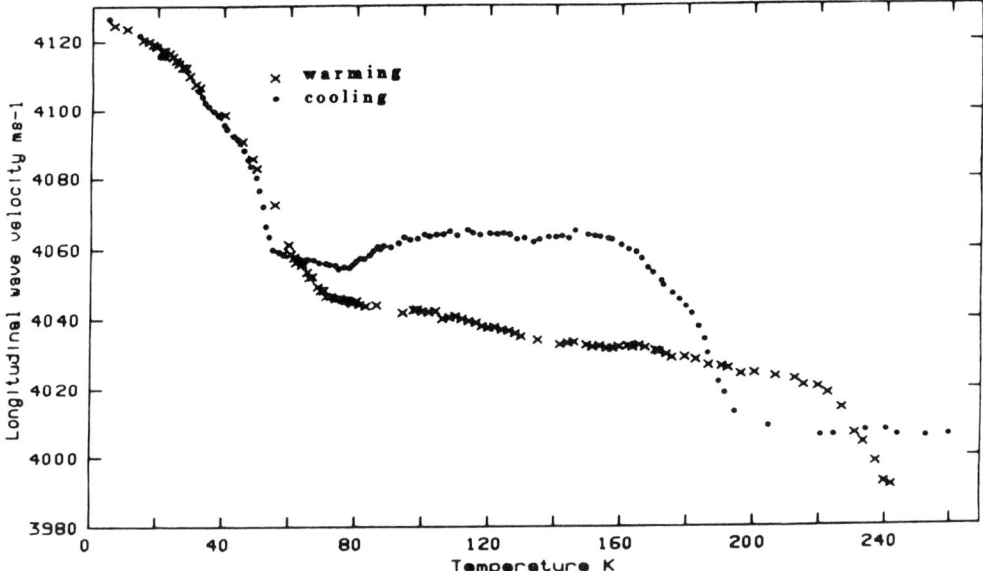

FIG. 1. The temperature dependence of 5 MHz longitudinal ultrasonic waves propagated in YBCO, 95% density, large grain sized ceramic (Almond et al., 1987).

hysteresis disappeared. These effects at these temperatures are evident in the data, Fig. 1, and in the other examples just discussed. Ledbetter and Kim suggested that the effects might be interpreted as being a consequence of a very broad phase transition between an unidentified high and low temperature phase.

The data shown in Fig. 2 (Bishop et al., 1987) is typical of the second group. It shows no evidence of hysteresis and is merely the familiar smooth enhancement of velocity at low temperatures perturbed by an inflection near T_c. The data was interpreted as showing an anomalous hardening of the bulk modulus below T_c that could not be explained by conventional superconductor theory. The effects are much smaller than those discussed above for the first group. Similar minor departures from the conventional velocity–temperature characteristic were reported by others (Horie et al., 1987; Bhattacharya et al., 1988).

The reason for the two very different groups of data is to be found in the early and very revealing paper of Ewert et al. (1987). In this, the authors reported extensive hysteresis in the data of one sample and none in a second. The difference between the samples was that the first had a very coarse grain structure, with grains up to 50 μm in diameter, whilst in the second grain size was only about 4 μm. The authors commented that a pronounced domain structure, twinning, was only evident in the grains of the large grain size sample, and they

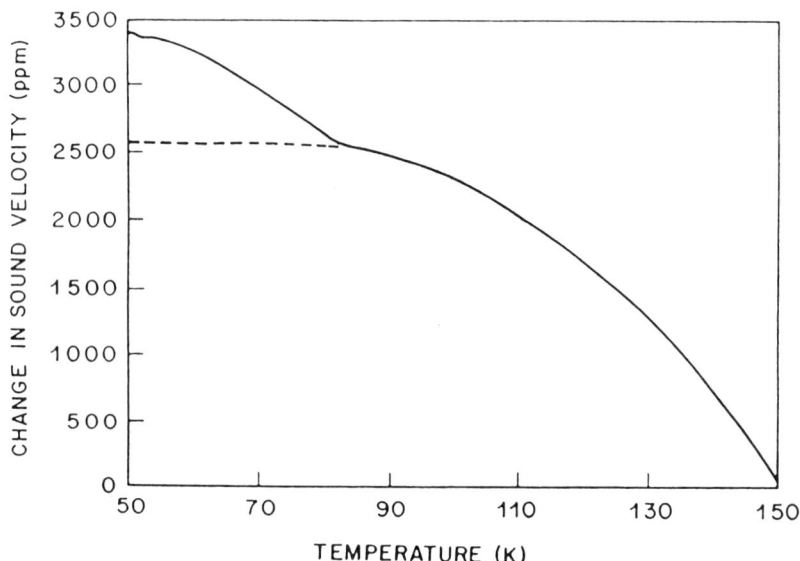

FIG. 2. The change in 10 MHz longitudinal sound velocity in polycrystalline YBCO (Bishop *et al.*, 1987).

attributed the hysteretic effects to this domain structure. They concluded that the results of the small grain size sample were a more faithful measure of the basic properties of the material and that the formation of the superconducting state had no effect on the elastic properties.

Muller *et al.* (1987a, 1987b, 1988) reported that the velocity hysteresis observed, again in a coarse-grained (~10 μm) sample, was cooling-rate–dependent and that it could be eradicated by several shock coolings into liquid nitrogen. They pointed out that similar behavior was found in a twinned ferroelastic crystal of $LiKSO_4$, where a first-order structural phase transition explains the data. Like Ledbetter and Kim (1988), they concluded that a similar phase transition would explain the results in YBCO and that it was one which influenced the formation of twins. They, alternatively, speculated that hysteresis effects may be due to internal strains which are enhanced on shock cooling to smear out the hysteresis effect.

2.2. Effect of Sample Oxygen Content

The early ultrasonic studies of YBCO were performed before the critical dependence of this material on oxygen concentration (Cava *et al.*, 1987; Alford *et al.*, 1988) had been appreciated. It is now known that the superconducting

9. Elastic Response of Polycrystalline Superconducting Oxides

properties (transition temperature, critical current density, etc.) of YBCO fall sharply with the deviation of oxygen content from the optimum value of O7. All the polycrystalline samples used in ultrasonic studies were produced by pressing and sintering YBCO powder. The established sintering temperature is about 950°C, and unfortunately, at this temperature the oxygen content drops stoichometrically below O6.5. It is essential to raise the oxygen content toward O7 by annealing, usually in oxygen, at about 400°C after the sintering process is completed. Sintering is, however, a process of densification, and consequently the reintroduction of oxygen is a slower process than its loss prior to sintering. For moderately dense samples (about 90% theoretical density), an annealing time of 48 hours is adequate to ensure high oxygen content throughout a sample. At higher densities, however, porosity becomes closed (Alford *et al.*, 1988), and oxygen access into the bulk of a sample becomes limited by the solid state diffusion rate of oxygen through the structure. It has been found (Yamamoto *et al.*, 1988) that for 98% density samples which were only 1 mm thick, annealing times of 21 days were required. Samples for conventional ultrasonic studies need to be at least 3 mm thick, and many workers in this field have used samples which were between 7 and 10 mm thick. Many of the groups (Almond *et al.*, 1987; Lang *et al.*, 1987; Ewert *et al.*, 1987; Ledbetter and Kim, 1988) reported using samples which were considerably more than 90% of theoretical density. Consequently, there must be some doubt about the oxygen contents of the bulks of these samples.

It has often been suggested in the past that a great advantage of ultrasonics over other investigative techniques is that it senses the bulk rather than being adversely affected by a sample's surface. This would appear to have become a disadvantage in the study of high-T_c superconductors where the interiors of high-density samples are the least likely to achieve optimum oxygen concentration. In addition, resistance temperature characteristics are dominated by the quality of the material at the sample surface, which has the highest oxygen content. It is, therefore, possible that the lack of correlation between resistive indications of superconductivity and ultrasonic phenomena may be traced, in some instances, to the material in the acoustic paths through the samples having significantly lower oxygen contents than that monitored by resistive measurements made at the surfaces.

2.3. INTERNAL STRESS

It was recognised at the outset by some (Almond *et al*, 1987; Muller *et al.*, 1987a) that internal stresses between the crystallites in polycrystalline ceramic samples may influence their net elastic properties. Such effects are common throughout ceramic science, and in extreme cases can lead to extensive internal cracking and thermal shock failure. The magnitude of the effect of internally generated stress on elastic properties can be gauged from measurements of the effect of externally applied pressure. A series of studies (Al-Kheffaji et al., 1989a; Cankurtaran et al., 1988, 1989) of the affect of hydrostatic pressure on the various elastic moduli of polycrystalline YBCO and $GdBa_2Cu_3O_{7-x}$ (GBCO) have revealed the bulk modulus to be particularly sensitive to pressure. The pressure dependence of the bulk modulus, dB/dP, was found to be about 50, whilst values for most other ionically bonded solids are found to be in the range 5 to 10. Consequently, it may be concluded that the elastic properties determined ultrasonically for a bulk sample will be a function of the various internal stresses generated, inter-granularly, within the sample. In addition, to further complicate matters, these internal stresses may vary with sample temperature to generate a spurious temperature dependence in the measured sound velocity.

Evidence of both the presence of internal stress and its variation with temperature has been found in optical microscopic studies. YBCO has an orthorhombic crystal structure, for oxygen contents between 6.5 and 7, which becomes heavily twinned to relieve elastic strain energy. The (110) twin boundaries are formed (Hewat et al., 1987; Van Tendezoo et al., 1987) when the material undergoes a structural phase transition from its high-temperature (~600°C) tetragonal phase to the lower-temperature orthorhombic phase. The resulting domain structure can be studied by polarised light microscopy. A micrograph of a dense (92%) sample of YBCO was reproduced in the paper of Ewert et al., (1987). The banded twinned structure was evident in most of the larger grains. It had a typically irregular structure that signifies the influence of stresses between the grains. These stresses vary with temperature because of the anisotropy of the expansion coefficients. Such effects can be seen to occur in a sample heated on a microscope hot stage. They are most pronounced in grains that are in contact with other grains of greatly differing crystallographic orientation, as evidenced by sharply differing colour contrast when viewed in polarised light.

2.3.1. Pseudoplasticity

Shen et al. (1988) reported direct tensile test measurements of the elastic properties of YBCO and found hysteresis that they attributed to the hysteretic movement of twin boundaries in response to an applied stress. They also found this hysteresis to disappear at low temperature like that, discussed previously, in some of the sound velocity studies. Since twin boundary movement produces an extensive distortion of the material, it may in some respects be likened to the familiar dislocation movement that results in ductility. Consequently, it has been

9. Elastic Response of Polycrystalline Superconducting Oxides

suggested (Shen *et al.*, 1988; Almond *et al.*, 1988) that YBCO is a "pseudoplastic" material. It has been noted (Almond *et al.*, 1988) that there is a marked similarity between the twinning in YBCO and that in the indium–thallium alloys (Guttman, 1950) that exhibit marked pseudoplasticity, producing almost rubber-like properties.

2.3.2. Heat Treatment

Like conventionally ductile materials, metals, the properties of YBCO have been found to be affected by heat treatment and thermal history; this has already been touched on in the remarks about the work of Muller *et al.* (1987b). A distinction must be made here between the alteration of sample oxygen content by annealing and the effect that the raised temperature has on other properties of the material. An explanation of the elastic hysteresis in YBCO is that the elastic response is modified by twin boundary movement, and that this is hysteretic because of the presence of pinning centres. An analogy may be drawn here with the explanation of magnetic hysteresis being caused by the influence of pinning centres on domain wall movement. The annealing process may, in addition to increasing sample oxygen content, also facilitate the unpinning of numerous twin boundaries. The sample, on its return to room temperature, may then have quite a different elastic domain structure, and hence a quite different array of internal stresses. Direct evidence of these effects was reported by Hatanaka and Sawada (1989). They showed that, in a single-crystal sample of YBCO, the domain pattern was readily affected by an externally applied stress when sample temperature was raised to 300°C. A nearly single-domain sample was formed at this temperature, and it was found to be possible to retain this structure at room temperature provided the sample was cooled under load. The authors commented that the coercive stress was comparatively large for this "ferroelastic" material.

The effect of heat treatment on the ultrasonic characteristics of YBCO was discussed by Almond *et al.* (1988). It was reported that the elastic moduli, derived from ultrasonic wave velocity measurements, of a dense YBCO sample were apparently increased by as much as 20% by the process of annealing at 400°C. Subsequently, the velocity–temperature characteristic shown in Fig. 3a was obtained, in sharp contrast to that obtained earlier in the same sample, Fig. 1. The second cooling characteristic was obtained after the sample had been left at room temperature for 24 hours. During that time the sound velocity partially recovered from the reduction caused by the first cooling in the manner shown in Fig. 3b. These phenomena were found to be reproducible and independent of atmosphere, showing that they were not attributable to oxygen concentration effects. The interpretation of the phenomena was that twin boundary pinning

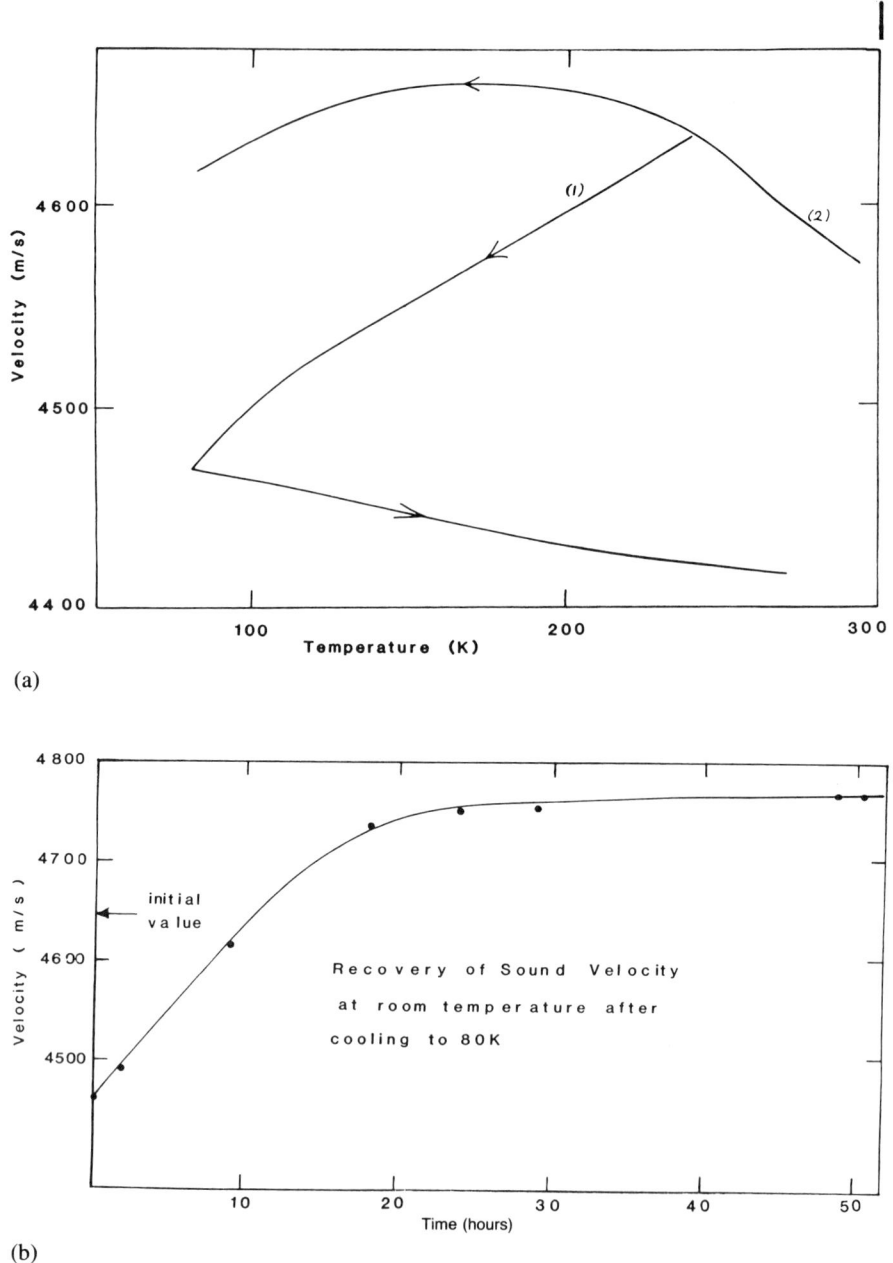

FIG. 3. (a) The temperature dependence of 10 MHz longitudinal ultrasonic waves in YBCO: (1) after annealing; (2) after first cooling–warming. See text. (b) The temporal recovery of 10 MHz longitudinal wave velocity in YBCO at room temperature after the first thermal cycle, (1) in (a) (Almond et al., 1988).

9. Elastic Response of Polycrystalline Superconducting Oxides

was relieved during annealing, producing an alteration in domain structure that was reflected in a change in measured modulus. On cooling, stresses in the sample induced twin boundary movement that became pinned at the low temperatures. The recovery monitored at room temperature was attributed to the delayed breakaway of twin boundaries from pinning centres. These results emphasise the alarming differences that can be produced in ultrasonic characteristics by heat treatment and subsequent cooling cycles. They have been attributed to modifications to the elastic domain structure of the material manifested in the ultrasonic characteristics via their sensitivity to the resulting alterations in internal stress.

2.4. POROSITY

Porosity is an inevitable feature of the most commonly used "shake and bake" technique for preparing polycrystalline YBCO samples. In the most dense samples studied ultrasonically, porosity has been as little as 3%, and in the least dense, as much as 40%. Thus, porosity alone is sufficient to explain a significant part of the variation in sound velocity measurements reported in the literature.

Horie *et al.* (1989) reported a systematic study of the dependence of the absolute value of sound velocity on porosity. They found that samples of $(RE)Ba_2CU_3O_{7-x}$, where the rare earths (RE) were Y, Ho, Dy, Er, Sm and Nd, varied in density from 80 to 95% and that sound velocity increased, over this range, linearly with density. Their estimate, by extrapolation, of the longitudinal wave velocity in a fully dense sample was 4.9 km/s. The magnitude of the influence of porosity was clearly illustrated by the finding that for an 80% dense $HoBa_2Cu_3O_{7-x}$ sample, longitudinal wave velocity fell to about 3.1 km/s.

An alternative theoretical analysis, based on the multiple scattering theory of Sayers and Smith (1982), was adopted by Cankurtaran *et al.* (1988). This theory was developed assuming a homogeneous medium containing spherical pores. A longitudinal wave velocity of about 4.8 km/s was obtained for the non-porous matrix of YBCO from the correction of experimental measurements of 82% and 95% dense samples. This value is in remarkably good agreement with the value of 4.9 km/s obtained by Horie *et al.* using the quite different method described above.

At first sight, porosity is an undesirable feature of an ultrasonic sample because not only does it affect the magnitudes of the sound velocities, but also it might be expected to seriously attenuate ultrasonic waves by scattering processes. An alternative view that has been proposed before by the author (Almond *et al.*, 1989) is that porosity is in other ways benificial. Firstly, connected porosity provides conduits for oxygen transport into the bulk of a sample. Then, during annealing, oxygen diffusion by the slow solid state diffusion process is initiated throughout the volume of the sample, rather than from the exposed outer surfaces alone. Hence, it is reasonable to expect a much more uniform oxygen stoichiometry in such porous samples. The second benificial feature of pores is their contribution to the relief of internal stress. In porous samples the crystallites are simply less constrained by their neighbours, and the stresses generated during sintering and subsequent cooling are reduced.

3. A Comparison of the Ultrasonic Characteristics of Superconducting and Non-superconducting Material

3.1. RATIONALE

Many of the observations cited in Section 2 were obtained at an early stage in the investigation of YBCO and may, collectively, be regarded as forming a pilot study. The lessons of this pilot study were that the ultrasonic characteristics are affected by many different phenomena, and that it is hard to distinguish effects associated with superconductivity from those associated with the material's other complex conventional properties. In earlier studies of lower–transition-temperature metallic superconductors, this type of difficulty was overcome by comparing characteristics before and after destroying the superconducting state with a magnetic field in excess of the critical field H_c or, in the case of type II superconductors, the upper critical field H_{c2}. For the high-T_c superconductors, however, this procedure can not be adopted because their H_{c2}s far exceed the fields of laboratory magnets. An alternative is to compare results obtained in the same sample with oxygen content, adjusted to make it firstly superconducting and then non-superconducting. However, in addition to altering the electronic properties, the removal of structurally incorporated oxygen also converts the crystal structure of the material from orthorhombic to tetragonal and it becomes antiferromagnetic. Consequently, this alternative is by no means as "clean" as the use of a magnetic field, and comparative results must be interpreted with caution.

3.2. COMPARISON OF SOUND VELOCITIES

Some of the results of a comparative study (Almond *et al.*, 1989) of the ultrasonic characteristics of superconducting and non-superconducting GBCO are shown in Fig. 4. The sample used for this work displayed excellent ultrasonic wave transmission characteristics despite being 22% porous. As mentioned in Section

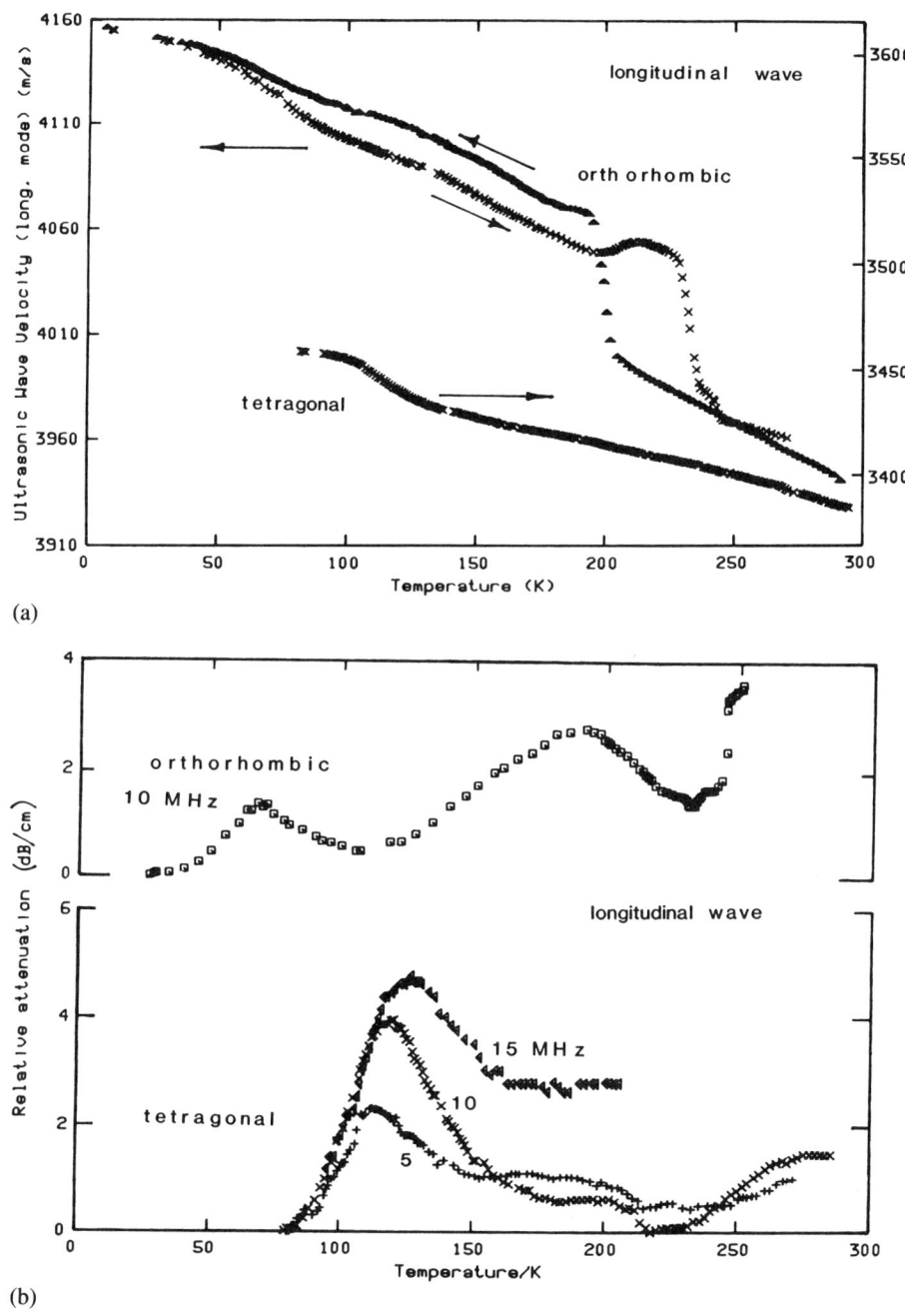

FIG. 4. The temperature dependences of (a) the longitudinal wave velocities and (b) the ultrasonic attenuations in both the orthorhombic and the tetragonal states of GBCO (Almond *et al.*, 1989).

2, this porosity may be beneficial in ensuring oxygen access to the centre of the sample. The sample was made superconducting by the usual process of annealing in an oxygen atmosphere at 400°C. After a programme of measurements, it was converted to a non-superconducting sample by removing some of its structurally incorporated oxygen. This was achieved by heating it in a vacuum to 700°C for 24 hours, a process known to reduce oxygen concentration to O6.4.

The dominant feature of the sound velocity characteristic of superconducting GBCO, Fig. 4a, is a steplike change at temperatures above 200 K. Hysteresis is found predominantly in the same temperature range. It is notable that these features are absent from the characteristics of the sample in its non-superconducting tetragonal form. Similar behaviour at temperatures above 200 K has been observed in other high-T_c superconductors, and these effects are discussed separately in Section 4.

At temperatures close to T_c, 90 K, there is a distinct inflection in the velocity–temperature characteristic of the superconducting sample data. This is similar to that observed by others, e.g., Bishop et al. (1987) (Fig. 2). However, it is particularly significant that an inflection also occurs in the velocity–temperature characteristic of the non-superconducting state data, though at a somewhat higher temperature. Obviously this phenomenon is not attributable to superconductivity, and its appearance throws doubt on whether its occurrence in the superconducting sample has anything to do with superconductivity either.

3.3. ANELASTIC RELAXATION IN THE TETRAGONAL STATE

The inflection in sound velocity found in the non-superconducting, tetragonal sample is a conventional anelastic effect. Ultrasonic attenuation measurements across the same temperature range, Fig. 4b, revealed a distinctive peak which displaced to higher temperatures with ultrasonic wave frequency. This is classical anelastic relaxation behaviour caused by an interaction of the acoustic wave with an entity within the sample which has a thermally activated relaxation time. Such effects are commonly observed in solids containing mobile atoms, ions, or molecules. They are accompanied by a low-temperature stiffening of elastic constants, and this is assumed to be the cause of the inflection in the velocity temperature characteristic, Fig. 4a. The attenuation peak is usually taken to occur where the conditions $\omega\tau = 1$ is satisfied, in which ω is the angular frequency of the acoustic wave and τ is the relaxation time. τ is conventionally found to fall with temperature as

$$\tau = \tau_0 \exp(E_a/kT), \qquad (1)$$

9. Elastic Response of Polycrystalline Superconducting Oxides   421

in which E_a is an activation energy, τ_0 is the inverse of the attempt frequency ν_0, k is the Boltzmann constant, and T is absolute temperature. Similar effects were reported by Cannelli *et al.* (1988a) in much lower-frequency (1.1 to 17.4 kHz) vibrating beam internal friction measurements of a similarly oxygen-outgassed sample of YBCO. Some of their data are reproduced here as Fig. 5.

A combined Arrhenius plot of the relaxation rates, τ^{-1}, (obtained assuming $\omega\tau = 1$ at each peak) indicated by the ultrasonic and internal friction data is shown in Fig. 6. It is evident that there is excellent agreement between the two data sets. It confirms that the two techniques were sensing the same phenomenon. This is important to note because the ultrasonic work was carried out on a GBCO sample, whereas the internal friction was of an YBCO sample. As in most of the properties of these materials, the particular rare earth appears here to have little influence on this mechanical relaxation phenomenon.

The ultrasonic and internal friction data, taken together, cover a frequency range of some four decades, and consequently provide a quite reliable parameterisation of the relaxation process involved. The line drawn through the data indicates this process to be controlled by a rather typical attempt frequency of about 10^{12} Hz and a particularly low activation energy of about 0.1 eV. A broadly similar conclusion was drawn by Cannelli *et al.* (1988a), based on their low-frequency data alone, and by Lemmens *et al.*, some of whose ultrasonic data is also included in the figure.

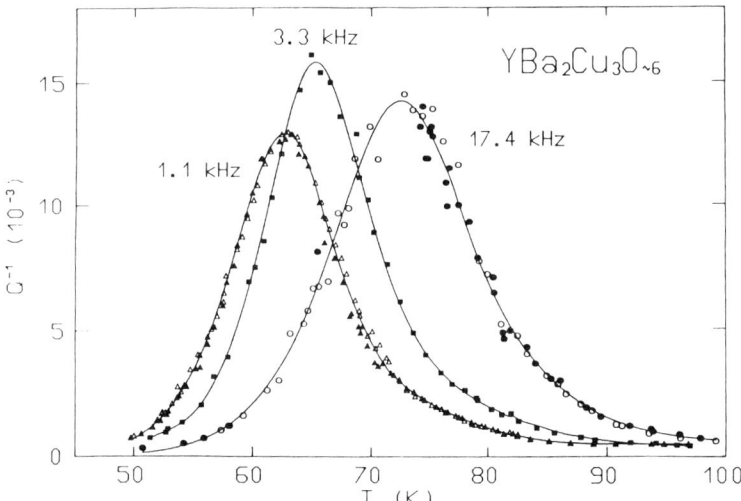

FIG. 5. The temperature dependence of the internal friction in an oxygen-outgassed sample of YBCO at low temperatures (Cannelli *et al.*, 1988a).

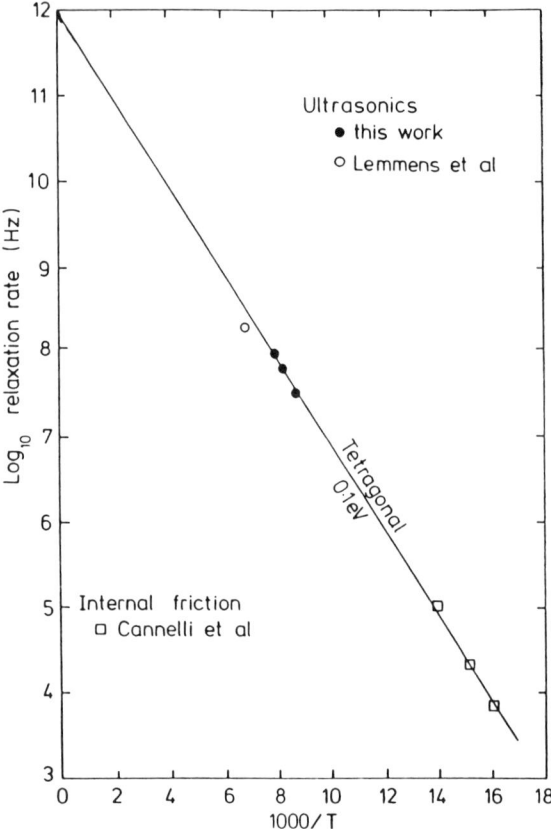

FIG. 6. An Arrhenius plot of the of the relaxation rates in oxygen-outgassed, tetragonal YBCO/GBCO.

Further confirmation of the preceding interpretation is found in the temperature dependence of the ultrasonic attenuation peaks. The simple single relaxation time expression for a mechanical absorption peak is

$$\alpha = \Delta\omega^2\tau/(1 + \omega^2\tau^2), \qquad (2)$$

in which Δ is the relaxation strength which will be discussed later. This expression is fitted to some of the ultrasonic attenuation data in Fig. 7, using the temperature dependence of the relaxation time, Eq. (1), indicated in Fig. 6. The fit is remarkably good. Peaks considerably broader than the prediction of the single relaxation time Debye expression are widely encountered in complex solids. It may be inferred from the sharpness of the peak that the entities concerned relax independently of each other.

9. Elastic Response of Polycrystalline Superconducting Oxides 423

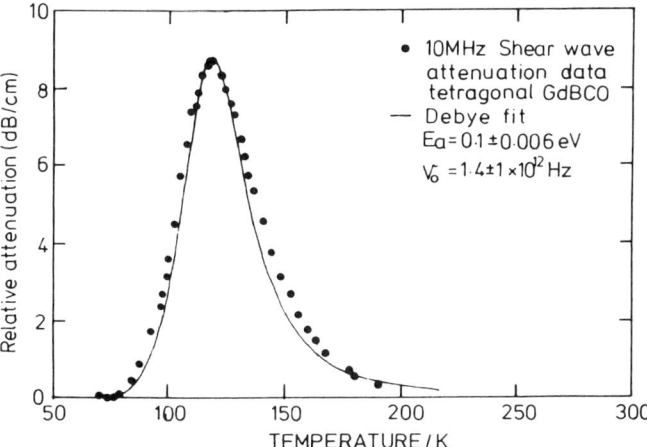

FIG. 7. A fit of the single–relaxation-time Debye expression, Eq. (2), to the ultrasonic attenuation in tetragonal GBCO, using the parameters shown in the figure.

3.4. ANELASTIC RELAXATION IN THE ORTHORHOMBIC, SUPERCONDUCTING STATE

The appearance of the preceding clear anelastic phenomena at low temperatures raises the possibility that similar anelastic phenomena may be responsible for some of the effects observed in superconducting samples. The key to the identification of conventional mechanical relaxation phenomena is the frequency dependence of the absorption peaks. This contrasts with phenomena genuinely associated with a phase transition, such as superconductivity, which have temperature dependences locked to the transition temperature. The ultrasonic attenuation of the superconducting sample exhibited two absorption peaks, Fig. 4b. The lower-temperature one of these can be seen to be associated with the velocity temperature inflection, Fig. 4a, that occurs in the vicinity of the critical temperature. Sun et al. (1988) reported frequency-dependent ultrasonic attenuation peaks at these low temperatures, and Bhattacharya et al. (1988) reported an attenuation peak at a similar temperature. The higher-temperature peak is similar to those reported by Ewert et al. (1987) and by Lemmens et al. (1989).

Again, instead of considering ultrasonic data in isolation, it proves instructive to examine data obtained by other lower-frequency techniques. There have been a wide range of measurements of mechanical relaxation, employing a variety of different techniques covering the frequency range 8 Hz to 100 kHz, in the superconducting form of YBCO. An example of 1.05 kHz measurements by Duran et al. (1988) is shown in Fig. 8. The pattern of these data is similar to

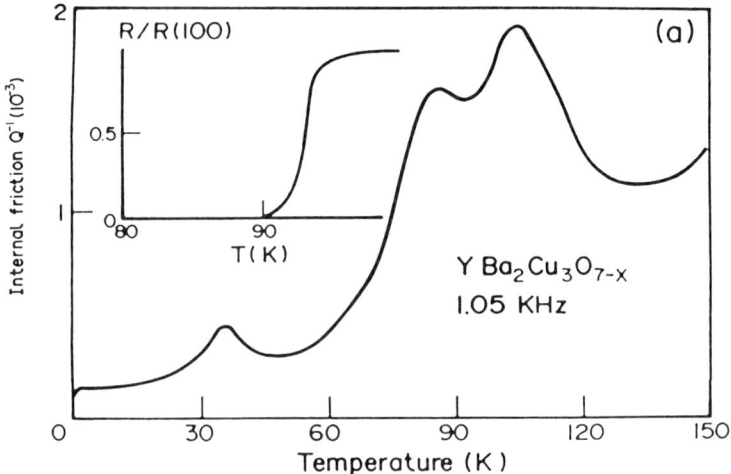

FIG. 8. The temperature dependence of the internal friction Q^{-1} in a superconducting, orthorhombic sample of YBCO (Duran et al., 1988).

that of the ultrasonic attenuation. It shows a lower-amplitude peak at low temperatures, and a larger-amplitude double peak at higher temperatures. It is notable that the various peak temperatures of these data are lower than those of the corresponding ultrasonic attenuation peaks, Fig. 4b, obtained at 10 MHz. The links between these and other mechanical relaxation studies are demonstrated in Fig. 9. This figure shows Arrhenius plots of relaxation rates indicated by the absorption peaks reported to have been found in internal friction studies performed at frequencies from 8 Hz to 100 kHz, as well as some obtained from ultrasonic work at megahertz frequencies. The higher-temperature double peaks show a systematic frequency dependence characteristic of two separate thermally activated processes controlled by 0.14 and 0.2 eV activation energies.

These characteristics converge at higher temperatures and relaxation rates, explaining the single peak obtained by the higher-frequency ultrasonic techniques. There is less evidence to confirm that the lower-temperature peak is also caused by a conventional anelastic mechanism, because the majority of the lower-frequency internal friction studies were not pursued to below liquid-nitrogen temperature. When they were (Duran et al., 1988; Mizubayashi et al., 1988), internal friction peaks were found which are consistent with the lower-temperature ultrasonic attenuation peak, Fig. 4b, being attributable to a further thermally activated process controlled by an activation energy of only 0.06 eV. In each case the extrapolated attempt frequency, which should be comparable with a lattice vibrational frequency (10^{12}–10^{13} Hz), has a reasonable value. This contributes significantly to the plausibility of the data relating to real physical pro-

9. Elastic Response of Polycrystalline Superconducting Oxides

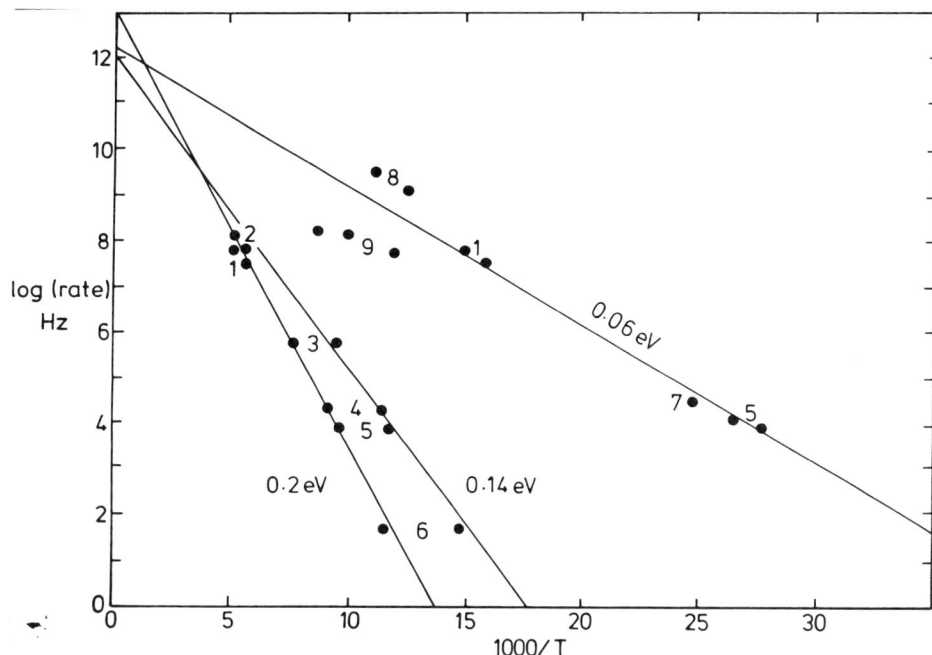

FIG. 9. Arrhenius plots of the relaxation rates in superconducting, orthorhombic YBCO or GDCO obtained from the mechanical loss peaks of (1) Almond *et al.* (1989); (2) Lemmens *et al.* (1989); (3) Cheng *et al.* (1988); (4) Cannelli *et al.* (1988a,b); (5) Duran *et al.* (1988); (6) Weller *et al.* (1989); (7) Mizubayashi *et al* (1988); (8) Saint-Paul and Henry (1989); (9) Sun *et al.* (1988).

cesses in the material. The controlling activation energies, however, are again particularly small.

The preceding analysis has produced a rationalisation of a large number of the mechanical relaxation studies of YBCO. Formerly, each of these was reported as a characterisation produced by the various techniques at low temperatures, and in most cases the suggestion was made that absorption peaks occuring at or near T_c might be related to the appearance of superconductivity. The double peaks occured within a few tens of degrees of T_c. This can now be seen to be coincidental (a consequence of $\omega\tau = 1$ being satisfied for the frequency employed at a temperature near T_c). Similarly, the appearance of the ultrasonic anomalies just below T_c would seem to be coincidental and not related directly to superconductivity. This analysis of mechanical relaxation in YBCO has been adopted by Kusz *et al.* (1990) and by Cannelli *et al.* (1990), who each have collected data covering a wide frequency range which they conclude must be explained by the presence of relaxation processes with activation energies very similar to those listed here.

3.5. RELAXATION STRENGTHS

The magnitudes of the absorption peaks provide indications of the concentrations of the entities involved in the mechanical relaxation phenomena. This additional information is helpful in deciding among a number of possible candidates for the physical origin of the entitles. Anelastic absorption peak amplitudes are determined by the relaxation strength Δ, Eq. (2), which is given by (Jackle et al., 1976)

$$\Delta = NB^2/(4\pi\rho v^3 kT), \qquad (3)$$

in which N is the concentration of entities, B is the elastic strain dependence of their site energy—a deformation potential, ρ is density, and v the velocity of sound. Since the terms in the denominator of this expression are all known, the absorption peak amplitudes provide a measure of the product NB^2. The values of NB^2 obtained from the ultrasonic attenuation peaks, Fig. 4b, are shown in Table I. The value of this parameter obtained in a similar study (Almond and West, 1988) of the ionic conductor sodium beta-alumina is also included for comparison. The entities responsible for the ultrasonic attenuation peaks in the latter material were the mobile sodium ions, whose concentration is well known. Consequently, for beta-alumina the deformation potential could be assessed. The value of 0.44 eV is quite typical; in fact, most solids that have been analyzed in this way yield values near to or less than 1 eV. To provide a conservative estimate of the concentrations of entities causing the absorption peaks in GBCO, an upper value of $B = 1$ eV was assumed. It should be noted that the NB^2 products obtained from the GBCO peaks are of similar magnitudes to that obtained from the peak in beta-alumina which was produced by the high concentration of mobile sodium ions in that material. The values of concentration, N,

TABLE I

ESTIMATES OF CONCENTRATION N OBTAINED FROM LONGITUDINAL WAVE ATTENUATION DATA (ALMOND ET AL. 1989) ASSUMING $B = 1$ eV[a]

	Peak temperature (K)	Activation energy (eV)	NB^2 (10^{20} eV^{2}cm^{-3})	B (eV)	N (10^{20} cm^{-3})
GdBa$_2$Cu$_3$O$_{7-x}$ orthorhombic	65	0.06	1.05	1	1.05
	185	0.14, 0.2	5.3	1	5.3
GdBa$_2$Cu$_3$O$_{7-x}$ tetragonal	120	0.1	3.0	1	3.0
Naβ-alumina[a]	235	0.16	8.4	0.44	43.0

[a]Naβ-alumina data (Almond and West, 1988) are included for comparison.

9. Elastic Response of Polycrystalline Superconducting Oxides

taken from the GBCO data indicate that the entities responsible for the absorption peaks are present in as many as 10% (or more if B is, in fact, less than 1 eV) of the unit cells of the material.

The high concentrations of the sources necessary to explain the strengths of the absorption peaks indicate an origin in intrinsic mechanisms. The peaks are far too large to have been caused by impurities, twin boundaries, or adsorbed gases. In addition, since a large peak is found in the oxygen-outgassed non-superconducting sample, polarons can be ruled out. The sample in this state is a semiconductor in which the carrier density is far too low to generate a polaron density approaching that necessary to explain the absorption peaks. There remains, however, oxygen, which is well known to be mobile and is hence a likely source of anelastic phenomena.

3.6. ANELASTICITY AND OXYGEN DIFFUSION

Oxygen diffusion in YBCO has been studied in depth because of the pivotal role oxygen content has in the determination of superconducting properties. Direct measurements (Tu *et al.*, 1988) have shown it to be controlled by an activation energy of 1 eV, an order of magnitude larger than the activation energies deduced from the low-temperature mechanical relaxation measurements. Xie *et al.* (1989) have completed a detailed study of oxygen diffusion employing classical internal friction techniques. They used an inverted torsional pendulum, at a frequency of about 1 Hz, and found an internal friction peak at a temperature of about 500 K, as expected for a process controlled by an activation energy of about 1 eV. This work shows very clearly the difference between the phenomena of concern here, observed at very much lower temperatures, and that which can be directly ascribed to the oxygen diffusion process.

Cannelli *et al.* (1990) have proposed that the 0.16 eV and 0.19 eV activation energy processes, associated with the low-temperature superconducting state internal friction peaks, are caused by a coupling to oxygen atoms in the Cu(1)–O(4) chains. There is evidence (Francois *et al.*, 1988) that these chains have a zigzag configuration in which O(4) atoms may occupy one of two equilibrium off-centre positions either side of the chain. An analysis indicates resulting chain segments to be of two distinct types that give rise to different shear deformations in the basal plane. Their relaxations are suggested to be the cause of one or more of the low–activation-energy processes. A similar suggestion has been made by Kusz *et al.* (1990). Cannelli *et al.* (1990) also consider that 0.1 eV is the correct activation energy for oxygen diffusion in tetragonal YBCO, and that one of the low-temperature peaks in orthorhombic YBCO may be associated with oxygen hopping within small oxygen-depleted islands.

3.7. MAGNETIC PHENOMENA

In addition to being a superconductor, YBCO exhibits a range of intrinsic magnetic properties. It is of interest to note that the associated exchange energy $J \sim$ 0.1 eV is similar in magnitude to the activation energies obtained in the low-temperature mechanical relaxation studies. Almond et al. (1990a) have performed an analysis of spin excitations on the copper–oxygen chains which revealed a spectrum of discrete excitation energies matching those obtained experimentally. It seems possible that the entities responsible for the mechanical relaxation peaks in both the superconducting and the non-superconducting forms of YBCO are magnetic in origin. It must be stressed, however, that this suggestion is somewhat speculative, based as it is only on the energies calculated for excitations that are theoretically expected to be present in YBCO. To date there is no clear evidence of the existence of these magnetic excitations in data from other techniques.

4. High-Temperature Anomalies

The early literature of YBCO contains numerous references to anomalies at temperatures close to 240 K. Chen et al. (1987) claimed to have detected a reverse ac Josephson effect, Ovshinsky et al. (1987) reported an electrical resistance drop, Bhargava et al. (1987) claimed to have found a T_c enhancement effect associated with the same temperature, and Laegreid et al. (1987) reported a specific-heat peak at this temperature. The step change in sound velocity shown here in Fig. 4a is a particularly clear example of the many elastic anomalies reported at temperatures close to 240 K. Laegreid and Fossheim (1988) suggested in light of measurements at 2.5 kHz, 3.5 kHz, and 10 MHz that the behaviour was caused by an intrinsic structural instability related to oxygen vacancies. Cannelli et al. (1988b) concluded that a phase transformation had been detected at 240 K. The frequency-independence of the phenomena found at this temperature supports the suggestion of a phase transformation, though there appears to be no reliable data to support it being a structural phase transformation.

Similar elastic anomalies have been reported in other high-T_c superconductors. Fukami et al. (1989) have found step changes and hysteresis very much like that shown in Fig. 4a in one of the bismuth-based superconductors. Al-Kheffaji et al. (1989) reported an anomaly in the electron-doped superconductor $ND_{1.85}Ce_{0.15}CuO_{4-y}$ at this temperature. Kusz et al. (1989) performed 200 Hz vibrating reed internal friction measurements of the basic non-superconducting compound CuO and found a series of distinctive absorption peaks, the strongest of which was at 215 K. They pointed out that CuO has a Neèl transition at 250

9. Elastic Response of Polycrystalline Superconducting Oxides 429

K and a further transition to the commensurate antiferromagnetic order at 213 K. They also noted the number of elastic anomalies reported in YBCO at similar temperatures and suggested that these results indicated that some of the properties of CuO were preserved in YBCO. The appearance of similar anomalies in the other CuO-based superconductors supports this contention. There is certainly evidence that the non-superconducting host compounds La_2CuO_4 (Vaknin et al., 1987) and Nd_2CuO_4 become antiferromagnetic below 220 K. That the elastic anomalies have been found in superconducting samples, which should not be antiferromagnetic, may be due to sample inhomogeneity. The problem of non-uniformity of doping throughout the bulk of a sample has already been discussed. In this context the elastic anomalies might signify the presence of an undoped, antiferromagnetic sample core. However, the complete absence of these elastic anomalies in the oxygen-outgassed tetragonal material would seem to be in conflict with this suggestion.

Almond et al. (1990b) reported finding a substantial resistivity drop in a very large grain size sample of YBCO at 164 K. They also reported finding an elastic anomaly, similar to those found at 240 K, at 170 K in an early low–oxygen-content YBCO sample. They suggested, in light of the phase diagram of de Fontaine et al. (1990), that both the 240 K and 164 K effects may be caused by the tetragonal-to-orthorhombic structural phase transition, occuring at these lower temperatures in low–oxygen-content samples.

5. Conclusions

In this review an attempt has been made to establish an overall understanding of the ultrasonic characteristics of YBCO. Instead of assuming that all the reported phenomena are caused by the material's superconductivity, the reverse approach of looking for conventional, non-superconducting explanations has been adopted. This approach seemed appropriate because of the great diversity of the data in the literature and because of the absence of comparable effects in single-crystal studies. Of the single-crystal studies, that of Golding et al. (1988) (and elsewhere in this volume) provides a notable benchmark. Their data of a very well-characterised superconducting single-crystal sample showed only the expected tiny (30 parts per million) thermodynamic discontinuity in sound velocity at T_c. In other single-crystal studies, Saint-Paul et al. (1989b) (15 MHz) estimate the discontinuity to be about 50 ppm, and the lower-frequency work of Hoen et al. (1988) and of Shi et al. (1989) indicate 90 and 190 ppm, respectively. There may be evidence here of stronger critical effects at lower frequencies, as has been found with other critical phenomena, and the possibility that the GHz data underestimates the thermodynamic anomaly.

The sensitivity of YBCO to its oxygen content is a factor which has been stressed throughout this review. It leads to serious concerns about the homogeneity and actual stoichiometry of the large bulk samples employed in most ultrasonic studies. It is likely that the majority of ultrasonic studies to be found reported in the literature were performed on samples with oxygen concentrations significantly different from the optimum O7. Magnetic measurements of Meissner fraction would seem valuable, but none appear to have been reported for ultrasonic samples. Similarly, structural data has been seldom reported, possibly because to be truly representative, its collection necessitates grinding to powder a valuable sample.

To date there is a shortage of data from well-characterised samples of known oxygen stoichiometry, and none comprehensively mapping the variation of ultrasonic characteristics with oxygen content. Such studies should prove fruitful for several reasons The anelastic absorption phenomena discussed in Section 3 are certainly composition-dependent, and their eventual explanation necessitates details of this dependence. The suggestion has been made that the almost ubiquitous 240 K phenomena are due to the co-existance of superconducting and antiferromagnetic phases or a structural phase transition. Again, composition-dependent studies should clarify this. The substantial lattice stiffening indicated by several early studies of sound velocity at low temperatures, e.g., Fig. 1, have not been satisfactorily explained, though subsequent work shows them not to be an intrinsic feature of well-characterised high–oxygen-content superconducting material. It may be significant that the temperature associated with these features, 65 K, corresponds with the lower T_c plateau obtained by Cava et al. (1987) in studies of the dependence of the superconducting properties of YBCO on oxygen content. There is evidence of a series of ordered oxygen defect states (Beyers et al., 1989) forming in reduced–oxygen-content material that have been associated with the lower T_c plateau. The ways in which these vary with temperature are not at all clear. There is, however, reason to expect their formation to affect elastic properties, and this provides further incentive to study oxygen-reduced samples.

Whilst the single-crystal results of Golding et al. (1988) bore no resemblance to the early data, e.g., Fig. 1, obtained in dense samples, that of Kim et al. (1990) was very reminiscent. They reported hysteresis, mode softening, and low-temperature stiffening, just like that shown in Fig. 1, for the longitudinal C_{33} mode, and attenuation peaks, like those shown in Fig. 4b, for the transverse C_{44} mode. Their single crystal had dimensions of $3 \times 3 \times 5$ mm, whilst that of Golding et al. was only $0.36 \times 0.27 \times 0.22$ mm, and these authors annealed their crystal for 22 days to establish oxygen homogeneity. Many of the comments about high-density polycrystalline samples would seem to apply to the large

9. Elastic Response of Polycrystalline Superconducting Oxides

single crystal studied by Kim *et al.* Although their work adds significantly the details of the acoustic responses of YBCO, it also confirms the presence of a number of phenomena that are yet to be understood. Perhaps most significant of these is the discovery of a weak attenuation peak at T_c that does seem to be associated with the superconducting state, and not with the anelastic relaxation phenomena that have been discussed at length here. A similar discovery has been reported by Sun *et al.* (1990) (and elsewhere in this volume) in a sample of Ta–Ca–Ba–Cu–O superconductor. It is not clear why others have not observed this effect, or how this new effect relates to the superconducting transition.

ACKNOWLEDGMENTS

The author is indebted to Professor G. A. Saunders, School of Physics, University of Bath, with whom he has worked closely since the first announcements of the high-T_c superconductors. The author is also grateful for the assistance of the numerous students and staff at the University of Bath that have contributed to the Group's studies of the ultrasonic characteristics of high-T_c superconductors.

IOP Publishing Ltd. is acknowledged for granting permission to the author to re-use Figs. 1, 3, and 4, and the author is grateful to J. Freestone for Fig. 7.

References

Alford, N. M., Clegg, W. J., Harmer, M. A., Birchall, J. D., Kendall, K., and Jones, D. H. (1988). *Nature* **332**, 58–59.
Alford *et al.* (1988). p. 444.
Al-Kheffaji, A., Cankurtaran, M., Saunders, G. A., Almond, D. P., Lambson, E. F., and Draper, R. C. G. (1989a). *Phil. Mag. B* **59**, 487–497.
Al-Kheffaji, A., Freestone, J., Almond, D. P., Saunders, G. A., and Jing Wang (1989b). *J. Phys.: Condens. Matter* **1**, 5993–5996.
Almond, D. P., and West, A. R. (1988). *Solid State Ionics* **26**, 265–278.
Almond D. P., Lambson, E. F., Saunders, G. A., and Wang Hong (1987). *J. Phys.* F**17**, L221–L224.
Almond, D. P., Saunders, G. A., and Lambson, E. F. (1988). *Supercond. Sci. Technol.* **1**, 163–166.
Almond, D. P., Qingxian Wang, Freestone, F., Lambson, E. F., Chapman, B., and Sanders, G. A. (1989). *J. Phys.: Condens. Matter* **1**, 6853–6864.
Almond, D. P., Long, M. W., and Saunders, G. A. (1990a). *J. Phys.: Condens. Matter* **2**, 4667–4674.
Almond, D. P., Chang Fanggao, Ford, P. J., and Saunders, G. A. (1990b). *Supercond. Sci. Technol.* **3**, 583–586.
Beyers, R., Ahn, B. T., Gorman, G., Lee, V. Y., Parkin, S. S. P., Ramirez, M. L., Roche, K. P., Vazquez, J. E., Gur, T. M., and Huggins, R. A. (1989). *Nature* **340**, 619–621.
Bhargava, R. N., Herko, S. P., and Osbourne, W. N. (1987). *Phys. Rev. Lett.* **59**, 1468–1470.
Bhattacharya, S., Higgins, M. J., Johnston, D. C., Stokes, J. P., Lewandowski, J. T., and Goshorn, D. P. (1988). *Phys. Rev.* B**37**, 5901–5904.

Bishop, D. J., Ramirez, A. P., Gammel, P. L., Batlogg, B., Rietman, E. A., Cava, R. J., and Millis, A. J. (1987). *Phys. Rev.* **B36**, 2408–2410.
Cankurtaran, M., Saunders, G. A., Willis, J. R., Al-Kheffaji, A., and Almond, D. P. (1988). *Phys. Rev.* **B39**, 2872–2875.
Cankurtaran, M., Saunders, G. A., Almond, D. P., Al-Kheffaji, A., Lambson, E. F., and Draper, R. C. J. (1989). *J. Phys.: Condens. Matter* **1**, 9067–9076.
Cannelli, G., Cantelli, R., and Cordero, F. (1988a). *Phys. Rev.* **B38**, 7200–7202.
Cannelli, G., Cantelli, R., Cordero, F., Ferretti, M. and Olcese, G. L. (1988b). *Europhys. Lett.* **6**, 271–276.
Cannelli, G., Cantelli, R., Cordero, F., Ferretti, M., and Verdini, L. (1990). *Phys. Rev.* **B42**, 7925–7930.
Cava, R. J., Batlogg, B., Chen, C. H., Rietman, E. A., Zahurak, S. M., and Werder, D. (1987). *Nature* **329**, 423–427.
Chen, J. T., Wenger, L. E., McEwan, C. J., and Logothetis, E. M. (1987). *Phys. Rev. Lett.* **58**, 2579–2582.
Cheng, X., Sun, L., Wang, Y., Shen, H., and Yu, Z. (1988). *J. Phys. C* **21**, 4603–4609.
de Fontaine, D., Ceder, G., and Asta, M. (1990). *Nature* **343**, 544–546.
Duran, C., Esquinazi, P., Fainstein, C., and Nunez Regueiro, M. (1988). *Solid State Commun.* **65**, 957–961.
Ewert, S., Guo, S., Lemmens, P., Stellmach, F., and Wynants, J. (1987). *Solid State Commun.* **64**, 1153–1156.
Francois, M., Junod, A., Yvon, K., Hewat, A. W., Capponi, J. J., Strobel, P., Marezio, M., and Fischer, P. (1988). *Solid State Commun.* **66**, 1117–1122.
Fukami, T., Youssef, A. A. A., Horie, Y., and Mase, S. (1989). *Physica* **161**, 34–38.
Golding, B., Haemmerle, W. H., Schneemeyer, L. F., and Waszcak, J. V. (1988). *IEEE Ultrasonics Symp. Proc.*, Vol. 2, p. 1079–1083.
Golding *et al.* (1989), p. 464, p. 466.
Guttman, L. (1950). *Trans. Am. Inst. Mining and Metall. Petrol Engrs.* **188**, 1472–1474.
Hatanaka, T., and Sawada, A. (1989). *Jap. J. Appl. Phys.* **28**, L794–L796.
Hewat, E. A., Dupuy, M., Bourret, A., Capponi, J. J., and Marezio, M. (1987). *Nature* **327**, 400–402.
Hoen, S., Bourne, L C., Choon M. Kim, and Zettl, A. (1988). *Phys. Rev.* **B38**, 11949–11951.
Horie, Y., Terashi, Y., Fukuda, H., Fukami, T., and Mase, S. (1987). *Solid State Commun.* **64**, 501–504.
Horie, Y., Terashi, Y., and Mase, S. (1989). *J. Phys. Soc. Japn,* **58**, 279–290.
Jackle, J., Picke, L., Arnold, W., and Hunklinger, S. (1976). *J. Non-cryst. Solids* **20**, 365–391.
Kim, T. J., Kowalewski, J., Assmus, W., and Grill, W. (1990). *Z. Phys. B* **78**, 207–212.
Kusz, B., Barczynski, R., Murawski, L., Gazda, M., Gzowski, O., Davoli, I., and Stizza, S. (1989). *Solid State Commun.* **72**, 97–99.
Kusz, B., Barczynski, R., Gazda, M., Pastuszak, R., Murawski, L., Gzowski, O., Davoli, I., and Stizza, S. (1990). *Solid State Commun.* **75**, 789–790.
Lang, M., Lechner, T., Reigel, S., Steglich, F., Weber, G., Kim, T. J., Luthi, B., Wolf, B., Rietschel, H., and Wilhelm, M. (1987). *Z. Phys. B* **69**, 459–463.
Lang *et al.* (1988). p. 442.
Laegreid, T., and Fossheim, K. (1988). *Europhys. Lett.* **6**, 81–88.
Laegreid, T., Fossheim, K., Sandvold, E., and Julsrud, S. (1987). *Nature* **330**, 637–638.
Ledbetter, H. M., and Kim S. A. (1988). *Phys. Rev.* **38**, 11857–11860.
Lemmens, P., Honnekes, C., Brakmann, M., Ewert, S., Comberg, A., and Passing, H. (1989). *Physica C*, Vol. 162-4 p. 452–453.
Mizubayashi, H., Takita, K., and Okuda, S. (1988). *Phys. Rev.* **B37**, 9777–9779.
Muller, V., De Groot, K., Maurer, D., Roth, C., Rieder, K. H., Eickenbusch, E., and Schollhorn, R. (1987a). *Japn. J. Appl. Phys.* **26**, supplement 26-3, 2139–2140.

Muller, V., De Groot, K., Maurer, D., Roth, C., and Rieder, K. H. (1987b). *Physica* **148B**, 296–297.
Muller, V., Maurer, D., Roth, Ch., Hucho, C., Winau, D., and De Groot, K. (1988). *Physica C* **153—155**, 280–281.
Ovshinsky, S. R., Young, R. T., Allred, D. D., DeMaggio, G., and van der Leeden, G. A. (1987). *Phys. Rev. Lett.* **58**, 2579–2581.
Saint-Paul, M., and Henry, J. Y. (1989). *Solid State Commun.* **72**, 685–687.
Saint-Paul, M., Tholence, J. L., Noel, H., Levet, J. C., Potel, M., and Gougeon, P. (1989b). *Solid State Commun.* **69**, 1161–1163.
Sayers, C. M., and Smith, R. L. (1982). *Ultrasonics* **20**, 201–205.
Shen, H., Wang, Y., Zhang, Z., Zhang, S., and Sun, L., (1987). *J. Phys. C,* **20**, L897–L892.
Shi, X. D., Yu, R. C., Wang, Z. Z., Ong, P. M., and Chaikin, P. M. (1989). *Phys. Rev.* **B39**, 827–830.
Sun, K. J., Winfree, W. P., Xu, M. F., Sarma, B. K., Levy, M., Caton, R., and Selim, R. (1988). *Phys. Rev.* **B38**, 11988–11991.
Sun, K. J., Winfree, W. P., Xu, M-F, Levy, M., Bimal K. Sarma, Singh, A. K., Osofsky, M. S., and Le Tourneau, V. M. (1990). *Phys. Rev.* **B42**, 2569–2572.
Testardi, R. L. (1973). In "Physical Acoustics," Vol. 10 (W. P. Mason and R. N. Thurston, Eds.). Academic Press, New York, p. 242.
Tu, K. N., Tsuei, C. C., Park, S. I., and Levi, A. (1988). *Phys. Rev.* **B38**, 772–775.
Vaknin, D., Sinha, S. K., Moncton, D. E., Johnston, D. C., Newsam, J. M., Safinya, C. R., and King, H. E. (1987). *Phys. Rev.* **B58**, 2802–2805.
Van Tendezoo, G., Zanbergen, H. W., and Amelinckx, S. (1987). *Solid State Commun.* **63**, 389–393.
Weller, M., Jaeger, H., Kaiser, G. and Schulze, K. (1989). *Physica C,* Vol. 162-4, p. 953–954.
Xie, X., Chen, T. G., and Wu, Z. L. (1989). *Phys. Rev.* **B40**, 4549–4556.
Yamamoto, T., Furusawa, T., Seto, H., Park, K., Hasegawa, T., Kishio, K., Kitazawa, K., and Fueki, K. (1988). *Supercond. Sci. Technol.* **1**, 153–159.

—10—
High T_c Superconductivity and Ultrasonics—Theoretical Aspects

VLADIMIR Z. KRESIN
Materials and Chemical Sciences Division, Lawrence Berkeley Laboratory, University of California, Berkeley

1. Introduction 435
2. Sound Attenuation and Conventional Superconductivity 436
   2.1. Ultrasound Attenuation; Temperature Dependence 436
   2.2. Energy Gap Anisotropy 437
   2.3. Low-Frequency Sound; Transverse Waves 439
   2.4. Sound Velocity 439
3. Exotic Superconductors (Organic Materials; Heavy Fermions) 440
4. High-T_c Oxides 441
   4.1. Normal and Superconducting Properties of Cuprates 442
   4.2. Lattice Stability; Sound Velocity 445
5. Ultrasonic Attenuation in High-T_c Oxides 446
   5.1. Analysis of the Normal Properties 446
   5.2. Superconducting Properties: Multigap Structure, Energy Gap Anisotropy ... 448
   Acknowledgment 452
   References 452

1. Introduction

Analysis of propagation and attenuation of sound waves is a powerful method of studying the properties of superconductors. It allows unique information to be obtained about their spectra. The recent discovery of high-temperature superconductivity (Bednorz and Muller, 1986; Wu et al., 1987; Maeda et al., 1988; Sheng and Hermann, 1988) has given birth to a new field: the physics of a new class of materials, namely the high-T_c oxides. The main goal of this article is to share some thoughts about acoustic study of the cuprates and, in particular, about the expectations and future results important for the theory of high T_c.

As is known, the success of the acoustic technique depends strongly on sample quality and size. Lately, there has been significant progress in preparation of bulk high-T_c superconductors, single crystals, thin films, etc. As a result, one can look forward to many new interesting results, obtained by ultrasonic study.

For this reason, we think that a discussion of future directions is of definite interest.

The structure of the paper is as follows. Since there are a lot of similarities between high-T_c oxides and conventional superconductors, we will describe briefly the major aspects of acoustic study of usual superconductors (Section 2). Exotic systems, such as organic superconductors, heavy fermions and the associated problems of ultrasonic investigations, are discussed in Section 3. Section 4 provides a description of the present status of the physics of high-T_c superconductivity. Section 5 contains a discussion of ultrasonic analysis of high-T_c oxides.

2. Sound Attenuation and Conventional Superconductivity

2.1. ULTRASOUND ATTENUATION; TEMPERATURE DEPENDENCE

The subject of ultrasound attenuation has been discussed in many reviews and papers (Geilikman and Kresin, 1974; Shepelev, 1969; Lynton, 1969). The most famous result was obtained in the original BCS paper (Bardeen et al., 1957); it describes the temperature dependence of ultrasonic attenuation:

$$\frac{\gamma_s}{\gamma_n} = \frac{2}{e^{\Delta/T} + 1} \tag{1}$$

Here γ_s and γ_n are the absorption coefficients in the superconducting and normal states, respectively, and Δ is the energy gap. This elegant expression correlates ultrasound attenuation with the major superconducting parameter, the energy gap. Equation (1) is valid if $\hbar\omega \ll 2\Delta$. The attenuation is due to the presence of quasi-particles excited thermally, but not by a sound quantum. The number of thermally excited quasi-particles decreases exponentially with temperature, in accordance with the Fermi distribution: $n = [\exp(E/T) + 1]^{-1}$, and consequently, Eq. (1) leads to the dependence $\gamma_s/\gamma_n \sim \exp(-\Delta/T)$, is $T \to 0$ K.

Equation (1) has played a very important role in the history of superconductivity, because it was seen as one of the major triumphs of the BCS theory. As was noted above, Eq. (1) directly relates the energy gap to the experimentally measured ratio γ_s/γ_n. Equation (1) is very simple, which makes ultrasound measurements very effective in determining the form of the function $\Delta(T)$. Figure 1 shows the theoretical curve obtained in BCS theory and the experimental data. One sees that there is excellent agreement.

Let us make several comments regarding Eq. (1). It describes the absorption of high-frequency ultrasound. This means that the sound wave period must be much shorter than the relaxation time ($\omega\tau \gg 1$, where ω is the sound frequency). Then the interaction of a wave with the electron system can be viewed as radiation

10. High T_c Superconductivity and Ultrasonics

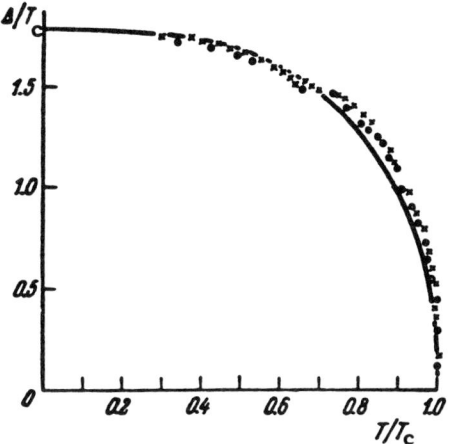

FIG. 1. Energy gap as a function of temperature.

and absorption of the quanta by electronic excitations. It turns out that Eq. (1) is valid if $ql \gg 1$ (l is the mean free path, and q is a wave vector). This condition is not as strong as $\omega\tau \gg 1$. It is clear that the condition $ql \gg 1$ is sufficient for neglecting the collisions. This absorption mechanism is analogous to Landau damping and actually represents absorption of sound waves by resonant electrons.

Secondly, Eq. (1) allows the determination of the shape of the function $\Delta(T)$. Near T_c the dependence $\Delta(T)$ has the form

$$\Delta(T) = aT_c\sqrt{1 - T/T_c}$$

According to the BCS theory, which has been developed in the weak coupling approximation, $a = 3.05$. Strong coupling effects lead to an increase in the value of the coefficient a (see Geilikman and Kresin, 1966; Geilikman et al., 1975). For example, for Pb, $a = 4$. In addition, near $T = 0$ K, $\Delta(0) > 3.52\,T_c$. Namely [7],

$$2\Delta(0) = 3.52\,T_c[1 + 5.3(T_c/\tilde{\Omega})^2 \ln(\tilde{\Omega}/T_c)], \qquad (2)$$

where $\tilde{\Omega}$ is the average phonon frequency. As a result, the function $\Delta(T)$ decreases more rapidly if the superconductor is characterized by strong coupling.

2.2. Energy Gap Anisotropy

The value of the energy gap depends on direction, that is, $\Delta = \Delta(\mathbf{p})$, where $\mathbf{p}$ is the pseudomomentum. This dependence is due to anisotropies of the Fermi surface and the phonon spectrum, and to the angle dependence of the electron–phonon interaction. Energy gap anisotropy has been studied by different methods (thermal conductivity, infrared spectroscopy, etc.), but ultrasound attenuation appears to be the most powerful method. Indeed, the attenuation is a bulk effect that, in addition, is very sensitive to the mutual orientation of wave propagation and the crystal axes.

The effect of energy gap anisotropy on the value of the absorption coefficient has been studied theoretically by Pokrovskii (1961). The attenuation is described by the imaginary part of the phonon Green's function, or by the imaginary part of the polarization operator:

$$\Pi(\mathbf{q}, i\omega_n) = \frac{T}{(2\pi)^3} \sum_{\omega_{n'}} \int d\,\mathbf{p}[G\,G' - F\,F']. \tag{3}$$

Here G and F are the usual and abnormal Green's functions that describe the superconducting pairing. We are not going to write out the general expression for the absorption coefficient obtained by Pokrovskii (1961) (the interested reader can find a detailed analysis in the original paper; see also Geilikman and Kresin, 1974). The general result has the simplest form in the temperature region $1 < T_c/T < (2/\alpha) \ln (v_F/u)$, where v_F is the Fermi velocity, u is the sound velocity, and α is the so-called anisotropy coefficient, which is equal to the ratio of the change of the energy gap on the Fermi surface to its minimum value. Then

$$\gamma^s(\mathbf{q}) \sim \omega(\mathbf{q}) e^{-\Delta_{\min}/T} \tag{4}$$

Here ω is the phonon frequency, and $\Delta_{\min}(\mathbf{n})$ is the minimum value of the energy gap on the stereographic projection of the Fermi surface onto the vector $\mathbf{n} = \mathbf{q}/q$. It is evident that the attenuation picture is highly anisotropic.

According to Eq. (4), absorption depends on the minimum value of the energy gap along the line $\mathbf{qv} = 0$. This result has an obvious physical meaning. As was mentioned above, absorption of high-frequency sound is a quantum electron–phonon process. Because of the inequality $u \ll v_F$, the equation describing conservation of energy, $\varepsilon_{\bar{p}+\bar{q}} - \varepsilon_\mathbf{p} = \omega_\mathbf{q}$, can be written in the form: $\vec{v}\,\vec{q} = \omega$, or, approximately $\mathbf{vq} = 0$. This is a natural condition for resonant absorption. Therefore, the main contribution comes from excitations with velocities perpendicular to the direction of wave propagation. Equation (4) contains a Boltzmann distribution of such electronic excitations. Note also that the experimental data on anisotropy of the attenuation can be used in order to reconstruct the dependence $\Delta(\mathbf{n})$ (Pokrovskii and Toponogov, 1961).

One should note, however, that despite the strong anisotropy of the Fermi surface and the electron–phonon interaction, the effect of energy gap anisotropy in conventional superconductors is not large, and their properties within a good accuracy are described by an isotropic model. This is due to the long coherence length. According to the Anderson theorem [10], the inequality $l \ll \xi_0$ (ξ_0 is the coherence length) results in gap averaging. Transitions between different parts of the Fermi surface caused by impurity scattering are the main mechanism of this averaging. As a result, the effect of energy gap anisotropy is depressed by the elastic scattering, although it still can be observed (see, for example, the review Shepelev, 1969).

The picture with the new high-T_c oxides is entirely different. This is connected with the short coherence length in these materials. We discuss this question later in Sections 4 and 5.

Another effect that also is due to deviation of the Fermi surface from a simple spherical shape is the effect of multiband structure. This effect does not play an essential role in conventional superconductivity, but is important in the physics of high T_c; we also discuss it in Section 4.

2.3. Low-Frequency Sound; Transverse Waves

So far we have been discussing the absorption of high-frequency sound, when interaction between the carriers and the acoustic wave can be treated as a direct quantum process. The picture is different if $ql \ll 1$; in this case, sound attenuation can provide interesting information about relaxation effects. If the sound frequency is small, then the effect of the sound field is to deform the lattice. As a result, the electron dispersion relation is modified. In this case it is necessary to analyze the Boltzmann equation for the electron distribution function and to take into account various relaxation mechanisms. The absorption coefficient turns out to be a function that decreases with temperature in the region $T < T_c$.

Sound can be absorbed not only by electronic excitations, but by phonons as well. As $T \to 0$, the number of electronic excitations decreases. As a result, the phonon mean free path grows, and sound attenuation increases. The picture is similar to that in the theory of thermal conductivity of superconductors (see the review Geilikman and Kresin, 1974).

An interesting direction is the study of attenuation of transverse sound waves (Morse and Bohn, 1959; Levy et al., 1963). For some systems, one observes a very sharp drop in absorption near T_c, although the effect is not universal. This drop is due to a significant contribution of the field associated with the transverse sound wave (for a more detailed discussion, see Geilikman and Kresin, 1974).

2.4. Sound Velocity

The superconducting transition also changes the phonon spectrum and, consequently, the sound velocity. In order to evaluate the latter change, one has to calculate the phonon Green's function and the polarization operator (see Eq.

(3)). The change caused by the superconducting transition is proportional to Δ^2. A more detailed investigation (Bardeen and Stephen, 1964) has shown that the phonon spectrum in conventional superconductors changes only by a small amount ($\sim \Delta^2/E_F^2$, where E_F is the Fermi energy). Usually $E_F \simeq 5\text{–}10$ eV, whereas $\Delta(0)$ is on the order of 1–2 meV. As a result, the velocity shift $\Delta u/u$ is very small. Even in superconductors with strong electron–phonon coupling (e.g., Pb), we are dealing with a change on the order of $10^{-4}\text{–}10^{-5}$.

Note that the smallness of E_F along with the large $\Delta(0)$ in the new high-T_c oxides leads to an entirely different situation (see Section 5).

3. Exotic Superconductors (Organic Materials; Heavy Fermions)

Organic superconductors. The effect of organic superconductivity was predicted by W. Little in 1964, but discovered only in 1980 (Parkin *et al.*, 1981). At present, there are many known organic superconductors, but they all can be divided into two groups: the (TMTSF)$_2$X (Parkin *et al.*, 1981) and the (BEDT–TTF)$_2$X (Williams and Carneiro, 1985) series (see the reviews Greene and Chaikin, 1984; Jerome and Cruezet, 1987; Ferraro and Williams). For example, the compound (BEDT–TTF)$_2$Cu(SCN)$_2$ has $T_c = 10.4$ K; we are thus dealing with materials with relatively high T_c. A new material, namely $\kappa = $ (ET)$_2$Cu[N(CN)$_2$]Br with higher T_c ($= 12$ K) has been obtained in the Argonne national Laboratory (Williams and Carneiro, 1985). The rapid progress in the field in the last 10 years is very impressive.

Lack of large crystals is an obstacle for ultrasound experiments on organic superconductors. But if this obstacle were overcome, it would be very interesting to carry out such studies. The important point is that these materials contain low-dimensional subsystems. There are many similarities between organic superconductors and the high-T_c oxides (see Section 4). Both families are characterized by small carrier concentration and high anisotropy.

Superconductors belonging to the (BEDT–TTF)$_2$X group have a layered structure; in addition, their in-plane transport properties are also anisotropic. The Fermi surface has a cylindrical shape whose cross-section is a stretched ellipse; as a result, the Fermi surface contains a lot of nesting states. Conventional superconductors are characterized by an exponential decrease of the heat capacity as T approaches zero. Organic superconductors display a power law, which may be related to the presence of nodes in the order parameter. Some investigators consider this to be a manifestation of an unconventional, magnetic nature of organic superconductivity. But an alternative explanation (Wolf and Kresin, 1991) is possible. The fact of the matter is that these materials are characterized by a relatively short coherence length (~ 50 Å). The mean free path is relatively large; as a result, the criterion $1 \gg \xi_0$ is satisfied. Therefore, one can observe energy gap anisotropy, and possibly an even stronger effect: multigap structure (the presence of two gaps was reported in the tunneling experiment (Hawley *et al.*, 1986). If the material has multigap structure, then the heat capacity is given by a sum of exponentials, and such a sum can look like a power law. The ultrasound technique is an ideal method to study multigap structure and energy gap anisotropy. One can expect that future ultrasound experiments will allow the resolution of the problem of the mechanisms present in organic superconductors.

Heavy fermions. Heavy fermions form another relatively young family of superconductors (see, for example, the reviews Steward, 1984; Ott, 1987). They were discovered in 1979 (Steglich *et al.*, 1979) and are characterized by uniquely large values of the effective mass. Ultrasound studies (Bishop *et al.*, 1984; Golding *et al.*, 1985), as well as heat capacity measurements, have shown noticeable deviations from the BCS behavior. In addition, a recent study of the heat capacity in the region near T_c (Fisher *et al.*, 1989) has shown that there are two transitions. This corresponds to a change in the symmetry of the order parameter. Probably, the origin of heavy-fermion superconductivity is due to the contribution of magnetic fluctuations. A maximum has been observed in ultrasound attenuation in URu_2Si_2 (Levy *et al.*, 1987) below the superconducting transition temperature. At the same time, the shift in sound velocity due to the superconducting transition is similar to that in usual BCS superconductors. One can look forward to many new, interesting results in the new field of heavy-fermion superconductivity.

4. High-T_c Oxides

The discovery of a new class of high-T_c oxides [1] and further developments [2] have generated unprecedented activity associated with study and characterization of these materials. More than 9,000 (!) papers concerned with different aspects of high-temperature superconductivity have been published in the past three years (1987–1989); see, for example, Wolf and Kresin (1987) and Ginsberg (1989). As a result, there has been significant progress in our understanding of the normal and superconducting properties of these materials, although, of course, many unresolved problems remain.

We are going to discuss those aspects of high-temperature superconductivity that can benefit from ultrasonic studies, but we start out by discussing briefly some normal and superconducting properties of the new materials. We will be following the analysis developed by the present author, jointly with S. A. Wolf

(NRL) and H. Morawitz (IBM) and described in papers (Kresin and Wolf, 1987a,b, 1988, 1990a; Kresin and Morowitz, 1988a,b, 1989).

4.1. Normal and Superconducting Properties of Cuprates

Normal state. The high-T_c oxides are doped materials. We are going to focus on the doped state, because this is the state that undergoes the superconducting transition. This state is metallic and can be described by the Fermi liquid theory. According to this theory (Landau, 1956; see e.g., Abrikosov (1988)), the low-lying excited states of highly correlated Fermi systems, such as the electron system in a metal, can be classified in a way similar to the Fermi gas. In other words, such quantities as Fermi energy, dispersion relation, Fermi surface, etc., have a direct physical meaning. Recent experimental photoemission data on cuprates (Arko et al., 1989) show the presence of a sharp Fermi edge. The very recent observation of De Haas–Van Alphen oscillations (Mueller et al., 1990) is a direct manifestation of the presence of a Fermi surface.

The new high-T_c materials are characterized by large anisotropy, which is manifested in the shape of the Fermi surface. As a first approximation, the Fermi surface in the La–Sr–Cu–O compound can be taken as cylindrically shaped; this corresponds to neglecting interlayer transitions. For such a Fermi surface, one can derive the following relations (Kresin and Wolf, 1987a,b, 1990a):

$$m^* = 3(\hbar^2/\pi)\,\kappa_B^{-2}\,d_c\,\gamma, \tag{5}$$

$$E_F = (\pi^2 \kappa_B^2/3)\,n/\gamma.$$

Here m^* is the effective mass, E_F is the Fermi energy, d_c is the interlayer distance, γ is the Sommerfeld constant, and n is the carrier concentration. The equations of (5) express the effective mass and the Fermi energy in terms of experimentally measured quantities. Therefore, based on the heat capacity and Hall effect data, one can evaluate the major normal parameters of La–Sr–Cu–O. Note that this material has a relatively simple band structure, and therefore the Hall effect data, which appear to be almost temperature-independent, can be used in order to estimate n. Employing the experimental data found in (Phillips et al., 1987; Panson et al., 1987; Kwak et al., 1987; Ong et al., 1987), one can evaluate the major parameters, as presented in Table I.

One can see from Table I that the material is characterized by uniquely small values of the Fermi energy (~ 0.1 eV) and the Fermi velocity ($\sim 8 \times 10^6$ cm s^{-1}). These values are much smaller than those in conventional metals. On the other hand, the Fermi momentum is comparable with its values in usual superconductors. This fact is important, because it provides a large phase space for pairing.

10. High T_c Superconductivity and Ultrasonics

TABLE I

COMPARISON OF NORMAL-STATE PROPERTIES OF
CONVENTIONAL METALS WITH $La_{1.8}Sr_{0.2}CuO_4$

Quantity	Conventioanl metals	$La_{1.9}Sr_{0.2}CuO_4$
m^*	$(1-15)\, m_e$	$5\, m_e$
k_F (cm^{-1})	10^8	3.5×10^7
v_F (cm s^{-1})	$(1-2) \times 10^8$	8×10^6
ε_F (eV)	5–10	0.1

The small values of E_F and v_F, along with the high anisotropy, are the key physical properties of the new high-T_c oxides.

The structure of the Fermi surface of the Y–Ba–Cu–O compound (see Fig. 2) is more complicated, which results from the presence of chain structure with its own set of carriers and energy gap (see Section 5). The Fermi surface contains a cylindrical part, together with a set of planes (Kresin and Wolf, 1987a,b, 1990a). The Fermi energy and the Fermi velocity also appear to be small (Kresin and Wolf, 1987a,b, 1990a; Kresin et al., 1988) (e.g., $E_F = 0.2$–0.25 eV).

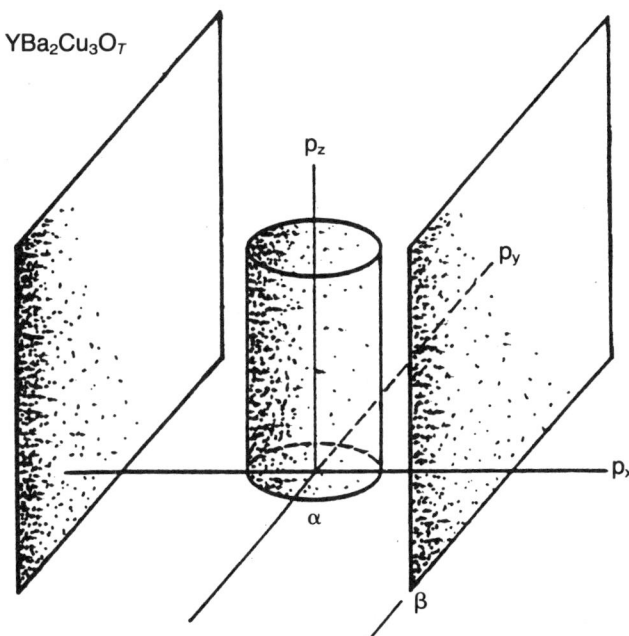

FIG. 2. Fermi surface for Y–Ba–Cu–O.

Therefore, the cuprates are metallic, but possess exotic values of the major parameters. This leads to unusual superconducting properties (see later sections).

Thus far, we have been discussing the properties of single-particle excitations in the cuprates. In addition, the spectrum of any metallic system, including the oxides, contains collective modes. Phonons are one example of a collective excitation. The phonon spectrum of the cuprates differs drastically from that in usual metals. According to neutron spectroscopy data (Maraki *et al.*, 1987; Ramizez *et al.*, 1987; Boni *et al.*, 1988), it contains low-frequency optical modes that are highly anharmonic. This anharmonicity is an important property of the materials (Muller, 1989; Morawitz and Kresin, 1989).

The carrier system also contains collective modes. From the theoretical point of view, one-particle excitations (electrons, holes) correspond to the poles of the one-particle Green's function, whereas collective modes are given by the poles of the two-particle Green's function. Plasmons, which can be viewed as a collective vibrational motion of the carriers relative to the lattice, are an example of such collective excitations. Plasmons in usual 3-D metals are characterized by the gap ω_0; in other words, their dispersion relation is $\omega = \omega_0 + aq^2$, where ω and q are the plasmon energy and momentum. Usually, ω_0 is large and is on the order of several eV. The structure of the plasmon spectrum in layered conductors, such as the cuprates, is entirely different (see the reviews Kresin and Morawitz, 1989). Namely, the dependence $\omega(\vec{q})$ is anisotropic (the z-axis has been chosen perpendicular to the layers) and forms an entire plasmon band. This band is restricted by two branches. The upper branch corresponds to $q_z = 0$ and resembles the usual bulk plasmon in conventional metals (although the value of ω_0 is much smaller, ~ 1 eV). The most interesting feature of the plasmon spectrum in the oxides is the presence of the lowest plasmon branch ($q_z = \pi/2$), which has an acoustic nature and can be termed the "electron sound." The plasmon spectrum in the cuprates does not have an energy gap and contains the low-frequency acoustic branch. One can prove that the density of states is peaked near the uppermost and the lowest branches; therefore, the plasmon spectrum can be approximated as a combination of these two branches. The slope of the lowest acoustic branch is on the order of v_F. It is important to recall that the value of the Fermi velocity is small (see Table I). Consequently, the acoustic plasmon branch can be viewed as an additional part of the total phonon spectrum.

We think that strong coupling to both the low-frequency phonon and plasmon modes (phonon–plasmon mechanism) is the origin of high-temperature superconductivity.

Superconducting properties. Let us turn to the problem of evaluating the superconducting parameters of the cuprates (Kresin and Wolf, 1988, 1990a).

10. High T_c Superconductivity and Ultrasonics

First of all, let us focus on the coherence length ξ_0. By definition, $\xi_0 = \hbar v_F/\pi\Delta(0)$, where $\Delta(0)$ is the energy gap. The energy gap is directly related to T_c, so that $2\Delta(0) = \alpha T_c$. The value of α depends on the strength of the coupling (in the weak coupling BCS approximation, $\alpha = 3.5$). According to strong coupling theory (Geilikman and Kresin, 1966), α is determined by the relation (2).

For La–Sr–Cu–O, $\tilde{\Omega} = 15$ meV (Maraki et al., 1987; Boni et al., 1988) and we obtain $2\Delta(0) \simeq 5T_c$. This is in good agreement with the tunneling data and with coupling strength evaluation (Kresin and Wolf, 1990a).

Using the Fermi velocity and the energy gap from Table I, we obtain $\xi_0 = 20$ Å. This strikingly small value of the coherence length is due to the small Fermi velocity and the large energy gap. It is a very important feature of the new materials.

Another important parameter is the ratio $\Delta(0)/E_F$. In conventional superconductors, this parameter is small ($\sim 10^{-4}$), which means that only a small number of states near the Fermi surface are involved in the pairing. The situation in the cuprates is different. The Fermi energy is small (see Table I), and as a result we have $\Delta(0)/E_F \simeq 0.1$. The large value of this ratio also implies that a significant fraction of the carriers are paired.

The parameter Δ/E_F is related to the size of the critical region; its large value provides an opportunity to observe critical behavior (Deutcher, 1987), similar to that in liquid HeII. It also leads to a noticeable shift in positron-annihilation lifetime. The parameter Δ/E_F also determines the change in sound velocity caused by the superconducting transition. This change is small in conventional superconductors (see Section 2), but has a greater value in the high-T_c materials.

In the next section, we discuss the sound velocity change observed experimentally, and then focus on possible future experiments where the powerful ultrasonic method will display its full potential.

4.2. LATTICE STABILITY; SOUND VELOCITY

The superconducting transition leads to a drastic change in the electron spectrum. In addition, one can in principle observe a change in the phonon spectrum; it is due to the coupling between electrons and the lattice. This change is manifested as a shift in sound velocity. However, in conventional superconductors this change is very small (see Section 2); even for lead, which is a strongly coupled superconductor, this ratio is on the order of 10^{-4}.

The situation is different in the high-T_c materials, because the parameter Δ/E_F is relatively large (e.g., for La–Sr–Cu–O, $\sim 10^{-1}$; see Section 3). As a result, one expects a noticeable change in sound velocity. An analysis of the phonon

Green's function in the presence of pairing (Bardeen and Stephen, 1964) leads to the expression

$$\frac{\Delta u}{u} \sim \frac{\Delta^2}{E_F^2} \ln \frac{E_F}{\Delta}.$$

A noticeable shift in sound velocity has been observed experimentally (see, for example, Bishop et al., (1964) and the review Poole et al., 1988. The shift increases with decreasing temperature and scales with the energy gap, which increases as $T \to 0$. One also observes changes in sound velocity at temperatures above T_c (see, for example, the review Poole et al., 1988). These changes are due to structural transitions; this information is important for studying the lattice dynamics.

Changing sound velocity is directly related to the problem of lattice stability, which is discussed briefly in (Kresin, 1987). According to adiabatic theory (Geilikman, 1971), an increase in the strength of the electron-photon coupling eventually leads to the appearance of an imaginary pole in the phonon Green's function, signifying lattice instability. This important problem has not been studied in detail, but it is clear that at some point, increasing the parameter $\bar{\Omega}/E_F$ further will result in lattice instability. This factor determines the upper achievable limit of T_c.

5. Ultrasonic Attenuation in High T_c Oxides

5.1. ANALYSIS OF THE NORMAL PROPERTIES

As was described above, the doped cuprates represent a Fermi liquid which is highly anisotropic and, in addition, is characterized by unusual parameter values. The Fermi energy and the Fermi velocity turn out to be uniquely small, whereas the Fermi momentum is relatively large. As for the Y–Ba–Cu–O compound, its Fermi surface contains several distinct parts. Because our understanding of high-temperature superconductivity depends in a crucial way on the information about normal parameters of the materials, it is very important to develop a detailed fermiology of the cuprates, similar to that for the common metals (see, for example Cracknell and Weng, 1973).

As is known, the Fermi surfaces of many metals have highly complicated structures. It is natural to assume that the Fermi surface of such a complex system as a high-T_c oxide also contains many peculiar features. A simple cylindrical model described in the previous section reflects the anisotropy of the material and allows estimation of the values of major parameters—but of course, the real shape is more complicated, and a detailed reconstruction should be carried out.

For example, interlayer transitions lead to deviations from the simple cylindrical shape to the appearance of warped structures.

The shapes of the Fermi surfaces of B_i and T1-based high-T_c compounds are unknown, as are the values of their major parameters. Needless to say, this problem is of great importance, because the Bi- and T1-based materials have the highest critical temperatures. The important question about the origin of the dependence of T_c on the number of layers per unit cell requires knowledge of the detailed fermiology of the material. The question of the influence of the multisheet structure on the Fermi surface is awaiting an answer.

In order to study these problems, it is necessary to use an experimental technique that would be very sensitive to the structure of the Fermi surface. From this point of view, the ultrasonic technique is very promising. Sound attenuation is very sensitive to the anisotropy of the Fermi surface. For example, according to the theoretical analysis in (Gokhfel'd, 1987), the absorption of a longitudinal acoustic wave that propagates perpendicular to the planes in a layered conductor (c-direction) is weaker than in a 3-D metal with the same carrier concentration.

Particularly promising is the study of attenuation in the presence of an external magnetic field. The present author, jointly with Kokotov (1978), proposed a method of reconstructing the shape of the Fermi curve for a size-quantizing film that forms a two-dimensional system in momentum space. Layered crystals, such as the copper oxides, are to a large extent similar to a size-quantizing system, because they are also characterized by quasi–2-D dynamics. For this reason, the method proposed in (Kresin and Kokotov, 1978) can be also used for fermiology of the cuprates in order to reconstruct the cross-section of the cylindrical Fermi surface formed by a plane perpendicular to the c-axis.

Consider in-plane propagation of an acoustic wave in the presence of a magnetic field perpendicular to the plane. Solution of the kinetic equation and evaluation of the attenuation coefficient leads to the following expression:

$$\gamma = \gamma_0 + \gamma_1 \cos\theta. \tag{6}$$

One can see from Eq. (6) that the absorption coefficient contains an oscillating part (the second term in Eq. (6), where $\theta = CD_\perp(e\lambda H)^{-1}$; λ is the wavelength, and $D_\perp$ is the projection of the extremal diameter of the cross-section perpendicular to the propagation direction). The absorption coefficient oscillates as a function of H^{-1} with the period given by $e\lambda(cD_\perp)^{-1}$. By varying the direction of sound-wave propagation, one can reconstruct the shape of the cross-section.

Note that high-T_c films can be used in order to carry out such an investigation. If we have a c-oriented film (the planes are parallel to the boundary of the film), then the magnetic field $\mathbf{H} \perp \mathbf{c}$.

As discovered by J. Torrance *et al.*, 1988 and then reproduced for various high-T_c systems, the critical temperature depends strongly on the carrier concentration. This dependence turns out to be non-monotonic. At first, T_c increases with the carrier concentration n, but then one observes a decrease in T_c followed by a superconductor–metal transition. A detailed analysis of the electron–phonon coupling leads to such a non-monotonic dependence (Kresin and Morawitz, 1990), which was observed many years ago in superconducting semiconductors. Inspired by this dependence, it would be interesting to study the dynamics of the fermiology, that is, the dependence of the shape of the Fermi surface and its parameters on the carrier concentration. This would allow a direct correspondence to be established between the values of these parameters and T_c. In addition, one should note that if $n \ll n_{max}$ (n_{max} corresponds to T_c), then the value of T_c is lower, and the criterion $ql \gg 1$ is easier to meet.

We have been analyzing the absorption of high-frequency ultrasound as a method of reconstructing the Fermi surface, that is, we have been concerned with single-particle excitations. In addition, of course, there are the collective excitations, and sound attenuation analysis can provide interesting information about their spectroscopy. For this purpose, an analysis of low-frequency acoustic waves looks more promising. As was noted above (see Section 2), propagation of low-frequency waves ($ql \ll 1$) is sensitive to the relaxation processes, and analysis requires the solution of the corresponding Boltzmann equation. Usually the electron–phonon interaction is the main relaxation mechanism. However, the layered structure of such materials as the cuprates results in the appearance of a peculiar collective acoustic plasmon branch ("electron" sound; see Section 4.1). Therefore, there is present an additional relaxation mechanism characterized by its own relaxation time. Indications of such an additional relaxation channel have been seen experimentally (Levy, 1989). It would certainly be very interesting to carry out a detailed theoretical and experimental investigation of this problem.

5.2. Superconducting Properties: Multigap Structure, Energy Gap Anistropy

In the preceding section, we discussed potential applications of the method of ultrasound attenuation to the analysis of the normal properties of the oxides. Here we are going to describe possible applications of this method to the spectroscopy of the superconducting state. Our main focus will be on the energy gap, its anisotropy, and the multigap structure. Ultrasound attenuation appears to be the most powerful technique of studying energy gap anisotropy in conventional superconductors (see Section 2). The smallness of the coherence length

10. High T_c Superconductivity and Ultrasonics

in the new high-T_c oxides should lead to even wider applications of the method to these materials.

In particular, we would like to emphasize the appearance of multigap structure in the cuprates. The properties of some oxides (e.g., Y–Ba–Cu–O) are greatly affected by the presence of overlapping energy bands and by their two-gap structure. But first, let us describe the main aspects of this unusual superconductivity.

The importance of multigap structure in the high-T_c oxides, as opposed to conventional superconductors, is due to the short coherence lengths in the former. At present, there are a number of experimental data on NMR, tunneling, and transport properties that indicate the presence of two energy gaps in Y–Ba–Cu–O.

The two-band model has been introduced in the theory of superconductivity initially by Suhl et al. (1959) and by Moskalenko (1959). A detailed analysis has been carried out by Geilikman, Zaitsev, and the present author (1966; 1973; see also Geilikman and Kresin, 1974).

Multigap structure is caused by the presence of overlapping energy bands; each band is characterized by its own energy gap. The effect of multigap structure is similar to that of energy gap anisotropy. In both cases we are dealing with a deviation of the Fermi surface from the spherical shape. Multiband structure implies the existence of several sections of the Fermi surface even for a fixed direction in momentum space. This effect is stronger than usual gap anisotropy and results in a larger spread of the values of the energy gaps.

Speaking of the cuprates, one should note that the La–Sr–Cu–O compound has a relatively simple band structure, and as a result, its properties can be adequately described by the one-band model. The situation is different for Y–Ba–Cu–O and Bi–S–Ca–Cu–O.

Consider two overlapping energy bands. Pairing is described by the order parameters $\Delta_1(\mathbf{p},\omega_n)$ and $\Delta_2(\mathbf{p}, \omega_n)$, where p is the crystal momentum, and $\omega_n = (2n + 1)\pi T$. They satisfy the following equations:

$$\Delta_i(\vec{\mathbf{p}},\omega_n) Z_i = T \sum_{\omega_{n'}} \sum_{\ell=1,2} \int d\mathbf{p}' \, \Gamma_{i\ell}(\mathbf{p},\omega_n; -\mathbf{p}, -\omega_n; \vec{p}', \omega_{n'}; -\vec{p}', -\omega_{n'})$$

$$\frac{\Delta(\vec{p}';\omega_{n'})}{\sqrt{\omega_{n'}^2 + \Delta^2(\vec{p}', \omega_n)}} \qquad (7)$$

$$i = \{1,2\}.$$

If we study the usual phonon mechanism, then $\Gamma_{i\ell} = \lambda_{i\ell} D$, where $D = D(\mathbf{q}, \omega_n - \omega_{n'})$ is the phonon thermodynamic Green's function: $D = \Omega^2(q)[\Omega^2(q) + (\omega_n - \omega_{n'})^2]^{-1}$, $\mathbf{q} = \mathbf{p}' - \mathbf{p}$, Ω is the phonon frequency; we assume summation over all phonon branches. As a result, we obtain the Eliashberg equation generalized to the case of several gaps.

The Hamiltonian can be written in the form

$$\hat{H} = \hat{H}_0 + \hat{H}_1 + \hat{H}_2 + \hat{H}_{12},$$

where $\hat{H}_0$ corresponds to the absence of the pairing interaction.

We shall focus on the two-band case (generalization to a greater number of overlapping bands is straightforward). The term H_{12} describes interband processes. The carrier from the first band can emit a virtual excitation (e.g., a phonon) and make a transition into the second band. Another carrier (also from the first band) can absorb the phonon and also make a transition into the second band, forming a Cooper pair with the first carrier. This interband process leads to a single T_c despite the presence of two gaps (both gaps are closed at the same T_c). T_c is described by the expression

$$T_c = 1.14 \, \tilde{\Omega} \exp(-1/\tilde{\lambda}), \tag{8}$$

where

$$\hat{\lambda} = \frac{1}{2} \{\lambda_{11} + \lambda_{22} + [(\lambda_{22} - \lambda_{11})^2 + 4\lambda_{12}\lambda_{21}]^{1/2}\}.$$

Note also that usually $p_{F1} \neq p_{F2}$; as a result, one can neglect interband pairing.

As is known, the ratio $\Delta(0)/T_c$ allows, in the one-gap theory, a conclusion to be made about the strength of the coupling. This can be seen directly from the expression (2). In the weak coupling limit, $T_c \ll \tilde{\Omega}$, and we have $A \equiv \Delta(0)/T_c = 3.52$. The effect of strong coupling is to increase A. T_c can be expressed directly in terms of λ, and, as a result, one can establish a one-to-one correspondence between $\Delta(0)/T_c$ and λ.

In the two-band case, we have three independent coupling constants λ_{11}, λ_{22}, λ_{12} ($\lambda_{21} = \lambda_{12} \nu_1/\nu_2$). Because of this, the question of correspondence between $\Delta_i(0)/T_c$ and $\lambda_{i\ell}$ is not straightforward and should be treated with considerable care.

Assume for concreteness that $\Delta_1 > \Delta_2$. Then (Kresin and Wolf, 1990a)

$$(2\Delta_1/T_c) > A_{BCS}; \quad (2\Delta_2/T_c) < A_{BCS}. \tag{10}$$

Note that $2\Delta_1/T_c$ exceeds A_{BCS}, but such a behavior does not indicate that strong coupling is present, as it would in the one-band case (see Eq. (2)). The deviation from the BCS value is caused by the multigap structure.

Consider the following example: $\lambda_{11} \gg \lambda_{12}, \lambda_{21}; \lambda_{22} \simeq 0$. This means that the superconducting state of the second band is due to interband transitions only. This model is realistic for the cuprates. In this case we obtain

$$\frac{\Delta_2}{\Delta_1} \simeq \frac{\lambda_{21}}{\lambda_{11}} \ll 1.$$

Therefore, the ratio Δ_2/Δ_1 is small; in conjunction with Eq. (10), this means that the energy gaps may noticeably differ from each other. The situation is different in the case of strong coupling. Then $\Delta_2/\Delta_1 \simeq \lambda_{21} (1 + \lambda_{11}) \lambda_{11}^{-1}$. Strong coupling tends to diminish the relative difference in the values of the energy gaps.

As was noted earlier, a short coherence length makes it possible to meet the criterion $\xi_0 < \ell$. As a result, it is perfectly realistic to observe multigap structure in the high-T_c oxides. Of course, this requires that different energy bands overlap. This is realized in Y–Ba–Cu–O, and probably in Bi–S–Ca–Cu–O. As for La–Sr–Cu–O, it is characterized by a relatively simple band structure (see, for example, the review by Pickett (1989), and its properties can be described by a one-gap model.

The situation with Y–Ba–Cu–O is entirely different, which is because of the presence of chains. As a result, we have two different energy bands, of a quasi–2-D and a quasi–1-D nature. The presence of such a two-band picture is manifested in the strong temperature dependence of the Hall effect and is supported by the band structure calculations (Pickett, 1989).

Therefore, one should be able to observe two energy gaps in the Y–Ba–Cu–O superconductor. There have been several experimental publications indicating the presence of the two-gap structure.

The most convincing evidence comes from NMR experiments (Warren *et al.*, 1987; Barrett *et al.*, 1988). A detailed investigation carried out by Barrett *et al.* (1988) which analyzed the temperature dependence of the Knight shift and clearly demonstrated the presence of two energy gaps. Therefore, the Y–Ba–Cu–O compound is characterized by two superconducting subsystems, and two different bands with different energy gaps. According to Barrett *et al.* (1990), their values are: $\Delta_1 \simeq 6T_c$ (plains), $\Delta_2 \simeq 3.5\, T_c$ (chains).

Ultrasound attenuation and multigap structure. Ultrasound attenuation is the ideal method for analyzing multigap structures and anisotropy. An evaluation of the attenuation coefficient in the presence of several gaps leads to the following result:

$$\frac{\gamma_s}{\gamma_n} = \sum_i \frac{f_i}{e^{\Delta_i/T} + 1}, \qquad (11)$$

where

$$f_i = \frac{2|\lambda^i_{\kappa\kappa'}|^2_F (m_i)^2}{\sum_i |\lambda^i_{\kappa\kappa'}|^2_F (m_i)^2} \quad (11')$$

(m_i are the effective masses, and the quantities $\lambda^{+-}_{kk'}$ describe the electron–lattice coupling). For the case of a single band, we obtain Eq. (1); if we assume that the matrix elements are equal, we obtain

$$f_i = 2(m_i)^2 \sum_i (m_i)^2$$

(see Kadanoff and Falko, 1964). When $T \to 0$, the main contribution to the absorption comes from the smallest gap. Consider a two-gap situation. Then we obtain (in the region $T \to 0$): $\gamma_s/\gamma_n \sim f_2 e^{-\Delta/T}$, and one can determine the values of Δ_2 and f_2. Then the measurements at higher temperatures allow (with the use of Eqs. (11), (11'), and the relation $\Sigma_i f_i = 2$) evaluation of Δ_1 and f_1. Note also that $\Delta_1/\Delta_{2|0} = \Delta_1/\Delta_{2|T_c}$, as has been shown by the author (1973).

The temperature dependence of ultrasound attenuation has been studied experimentally (Xu et al., 1988). An attenuation maximum was observed, which is probably due to fluctuations. In addition, the nonexponential decrease of attenuation observed by the authors may be connected with the multigap structure described here: As was noted above, a power law can be matched by a sum of exponentials.

Thus, ultrasound measurements can provide a wealth of important information about the spectra of the high-T_c superconductors. Recent progress in material preparation makes further progress in this field likely in the near future.

ACKNOWLEDGMENT

The author is grateful to Prof. M. Levy for many valuable discussions.

References

Abrikosov, A. (1988). "Fundamentals of the Theory of Metals." North-Holland, Amsterdam.
Anderson, P. (1959). *J. Phys. Chem. Sol.* **11**, 26.
Arko, A., List, R., Cheong, S., Fisk, Z., Thompson, J., Olson, C., Yang, A., Liu, R., Cu, C., Veal, B., Liu, J., Paulikas, A., Vandervoort, K., Claus, M., Campusano, J., Schreib, J., and Shinn, N. (1989). *Phys. Rev.* **B40**, 2268.
Bardeen, J., Cooper, L., and Schrieffer, J. (1957). *Phys. Rev.* **108**, 1175.
Bardeen, J., and Stephen, M. (1964). *Phys. Rev.* **136**, 1485.
Barrett, S., Durand, D., Dennington, C., Slichter, C., Friedman, T., Rice, J., and Ginsberg, D. (1990). *Phys. Rev.* **B41**, 6283.
Bednorz, J., and Muller, K. (1986). *Z. Phys.* **B64**, 189.

Bishop, D., Varma, C., Batlogg, B., Bucher, E., Fisk, Z., and Smith, J. (1984). *Phys. Rev. Lett.* **53**, 1009.
Bishop, D., Gammel, P., Ramirez, A., Batlogg, B., Caval, R., and Millis, A. (1987). *In* "Novel Superconductivity" (S. Wolf and V. Kresin, eds.). Plenum, New York, p. 659.
Boni, P., Axe, J., Shirane, G., Birgeneau, R., Gaffe, D., Jenssen, H., Kastner, M., Peters, C., Picone, P., and Thirston, T. (1988). *Phys. Rev.* **B38**, 185.
Cracknell, A., and Weng, K. (1973). "The Fermi Surface." Clarendon, Oxford.
Deutcher, G. (1987). *In* "Novel Superconductivity" (S. Wolf and V. Kresin, eds.). Plenum, New York, p. 293.
Ferraro, J., and Williams, J. (1987). "Introduction to Synthetic Electrical Conductors." Academic Press, Orlando.
Fisher, R., Kim, S., Woodfield, B., Phillips, N., Taillefer, L., Kesselbach, K., Flouqu, J., Giorgi, A., and Smith, J. (1989). *Phys. Rev. Lett.* **62**, 1411.
Geilikman, B. (1971). *J. Low Temp. Phys.* **4**, 189.
Geilikman, B., and Kresin, V. (1966). *Sov. Phys.—Solid State* **7**, 2659.
Geilikman, B., and Kresin, V. (1972). *Phys. Lett.* **40A**, 123.
Geilikman, B., and Kresin, V. (1974). "Kinetic and Non-Stationary Phenomena in Superconductors." Wiley, New York.
Geilikman, B., Kresin, V., and Masharov, N. (1975). *J. Low Temp. Phys.* **18**, 3241.
Geilikman, B., Zaitsev, R., and Kresin, V. (1966). *Sov. Phys.—Solid State* **9**, 642.
Ginsberg, D., ed. (1989). "Physical Properties of High Temperature Superconductors." World, Singapore, pp. I,II.
Gokhfel'd, V. (1987). *Sov. Phys.—Low Temp.* **12**, 661.
Golding, B., Bishop, D., Batlogg, B., Haemmerle, W., Fisk, Z., Smith, J., and Ott, H. (1985). *Phys. Rev. Lett.* **55**, 2479.
Greene, R., and Chaikin, P. (1984). *Physica* **126B**, 431.
Hawley, M., Grey, K., Terris, B., Wang, H., Carlson, K., and Williams, J. (1986). *Phys. Rev. Lett.* **57**, 629.
Jerome, D., and Creuzet, F. (1987). *In* "Novel Superconductivity" (S. Wolf and V. Kresin, eds.). Plenum, New York, p. 103.
Kadanoff, L., and Falko, I. (1964). *Phys. Rev.* **136** A1170.
Kresin, V. (1971). *J. Low Temp. Phys.* **5**, 565.
Kresin, V. (1973). *J. Low Temp. Phys.* **11**, 519.
Kresin, V. (1987). *Solid State Comm.* 63, 725.
Kresin, V., Deutcher, G., and Wolf, S. (1988). *J. of Superconductivity* **1**, 327.
Kresin, V., and Kokotov, B. (1978). *Sov. Phys.—JETP* **48**, 537.
Kresin, V., and Morawitz, H. (1988a). *Phys. Rev.* **B37**, 7854.
Kresin, V., and Morawitz, H. (1988b). *J. of Superconductivity* **1**, 89.
Kresin, V., and Morawitz, H. (1989). *J. Opt. Soc. Am.* **B6**, 490.
Kresin, V., and Morawitz, H. (1990). *Solid State Comm.* **11**, 1203.
Kresin, V., and Wolf, S. (1987a). *Solid State Comm.* **63**, 1141.
Kresin, V., and Wolf, S. (1987b). *In* "Novel Superconductivity" (S. Wolf and V. Kresin, eds.). Plenum, New York, p. 287.
Kresin, V., and Wolf S. (1988). *J. of Superconductivity* **1**, 143.
Kresin, V., and Wolf, S. (1990a). *Phys. Rev.* **B41**, 4278.
Kresin, V., and Wolf, S. (1990b). *Physica.* **C169**, 476.
Kwak, W., Crabtree, G., Hinks, D., Capone, D., Jorgensen, J., and Zhang, K. (1987). *Phys. Rev.* **B35**, 5343.
Landau, L. (1956). *Sov. Phys.—JETP* **3**, 920.
Levy, M. (1988). Personal communication.
Levy, M., Kagiwada, R., Rudnick, I. (1963). *Phys. Rev.* **132**, 2039.

Levy, M., Schenmstrom, A., Sun, K., and Sarma, B. (1987). *In* "Novel Superconductivity" (S. Wolf and V. Kresin, eds.). Plenum, New York, p. 265.
Little, W. A. (1964). *Phys. Rev.* **134**, A1416.
Lynton, E. (1969). "Superconductivity." Methuen, London.
Maeda, H., Tanaka, Y., Fukutami, M., and Asano, T. (1988). *J. Appl. Phys. Lett.* **27**, 1209.
Maraki, A. (1987). *Jpn. Appl. Phys. Lett.* **20**, L405.
Morawitz, H., and Kresin, V. (1989). *Physica* **C162–164**, 1471.
Morse, R., and Bohm, H. (1959). *J. Acous. Soc. Am.* **31**, 1523.
Moskalenko, V. (1959). *Fiz. Metal. Metallov.* **8**, 503.
Mueller, F., Fowler, C., Freeman, B., Hults, W., King, J., and Smith, J. (1990). *In* "Electronic Structures in the 1990s." Ban Honnet, Germany.
Muller, K. (1989). *Z. F. Phys.* **B79**, 143.
Ong, N., Wang, Z., Clayhold, J., Tarascon, J., Greene, L., and McKinnon, W. (1987). *Phys. Rev.* **B35**, 8807.
Ott, H. (1987). *In* "Novel Superconductivity" (S. Wolf and V. Kresin, eds.). Plenum, New York, p. 18.
Panson, A., Wagner, G., Braginski, A., Gavaler, J., Janocko, M., Pohl, H., and Talvaccio, J. (1987). *Appl. Phys. Lett.* **50**, 1104.
Parkin, S., Ribault, ., Jerome, D., and Bechgaard, K. (1981). *J. Phys.* **C14**, 5305.
Phillips, N., Fisher, R., Lacy, S., Marcenat, C., Olsen, J., Ham, W., and Stacy, A. (1987). *In* "Novel Superconductivity" (S. Wolf and V. Kresin, eds.). Plenum, New York, p. 739.
Pickett, W. (1989). *Rev. Mod. Phys.* **61**, 433.
Pokrovskii, V. (1961). *Sov. Phys.—JETP* **13**, 628.
Pokrovskii, V., and Toponogov, V. (1961). *Sov. Phys.—JETP* **13**, 785.
Poole, C., Jr., Datta, T., and Farach, H. (1988). "Cooper Oxide Semiconductors." Wiley, New York.
Ramirez, A., Batlogg, B., Aeppli, G., Cava, R., Rietman, E., and Goldman, A. (1987). *Phys. Rev.* **B35**, 8883.
Sheng, Z., and Hermann, A. (1988). *Nature* **332**, 138.
Shepelev, A. (1969). *Sov. Phys.—Usp.* **11**, 690.
Steglich, F., Aarts, J., Bredl, C., Lieke, W., Meschede, D., Franz, W., and Schafer, H. (1979). *Phys. Rev. Lett.* **43**, 1892.
Steward, G. (1984). *Rev. Mod. Phys.* **56**, 755.
Suhl, H., Mattias, B., and Walker, L. (1959). *Phys. Rev. Lett.* **3**, 552.
Suzuki, M. (1989). *Phys. Rev.* **B39**, 2312.
Torrance, J., Tokura, Y., Nazzal, A., Bezinge, A., Huang, T., and Parkin, S. (1988). *Phys. Rev. Lett.* **61**, 1127.
Warren, W., Walstedt, R., Brennert, G., Espinosa, G., and Remeika, J. (1987). *Phys. Rev. Lett.* **59**, 1860.
Williams, J., and Carneiro, K. (1985). *Adv. Inorganic Chemistry and Radiochemistry* **29**, 249.
Wolf, S., and Kresin, V., eds. (1987). "Novel Superconductivity." Plenum, New York.
Wolf, S., and Kresin, V. (1991). *In* "Organic Superconductors" (V. Kresin and W. A. Little, eds.). Plenum, New York.
Wu, M., Ashburn, J., Torng, C., Hor, P., Meng, R., Gao, L., Huang, Z., and Chu, C. (1987). *Phys. Rev. Lett.* **58**, 908.
Xu, M., Baum, H., Schenstrom, A., Sarma, B., Levy, M., Sun, K., Toth, L., Wolf, S., and Gubser, D. (1988). *Phys. Rev.* **B37**, 3675.

Index

A

Ac magnetic susceptibility, temperature-dependent
  of $Er_{0.187}Ho_{0.813}Rh_4B_4$ at constant fields, 201
  of $Er_{0.6}Ho_{0.4}Rh_4B_4$ at zero field, 199
Activation energy, 249, 262, 263, 289, 297
Anelastic relaxation, 420–428
  activation energies, 421–425, 427
  magnetic interpretation, 428
  orthorhombic YBCO, 423–425
  oxygen diffusion, 427
  relaxation strengths, 426
  tetragonal YBCO, 420–423
Anisotropy of attenuation in magnetic superconductor, 209–211
Attenuation
  anisotropy of attenuation in magnetic superconductor, 209–211
  anomaly 203
    at magnetic phase transition, 201
    spin–phonon interactions, 203
    at superconducting transition, 193, 194
  in $Ba_{1-x}K_xBiO_3$
    longitudinal 287
    transverse 282, 288
  in BiSrCaCuO
    curve, 278
    peak at T_c, 278
  in $ErBa_2Cu_3O_7$
    curve, 270
  frequency-dependence
    of $Er_{0.187}Ho_{0.813}Rh_4B_4$, 219
    of $Er_{0.705}Ho_{0.295}Rh_4B_4$, 216
    of $HoRh_4B_4$, 219
    measurement, 197
  in $GdBa_2Cu_3O_7$
    curves, 273
  in $La_{2-x}Sr_xCuO_4$, 238–243
    peaks, 242
  of $ErRh_4B_4$, 210, 211
  of $Er_{0.4}Ho_{0.6}Rh_4B_4$, 200
  of $Er_{0.705}Ho_{0.295}Rh_4B_4$, 200
  of $Er_{0.088}Ho_{0.912}Rh_4B_4$, at high magnetic fields, 227
  in Nb_3Sn
  in magnetic field, 19, 20
  relaxation-type (in $Er_{1-x}Ho_xRh_4B_4$), 218–227
    broad maximum, 196, 197
  in $HoRh_4B_4$, at different frequencies, 219
    relaxation attenuation equation, 219, 222
    residual, 7
  in Ta, 9
  in $T\ell BaCaCuO$
    curves, 277
    peaks at T_c, 297
  in $YBa_2Cu_3O_7$
    anistropy, 254–264
    peaks, relaxation, 256, 257, 258, 261, 262, 297
  temperature-dependence, at constant magnetic fields
    common behavior, 196
    of $ErRh_4B_4$, low temperature, 203
    of $Er_{0.4}Ho_{0.6}Rh_4B_4$, 199
    of $Er_{0.705}Ho_{0.295}Rh_4B_4$, low temperature, 204
    at zero magnetic field, 195, 196
Attenuation anomaly, 203
  at magnetic phase transition, 201
  at superconducting transition, 193, 194
Attenuation peaks, 228, 242, 256, 257, 258, 261, 262, 297, 419
Attenuation signature, 290
Axial state, 10

B

$Ba_{1-x}K_xBiO_3$, 280–289
  attenuation, longitudinal, 287

attenuation, transverse, 282, 288
  crystal structure, 295
  elastic constants, 295
  magnetic field, 284
  phase diagram, 281
  velocity, longitudinal magnetic field, 288
  velocity, transverse magnetic field, 283
    magnetic field, 285
BCS attenuation
  isotropic energy gap, 10
  longitudinal, 8
  transverse, 6
BiSrCaCuO, 274–280
  attenuation curve, 278
    peak at T_c, 278
  crystal structure, 294
  elastic constants, 296
  superconducting fluctuations, 290
  velocity curve, 279
Boltzman transport equation, 3, 7

C

Coexistence of magnetic order and superconductivity, 192
  in $Er_{0.705}Ho_{0.295}Rh_4B_4$, 208, 216, 218
Coherence length, 13
  effective, 16
    clean type II, 16
    dirty type II, 16
Compressibility coefficient, 253
Conductivity
  electrical, 246
  thermal, 251
Conventional superconductors, attenuation, 1
  magnetic field dependence, 13
    clean type II, 19
    dirty type II, 16
  temperature dependence, 2
    longitudinal waves, 4
    transverse waves, 4
Conventional ultrasound pulse-echo technique
  difficulties, 384–385
  measurement, 383–385
  sample preparation, 383–385
Cooper pairs, 262
Critical phonons, 262
Crystal structure, high T_c superconductors, 291–295

Crystalline electric fields (CEF), 220
  in $Er_{1-x}Ho_xRh_4B_4$, 206
  in magnetic superconductor, 219
  screened, superconducting currents, 206
CuO planes, distortions, 289, 290
Current density
  electron, 8
  lattice, 8

D

Density, high T_c superconductors, 295–296
Density gradients, 7

E

Effective relaxation time, 251
Elastic constants
  $Ba_{1-x}K_xBiO_3$, 295
  BiSrCaCuO, 296
  $ErBa_2Cu_3O_7$, 295
  La_2CuO_4, 398
  $La_{1.86}Sr_{0.14}CuO_4$, 398–399
  $La_{2-x}Sr_xCuO_4$, 239, 295
  Pulse-echo technique, measurement by, 383–385
  quartz, 396, 398
  resonant ultrasound, measurement by, 393–396
  TℓBaCaCuO, 296
  $YBa_2Cu_3O_7$, 295
Elastic constants, high T_c superconductors, 295–296
Electromagnetic interaction
  in $Er_{1-x}Ho_xRh_4B_4$, 211
  in magnetic superconductor, 192, 208
Electron–phonon interaction, 2, 246
Electron–phonon interaction, magnetic superconductor
  BCS theory, 193, 198, 207
  order of, 228
  strong, 193
Electron reservoir, between CuO planes, 262, 290
Energy-level crossing, at high magnetic field, 229

Index

Enhanced ultrasonic attenuation,
  $Er_{1-x}Ho_xRh_4B_4$, in superconducting state, 198–207
$ErBa_2Cu_3O_7$, 270–274
  attenuation curve, 270
  crystal structure, 292
  elastic constants, 295
  velocity curve, 271
$Er_{1-x}Ho_xRh_4B_4$
  phase transition temperature, T_{c1}, T_{c2}, T_m, 194
  physical dimensions, 194
  sound velocity, 194
  synthesis, 194

F

Free electron, attenuation model, 246
Frequency-dependent
  attenuation
    of $Er_{0.187}Ho_{0.813}Rh_4B_4$, 219
    of $Er_{0.705}Ho_{0.295}Rh_4B_4$, 216
    of $HoRh_4B_4$, 219
  measurement, 197

G

$GdBa_2Cu_3O_7$, 270–274
  attenuation curves, 273
  crystal structure, 292
  velocity, 274
Ginsburg–Landau parameter, 13
  type I superconductor, 16
  type II superconductor, 16
Glasslike solids, 250
Gruneisen constant, 222

H

Heat treatment, 244, 245, 248, 276, 415
Heavy Fermion superconductor, 10
High-temperature anomalies, 419, 428
Hysteresis, 270, 410, 414, 419

I

Internal stress, 413

Interplay
  ferromagnetic order and superconductivity, 193
  magnetic ions and superconducting current, 207

L

La_2CuO_4, single crystal
  elastic constants, 398
  structure phase transition, 400
  temperature dependance, 399
  ultrasound velocity minimum, 400–401
$La_{1.86}Sr_{0.14}CuO_4$, single crystal
  elastic constants, 398–399
  temperature dependence, 400, 405–406
  orthorhombic-tetragonal structural transition, 403–406
  self-consistent phonon approximation, 404–406
$La_{2-x}Sr_xCuO_4$, 238
  attenuation, 238–243
  attenuation peaks, 242
  crystal structure, 292
  elastic constants, 239, 295
  internal friction, 241
  phase transition, 240
  velocity, 238–243
Linear dependence, attenuation, dirty type II, 14
London penetration length, 13

M

Magnetic-field dependent attenuation
  of $ErRh_4B_4$, 210, 211
  of $Er_{0.4}Ho_{0.6}Rh_4B_4$, 200
  of $Er_{0.705}Ho_{0.295}Rh_4B_4$, 200
  of $Er_{0.888}Ho_{0.912}Rh_4B_4$, at high magnetic fields, 227
Magnetic flux quantization, 212, 213
Magnetic superconductor
  antiferromagnetic, 192
  ferromagnetic, 192
  magnetic impurity, 191
Magnetization, 13
  type I superconductor, 13, 14
  type II superconductor, 13, 14
Magnetoelastic interaction (coupling), 228

458 Index

Mean-field behavior, HoRh$_4$B$_4$, 192, 204
Modulated magnetic order in superconducting state, 192, 208

N

Nb$_3$Sn
  attenuation, magnetic field, 19, 20

O

Order parameter
  clean type II, 16–18
  dirt type II, 16–18
Oxygen content, 412

P

Phase diagram
  type I superconductor, 15
  type II superconductor, 15
Phase diagram of H vs. T Er$_{0.705}$Ho$_{0.295}$Rh$_4$B$_4$, 217
Phonon drag, 4
Phonon–phonon interaction, 227
Polar state, 10
Porosity, 417
Pseudoplasticity, 417

Q

Quartz, elastic constants, 396, 398

R

Reentrant superconductor, 192
Relaxation mechanism, of Er$_{1-x}$Ho$_x$Rh$_4$B$_4$
  relaxation time, 219
    effective, 223–227
    two-level model, 220–223
Relaxation process, 249
Relaxation-type attenuation, of Er$_{1-x}$Ho$_x$Rh$_4$B$_4$, 218–227
  broad maximum, 196, 197
  of HoRh$_4$B$_4$, at different frequencies, 219
  relaxation attenuation equation, 219, 222

Residual attenuation, 7
Resonant ultrasound technique
  advantages, over pulse-echo technique, 385–386
  computation algorithm, 395–396
  development, 382–396
  measurement procedure
    phase sensitive technique, 385, 402–403
    frequency modulation technique, 391–392
  sample preparation, 386, 390–391
  theoretical model, 394
  transducer, requirements for, 386, 388

S

Sample preparation,
  pulse-echo measurements, requirements, 384, 385
  resonant ultrasound technique, requirements, 390–391
Schottky-type, heat capacity (specific heat)
  two-level magnetic system, 221
Second-order phase transition, superconducting state, 1, 253
  Ehrenfest equation, 253
  Meissner–Ochsenfeld effect, 1
Skin depth, 4
Sound velocity, 238–243, 252–254, 264–271, 274, 278, 279, 283, 285, 288, 414, 415, 419
Spin fluctuation, 192, 194, 202, 205
  suppressed, 205
Spin–phonon interaction, 201–204
  attenuation anomaly, 203
  energy dissipation of sound, random force, 202
  equation, 203
Square root dependence, attenuation, clean type II, 19, 20
Staggered magnetic susceptibility, 203, 208
  relationship with attenuation, 203
Structural phase transition
  in high T_c superconductors, 381, 403–406
  in La$_2$CuO$_4$, 400
  in LaSrCuO$_4$, 402–406
Superconducting fluctuations, 290
Superconducting screening, 7
  transverse wave attenuation, 4

Index

Superconductivity, with antiferromagnetism, 192, 193

T

Ta, 9
  attenuation, 9
  energy gap, 9
Temperature-dependent, attenuation
  common behavior, 196
  at constant magnetic fields
    of $ErRh_4B_4$, low temperature, 203
    of $Er_{0.4}Ho_{0.6}Rh_4B_4$, 199
    of $Er_{0.705}Ho_{0.295}Rh_4B_4$, low temperature, 204
  at zero magnetic field, $Er_{1-x}Ho_xRh_4B_4$, 195, 196, 198
Thermal expansion coefficient, 253
Thermal phonon, Barett's model, 221
TℓBaCaCuO, 274–280
  attenuation curves, 277
    peaks at T_c, 297
  crystal structure, 293
  elastic constants, 296
  superconducting fluctuations, 290
  velocity curves, 278
Transducers
  diamond lithium niobate composite, 402, 403
  polyvinylidence fluoride film
    comparison with quartz, 388
    fabrication for use with small samples, 389
    ringing, problems with, 384, 386
Transition rate, of electron, 220
Twinning, 414

Two-energy-level system, 248, 251
Type II-I superconductors, 212–215
  $Er_{0.705}Ho_{0.295}Rh_4B_4$, 215
  magnetization curve, 214

Y

$YBa_2Cu_3O_7$
  crystal structure, 292
  elastic constants, 295
  oriented, 254–271
    activation energy, 262, 263, 289, 297
    attenuation, anisotropy, 254–264
      peaks, relaxation, 256, 257, 258, 261, 262, 297
    critical current density, anisotropy, 255
    elastic constants, 297
    magnetization, anisotropy, 255
  sinter forged (*see also* oriented), 255–271
  sintered, ordinary, 243–254
    attenuation, 243–251
      activation energy, 249
      peak at T_c, 276
      peaks maxima, 244, 245
      relaxation peaks/maxima, 248
    superconducting fluctuations, 290
    velocity, 252–254
      discontinuity, 252, 253
    velocity, anisotropy, 264–271
      effect of hysteresis on T_c, 270–271
      hysteresis curves, 265, 266, 268, 270

Z

Zeeman effect, 227

Contents of Volumes in this Series

Volume I, Part A—Methods and Devices

Wave Propagation in Fluids and Normal Solids—*R. N. Thurston*
Guided Wave Propagation in Elongated Cylinders and Plates—*T. R. Meeker and A. H. Meitzler*
Piezoelectric and Piezomagnetic Materials and Their Function in Transducers—*Don A. Berlincourt, Daniel R. Curran, and Hans Jaffe*
Ultrasonic Methods for Measuring the Mechanical Properties of Liquids and Solids—*H. J. McSkimin*
Use of Piezoelectric Crystals and Mechanical Resonators in Filters and Oscillators—*Warren P. Mason*
Guided Wave Ultrasonic Delay Lines—*John E. May, Jr.*
Multiple Reflection Ultrasonic Delay Lines—*Warren P. Mason*

Volume I, Part B—Methods and Devices

The Use of High- and Low-Amplitude Ultrasonic Waves for Inspection and Processing—*Benson Carlin*
Physics of Acoustic Cavitation in Liquids—*H. G. Flynn*
Semiconductor Transducers—General Considerations—*Warren P. Mason*
Use of Semiconductor Transducers in Measuring Strains, Accelerations, and Displacements—*R. N. Thurston*
Use of p-n Junction Semiconductor Transducers in Pressure and Strain Measurements—*M. E. Sikorski*

The Depletion Layer and Other High-Frequency Transducers Using Fundamental Modes—*D. L. White*
The Design of Resonant Vibrators—*Edward Eisner*

Volume II, Part A—Properties of Gases, Liquids, and Solutions

Transmission of Sound Waves in Gases at Very Low Pressure—*Martin Greenspan*
Phenomenological Theory of the Relaxation Phenomena in Gases—*H. J. Bauer*
Relaxation Processes in Gases—*H. O. Kneser*
Thermal Relaxation in Liquids—*John Lamb*
Structural and Shear Relaxation in Liquids—*T. A. Litovitz and C. M. Davis*
The Propagation of Ultrasonic Waves in Electrolytic Solutions—*John Stuehr and Ernest Yeager*

Volume II, Part B—Properties of Polymers and Nonlinear Acoustics

Relaxations in Polymer Solutions, Liquids and Gels—*W. Philoppoff*
Relaxation Spectra and Relaxation Processes in Solid Polymers and Glasses—*I. L. Hopkins and C. R. Kurkjian*
Volume Relaxations in Amorphous Polymers—*Robert S. Marvin and John E. McKinney*
Nonlinear Acoustics—*Robert T. Beyer*
Acoustic Streaming—*Wesley LeMars Nyborg*
Use of Light Diffraction in Measuring the Parameter of Nonlinearity of Liquids and

the Photoelastic Constants of Solids—
L. E. Hargrove and K. Achyuthan

Attenuation of Elastic Waves in the Earth—*L. Knopoff*

Volume III, Part A—Effect of Imperfections

Anelasticity and Internal Friction Due to Point Defects in Crystals—*B. S. Berry and A. S. Nowick*
Determination of the Diffusion Coefficient of Impurities by Anelastic Methods—*Charles Wert*
Bordoni Peak in Face-Centered Cubic Metals—*D. H. Niblett*
Dislocation Relaxations in Face-Centered Cubic Transition Metals—*R. H. Chambers*
Ultrasonic Methods in the Study of Plastic Deformation—*Rohn Truell, Charles Elbaum, and Akira Hikata*
Internal Friction and Basic Fatigue Mechanisms in Body-Centered Cubic Metals, Mainly Iron and Carbon Steels—*W. J. Bratina*
Use of Anelasticity in Investigating Radiation Damage and the Diffusion of Point Defects—*Donald O. Thompson and Victor K. Paré*
Kinks in Dislocation Lines and Their Effects on the Internal Friction in Crystals—*Alfred Seeger and Peter Schiller*

Volume III, Part B—Lattice Dynamics

Use of Sound Velocity Measurements in Determining the Debye Temperature of Solids—*George A. Alers*
Determination and Some Uses of Isotropic Elastic Constants of Polycrystalline Aggregates Using Single-Crystal Data—*O. L. Anderson*
The Effect of Light on Alkali Halide Crystals—*Robert B. Gordon*
Magnetoelastic Interactions in Ferromagnetic Insulators—*R. C. LeCraw and R. L. Comstock*
Effect of Thermal and Phonon Processes on Ultrasonic Attenuation—*P. G. Klemens*
Effects of Impurities and Phonon Processes on the Ultrasonic Attenuation of Germanium, Crystal Quartz, and Silicon—*Warren P. Mason*

Volume IV, Part A—Applications to Quantum and Solid State Physics

Transmission and Amplification of Acoustic Waves in Piezoelectric Semiconductors—*J. H. McFee*
Paramagnetic Spin-Phonon Interaction in Crystals—*Edmund B. Tucker*
Interaction of Acoustic Waves with Nuclear Spins in Solids—*D. I. Bolef*
Resonance Absorption—*Leonard N. Liebermann*
Fabrication of Vapor-Deposited Thin Film Piezoelectric Transducers for the Study of Phonon Behavior in Dielectric Materials at Microwave Frequencies—*J. de Klerk*
The Vibrating String Model of Dislocation Damping—*A. V. Granato and K. Lücke*
The Measurement of Very Small Sound Velocity Changes and Their Uses in the Study of Solids—*G. A. Alers*
Acoustic Wave and Dislocation Damping in Normal and Superconducting Metals and in Doped Semiconductors—*Warren P. Mason*
Ultrasonics and the Fermi Surfaces of the Monovalent Metals—*J. Roger Peverley*

Volume IV, Part B—Applications to Quantum and Solid State Physics

Oscillatory Magnetoacoustic Phenomena in Metals—*B. W. Roberts*
Transmission of Sound in Molten Metals—*G. M. B. Webber and R. W. B. Stephens*
Acoustic and Plasma Waves in Ionized Gases—*G. M. Sessler*
Relaxation and Resonance of Markovian Systems—*Roger Cerf*
Magnetoelastic Properties of Yttrium-Iron Garnet—*Walter Strauss*
Ultrasonic Attenuation Caused by Scattering in Polycrystalline Media—*Emmanuel P. Papadakis*
Sound Velocities in Rocks and Minerals: Experimental Methods, Extrapolations to Very High Pressure, and Results—*Orson L. Anderson and Robert C. Liebermann*

Contents of Volumes in this Series

Volume V

Acoustic Wave Propagation in High Magnetic Fields—*Y. Shapira*

Impurities and Anelasticity in Crystalline Quartz—*David B. Fraser*

Observation of Resonant Vibrations and Defect Structure in Single Crystals by X-ray Diffraction Topography—*W. J. Spencer*

Wave Packet Propagation and Frequency-Dependent Internal Friction—*M. Elices and F. Garcia-Moliner*

Coherent Elastic Wave Propagation in Quartz at Ultramicrowave Frequencies—*John Ilukor and E. H. Jacobsen*

Heat Pulse Transmission—*R. J. von Gutfeld*

Volume VI

Light Scattering as a Probe of Phonons and Other Excitations—*Paul A. Fleury*

Acoustic Properties of Materials of the Perovskite Structure—*Harrison H. Barrett*

Properties of Elastic Surface Waves—*G. W. Farnell*

Dynamic Shear Properties of Solvents and Polystyrene Solutions from 20 to 300 MHz—*R. S. Moore and J. H. McSkimin*

The Propagation of Sound in Condensed Helium—*S. G. Eckstein, Y. Eckstein, J. B. Ketterson, and J. H. Vignos*

Volume VII

Ultrasonic Attenuation in Superconductors: Magnetic Field Effects—*M. Gottlieb, M. Garbury, and C. K. Jones*

Ultrasonic Investigation of Phase Transitions and Critical Points—*Carl W. Garland*

Ultrasonic Attenuation in Normal Metals and Superconductors: Fermi-Surface Effects—*J. A. Rayne and C. K. Jones*

Excitation, Detection, and Attenuation of High-Frequency Elastic Surface Waves—*K. Dransfeld and E. Salzmann*

Interaction of Light and Ultrasound: Phenomena and Applications Spin-Phonon Spectrometer—*R. W. Damon, W. T. Maloney, and D. H. McMahon*

Volume VIII

Spin–Phonon Spectrometer—*Charles H. Anderson and Edward S. Sabisky*

Landau Quantum Oscillations of the Velocity of Sound and the Strain Dependence of the Fermi Surface—*L. R. Testardi and J. H. Condon*

High-Frequency Continuous Wave Ultrasonics—*D. I. Bolef and J. G. Miller*

Ultrasonic Measurements at Very High Pressures—*P. Heydemann*

Third-Order Elastic Constants and Thermal Equilibrium Properties of Solids—*J. Holder and A. V. Granatb*

Interactions of Sound Waves with Thermal Phonons in Dielectric Crystals—*Humphrey J. Maris*

Internal Friction at Low Frequencies Due to Dislocations: Applications to Metals and Rock Mechanics—*Warren P. Mason*

Volume IX

Difference in Electron Drag Stresses on Dislocation Motion in the Normal and the Superconducting States for Type I and Type II Superconductors—*M. Suenaga and J. M. Galligan*

Elastic Wave Propagation in Thin Layers—*G. W. Farnell and E. L. Adler*

Solid State Control Elements Operating on Piezoelectric Principles—*F. L. N-Nagy and G. C. Joyce*

Monolithic Crystal Filters—*W. J. Spencer*

Design and Technology of Piezoelectric Transducers for Frequencies Above 100 MHz—*E. K. Sittig*

Volume X

Surface Waves in Acoustics—*H. Überall*

Observation of Acoustic Radiation from Plane and Curved Surfaces—*Werner G. Neubauer*

Electromagnetic Generation of Ultrasonic Waves—*E. Roland Dobbs*
Elastic Behavior and Structural Instability of High-Temperature A-15 Structure Superconductors—*Louis R. Testardi*
Acoustic Holography—*Winston E. Kock*

Volume XI

Third Sound in Superfluid Helium Films—*David J. Bergman*
Physical Acoustics and the Method of Matched Asymptotic Expansions—*M. B. Lesser and D. G. Crighton*
Ultrasonic Diffraction from Single Apertures with Application to Pulse Measurements and Crystal Physics—*Emmanuel P. Papadakis*
Elastic Surface Wave Devices—*J. de Klerk*
Nonlinear Effects in Piezoelectric Quartz Crystals—*J. J. Gagnepain and R. Besson*
Acoustic Emission—*Arthur E. Lord, Jr.*

Volume XII

The Anomalous Elastic Properties of Materials Undergoing Cooperative Jahn-Teller Phase Transitions—*R. L. Melcher*
Superconducting Tunneling Junctions as Phonon Generators and Detectors—*W. Eisenmenger*
Ultrasonic Properties of Glasses at Low Temperatures—*S. Hunklinger and W. Arnold*
Acoustical Response of Submerged Elastic Structures Obtained Through Integral Transforms—*H. Überall and H. Huang*
Ultrasonic Velocity and Attenuation: Measurement Methods with Scientific and Industrial Applications—*Emmanuel P. Papadakis*

Volume XIII

Anelasticity: An Introduction—*A. S. Nowick*
Structural Instability of A-15 Superconductors—*L. R. Testardi*
Plate Modes in Surface Acoustic Waves Devices—*R. S. Wagers*
Anisotropic Surface Acoustic Wave Diffraction—*Thomas L. Szabo*
Doubly Rotated Thickness Mode Plate Vibrators—*Arthur Ballato*
The Generalized Ray Theory and Transient Responses of Layered Elastic Solids—*Yih-Hsing Pao and Ralph R. Gajewski*

Volume XIV

Acoustic Microscopy—*Ross A. Lemons and Calvin F. Quate*
Sound Propagation in Liquid Crystals—*K. Miyano and J. B. Ketterson*
Electromagnetic-Ultrasound Transducers: Principles, Practice, and Applications—*H. M. Frost*
Ultrasonic Transducers for Materials Testing and Their Characterizations—*Wolfgang Sachse and Nelson N. Hsu*
Ultrasonic Flowmeters—*Lawrence C. Lynnworth*

Volume XV

A History of Ultrasonics—*Karl F. Graff*
Circuit-Model Analysis and Design of Interdigital Transducers for Surface Acoustic Wave Devices—*W. Richard Smith*
Theory of Resonance Scattering—*Lawrence Flax, Guillermo C. Gaunaurd, and Herbert Überall*
Acoustic Emission—An Update—*Arthur E. Lord, Jr.*

Volume XVI

Relaxation Processes in Sound Propagation in Fluids: A Historical Survey—*R. Bruce Lindsay*
Acoustic Vibrational Modes in Quartz Crystals: Their Frequency, Amplitude, and Shape Determination—*Harish Bahadur and R. Parshad*
Electron and Phonon Drag on Mobile Dislocations in Metals at Low Temperatures—*J. M. Galligan*
Two-Pulse Phonon Echoes in Solid-State Acoustics—*K. Fossheim and R. M. Holt*
Dynamic Polarization Echoes in Powdered Materials—*Koji Kajimura*
Memory Echoes in Powders—*R. L. Melcher and N. S. Shiren*

Fiber Optic Acoustic Transduction—*J. A. Bucaro, N. Lagakos, J. H. Cole, and T. G. Giallorenzi*

Volume XVII

Determination of Third-Order Elastic Constants from Ultrasonic Harmonic Generation Measurements—*M. A. Breazeale and Jacob Philip*
Acoustoelasticity and Ultrasonic Measurements of Residual Stresses—*Yih-Hsing Pao, Wolfgang Sachse, and Hidekazu Fukuoka*
Absorption of Sound by the Atmosphere—*H. E. Bass, L. C. Sutherland, Joe Piercy, and Landon Evans*
Statistical Properties of Random Wave Fields—*Karl Joachim Ebeling*

Volume XVIII

Number-Theoretic Phase Arrays and Diffraction Gratings with Broad Radiation (Scattering) Characteristics—*Manfred R. Schroeder*
Ultrasonic Generation by Pulsed Lasers—*D. A. Hutchins*
Electron Beam-Acoustic Imaging—*G. S. Cargill III*
Theory of Photothermal and Photoacoustic Effects in Condensed Matter—*F. Alan McDonald and Grover C. Wetsel, Jr.*
Opto/Photoacoustics: Vibrational Relaxation: Theory and Experiment—*M. Rebelo da Silva and F. Lepoutre*
Analytical Applications of Photoacoustic Spectroscopy to Condensed Phase Substances—*Tsuguo Sawada and Takehiko Kitamori*
Imaging with Optically Generated Thermal Wave—*G. Busse*

Volume XIX, Ultrasonic Measurement Methods

Radiated Fields of Ultrasonic Transducers—*D. A. Hutchins and G. Hayward*

The Measurement of Ultrasonic Velocity—*Emmanuel P. Papadakis*
The Measurement of Ultrasonic Attenuation—*Emmanuel P. Papadakis*
Physical Principles of Measurements with EMAT Transducers—*R. B. Thompson*
Optical Detection of Ultrasound—*James W. Wagner*
Measuring the Electrical Characteristics of Piezoelectric Devices—*Warren L. Smith*
Photoelastic Visualization and Theoretical Analyses of Scatterings of Ultrasonic Pulses in Solids—*C. F. Ying*

Volume XX, Ultrasonic Attenuation in Conventional Superconductors

Sound Propagation in Superfluid 3He—*J. B. Ketterson, S. Adenwalla, Z. Zhao and Bimal K. Sarma*
Sound Propagation in the Heavy Fermion Superconductors—*Bimal K. Sarma, Moises Levy, S. Adenwalla and J. B. Ketterson*
Ultrasonic Attenuation in the Magnetic Superconducting System $Er_{1-x}Ho_xRh_4B_4$—*Keun J. Sun and Moises Levy*
Ultrasonic Propagation in Sintered High T_c Superconductors—*Moises Levy, Min-Feng Xu, Bimal K. Sarma and Keun-Jenn Sun*
Sound Velocity Studies of Ceramic High-Temperature Superconductors—*S. Bhattacharya*
Acoustic Studies of Single Crystal High-T_c Superconductors—*Brage Golding*
Ultrasonic Measurements of Elastic Constants in Single Crystals of La_2CuO_4—*J. D. Maynard, M. J. McKenna, A. Migliori and William M. Visscher*
A Rationalisation of the Diversity in the Elastic Response of Polycrystalline Superconducting Oxides—*D. P. Almond*
High-T_c Superconductivity and Ultrasonics—Theoretical Aspects—*Vladimir Z. Kresin*